C.-J. Winter, R. L. Sizmann,
L. L. Vant-Hull (Eds.)

Solar Power Plants

Fundamentals, Technology, Systems, Economics

With 230 Figures

Springer-Verlag
Berlin Heidelberg New York
London Paris Tokyo
Hong Kong Barcelona Budapest

Prof. Dr.-Ing. C.-J. Winter
Deutsche Forschungsanstalt
für Luft- und Raumfahrt (DLR)
Pfaffenwaldring 38-40
W-7000 Stuttgart 80
Germany

Prof. Dr. rer. nat. Rudolf L. Sizmann
Sektion Physik
Ludwig-Maximilians-Universität München
Amalienstraße 54
W-8000 München 40
Germany

Dr. Lorin L. Vant-Hull
Prof. of Physics
Energy Laboratory
University of Houston
4800 Calhoun Road
Houston, TX 77004
USA

ISBN 3-540-18897-5 Springer-Verlag Berlin Heidelberg NewYork
ISBN 0-387-18897-5 Springer-Verlag NewYork Berlin Heidelberg

This work is subject to copyright. All rights are reserved, whether the whole or part of the material is concerned, specifically the rights of translation, reprinting, re-use of illustrations, recitation, broadcasting, reproduction on microfilms or in other ways, and storage in data banks. Duplication of this publication or parts thereof is only permitted under the provisions of the German Copyright Law of September 9, 1965, in its current version and a copyright fee must always be paid. Violations fall under the prosecution act of the German Copyright Law.

© Springer-Verlag Berlin, Heidelberg 1991
Printed in the United States of America

The use of registered names, trademarks, etc. in this publication dies not imply, even in the absence of a specific statement, that such names are exempt from the relevant protective laws and regulations and therefor free for general use.

2161/3020-543210 – Printed on acid-free paper

Preface

In less than 20 years solar power has developed from a sub-kilowatt novelty to a system of multi-megawatt grid-connected power stations. Pilot or demonstration electric generating plants representing several different direct solar technologies (solar tower, distributed receiver dish and trough, and photovoltaics) have been operated and have delivered power at megawatt levels to the electrical grid. In addition cogeneration and process heat operation have been demonstrated in the field, and laboratory demonstrations of direct photo- and thermal-catalytic processes have been successful. These successes, in light of present awareness of the environmental threat associated with conventional energy technology (greenhouse effect from CO_2, acid rain from SO_2 and NO_x, waste disposal and radioactive emissions from both coal and nuclear plants), should be propelling solar power plants into rapid implementation. In fact this is not happening.

The situation at the close of the 1980's was not favorable to solar energy utilization. Rational energy usage coupled with excess production capacity in the OPEC nations resulted in an oil glut, with the result that energy was abundant. The utility companies, with few exceptions, had overcapacities and those with inadequate generating capacity prefered to purchase their excess requirements elsewhere instead of installing expensive new plants. Although the environmental problems are worsening, the situation does not yet appear to be dramatic. Nonetheless time is running out while polluting technologies escape their responsibilities to the environment and the widespread use of solar energy (which is almost entirely benign in pollution) waits on the sidelines. Meanwhile, the population of developing countries is increasing drastically. They need energy for industrialization, not wood or dung but commercial energy supplies, which they cannot afford. They are potential users of advanced solar energy technology, especially those in the globe's sun-belt, since they already have high quantities of irradiation as primary energy in their countries, and can profit from an energy supply which is free of charge and devoid of environmental problems.

Solar energy technology is in general still lacking commercial mass production and it is not readily available to the power industry; consequently it is dependent on government support for its further development, and on government inducements to promote its utilization. But governments and their priorities change from time to time. Such conditions place solar energy development at risk, since the time required for technology development is so much longer than the usual elective life of governments. What solar energy really needs is a long-term commitment of the world community of industrialized countries in the north as well as the well insolated developing countries in the south. Solar energy, truly, is a universal resource and, consequently, requires an international consensus to foster its development.

The American and European authors of this book have undertaken a commitment to solar power as a viable energy supply for much of the world's population. While not the native tongue of all authors or readers, English has been chosen as a neutral language of science. Our apologies for any resulting inconvenience. In an effort to assist readers of varied backgrounds, most terms are defined on first use. A glossary of terms, abbreviations and acronyms is included as an appendix, along with solar related definitions of each term.

The authors are scientists and engineers selected from academia and national laboratories on the basis of their expertise and their significant contributions to the field. Industry is represented by only one author because, unfortunately, solar power plants are not yet, to a significant extent, in the hands of the power plant industry. It is this very situation that inspired the writing of this book. In it we aim to speak openly of merits and deficiencies, of successes and failures encountered to date, of unforeseen developments and over-optimistic expectations, of open questions concerning solar energy conversion in general and solar power plants in particular.

The authors and editors of this book join together to foster a long term political commitment to renewable solar energy. To achieve this end we feel it is essential to apprise those active in the political arena, either as voters or as politicians and administrators, of the subject matter and of the issues involved. Thus, the potential of solar power and the requirements to achieve that potential are addressed in Chaps. 1 and 10. Our objective has been to make these chapters as accessible as possible to an educated lay reader, while providing firm support to any technical and economic assumptions in the intervening chapters.

Chapters 2 and 3 provide the basic physics applicable to solar power plants. This material provides the thermodynamic and optical background driving the engineering and technical designs which follow. To satisfy the needs of the engaged student or engineer, these chapters contain considerable mathematical and physical detail. The lay reader should feel free to skim (or even skip) the more difficult passages. References to this material in subsequent chapters are usually for technical support, rather than for comprehensibility. In addition, the glossary provides less familiar technical definitions of terms or quantities required in other chapters.

Chapters 4–9 are the technical substance of the book. These chapters provide a snapshot of the status of the primary solar power technologies. They characterize installations throughout the world representative of the current state-of-the-art in engineering. For the engineer or student, they assess the performance of these installations and provide sufficient background and supporting information so that the interested reader can also participate in future developments.

This book has a place on the desk of the student, scientist, or engineer interested in advancing the state of solar power. It also has a place in the hand, and in the bookcase of the voter or politician interested in providing a safe, renewable resource base for the world and all its children, including their own. We invite you to join us in this endeavor.

This preface would not be complete without an expression of gratitude to many: to the authors for four years of labor, scientific exchange and animated correspondence; to the assistants and secretaries who helped to prepare the manuscript and figures; to DLR which generously provided resources to help our efforts along, and of course to the Springer publishing house which handled design and layout of the book with customary thoroughness and care, and included this text in its world-famous series of books. May we be forgiven if we mention only one person by name, Hansmartin P. Hertlein; without his unremitting coordination, drive and beneficial impatience behind the scenes the book would never have appeared.

Gerald W. Braun, USA, and Claude Etievant, France, kindly read through the manuscript and supplied valuable comments; many thanks for their efforts!

Stuttgart/Houston/Munich
September 1990

Carl-Jochen Winter
Lorin L. Vant-Hull
Rudolf Sizmann

Table of Contents

1 The Energy Heptagon   1
By C.-J. Winter
Bibliography 16

2 Solar Radiation Conversion   17
By R. Sizmann, with contributions by P. Köpke and R. Busen
- 2.1 Introduction 17
- 2.2 Solar Radiant Flux 17
  - 2.2.1 Modulation Through Revolution and Rotation 20
  - 2.2.2 Beam Radiation on Tilted Surfaces 26
  - 2.2.3 Terrestrial Solar Radiation 27
  - 2.2.4 Beam Radiation and Clouds 29
  - 2.2.5 Diffuse and Global Radiation 29
  - 2.2.6 Spectral Direct and Diffuse Radiation 33
- 2.3 Thermodynamic Quality of Solar Radiation 35
  - 2.3.1 Measure of Quality 35
  - 2.3.2 Quality of Radiation 36
  - 2.3.3 Standard Spectra 38
  - 2.3.4 Quality of Solar Irradiance 40
- 2.4 Concentration of Radiation 41
- 2.5 Conversion to Heat 44
  - 2.5.1 Process Heat and Concentrated Radiation 46
  - 2.5.2 Selective Absorption-Transmission 48
  - 2.5.3 Yield of Process Heat 49
  - 2.5.4 Simultaneous Concentration and Selective Absorption 53
- 2.6 Conversion of Radiation to Electrical Energy 54
  - 2.6.1 Photoionization 54
  - 2.6.2 Photovoltaics 55
  - 2.6.3 Ideal Photocell 57
  - 2.6.4 Ideal Solar Cell Equation 58
  - 2.6.5 Parameters of Solar Cells 59
  - 2.6.6 Maximum Photovoltaic Efficiencies 63
  - 2.6.7 Spectral Matching of Solar Cell Devices 67
  - 2.6.8 Tandem Solar Cells 68
- 2.7 Photochemical Conversion 70
  - 2.7.1 Equation of Ideal Photochemical Processes 71
  - 2.7.2 Maximum Yield in Photochemical Processes 74
  - 2.7.3 $h\nu$-, eV-, and kT- Reaction Paths 75

2.8 Appendix 1: Measurement of Solar Radiation 76
    2.8.1 Introduction 76
    2.8.2 Basic Quantities and Instrumentation 76
    2.8.3 Detectors, Windows, Filters 77
    2.8.4 Description of Instruments 77
2.9 Appendix 2: Frequently Used Symbols 80
Bibliography 81

3 Concentrator Optics   84
By L. L. Vant-Hull

3.1 Introduction 84
3.2 Basic Optics 84
3.3 Concentration Optics 87
    3.3.1 Concepts of Concentrator Optics 87
    3.3.2 Solar and Circumsolar Brightness Distribution – Sunshape 88
    3.3.3 The Degraded Sun 90
    3.3.4 The Error Function for the Concentrator 92
    3.3.5 Flux Density and Concentration 95
    3.3.6 Cassegranian Optics 96
3.4 Ideal Concentrators 98
    3.4.1 Conceptual Framework 98
    3.4.2 Compound Parabolic Concentrator 99
    3.4.3 Flow-line or Trumpet Concentrators 102
    3.4.4 Conical Flux Density Redirector 103
3.5 Parabolic Geometries 104
    3.5.1 General Considerations 104
    3.5.2 Geometric Concentration Ratio 106
    3.5.3 Local Concentration Ratio: Flux Density Distribution 109
    3.5.4 The Iso-Intensity Problem 112
3.6 Other Concentrating Geometries 113
    3.6.1 Introduction 113
    3.6.2 The Hemispherical Bowl Concentrator 113
    3.6.3 The Line Focus Fixed Mirror Collector 113
    3.6.4 Tracking Facet Distributed Receiver Systems 114
    3.6.5 Fresnel Reflectors 115
    3.6.6 Fresnel Lenses 117
    3.6.7 Other Optical Configurations 119
3.7 Central Receivers 119
    3.7.1 Introduction 119
    3.7.2 Scaling Relationships 119
    3.7.3 Shading and Blocking Calculations 120
    3.7.4 Flux Density Distribution at the Receiver 120
3.8 Design Issues and Constraints 122
    3.8.1 Preliminary System Level Considerations 122
    3.8.2 System Optimization 125
    3.8.3 System Performance 128
    3.8.4 Layout 128
3.9 System Sizing 129
3.10 Appendix: Frequently Used Symbols 130
Bibliography 131

4 Aspects of Solar Power Plant Engineering 134
By W. Grasse, H. P. Hertlein, and C.-J. Winter
4.1 Introduction 134
4.2 Solar and Conventional Power Plants: Similarities and Differences 134
4.3 Engineering Aspects 136
    4.3.1 Collection 136
    4.3.2 Energy Conversion 139
    4.3.3 Characterization and Physical Properties of Solar Power Plants 142
4.4 Design Aspects 153
    4.4.1 Terminology 154
    4.4.2 Factors Influencing Power and Energy Performance 155
    4.4.3 Design Objectives 157
    4.4.4 Design Process and Parameters 158
Bibliography 161

5 Thermal Receivers 163
By M. Becker and L. L. Vant-Hull
5.1 Introduction 163
5.2 Principles and Concepts for Energy Transfer 164
5.3 Thermal and Thermodynamic Basis for Receiver Design 167
5.4 Physical Interactions 174
5.5 Engineering Methods of Computation 177
5.6 Receiver Designs 179
    5.6.1 Tube Receiver Concept (Central Receiver) 179
    5.6.2 Tube Receiver Concept (Parabolic Trough) 184
    5.6.3 Volumetric Receiver Concept 185
    5.6.4 Direct Absorption Receiver Concept 189
5.7 Relationships Between Design and Type of Application 190
5.8 Status and Prospects 191
5.9 Measurement Techniques 192
5.10 Receiver Loss Calculation Examples 194
5.11 Appendix: Frequently Used Terms 196
Bibliography 197

6 Thermal Storage for Solar Power Plants 199
By M. A. Geyer
6.1 Impact of Storage on Solar Power Plants 199
    6.1.1 Capacity Factor and Solar Multiple 199
    6.1.2 Optimization of Solar Multiple and Storage Capacity 201
6.2 Media for Thermal Storage 203
    6.2.1 Sensible Heat Storage Media 203
    6.2.2 Latent Heat Storage Media 204
    6.2.3 Chemical Storage Media 204
    6.2.4 Single Versus Dual Medium Concepts 204
6.3 State-of-the-art of Thermal Storage for Solar Power Plants 205
    6.3.1 Thermal Storage for Oil-Cooled Solar Plants 206
    6.3.2 Thermal Storage for Steam-Cooled Solar Plants 207
    6.3.3 Thermal Storage for Molten Salt-Cooled Solar Plants 207
    6.3.4 Thermal Storage for Sodium-Cooled Solar Plants 210
    6.3.5 Thermal Storage for Gas-Cooled Solar Plants 211

	6.4	Appendix: Frequently Used Symbols	212
	Bibliography		213

7 Thermal Solar Power Plants Experience 215
By W. Grasse, H. P. Hertlein, and C.-J. Winter, with contributions by G.W. Braun

- 7.1 Introduction . . . 215
- 7.2 Farm Solar Power Plants with Line-Focussing Collectors . . . 215
  - 7.2.1 Plant Configurations . . . 216
  - 7.2.2 System Examples . . . 219
  - 7.2.3 Collector Subsystem . . . 223
  - 7.2.4 Plant Performance Characteristics . . . 225
  - 7.2.5 Technical and Operational Potential . . . 226
- 7.3 Farm Solar Power Plants with Point-Focussing Collectors . . . 229
  - 7.3.1 Plant Configurations . . . 229
  - 7.3.2 System Examples . . . 231
  - 7.3.3 Plant Performance Characteristics . . . 235
  - 7.3.4 Technological and Operational Potential . . . 235
- 7.4 Central Receiver Solar Power Plants with Heliostat Fields . . . 237
  - 7.4.1 Plant Configurations . . . 237
  - 7.4.2 System Examples . . . 241
  - 7.4.3 Heliostat and Heliostat Field . . . 244
  - 7.4.4 Plant Performance . . . 246
  - 7.4.5 Plant Performance Characteristics . . . 247
  - 7.4.6 Technological and Operational Potential . . . 249
- 7.5 Individual Dish Solar Power Plants . . . 250
  - 7.5.1 Configuration and Technology . . . 250
  - 7.5.2 Dish/Stirling Examples . . . 254
  - 7.5.3 Plant Performance . . . 255
  - 7.5.4 Plant Characteristics . . . 257
  - 7.5.5 Technological and Operational Potential . . . 259
- 7.6 Comparison of Thermal Solar Power Plants . . . 259
  - 7.6.1 Performance Comparison . . . 259
  - 7.6.2 Long-term Operating Histories . . . 261
- 7.7 Test Sites for Solar-Thermal R&D . . . 262
- 7.8 Thermal Solar Power Plant Modelling and Calculation Codes . . . 263
  - 7.8.1 Performance Models . . . 263
  - 7.8.2 Economic Analysis Models . . . 267
- 7.9 Appendix: Solar Thermal Facility Data . . . 271
- Bibliography . . . 280

8 Photovoltaic Power Stations 283
By W. Bloss, H. P. Hertlein, W. Knaupp, S. Nann, and F. Pfisterer

- 8.1 Introduction . . . 283
- 8.2 Technical Aspects of Solar Cells . . . 284
  - 8.2.1 IV-Characteristic of Solar Cells . . . 284
  - 8.2.2 Temperature Effects . . . 285
  - 8.2.3 Radiation Absorption and Material Selection . . . 286
  - 8.2.4 Tandem Systems . . . 287
- 8.3 Status of Solar Cell Development . . . 288
  - 8.3.1 Crystalline Silicon Solar Cells . . . 289

	8.3.2	Amorphous Silicon Thin Film Solar Cells 291
	8.3.3	Polycrystalline Thin Film Solar Cells 292
	8.3.4	Concentrator Cells 292
	8.3.5	Tandem Solar Cells 292
8.4	Photovoltaic Modules 293	
	8.4.1	Status of Non-Concentrator Module Technology 293
	8.4.2	Module Design and Interconnection of Cells 294
8.5	Power Conditioning Systems 297	
	8.5.1	DC-DC Converter, Maximum-Power-Point Tracking 297
	8.5.2	DC-AC Inverter 299
	8.5.3	Batteries and Charge Regulators 299
8.6	Supporting Structures 300	
	8.6.1	Basic Design Considerations 301
	8.6.2	Review of Selected Support Structures 301
	8.6.3	Support Structures for Tracking Arrays 303
8.7	Tracking and Concentrating Systems 304	
	8.7.1	Fresnel Modules 305
	8.7.2	V-Trough Concentrator 306
	8.7.3	Parabolic Geometries 307
	8.7.4	Further Concentrator Concepts 307
	8.7.5	Perspectives of Tracking and Concentrating Systems 307
8.8	Design Considerations for Grid-Connected Power Plants 309	
	8.8.1	Site and System Selection 309
	8.8.2	Electrical Circuit Design Aspects 311
	8.8.3	Plant Monitoring 312
8.9	PV Plant Operating Experience 314	
	8.9.1	The Experience Base 314
	8.9.2	Operating Experience 314
	8.9.3	Summary and Conclusions 323
8.10	Photovoltaic Solar Systems Modelling and Calculation Codes 324	
8.11	Appendix: Frequently Used Symbols 327	
8.12	Appendix: Photovoltaic Facility Data 327	
	Bibliography 332	

9 Solar Fuels and Chemicals, Solar Hydrogen   336
By M. Fischer and R. Tamme

9.1 Introduction 336
9.2 Endothermal Chemical Processes Coupled with Solar Energy 337
9.3 Receiver-Reactors for Solar Chemical Applications 339
9.4 High Temperature Processes for Fuels and Chemicals Production 343
9.5 Additional Chemical Processing Using Solar Energy 347
9.6 Steam/Carbondioxide Reforming of Methane – A Candidate Process 348
9.7 High Temperature Processes by Direct Absorption of Solar Radiation 352
9.8 Electrolytic Production of Hydrogen with Photovoltaic and Solar Thermal
    Power Plants 354
    9.8.1 Electrolytic Production of Hydrogen with Photovoltaic Systems 354
    9.8.2 Electrolytic Production of Hydrogen with Thermal Solar Power Plants 361
Bibliography 364

10 Cost Analysis of Solar Power Plants   367
By H. P. Hertlein, H. Klaiss, and J. Nitsch
   10.1 SPP Technologies in Comparison 367
   10.2 Investment, Operating and Maintenance Cost 372
       10.2.1 Parabolic Trough Solar Power Plants 373
       10.2.2 Central Receiver (Tower) Solar Power Plants 375
       10.2.3 Dish/Stirling Units 380
       10.2.4 Photovoltaic Solar Power Plants 382
   10.3 Power Plant Cost Analysis and Comparison 386
       10.3.1 Conventional Power Plant Generating Costs 387
       10.3.2 Solar Power Plant Generating Costs 389
       10.3.3 Sensitivity Analysis of SPP Generating Costs 392
       10.3.4 Social Costs 396
   10.4 Market Considerations 399
       10.4.1 Introduction 399
       10.4.2 Costs 400
       10.4.3 Value 400
       10.4.4 Financing 403
       10.4.5 Risk 404
       10.4.6 Market Potential and Outlook 405
   Bibliography 407

Appendix A: Glossary of Terms   410
   A.1 Solar Resource Terminology 410
   A.2 Solar Thermal Terminology 411
   A.3 Photovoltaic Terminology 414
   A.4 Financial Terminology 415

Appendix B: Abbreviations and Acronyms   417
   B.1 Radiation, Solar 417
   B.2 Thermal 417
   B.3 Photovoltaic 418
   B.4 Cost/Economic 418
   B.5 Acronyms 419

Subject Index   420

1 The Energy Heptagon

C.-J. Winter[1]

Never in the history of mankind has a single source of energy provided all the energy services needed for the support of human life and development. The ever-increasing energy demands of a growing world population have required the use of all available energy sources. Newly emerging types of energy never fully replaced the previously existing ones, although they did alter their relative share and their individual importance.

From the time men appeared on Earth until the start of industrialization in Europe, it was solar energy which in various forms served mankind by providing heat for cooking and heating, wind for transportation on rivers and seas, and power for grinding or pumping water. Solar radiation, wind, running water, tides and combustible biomass were the first sources of renewable or solar energy to be utilized (Fig. 1.1).

In the second half of the 18th century, coal came into use and helped make possible the industrialization of the world. In the second half of the 19th century, only about 100 years ago, mineral oil started to be widely exploited, not only providing a lighting source, but also making possible the individual long distance transport of large numbers of people. Today several hundred million motor vehicles are in use worldwide, with a correspondingly huge market impact. Again a number of decades later, and in Europe hardly more than three to four decades ago, natural gas began to replace coal-derived coke-oven or city gas in larger quantities. All of these resources are fossil fuels, ultimately derived from solar energy.

The geothermal resource associated with hot aquifers is being relatively minimally exploited in several geothermal generating stations with a total capacity of not more than a few thousand megawatts. Use of tidal power is even more limited, amounting to a few hundred megawatts worldwide. Lately nuclear energy has been added to the list. Half a century ago, the first experimental nuclear chain reaction took place; today, nuclear energy provides 5% of the world's primary energy needs from some 400–500 reactors.

A renaissance of interest in solar energy began in the early 1970s, and this for three predominant reasons. First, the oil crises of that decade brought back to mind that the raw materials available in the Earth's crust for fossil and nuclear energy are finite and, moreover, so concentrated in a limited number of countries as to predestine, by geographical happenstance, oligopolistic potentials. The second reason is that the liberal use of fossil fuels is causing irreversible ecological damage at an increasing rate to nature in general, to mankind, fauna and flora. In addition, irreplaceable cultural monuments of a shared human heritage are being destroyed. The complex relationship between the environment and the industrial and agricultural behavior of modern man is not yet well understood. Much remains still hidden because of the long time scale, extending from decades to centuries, against which nature's reactions need to be measured. The urgency of this situation becomes evident with the realization that 90% or more of the world's energy consumed today is of carbon-containing

---

[1] Carl-Jochen Winter, Deutsche Forschungsanstalt für Luft- und Raumfahrt (DLR), Pfaffenwaldring 38-40, D-7000 Stuttgart 80

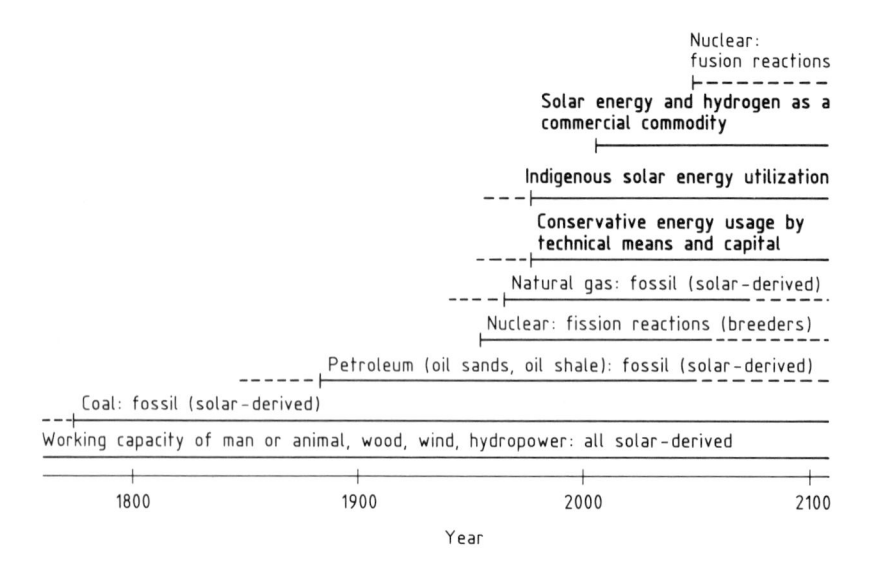

Fig. 1.1. History of the world energy economy (qualitative).

fossil energy, be it coal, oil, or natural gas, be it wood, or agricultural and forestry waste products. Invariably their utilization involves combustion processes with air as oxidizer. The resulting oxidation reactants are being unavoidably released into the environment: CO, CO_2, SO_2, NO_x, residual C_nH_m, toxicants and heavy metals, as well as dust, and occasionally soot and particulates, or, if desulfurizing or denoxing equipment is used, gypsum or surplus NH_3. Only CO_2 release from biomass combustion is environmentally neutral because the CO_2 which is released was recently taken from the environment when the biomass was formed.

The third and last reason for the renaissance of interest in solar energy is the fact that nuclear energy is only reluctantly tolerated in many countries and restricted by moratoria or quasi-moratoria in others. In the United States not a single nuclear reactor has been ordered since 1978, and several orders have been cancelled or are under renegotiation. In Europe, the use of nuclear power has been renounced in Denmark, Greece, Luxembourg and Portugal; the Netherlands have a de facto moratorium, and parliaments in Italy and Sweden have approved a ban. The major users of nuclear power are Belgium, France, Germany, and the United Kingdom, with France being the most confident. Worldwide, most of the nuclear reactors in operation are located in the industrial countries of the North. Some developing countries have committed themselves to the use of nuclear energy, but by far not to the extent previously planned. Further, if the intended transfer to breeder-type reactors cannot be successfully accomplished (nowhere has this been achieved so far), the light-water reactors now in operation must be shut down at the very latest when the Earth's known recoverable reserves and any additionally suspected resources of fissionable material are depleted. Failure to resolve the radioactive waste disposal problem may provide an even more serious limitation.

Solar energy usage has two fundamental facets. The first one comprises a variety of local applications characterized by collection, conversion and consumption of the solar energy *on site*. To this group belong passive solar energy usage in buildings, heat production by solar radiation collectors, photovoltaic arrays for electricity generation, ambient heat use in heat pumps, and the conversion of wind, hydropower, or biomass into electrical energy, heat and gaseous or liquid fuels. What these *solar* energy technologies have in common is that they are not restricted only to those regions of the world that offer ideal solar conditions. They are, indeed, capable of delivering solar-derived secondary energies in almost all climates. However, depending on such factors as irradiance, wind profiles, precipitation zones, etc., the amount of natural solar energy available may differ from place to place by factors of two to three, annually averaged. For example, the best average irradiation conditions on Earth, found on the Arabian peninsula or in the North American Southwest, provide $\approx$ 2,500 kWh/m^{2}a, while the annual average for central Europe is only 1,000 kWh/m^{2}a.

The *off-site* facet comprises thermal solar power plants with their need for high direct irradiation, photovoltaic solar power plants, large hydropower complexes and ocean thermal energy conversion (OTEC) systems, among others. All belong to the category of solar energy *converters* whose common feature is the necessity to deliver secondary energy over distances up to several thousand kilometers because their primary users rarely (and in the case of the OTEC systems hardly ever) reside in the vicinity of the plant. Heat and electricity are neither storable in the quantities needed for national markets, nor can they be transported with acceptable loss over the intercontinental distances mentioned (ignoring not yet commercially available superconducting bulk electricity transmission). This can be contrasted with the potential of hydrogen as a carrier of chemical secondary energy. Produced by dissociation of water with the help of solar process heat and solar-derived electricity, hydrogen can be stored in very large quantities over very long periods of time [24]. It can be transported in gaseous form through pipelines, or shipped in liquid form in cryogenic tankers to energy-intensive agglomerates. There hydrogen can be used as a raw material in the chemical industry, be oxidized, either with pure oxygen or with air as oxidizer, supplying heat or electricity to the local economy and providing energy for surface, air or sea transportation. Of course, within several thousand kilometers of a solar plant, high voltage direct current (HVDC) transmission and direct use of solar generated electricity is a reasonable alternative, as long as this fluctuating energy source can be fed into and tolerated by electricity grids without severe stability and control problems.

With the example of hydrogen, the *cyclic character* of solar energy becomes obvious: Solar radiation from the Sun, after end use, is returned to space in the form of heat at ambient temperature. Simultaneously, the water from the global water inventory used for electrolysis is given back, after splitting and recombining, to that same inventory without loss in quantity or quality. Generating and using solar hydrogen energy involves the removal from the natural cycle of no more than about 10% per unit area (proportional to the efficiency of solar power conversion into hydrogen) of the annual solar offer to that area from the natural cycle, its storage and transportation from, say, the Sahara over thousands of kilometers to countries of heavy energy usage in the northern hemisphere of the globe, and its diffusion into the environment after utilization, from where it is reradiated into space: *an environmentally closed cycle*. Before solar energy, or any energy for that matter, is put to use in substantial amounts, at each conversion stage where energy is generated, converted, transported, stored or used, one precondition must be fulfilled: Energy should be used efficiently! In this sense the efficient usage of energy is an additional energy '*resource*', one which does not require any energy raw materials and which can be unlocked by technical means and the investment of capital alone [11]. The importance of rational energy use can be deduced from the different amounts of energy needed for the production of the per capita GNP in the European countries as contrasted with Canada or the U.S. The relationship is approximately 1:2, each inhabitant of West Germany or Switzerland needing only 5–6 tce (tons of coal equivalent) per capita for the production of his part of the GNP, while Canadians or Americans require 10–12 tce per capita (Fig. 1.2), without major differences in quality of life standards between the two continents.

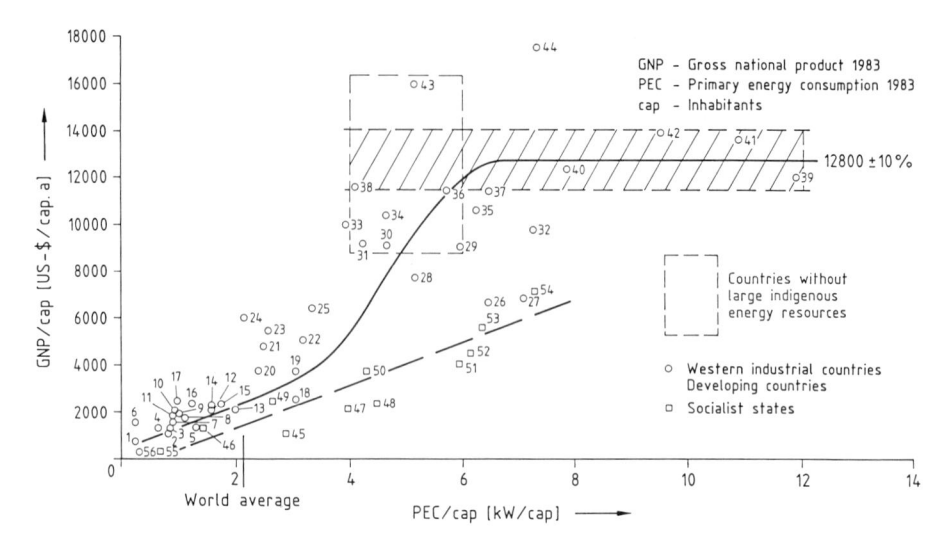

Fig. 1.2. Energy intensities of countries; 1:Ivory Coast; 2:Costa Rica; 3:Turkey; 4:Tunisia; 5:Jamaica; 6:Paraguay; 7:Ecuador; 8:Jordan; 9:Malaysia; 10:Chile; 11:Brazil; 12:South Korea; 13:Argentina; 14:Portugal; 15:Mexico; 16:Algeria; 17:Uruguay; 18:South Africa; 19:Venezuela; 20:Greece; 21:Spain; 22:Ireland; 23:Israel; 24:Hong Kong; 25:Italy; 26:Singapore; 27:Trinidad & Tobago; 28:New Zealand; 29:Belgium; 30:United Kingdom; 31:Austria; 32:Netherlands; 33:Japan; 34:France; 35:Finland; 36:West Germany; 37:Australia; 38:Denmark; 39:Canada; 40:Sweden; 41:Norway; 42:USA; 43:Switzerland; 44:Kuwait; 45:North Korea; 46:Cuba; 47:Hungary; 48:Romania; 49:Yugoslavia; 50:Poland; 51:Bulgaria; 52:USSR; 53:Czechoslovakia; 54:East Germany; 55:PR of China; 56:India [8,1].

A potential global *energy heptagon* may thus consist of *seven* different energy resources: (1) coal, (2) oil (3) natural gas, (4) nuclear fission, (5) efficient energy usage, (6) indigenous solar energy utilization, and (7) off-site solar electricity or solar hydrogen as secondary solar energy carriers which can perpetuate the world energy trade in a not-too-distant prospective post-fossil energy era. Coal, oil, natural gas and biomass usage account for 88% (1987) of world energy consumption, the rest being supplied by hydropower (7%) and nuclear energy (5%). Hardly quantifiable, but amounting to thousands of times this total, is the additional solar energy which provides food, oxygen, and a livable environment for all.

Coal, mineral oil, natural gas and nuclear fission are *open systems*: they take something irreplaceable from somewhere out of the Earth's crust and return it elsewhere to the geosphere in a chemically or isotopically altered form, sometimes poisonous, in nuclear systems unavoidably radioactive, in fossil systems always in combination with the removal of oxygen from and the release of CO_2 into the atmosphere. In both cases conversion occurs in conjunction with an additional transfer of heat into the environment amounting to the heat content of the total primary energy introduced.

The remaining three energies of the heptagon, efficient energy usage, indigenous solar energy utilization and off-site solar electricity or solar hydrogen energy, are fundamentally different: none of them depends on energy raw materials nor do they release any pollutants related to energy raw materials (Fig. 1.3). Solar radiation as the 'raw material' for primary solar energy is free of charge and ubiquitous, and water is very nearly so. For their deployment, these energies require only the investment of technology and capital. Consequently, efficient energy usage, indigenous solar energy utilization and off-site solar electricity or solar hydrogen energy are predestined for development by the energy-poor but technologically-skilled and capital-rich industrialized countries. Such countries not only have an opportunity to reduce or in some cases to free themselves from their dependence on an unwelcome and heavy energy raw materials import burden, but also have a chance to take the lead in developing and introducing the needed technologies for both domestic and foreign markets.

Since nothing is solely advantageous, it is only fair to mention also the difficult aspects of solar energy utilization. Because solar energy flux densities are low compared to the excellent densities of fossil energies, the material requirements are high, large areas of land must be reserved, and the time spans for investment returns can be long. Figures 1.4, 1.5, 1.6 and 1.7 relate to these issues.

The land requirements for solar systems are, indeed, higher than those for conventional systems by 2–3 orders of magnitude (Fig. 1.4). Unfortunately, engineers cannot substantially improve this situation because *irradiation* is, in essence, a natural location-specific parameter. On the other hand, the unavailability of land is rarely a real restriction. A thought exercise (Fig. 1.5) suggests that under very conservative assumptions only 0.5% of the otherwise unused global land area would suffice to supply the total worldwide demand for end-use energy of approximately 8×10^9 tce/a (1988) with solar energy or solar hydrogen energy, based on existing technologies and efficiencies – a thought exercise, *nota bene!*

On the other hand, engineers are quite capable of tackling the other consequence of low solar energy density, namely its high material intensity [2,23]. Figure 1.6 shows that the difference in material intensity between solar and conventional systems is only 1–1.5 orders of magnitude, and it is not unrealistic to expect that future developments will close the gap further. Through successful research and development work the *solar-specific* items of solar conversion systems (e.g. heliostats, photovoltaic cells) have already been substantially improved and will continue to be improved. Conventional systems, in contrast, must accommodate on an increasing scale both resource depletion and precautions against environmental pollution. A convincing example is the potential necessity for CO_2 containment, for which a technically sound, commercially acceptable solution has yet to be found.

Energy-related amortization times (energy pay-back times) relate to the time spans over which an energy conversion system must be operated in order to 'pay back', or to regain, the energy which was invested in its construction, its lifelong operation and its eventual recycling. Figure 1.7 provides insight into energy pay-back times and energy gain factors[2] (expressing the excess of energy supplied by the system during its lifetime over the energy expended for its construction, operation and recycling) for fossil, nuclear and solar systems. It is self-evident that the energy amortization times of solar power plants can be up to one order of magnitude higher than those of conventional plants. This, however, is only one side of the coin; the other shows that conventional fossil or nuclear systems dependent on energy raw material must pay during their entire lifetime for their daily primary energy supply as well as for the safe disposal of residuals and pollutants over very long periods. It is again trivial to state that the corresponding expenses for solar systems are negligible.

When all aspects are combined for an overall energy gain factor which reflects the current development status, the following exegesis seems appropriate: hydropower and windpower plants, on one end of the spectrum, rank the highest and, in all probability, will continue to do so. Coal-fired power plants, on the other end, rank lowest from an energetic viewpoint, and taking all presumptive environmental penalties into consideration (CO_2!) is reasonable to expect that this trend will become worse in the future. Solar plants, whose current level of maturity has been reached after only 10–15 years of development, and existing nuclear power plants compare reasonably well and can be located in the middle of the energy gain

---

[2] It should be noted that from the standpoint of a precise definition, only solar conversion systems have an energy amplification factor > 1; the others are depleters of non-renewable resources.

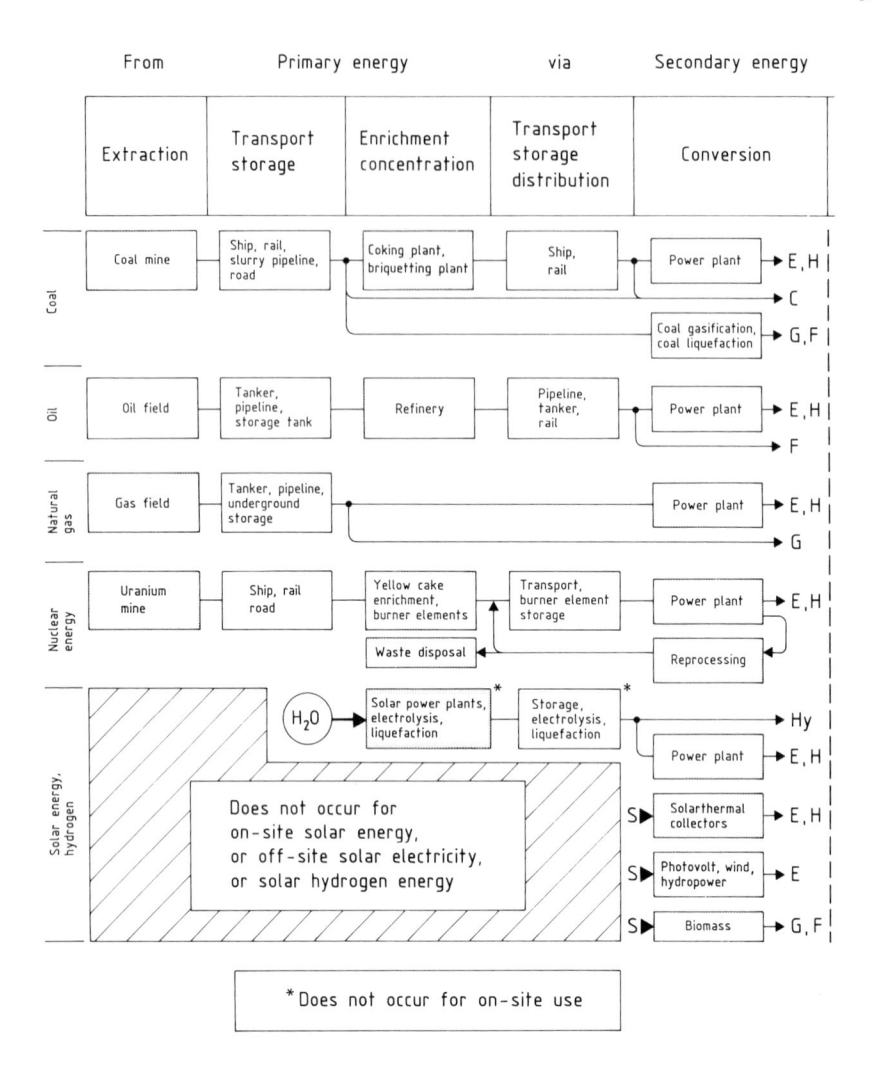

Fig. 1.3. Energy conversion chains.

spectrum. Considering the certainty of further improvements in the material intensities of solar power plants on the one hand, and the increasing environmental constraints and rigid safety measures in the entire nuclear conversion chain on the other, the argument is tenable that energy gain factors will altogether cease to be a significant penalizing criterion for solar systems. It may not even be far-fetched to surmise that, in the future, *solar cyclic energy systems* are *bound* to be energetically superior to any eventual environmentally-acceptable version of our current non-cyclic, open, environmentally-polluting fossil or nuclear systems which are, without exception and unavoidably, in discord with nature.

This book, 'Solar Power Plants', is to be seen in this context. So far, the equivalent of about two to three billion U.S. dollars has been spent worldwide in those countries involved in solar energy technologies – in France, Germany, Israel, Italy, Japan, Spain, Switzerland and,

1 The Energy Heptagon

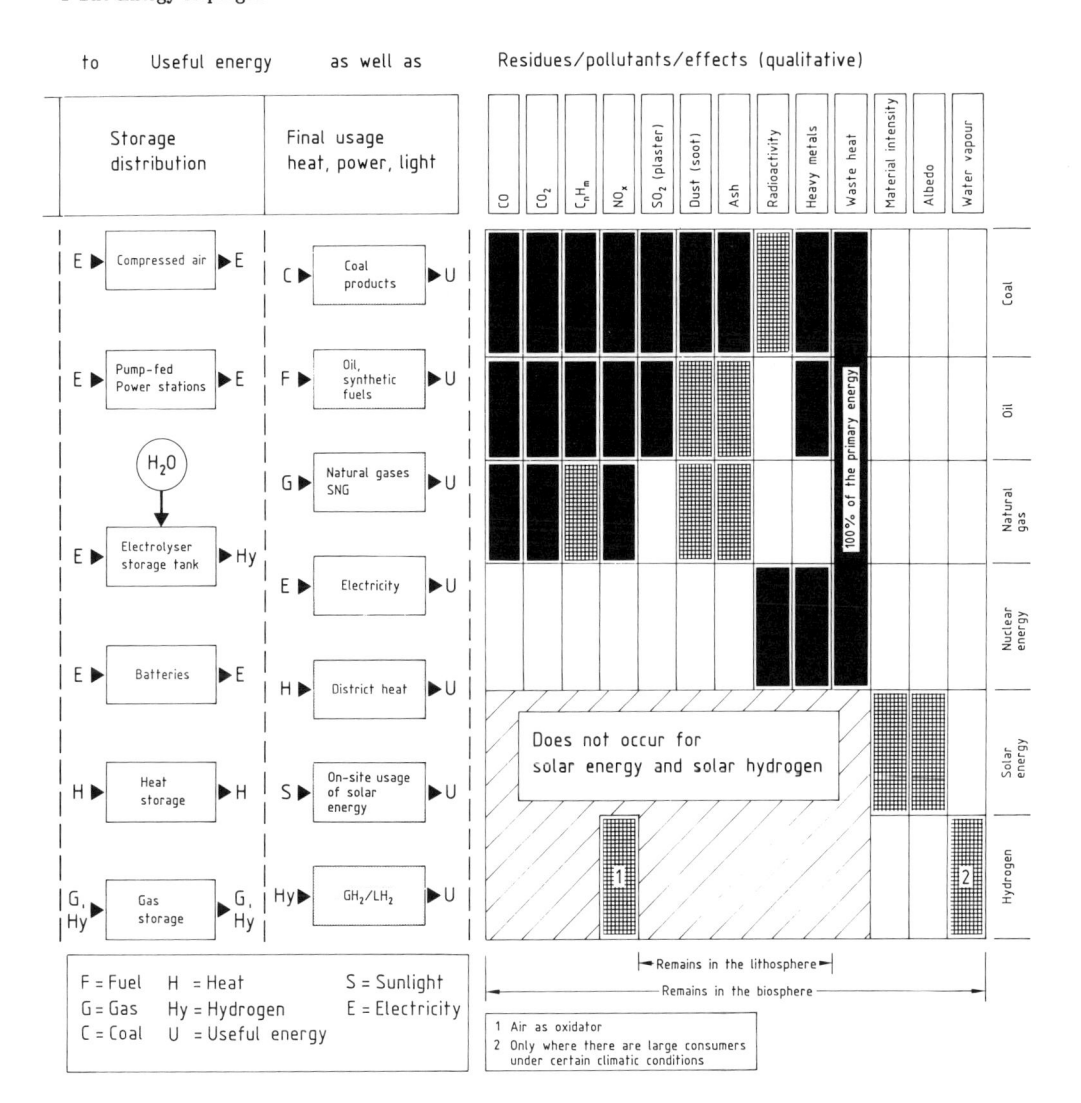

Fig. 1.3. (Continuation).

with the largest accumulated budget so far, in the U.S.. The knowledge shared on thermal and photovoltaic solar power conversion reflects an accumulation of several thousand man-years of research and experience pursued over 10–15 years. Positive as well as negative results from seven experimental solar tower power plants, distributed worldwide and altogether representing about 20 MW$_e$, has been made available. More than seven solar line-focussing farm plants with an accumulated installed capacity of some 200 MW$_e$ (1988) have been evaluated, as have quite a few point-focussing parabolic dishes. Similarly, the results of a representative number of photovoltaic stations have been incorporated for comparison, including the world's two largest single installations, both located in the U.S. and providing 6.5 and 1 MW$_e$, respectively. Today, world photovoltaic production capacity lies between 40 and 50 MW$_e$/a.

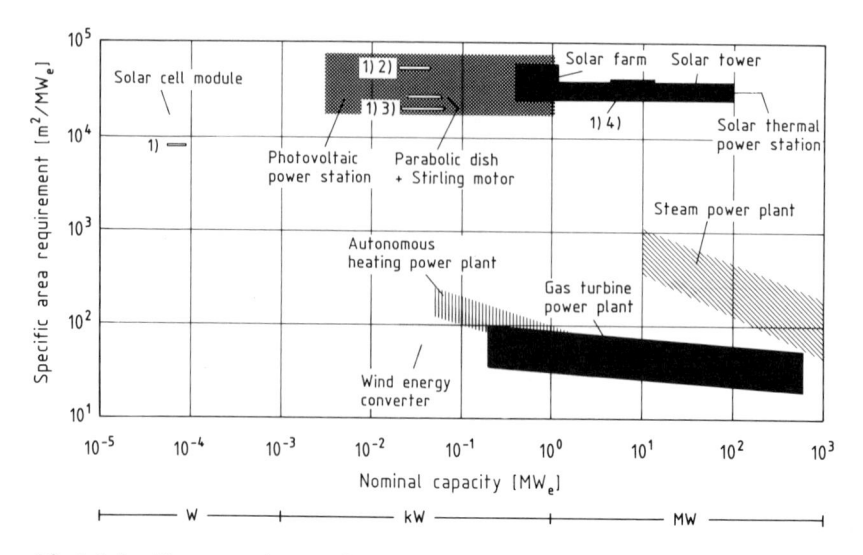

Fig. 1.4. Specific-area requirements for power plants [15,21]. 1) from [21] 2) with tracking; 3) without tracking; 4) without storage system · 0.5; 5) without coal mining.

Preconditions:

Global Irradiation:
> 2,300 kWh/m² ('very good')
> 2,000 kWh/m² ('good')

Surface structure:
flat, covered with gravel and stone fragments < 100 mm

Morphodynamics:
no sandstorms; no heavy rainfall

Vegetation:
none or very poor

1 EJ = 34.14 · 10⁶ tce

Fig. 1.5. Land areas available and suitable for solar hydrogen production [16]. 1,380 km · 1,380 km = 1.9 · 10⁶ km² = 5% of global desert area = 1.3% of global land area.

Modern research and development of solar power plant technology has only been going on for 10–15 years, too short a time to expect that solar plants can already provide a significant amount of useful energy services. Even when the intended contribution to the existing energy supply system is only 10%, the adoption of a 'novel' form of energy usually requires from several decades to half a century, as was demonstrated by nuclear energy, the most recent example. The first chain reaction on a laboratory scale took place in 1938 in Berlin, and half a

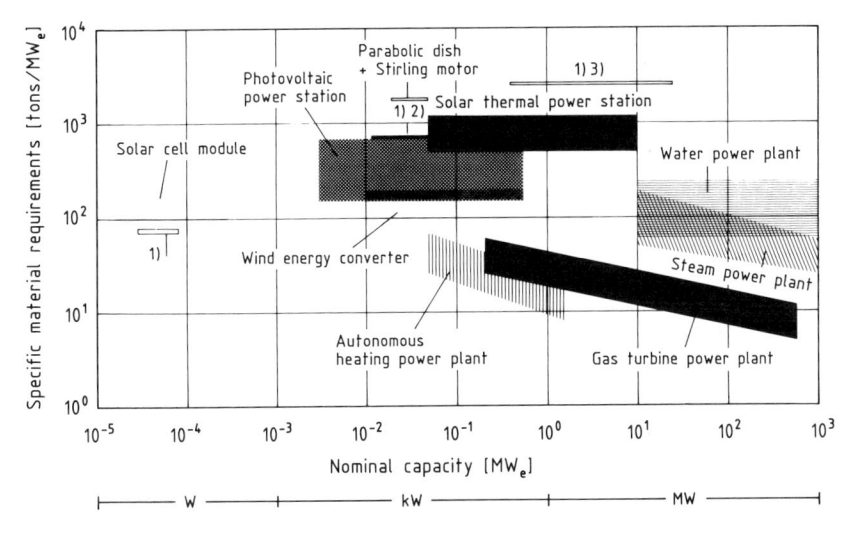

Fig. 1.6. Specific-material requirements for power plants [15,21]. 1) from [21]; 2) without tracking; 3) without storage system · 0.5.

century later nuclear energy still provides no more than 5% of the worldwide primary energy equivalent. The time constants of energy resource development have been of such duration in the past – 'come feast or famine'. Why should it be different with solar energy?

While there are certain similarities, there are also substantial differences between solar and other forms of energy:

- In the area of *ecology* and *safety*. There is absolutely no need whatsoever for a 'solar containment', whereas nuclear containment is an absolute necessity and 'fossil containment' will become increasingly important. Except for certain photovoltaic manufacturing processes, solar power plants are principally not burdened with toxicity or radioactivity; hence, their *inherent* safety is very high.
- In the area of *investment costs* and *financing*. The investment costs of solar plants are comparatively high, but a number of compensating advantages must also be taken into account. Plant construction times are short; 12–18 months for trough plants of a capacity of 30–80 MW_e is realistic according to reports. Consequently, the prefinancing needs of a plant prior to its first power delivery are moderate. Also, almost the entire capital outlay is due over the period of plant construction, i.e. follow-up costs for maintenance and repair are small and calculable. There are no expenses for energy raw materials or for dealing with pollutants. The non-contaminating construction material of modular design is easy to recycle. Since the classic bulk materials for solar plant construction (steel, concrete, aluminum, glass or silicon) are common in the machine-building and electrical industries, recycling difficulties may only be expected with the relatively small quantities needed of 'exotic' plastics or compounds. And of course, in hybrid plants which use certain amounts of natural gas, its consumption and atmospheric emissions have to be taken into consideration.
- In the area of *competitive markets*. Fossil and nuclear plants have incurred to date a number of social and environmental costs which are not, or at least not fully, internalized in the product's bill [9,13]. Relative to the solar kilowatt-hour, the price for the fossil or nuclear thermal or electrical kilowatt-hour is artificially low because the related external costs are

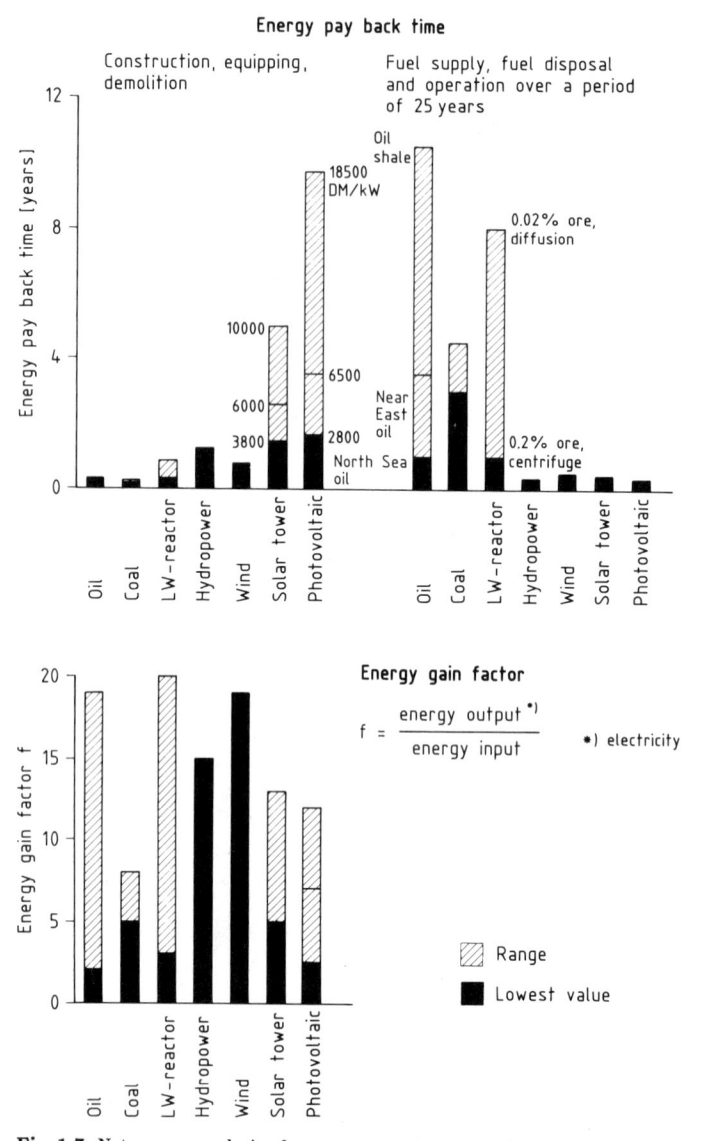

Fig. 1.7. Net energy analysis of energy conversion chains [4,6,10,12,17,18,19,21,22].

borne by the national economy, i.e. ultimately by the tax-paying public. In contrast, since very few external costs are known to occur for solar energy, it is at an unfair disadvantage in the energy market, having to face unequal competitive positions. Nevertheless, two developments will serve to promote solar energy in the future: the continuing technical developments and advances reported on in this book, and the concerted drive towards full internalization of all external costs. For example, a 1988 study carried out for the Commission of the European Communities [14] estimated the external costs for electricity production in Germany to be 0.02–0.045 $/kWh$_e$ (1987)[3] for fossil plants, and to be 0.05–0.105 $/kWh$_e$(1987) for nuclear plants. The weighted average of 0.025–0.06 $(1987)/kWh$_e$ almost corresponds to the present cost of power. In other words, if all external costs were to be internalized, the price of the product would double. This fact underlines how badly an *energy heptagon* is needed; the inclusion of solar energy is indeed a great leap forward!

To date some 300–500 MW$_e$a (1988) have been accumulated in the operation of thermal solar power plants and, although difficult to assess because of their worldwide dissemination, perhaps 100–200 MW$_e$a in photovoltaic capacity. These quantities are negligible compared to the fossil or nuclear plants, but rather satisfying if the short period of development is taken into account. At any rate, what is urgently needed for the future success of solar energy technologies in general and of solar power plants in particular, without compromise, is research and development financing. Without it, most efforts must ultimately be in vain. Of equal importance, however, is *continuity* of effort, instead of on-again/off-again attitudes and lack of foresight.

Solar power plants are unique in their dependence upon the availability of the solar resource and its quality. While most photovoltaic plants operate on total sunlight, thermal solar power plants utilize the high intensity direct irradiation, and hence can achieve high temperature as a result of the concentration of sunlight. Proper understanding of these alternatives and of the characteristics of sunlight itself requires considerable insight into the physics involved. The apparent motion and size of the Sun, atmospheric absorption, and cloud-free lines of sight are a few of these concepts, introduced in Chap. 2. In addition, thermodynamic arguments reveal that solar radiation provides a high quality 'fuel'. Detailed analysis shows that this solar 'fuel' can be utilized effectively to produce heat (to thousands of Kelvins), to excite photovoltaic cells (producing 0.3 to 3.0 volts potential difference) and to drive chemical reactions (either by delivering heat or photons to the reaction site).

It is not difficult to achieve the high concentration of sunlight. The basic laws of optics provide all the required theory. However, its large-scale realization in a cost-effective manner remains a challenge. The special mindset described in Chap. 3 leads to the definition of 'concentration optics', where image formation is abandoned in order to achieve the required concentration and collection of sunlight at lowest cost. The many possible optical configurations are conveniently divided into central receiver or distributed receiver systems. Central receiver systems use a multitude of large Sun-tracking mirrors (heliostats) to focus sunlight onto a single elevated central receiver, producing 10s or 100s of MW of heat. This thermal energy is used in a nearby plant as process heat or to generate electricity. In contrast, distributed receiver systems consist of a multitude of modules, each of which concentrates and transforms the sunlight to heat (or possibly to electricity or a chemical energy carrier). A 'farm' of hundreds or thousands of identical modules is required to produce commercial quantities of energy.

Distributed receivers are broadly classified as line focus or point focus systems. Line focus systems use cylindrical optical elements, require tracking of the Sun in one axis, and produce relatively low concentrations (10×–100×). Because of this low concentration to achieve even moderate temperatures, selectively absorbing receivers (low radiation loss) and concentric glass covers over the receiver (to suppress convection losses) are required. The linear modules can be easily connected end-to-end to form 'delta T strings' to achieve cost effective collection of heat at a centrally located site. In contrast, point focus collectors must track in two axes, but can achieve high concentration (100×–10,000×) and so are usually used to generate higher temperatures. A specialized photovoltaic array, a chemical reactor, or a heat engine may be mounted directly at the focal point to produce an easily collected chemical energy

---

[3] 2 DM ≃ 1 US$.

carrier. The alternative of collecting heat at high temperature from a multitude of individual tracking modules is less promising.

A discussion of the characteristics of the principal types of solar plants, as well as issues relating to the unique qualities and requirements of solar power plants, is presented in more detail in Chap. 4. Just as it is unreasonable to operate a large coal or nuclear plant in a daily peaking mode, solar plants cannot easily satisfy the steady requirements of base load power. On the other hand, since part of the peak load demand is solar-induced in some countries, for example the need for air conditioning in the U.S., solar plants provide a good match. This interaction and the role of storage and capacity factors in plant design are important features unique to solar plants.

Following Chap. 4 and as suggested by the physical concepts discussed in Chap. 2, the three conversion modes for the primary solar resource are discussed in more detail. Chapters 5, 6 and 7 concentrate on the conversion of sunlight into thermal energy and its use as process heat or as thermocynamic cycle input to run generators for producing 50/60 cycle alternating current in grid connection. Chapter 8 is devoted to photovoltaic solar power plants, and Chap. 9 presents a more experimental approach where concentrated sunlight is used to drive directly a chemical reaction, either by providing heat in a reaction zone, or by activating quantum processes using solar photons.

The final chapter discusses some of the economic factors introduced already in this chapter. A conservative estimate shows that full accounting of the costs for damages to the environment would typically double the cost of energy as delivered by conventional power plants. While solar plants are materials intensive, the fast energy payback assures that the environmental cost of constructing a solar plant is rapidly recovered. The costs of many of the existing experimental and pilot-scale solar plants are compiled and reasonable projections for future commercial-scale plants are presented. Based on this material it becomes clear that mature solar plants could compete with fossil or nuclear plants even today if social costs were reasonably accounted for. Chapter 10 argues that both fuel and social costs of fossil and nuclear plants are destined to go up, while advances in solar plant technology and increased production rates should lead to cost reductions. Consequently, once the questions of technology, economy and reliability have been resolved to the satisfaction of the utility managers and regulators, solar electric plants should have good market prospects where solar resource availability and electricity demand coincide, for instance in the U.S. and in Australia. In addition, solar-derived hydrogen can serve as a secondary energy carrier connecting resource-rich regions such as the Sahara with high-demand regions such as Europe.

Returning now to the three conversion modes discussed in detail in Chaps. 5–9, the question naturally arises, 'Which one is preferable?' While the three options are at substantially different stages in their development and evolution, some basic facts and several observations may provide guidance in such a decision. Note, however, that many of the statements made refer to a technology field yet evolving, so the decision made may well depend more on objectives, or on the time period when decisions have to be made.

Thermal Solar Plants

- They require concentrated direct irradiation and, therefore, can operate effectively only in the Earth's equatorial belt of ± 30–$40\,°$ N/S where direct sunlight is at a maximum.
- They deliver heat of medium or elevated temperature in the range of 100–1,000–(3,200) °C which can be utilized as process heat, converted into electricity, or serve in endothermic chemical reactions (solar chemistry).
- Their operation as heat-power (cogeneration) units is possible.

- Potential unit capacity is between about 100 kW$_t$ for single parabolic dishes and several hundred MW$_e$ for solar tower power plants, with parabolic trough plants situated in between. Larger capacities are possible when single modules are assembled in farm combinations.
- Solar heat can be stored for future use. Hence thermal solar plants are operable, if necessary, even before sunrise or after sunset (even 24-hour operation has been achieved). In addition, auxiliary fossil firing can be used as backup.
- Thermal solar power plants are naturally suited to be either peak-load power stations or intermediate-load power stations, with typical capacity factors of $\leq 40\%$ ($\simeq$3,000–3,500 h/a).
- One can distinguish between more or less conventional plant components (piping, valves, heat-exchangers, turbines, generators, etc.) and so-called solar-specific items such as heliostats or receivers. Due to very successful development efforts, the cost of the solar-specific items has been reduced from the original 60–70% (1975) of total plant cost to 30–40% (1988). Around 1975, heliostat installations cost $\approx 1,000$ \$/m^2, while the present price is less than 200 \$/m^2. The significant decrease in the cost of solar-specific plant items means that today a thermal solar plant requires a 40% solar-specific investment, and the remaining 60% is almost comparable in cost to a coal-fired power station (without its pollution abatement equipment or coal supply).
- The annual overall average efficiency of today's solar electrical power plant is typically $\leq 15\%$; potential efficiencies may be as high as ≤ 20–25%.
- Specific solar energy accumulations are typically 280–300 kWh$_e$/m^{2}a.
- Investment costs for an n-th (n $\geq$ 3) tower solar power plant of 100 MW$_e$ rating and providing a 38% annual capacity factor are in the range of 2,200–3,000 \$/kW$_e$ (1987), and are expected to decrease further. These figures include six hours of heat storage capacity and an oversize heliostat field to charge storage daily [9].
- With a 38% capacity factor, generating costs of 0.08–0.11 \$/kWh$_e$ (1987) are achievable [9].
- The accumulated operational time for thermal solar plants is approximately 800–1,000 MW$_e$a (1989).

Photovoltaic Solar Plants

- They use the global, and thus both diffuse and direct, irradiation and, therefore, are seldom subject to geographical restrictions. In principle, they can be installed all over the globe. Alternatively, concentration of direct beam irradiation can reduce the number of expensive photovoltaic cells required. This alternative gives up the advantage described above and adds the cost of the concentrating tracking system.
- They generate low voltage DC, which can be converted into AC voltage and transformed to any voltage level.
- Plant capacities, in principle, are unlimited and may range from a few kW$_e$ to potentially 1,000 MW$_e$ or even more. The primary interest in this book is in plants larger than 10 kW$_e$ which are grid connected or used to produce a transportable chemical energy carrier.
- Bulk electricity (GWh$_e$) storage is physically impossible and energy storage in batteries or flywheels, etc. is economically unattractive; the operating time of photovoltaic power plants is therefore essentially equal to the length of the sunshine period. Consequently, the maximum capacity factors are < 30%, equivalent to $\approx 2,500$ h/a.
- Annually averaged overall power plant efficiencies are presently in the range of 6–8%, and may possibly reach 20–25% (there are indications [5,7] that photovoltaic power plants may eventually achieve higher efficiencies than thermal solar plants, but complex cell structures will be required.)

- A typical value for the specific energy accumulation in central Europe is 140 kWh_e/m^2a.
- Present specific overall power plant investment costs range between 7,500 and 10,000 $(1987)/kW_e$, with decreasing tendency but exclusive of cost for storage.
- Potentially achievable power costs in Germany are currently 0.30–0.50 $/kWh_e$ (1987) for a sun-tracking plant.
- To date, the accumulated operational time for photovoltaic plants is approximately 100–200 MW_ea (1988).

In concluding this brief comparison it can be said that for sites with sufficiently high direct irradiation, thermal solar power plants will be the forerunners today as well as 'tomorrow.' Considering, however, the development potentials of both and their dissimilar economy-of-scale characteristics, it is quite possible that the photovoltaic plants will not only catch up in electricity production, but may perhaps in the long run even leave the thermal plants behind. The dominant advantages of thermal plants, however, provide them a second avenue for competition: process heat production, heat-power (cogeneration) combinations, 'solar captive energy economy', large capacity factors via thermal storage, and, last but not least, solar-thermally induced chemistry are only possible with thermal solar plants.

Conclusion: at present, only in the United States are the three necessary ingredients for solar power plants – Sun, technical skill and capital – found together. Under the conditions prevailing in a moderately insolated industrialized country in Central Europe, thermal solar power plants are an item only for the export list. Fortunately, however, potential sites for photovoltaic plants can be found all over the globe.

Solar Chemical Plants

Although solar chemistry has not yet reached the point where chemical reactors are coupled to solar power plants, it will nevertheless be addressed because of the prospect that highly concentrated solar radiation possesses properties specifically favoring solar endothermic chemistry.

- Temperatures of $3,000\,°C$ can be achieved with fully tracking parabolic dish concentrators, and of 600–$1,000\,°C$ with central receiver systems; high enough for nearly any chemical reaction.
- This high quality heat can be delivered directly to the reaction zone with relatively little of the environmental pollution associated with other sources.
- The solar heat delivery is thermodynamically superior to processes involving heat exchange from fossil burners.
- Solar heat can be collected at temperatures required by nearly any industrial process (Fig. 1.8), again increasing thermodynamic efficiency.
- Photon initiated and photo-catalytic processes have been identified which show promise for new, solar beneficial and solar specific processes.
- Many of these solar-driven processes proceed at much lower temperatures than do conventional thermally driven reactions, leading to the promise of interesting new product streams.
- All solar chemical plants must be located in close proximity to the collection area.
- The solar photon process must occur in the receiver of the solar plant.
- Such photon processes are, of course, sun-following, and are subject to cloud and nighttime interruptions.
- Chemical plants are currently designed to operate with steady energy and temperature inputs. Plant designs and chemical reactions without these requirements are under development.

1 The Energy Heptagon   15

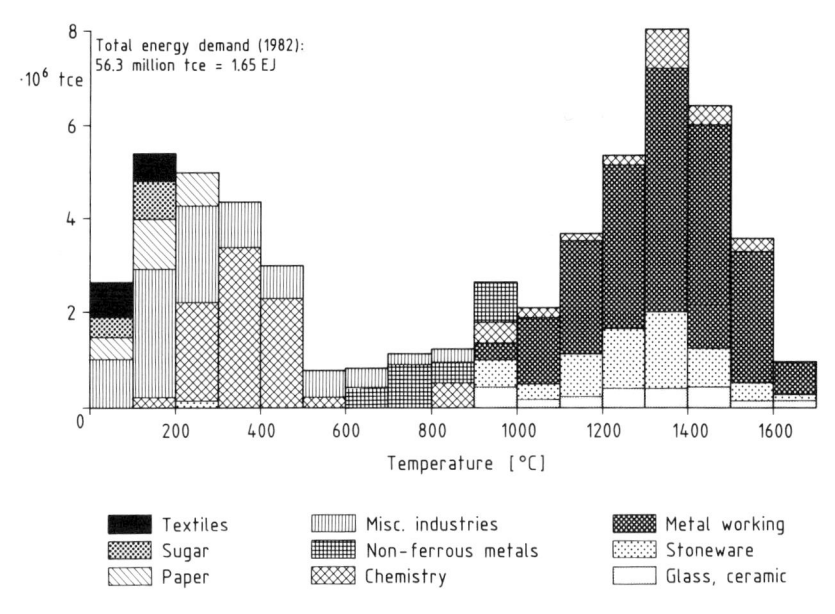

Fig. 1.8. Typical example of temperature requirement for industrial process heat in the Federal Republic of Germany 1982 [3].

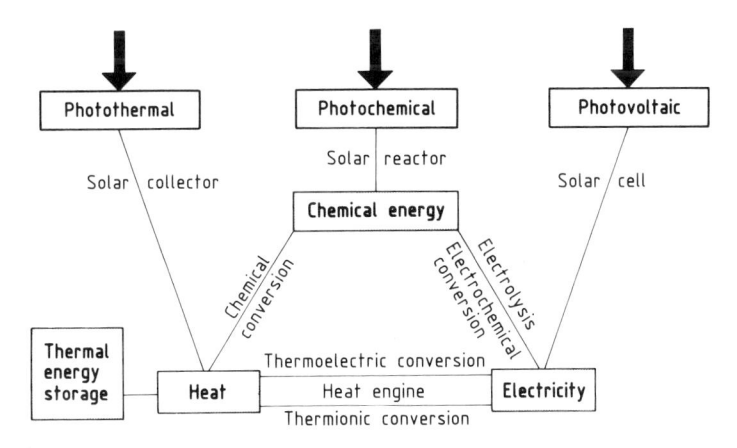

Fig. 1.9. Conversion of solar energy (redrawn from [20]).

- An endothermic transfer process might use sensible heat absorbed in a falling film of salt or particulates, stored in a buffer, and delivered to the process as needed.

In summary, without doubt thermal solar or photovoltaic electricity generation and solar process heat production are the first and second goals, but solar chemistry may, in the more distant future, become the third goal of large-scale solar energy conversion (Fig. 1.9). Furthermore, considering the as yet still moderate but nevertheless promising amount of knowledge about solar chemistry, it seems worth promoting this innovative, challenging field which will add a whole new dimension to the solar industry.

To come to a close: solar power plants have so gained in importance that it is not at all unrealistic to expect them to become a significant part of man's *energy supply heptagon*, a cyclic, non-polluting part which does not require energy raw materials. What is still lacking is continuing progress in research and development; may this book make a contribution to that end.

Bibliography

[1] In *Statistik der Energiewirtschaft*, Essen (D): Vereinigung der Industriellen Kraftwirtschaft (VIK)
[2] *Hydrogen Energy Technology* (in German). Volume 602 of *VDI-Berichte*, Düsseldorf (D): VDI-Verlag, 1987
[3] Information of the Forschungsstelle für Energiewirtschaft (FfE). TU München, 1984
[4] Private Communication. Sandia National Laboratories, Livermore/CA, 1981
[5] Program on Energy Research and Energy Technologies (in German). In *Statusreport 1987: Photovoltaik*, Projektleitung Biologie, Energie, Ökologie (BEO), KFA Jülich, Bonn (D): BMFT, 1987
[6] Aulich, H.: Energy Pay-Back Time – An Economic Criterion for Photovoltaics (in German). *Sonnenenergie*, 6 (1986) 14–17
[7] Bauer, G. H.: Photovoltaic Electricity Generation. In *Hydrogen as an Energy Carrier – Technologies, Systems, Economy*, Berlin, Heidelberg, New York: Springer, 1988
[8] Bockris, J. O.; Dandapani, B.: The Dependency of the Average Income within Countries upon the Energy Input to Run Them. *Hydrogen Energy*, 12 (1987) 439–444
[9] Couch, W. A.: Summary – Central Receiver Utility Study Activities. In *Proc. Annual Solar Thermal Technology Research and Development Conference*, Sandia National Laboratories, Albuquerque/NM, 1989
[10] Enger, R. C.; Weichel, H.: Solar Electricity Generating System Resource Requirements. *Solar Energy*, 23 (1979) 255–261
[11] Goldenberg, J.; Johansson, T. B.; Reddy, A. K. N.; Williams, R. H.: *Energy for a Sustainable World*. New York: J. Wiley & Sons, 1988
[12] Heinloth, K.: *Energy* (in German). Stuttgart (D): Teubner, 1983
[13] Hohenwarter, D. J.: The Relevance of Net Energy Analysis to Solar Energy Planning. In *Congress of the International Solar Energy Society*, Montreal: 1985
[14] Hohmeyer, O.: *Social Costs of Energy Consumption*. Berlin, Heidelberg, New York: Springer, 1988
[15] Jensch, W.: *Comparison of Energy Supply Systems with Different Degrees of Centralization* (in German). Volume 22 of *FfE-Schriftenreihe*, TU München, 1988
[16] Klaiss, H.; Nitsch, J.: Solar Hydrogen – Its Importance and Limits. In *Proc. ISES Solar World Congress, Hamburg 1987*, Oxford (UK): Pergamon Press, 1988
[17] Meyers, A. C.; Vant-Hull, L. L.: The Net Energy Analysis of the 100 MW_e Commercial Solar Tower. In *Proc. 1978 Annual Meeting, Denver/CO: Solar Diversification*, Boer, K. W.; Franta, G. E. (Ed.), pp. 768–792, Newark/DE: American Section of ISES, 1978
[18] Moraw, G.; et al.: Energy Investments in Nuclear and Solar Power Plants. *Nuclear Technology*, 33 (1987) 174–183
[19] Rotty, R. M.; et al.: *Net Energy from Nuclear Power*. Volume IEA 75-3 of *Report of the Institute for Energy Analysis*, Oak Ridge Associated Universities, 1975
[20] Sizmann, R.: Process Heat Applications. In *Proc. "Solar Energy '85" Summer School, Igls/Austria*, pp. 93–100, Köln (D): ASA/DLR, 1985
[21] Voigt, C.: Material and Energy Requirements of Solar Hydrogen Plants. *Hydrogen Energy*, 9 (1988) 491–500
[22] Wagner, H. J.: Energy Input for the Construction and Operation of Energy Supply Facilities (in German). In *7. Hochschultage Energie*, Universität Essen, 1986
[23] Wagner, H. J.: *Energy Input for the Construction and Operation of Several Energy Supply Technologies* (in German). Volume JÜL-1561 of *Berichte der KfA Jülich*, KfA Jülich, 1978
[24] Winter, C.-J.; Nitsch, J. (Ed.): *Hydrogen as an Energy Carrier – Technologies, Systems, Economy*. Berlin, Heidelberg, New York: Springer, 1988

2 Solar Radiation Conversion

R. Sizmann, with contributions by P. Köpke and R. Busen [1]

2.1 Introduction

Solar radiation incident upon the Earth is the primary energy source by which the life of mankind has developed. In this chapter basic concepts of the *conversion* of solar radiation for its use in the present day's energy economy are considered: to heat, to electric energy, to chemical energy carriers. The emphasis will be on the upper limits of conversion yields, derived from laws of thermodynamics. Such upper limits expose the boundaries of how much can at most be achieved from incident solar radiation. They are a guidance to appreciation of what has already been realized in practice and what still is the potential for further development. Though, in quoting Sadi Carnot [10], whose renowned 'Carnot factor' is in thermodynamics synonymous with ideal reversible processes, i.e. with best possible efficiency:

> *The attempts made to attain this result would be far more hurtful than useful if they caused other important considerations to be neglected.*

The chapter begins with a review of the influence of Sun-Earth astronomy on extraterrestrial solar radiant flux. A survey is presented of extinction processes in the atmosphere and the meteorology of cloudiness; both elements reduce the terrestrially available direct and diffuse solar flux.

2.2 Solar Radiant Flux

Matter emits incoherent electromagnetic radiation, usually referred to as thermal or heat radiation. The strength of flux through a unit surface area (referred to as *radiosity* or *radiant exitance*), M, depends on temperature and materials properties, in particular on surface properties. However, by laws of thermodynamics there exists an upper limit M_b of radiosity which is independent of the material and dependent solely upon the absolute temperature T,

$$M_b = \sigma T^4 \tag{2.1}$$

The universal factor σ for emission in the hemisphere, the Stefan-Boltzmann constant, contains fundamental constants of nature [11],

$$\sigma = \frac{2\pi^5 k^4}{15 c^2 h^3} = 5.67051 \cdot 10^{-8} \text{ Wm}^{-2}\text{K}^{-4} \tag{2.2}$$

---

[1] Rudolf Sizmann, Sektion Physik der Ludwig-Maximilians-Universität München, Amalienstr. 54, D-8000 München 40

k	=	Boltzmann's constant	=	$1.380658 \cdot 10^{-23}$	WsK^{-1}
h	=	Planck's constant	=	$6.6260755 \cdot 10^{-34}$	Ws^2
c	=	Vacuum light velocity	=	$2.9979258 \cdot 10^8$	ms^{-1}

Generally, radiosity depends on material properties and hence is lower than the upper limit M_b,

$$M = \bar{\epsilon}\sigma M_b = \bar{\epsilon}\sigma T^4. \tag{2.3}$$

Here $\bar{\epsilon}(T) \leq 1$ is the *emittance* ϵ, averaged over the spectral and angular distribution of unpolarized emitted radiation of the body at temperature T. The limiting case of maximum emittance, $\epsilon = 1$, also implies maximum *absorptance*, $\alpha = 1$. This is a consequence of Kirchhoff's law of equivalence of emittance and absorptance. Hence, a body with highest emittance $\epsilon = 1$ is at the same time the best absorber possible, $\alpha = 1$; any incident radiation becomes completely absorbed: the body appears to be perfectly black. For that reason radiosity with $\epsilon = 1$ is labelled *black body radiation*.

Assuming the Sun to be a spherical black body emitter, its radiant flux Φ_S is

$$\Phi_S = 4\pi R_S^2 M_S \quad \text{or} \quad \Phi_S = 4\pi R_S^2 \sigma T_S^4, \tag{2.4}$$

where $R_S = 6.96 \cdot 10^8$ m is the Sun's radius and T_S its surface temperature (photosphere temperature). By consequence of energy conservation, this flux passes through any imaginary external spherical surface concentric to the Sun (Fig. 2.1). In particular, Φ_S passes through a surface of radius D_{ES}, the distance between Earth and Sun, $D_{ES} = 1.496 \cdot 10^{11}$ m. The flux density observed at distance D_{ES} is called the (Earth's) *solar constant* E_{sc}

$$4\pi R_S^2 M_S = 4\pi D_{ES}^2 E_{sc}. \tag{2.5}$$

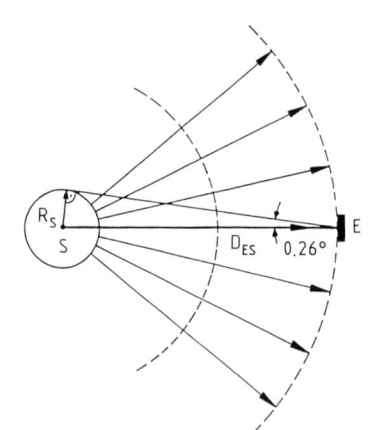

Fig. 2.1. Sun-Earth geometry. Geometry of radiosity M_S at the Sun's spherical surface of radius R_S and the solar constant E_{sc} observed at distance D_{ES} between Earth and Sun.

The solar constant is (extra)terrestrially directly accessible to measurement. Therefore, M_S and T_S can be calculated

$$M_S = \frac{E_{sc}}{f_s} \quad \text{and} \quad T_S = \left(\frac{E_{sc}}{\sigma f_s}\right)^{1/4}. \tag{2.6}$$

The ratio $f_s = (R_s/D_{ES})^2 = 2.165 \cdot 10^{-5}$ is an important number in Sun-Earth astronomy. The numerical value of E_{sc} at present accepted as most reliable is [20,21]

$$E_{sc} = 1,367.0 \pm 0.1 \text{ Wm}^{-2}. \tag{2.7}$$

2.2 Solar Radiant Flux

It can be used to calculate

$$M_S = 63.2 \text{ MWm}^{-2} \quad \text{and} \quad T_S = 5,777 \text{ K}. \tag{2.8}$$

The emitted power is in balance with the energy produced in the core of the Sun. There, at about $15 \cdot 10^6$ K the gross nuclear fusion reaction $4^1\text{H} \rightarrow {}^4\text{He}$ supplies $6.2 \cdot 10^8$ MJ per kg of hydrogen.

Several remarks are necessary.
- The assumption of the Sun being a black body emitter is an approximation only. A spherical black body emitter would appear as a disk uniform in brightness (Lambert's law). In fact, the disk of the Sun is darker near the rim than at the center (limb darkening) [55].
- The spectrum emitted by a black body of temperature T follows a thermodynamic relationship between spectral radiance L_λ and wavelength λ: Planck's equation of *spectral radiance*

$$L_\lambda(\lambda, T) = \frac{2hc^2}{\lambda^5} \frac{1}{\exp(hc/\lambda kT) - 1}. \tag{2.9}$$

The radiance L expresses a radiant flux $d^2\Phi$ related to an emitter surface dA and a solid angle $d\Omega$ in direction Ω with angle θ to the normal of dA

$$L = \frac{d^2\Phi}{dA \cos\theta d\Omega}. \tag{2.10}$$

The index λ denotes a spectral component of radiance, $L_\lambda = dL/d\lambda$

$$L_\lambda = \frac{d^3\Phi}{dA \cos\theta d\Omega d\lambda}. \tag{2.11}$$

The SI-units are for L Wm^{-2}sr^{-1} and for L_λ Wm^{-2}μm^{-1}sr^{-1}.
In solar energy utilization the radiant flux density, in particular the *irradiance* $E = d\Phi/dA$, in Wm^{-2}, has primary significance. Its spectral component is

$$E_\lambda = \int_\Omega L_\lambda \cos\theta \, d\Omega. \tag{2.12}$$

Ω is the solid angle in which the radiation penetrates the receiver area. In case of radial symmetry of Ω being a cone of angular aperture θ around the normal to the receiver area, it follows

$$E_\lambda = \int_0^\theta \int_0^{2\pi} L_\lambda \cos\theta \sin\theta \, d\theta d\varphi = \pi L_\lambda \sin^2\theta. \tag{2.13}$$

For hemispherical irradiance, $\theta = 90°$, it follows $E_\lambda(\lambda) = \pi L_\lambda(\lambda)$.
By summing over all wavelengths the integral irradiance is obtained

$$E = \int_0^\infty E_\lambda(\lambda) d\lambda = \int_0^\infty \pi L_\lambda(\lambda, T) d\lambda = \sigma T^4. \tag{2.14}$$

Table 2.1 summarizes definitions and relations of radiation quantities.
- In reality the Sun's spectrum exhibits several peaks of enhanced emission (Fig. 2.2). Thus, the M_S and T_S calculated from the measured solar constant are attributed to a fictitious black body emitter of equivalent total emissive power; Planck's spectral distribution corresponding to the calculated $T_S = 5,777$ K is only an approximation to the true solar spectrum.
- The solar constant is in fact not a constant. The distance between Sun and Earth varies periodically over the year due to the ellipticity of the Earth's orbit. The ensuing variation in E_{sc} is approximately

$$\Delta E_{sc}/E_{sc} = 0.034 \cdot \sin\left(\frac{2\pi(82.8 - N)}{365.25}\right). \tag{2.15}$$

N is the day number in a leap year cycle, $N = 1$ for January 1st of the leap year.

Table 2.1. Radiation quantities – definitions, relations, symbols, units.

Radiant Flux, Radiant Power Φ W (Watt)
Radiant energy of any origin passing in unit time through an area A. In a radiation field Φ generally depends on time, location $\mathbf{r}$, size, shape and orientation of A.

Radiant Flux Density E or M Wm^{-2}
Radiant flux of any origin *crossing* an area element.

Irradiance E Wm^{-2}
At a field point $\mathbf{r}$ radiant flux $d\Phi$ *incident* onto an elemental plane dA of arbitrary orientation Ω_A. Only the radiation is counted which comes from those directions within Ω, where the cosine of the angle between Ω and Ω_A is positive. E depends on $\mathbf{r}$ and Ω_A

$$E = \frac{d\Phi}{dA} \quad \text{and} \quad \Phi = \int E \, dA.$$

Radiosity, Radiant Exitance M Wm^{-2}
is the counterpart of irradiance E: radiant flux $d\Phi$ *emerging* from an elemental plane dA of arbitrary orientation Ω_A. Radiation is only counted which comes from those directions Ω, where the cosine of the angle between Ω and Ω_A is positive. M depends on $\mathbf{r}$ and Ω_A

$$M = \frac{d\Phi}{dA} \quad \text{and} \quad \Phi = \int M \, dA.$$

Radiance L Wm^{-2}sr^{-1}
At a field point $\mathbf{r}$ radiant flux emerges from an elemental plane dA of orientation Ω_A and along direction Ω in an elemental solid angle $d\Omega$. Radiance is defined by

$$L = \frac{d^2\Phi}{dA \cos\theta \, d\Omega} = \frac{d^2\Phi}{dA \, d\Omega^*}.$$

θ is the angle between Ω and Ω_A. By definition the *projected solid angle* is

$$d\Omega^* = \cos\theta \, d\Omega.$$

The product $dA \, d\Omega^*$ is an element of phase space, equivalent to $dU = dx \, dy \, dp_x \, dp_y$ in rectangular coordinates. p_x and p_y are the (optical) direction cosines of the ray. U is the *Etendue* or *throughput*, which is maintained in an optical system with conservation of radiant flux and its spectral distribution.

Radiant Intensity I Wsr^{-1}
A detector sensitive in a solid angle $d\Omega$ receives flux $d\Phi$ from the direction $\Omega(\theta, \phi)$. The radiant intensity of a 'point' emitter is defined by

$$I(\Omega) = \frac{d\Phi}{d\Omega}.$$

Its relation to the radiance L of the emitter of surface area A is $I(\mathbf{r}, \Omega) = \int_A L(\mathbf{r}, \Omega) \cos\theta_s \, dA$, where θ_s is the angle between Ω and Ω_A of surface dA at $\mathbf{r}$.

2.2.1 Modulation Through Revolution and Rotation

The extraterrestrial radiation is the permanent and rather constant input for the solar irradiance available terrestrially. In passing through the atmosphere the radiation becomes attenuated by complex and stochastically varying extinction processes. Another, yet predictable peculiarity of terrestrial irradiance is due to the apparent motion of the Sun. The angle subtended by the line of sight to the Sun and the horizon defines the solar elevation angle A. The elevation varies from sunrise (A is zero) via noon (A assumes a maximum) to sunset (A is again zero). The shortest path through the atmosphere and hence the case of least extinction occurs with the Sun overhead (in zenith). For $A \leq 90°$, the pathlength through the

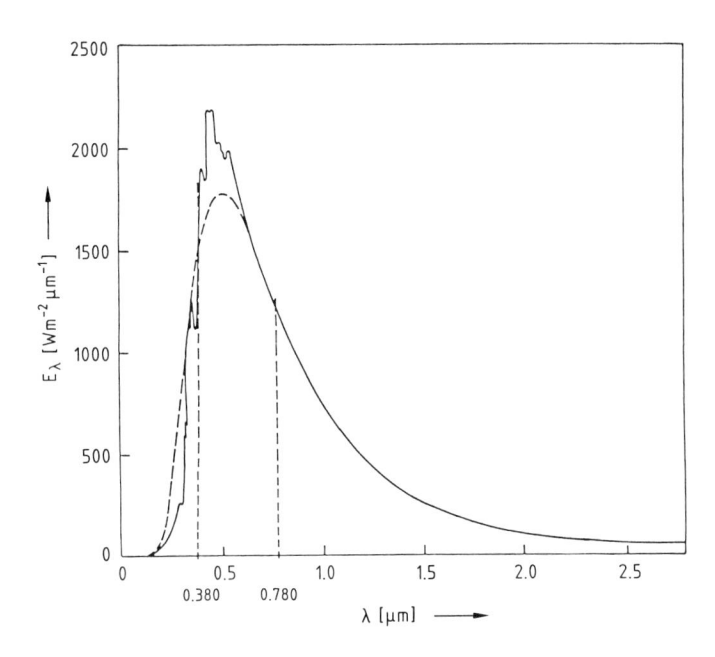

Fig. 2.2. Spectral distributions. Observed extraterrestrial solar spectrum together with smooth Planck's spectrum of $T = 5{,}777$ K, reduced in height by $f_s = 2.165 \cdot 10^{-5}$.

atmosphere is with reference to that shortest path geometrically increased by a factor

$$AM = \frac{1}{\sin A} = \frac{1}{\cos Z}, \qquad (2.16)$$

termed the *relative air mass*. $Z = 90° - A$, complementary to the elevation A, is the *solar zenith angle (zenith distance)* (Fig. 2.3).

The simple relationship $AM = 1/\sin A$ holds true only for a planar atmosphere; it remains an approximation better than 1% in the curved atmosphere of the Earth if $A > 20°$. For smaller elevation angles the finite radius of the Earth, $R_E = 6{,}370$ km, limits geometrically the relative air mass. For an elevation lower than $5°$, refraction becomes noticeable owing to the density gradient in the Earth's atmosphere. Kasten's relationship for relative air mass accounts for these effects [29]:

$$AM = \frac{1}{\sin A + 0.50572(6.07995° + A)^{-1.6364}}. \qquad (2.17)$$

with A, the apparent elevation angle in degrees.

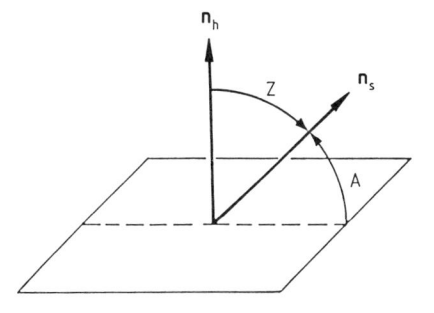

Fig. 2.3. Direction vectors. Zenith distance Z and elevation angle A of the Sun's position. $\mathbf{n}_s$ is the direction vector to the Sun and $\mathbf{n}_h$ is the orientation vector of the horizontal plane in consideration.

The direct or beam irradiance E_b on a horizontal receiver plane is, because of the geometrical projection, smaller than the incoming beam flux density $E_\perp$, see Fig. 2.3

$$E_b = E_\perp \sin A = E_\perp \cos Z = E_\perp \, \mathbf{n}_h \cdot \mathbf{n}_s. \tag{2.18}$$

If attenuation through the atmosphere were absent, $E_\perp$ would be equal to the solar constant E_{sc}.

The position of the Sun can be fixed in a geocentric reference frame by assigning two angular coordinates δ and ω (Fig. 2.4). The declination δ is measured from the celestial equator along the great circle passing through the Sun and the celestial north pole. This particular great circle is the horary circle of the Sun. The hour angle ω is measured along the celestial equator from the local meridian towards the horary circle of the Sun. The Sun's declination δ is limited to the range of $-23.44° \leq \delta \leq 23.44°$ (north being positive). The hour angle ω can vary between 0 and 360°, with $\omega = 0$ at solar noon and $\omega > 0$ after noon.

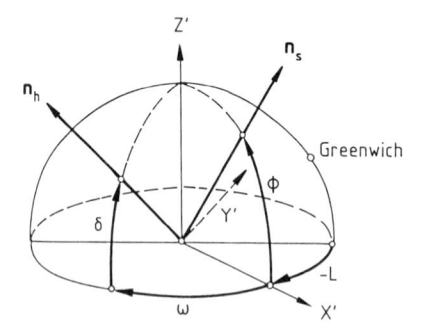

Fig. 2.4. Geocentric frame. Geocentric Cartesian coordinate frame X', Y', Z' with direction vectors $\mathbf{n}_h$ and $\mathbf{n}_s$ for the position of receiver plane and Sun, respectively. Corresponding polar angles are declination δ and geographic latitude ϕ. Azimuthal angles are hour angle ω counted from the X'-axis and L, the azimuthal local geographic length (west) of the Greenwich meridian. The X'-axis is located at L, the Z'-axis is parallel to the Earth's polar axis.

The direction vector $\mathbf{n}_s$ in terms of δ and ω in the geocentric frame is related to vector components x', y', z' in a rectangular Cartesian frame (Fig. 2.4) in the plane of the celestial equator and passes through the local meridian. It follows

$$\mathbf{n}_s(x'\ y'\ z') = \begin{pmatrix} \cos\delta \cos\omega \\ -\cos\delta \sin\omega \\ \sin\delta \end{pmatrix}. \tag{2.19}$$

In this particular Cartesian frame the orientation vector $\mathbf{n}_h$ of a horizontal receiver plane is

$$\mathbf{n}_h(x'\ y'\ z') = \begin{pmatrix} \cos\phi \\ 0 \\ \sin\phi \end{pmatrix}. \tag{2.20}$$

ϕ is the geographic latitude (Fig. 2.4). By means of these relations the elevation angle A in dependence on declination δ, hour angle ω, and latitude ϕ can be calculated

$$\sin A = \mathbf{n}_h \cdot \mathbf{n}_s = \cos\delta \cos\phi \cos\omega + \sin\delta \sin\phi. \tag{2.21}$$

Accurate values of the solar declination with resolution in hour intervals are published in Nautical Almanac and Astronomical Year Book [1]. An approximation valid within ±0.3° is

2.2 Solar Radiant Flux

(in degrees)
$$\delta = 23.44 \cdot \sin\left[P \cdot (N - 82.3 + 1.93 \cdot \sin\left[P \cdot (N - 2.4)\right])\right] \tag{2.22}$$

with the period $P = 2\pi/365.25$. N is the number of the day in a leap year cycle, i.e. $N = 1$ on January 1st of a leap year; the maximum of N is 1,461. Special day numbers N are

Year number	1	2	3	4
spring equinox	80	445	810	1,175
summer solstice	173	538	903	1,268
autumnal equinox	266	631	996	1,362
winter solstice	356	721	1,086	1,451

An approximation of the hour angle ω, accurate within 10 seconds if daily adjusted to solar noon $t_s = 0$ is (in degrees)
$$\omega = t_s \cdot 360/24, \tag{2.23}$$
where t_s is the so-called actual *local solar time* (in hours); $t_s = 0$ at noon, < 0 before and > 0 after noon. Its relation to the *local standard time* LST is (in hours)

$$t_s = LST - 12 + \frac{24}{360}(L_o - L) - EOT, \tag{2.24}$$

where L is the local geographic length (west) with L_o the geographic length referring to LST (zone time). With daylight saving time, 1 hour must be subtracted from the actual local time to obtain LST. L_o for various time zones are listed in Tab. 2.2.

Table 2.2. Standard time zones with the geographic length L_o (in degrees west) referring to Local Standard Time LST (Zone Time).

Time zone	L_o	Time zone	L_o
Universal Time Coordinate (UTC)	0	Central Standard Time	90
Central European Time	−15	Mountain Standard Time	105
South Atlantic Standard Time	45	Pacific Standard Time	120
Atlantic Standard Time	60	Yukon Standard Time	135
Eastern Standard Time	75	Alaska-Hawaii Standard Time	150

The third term in (2.24), the Equation of Time EOT, corrects for the difference between mean solar time and actual solar time. Mean solar time is a convention based on a fictitious Sun which is assumed to proceed with constant angular velocity in a circular orbit. Actual solar time is related to the real motion of the Earth around the Sun. An approximation accurate to within 10 seconds is (in hours)

$$EOT = 0.1276 \sin 0.9856(N - 3) + 0.1644 \sin 1.9713(N - 81). \tag{2.25}$$

For practical purposes in solar applications not the geocentric frame X', Y', Z' of Fig. 2.4 but a coordinate frame X, Y, Z referring to the *local horizontal plane* is used. Its X-axis is pointing to north, the Y-axis is pointing to west, the Z-axis lies along $\mathbf{n}_h$, the local horizontal plane direction. The transformation is accomplished by a matrix $\mathbf{R}$ for rotation of the x', y', z'-components into x, y, z-components,

$$\mathbf{n}_s(x\,y\,z) = \mathbf{R} \cdot \mathbf{n}_s(x'\,y'\,z') \tag{2.26}$$

with

$$\mathbf{R} = \begin{pmatrix} -\sin\phi & 0 & \cos\phi \\ 0 & -1 & 0 \\ \cos\phi & 0 & \sin\phi \end{pmatrix}. \tag{2.27}$$

It follows

$$\mathbf{n}_s(x,y,z) = \begin{pmatrix} -\cos\delta\sin\phi\cos\omega + \sin\delta\cos\phi \\ \cos\delta\sin\omega \\ \cos\delta\cos\phi\cos\omega + \sin\delta\sin\phi \end{pmatrix} \tag{2.28}$$

which gives the $x\,y\,z$-components of $\mathbf{n}_s$ as function of δ, ω, and ϕ.

This completes the information necessary for evaluating the local time dependence of the elevation angle A. In the following, several important cases are considered.

- Sunrise and sunset
  Neglecting refraction, A for the center of the Sun is zero at sunrise and again zero at sunset. Hence, the hour angles ω_s of sunrise and sunset are given by $\mathbf{n}_h \cdot \mathbf{n}_s = 0$ or

$$\cos\omega_s = -\tan\delta\tan\phi \tag{2.29}$$

  $\omega_s < 0$ for sunrise but > 0 at sunset. In local standard time LST we obtain

$$LST = 12 + \frac{24}{360}\cos^{-1}(-\tan\delta\tan\phi) + \frac{24}{360}(L - L_o) + EOT. \tag{2.30}$$

  The position vectors of sunrise and sunset for the center of the Sun are obtained by inserting ω_s in the expression for $\mathbf{n}_s$

$$\mathbf{n}_s = \begin{pmatrix} \sin\delta/\cos\phi \\ -(\cos^2\phi - \sin^2\delta)^{1/2}/\cos\phi \\ 0 \end{pmatrix}. \tag{2.31}$$

  For sunset the y-component changes sign.
  Example: Calculate time and position of sunrise and sunset on March 16 ($N = 441$ in the year following a leap year) for Munich, $\phi = 48.14°$, $L = -11.57°$ west and $L_o = -15°$ west (Central European Standard Time Zone). From the formulae given the declination angle is $\delta = -1.90°$, $EOT = -0.176$ (hours). This yields a local standard time LST of sunrise of $6^h 13^{min}$. Sunset is at $17^h 56^{min}$. The position of sunrise at the horizon is given by $\tan\gamma_s = -(\cos^2\phi - \sin^2\delta)^{1/2}/\sin\delta$. Numerically it is $\gamma_s = 87.3°$ east of north, i.e. 2.7° to the north of east.
- Daily sunshine hours
  The time interval between sunrise and sunset or the *daily maximum of sunshine hours* is

$$t_d = 2 \cdot \frac{24}{360}|\omega_s| = \frac{2}{15}|\arccos(-\tan\delta\tan\phi)|. \tag{2.32}$$

  Example: Calculate the durations of the shortest and longest day in Munich (i.e. t_d for declinations $\delta = \pm 23.44°$).
  The shortest day: $t_d = 8.14$ h; the Sun rises at 36.6° north of east.
  The longest day: $t_d = 15.9$ h; the Sun rises at 36.6° south of east.
- Daily sums of extraterrestrial solar irradiance
  The daily variation of solar declination changes the daily sum of irradiance for given geographic position. Figure 2.5 is a graphical presentation of daily extraterrestrial solar radiation input on a unit horizontal area in dependence on latitude and day of the year.
- Monthly maximum sunshine hours
  Table 2.3 lists the sum over the day lengths $t_d(N)$ per month for various geographic latitudes.
- Yearly maximum sunshine hours
  A global average of yearly sunshine hours t_a is easily estimated

$$t_a = 12 \text{ hours/day} \cdot 365.25 \text{ days/a} = 4{,}383 \text{ h/a}. \tag{2.33}$$

However, as seen in the last column of Tab. 2.3, due to eccentricity of the orbit and the 23.44° inclination of the axis of the Earth to the ecliptic there is a clear dependence of t_a on latitude ϕ.

2.2 Solar Radiant Flux

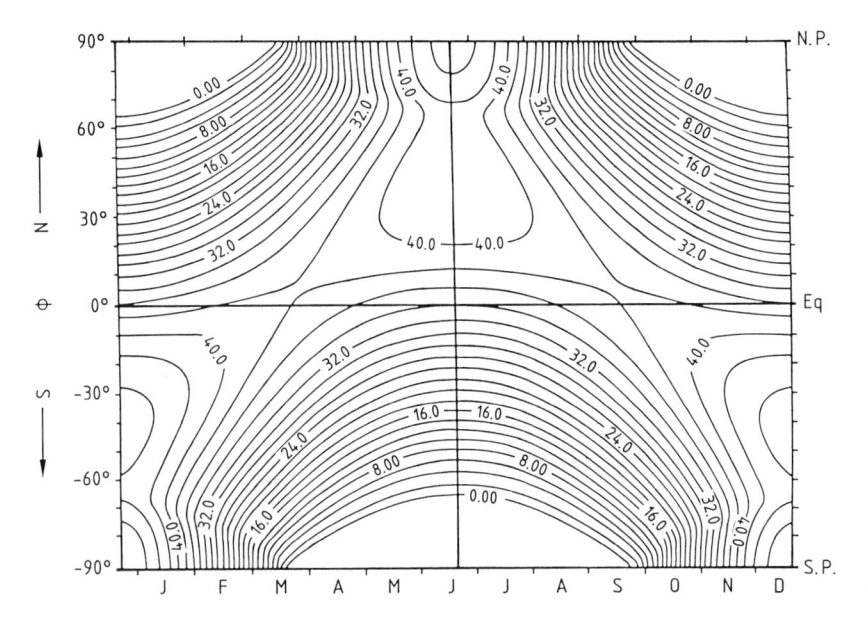

Fig. 2.5. Daily extraterrestrial solar radiation. Contour lines of daily extraterrestrial input of solar radiation on unit area of a horizontal plane in MJm^{-2} showing dependence on latitude and day of the year.

Table 2.3. Maximum monthly and yearly sunshine hours dependent on degrees latitude ϕ.

ϕ	Jan	Feb	Mar	Apr	May	Jun	Jul	Aug	Sep	Oct	Nov	Dec	Year
90	0	0	258	720	744	720	744	744	546	0	0	0	4,476
80	0	18	323	632	744	720	744	734	434	120	0	0	4,469
70	27	188	350	474	670	720	727	551	393	268	84	0	4,453
60	201	250	358	429	522	550	547	476	381	308	219	176	4,419
50	261	279	363	407	471	482	486	442	374	329	267	246	4,407
40	295	297	365	393	441	444	451	421	370	342	295	286	4,399
30	320	310	367	383	419	417	425	406	367	351	316	313	4,394
20	339	321	369	374	401	396	406	393	364	359	332	335	4,390
10	356	330	371	367	386	377	388	382	362	366	347	354	4,386
0	372	339	372	360	372	360	372	372	360	372	360	372	4,383
-10	388	348	373	353	358	343	356	362	358	378	373	390	4,380
-20	405	357	375	346	343	324	338	351	356	385	388	409	4,376
-30	424	368	377	337	325	303	318	338	353	393	404	431	4,372
-40	449	381	379	327	303	276	293	323	350	402	425	458	4,367
-50	483	399	381	313	273	238	258	302	346	415	453	497	4,359
-60	543	428	386	291	222	170	197	268	339	436	501	568	4,347
-70	717	490	394	246	74	0	17	193	327	476	636	744	4,313
-80	744	660	421	88	0	0	0	10	286	624	720	744	4,297
-90	744	678	486	0	0	0	0	0	174	744	720	744	4,290

2.2.2 Beam Radiation on Tilted Surfaces

Two convenient angular parameters of a tilted plane are to be defined first:

- the *surface tilt* β;   • the *surface azimuth* γ.

Then the incidence angle Θ of a solar beam onto the tilted plane and its time dependence $\Theta(t)$ can be derived. Finally, in Sect. 2.2.3, by including diffuse radiation, the global (i.e. direct *and* diffuse) irradiance of a tilted plane shall be calculated.

The surface tilt β is the angle subtended by the direction vectors $\mathbf{n}_h$ and $\mathbf{n}_\beta$ of the plane when it is horizontal and in tilted position, respectively (Fig. 2.6). The range of β lies between 0 and $\pm 90°$. Tilting towards the equator is counted positive.

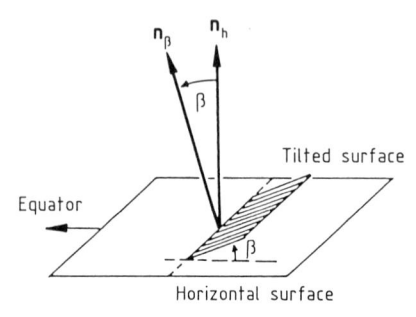

Fig. 2.6. Surface tilt angle β. Definition of surface tilt angle β. The direction vector of the horizontal plane is $\mathbf{n}_h$; the direction vector of the receiver plane is $\mathbf{n}_\beta$. Tilting performed towards the equator is counted positive.

The γ of a plane tilted with β is the angle between the plane subtended by ($\mathbf{n}_h$, $\mathbf{n}_{\beta\gamma}$) and the plane of the local meridian. γ is easy to determine: imagine the plane being tilted vertically ($\beta = 90°$); then γ is the angle between $\mathbf{n}_{\beta\gamma}$ ($\beta = 90°$) and the north-south line. γ is counted clockwise from north (to east, south, west) and runs from 0 to 360° (Fig. 2.7). There is another convention, which is often used: $\gamma = 0$ if $\mathbf{n}_\beta$ ($\beta = 90°$) points to south; $\gamma > 0$ in turning from south to east and $\gamma < 0$ in turning from south to west. Θ, the angle of incidence of a solar beam on a tilted plane is determined by

$$\cos \Theta = \mathbf{n}_{\beta\gamma} \cdot \mathbf{n}_s. \tag{2.34}$$

$\mathbf{n}_s$, the direction vector towards the Sun, has already been written in x, y, z-components of the local system. As is obvious from geometry the direction vector of the tilted plane in the X, Y, Z-frame is

$$\mathbf{n}_{\beta\gamma}(x\,y\,z) = \begin{pmatrix} \sin\beta \cos\gamma \\ -\sin\beta \sin\gamma \\ \cos\beta \end{pmatrix}. \tag{2.35}$$

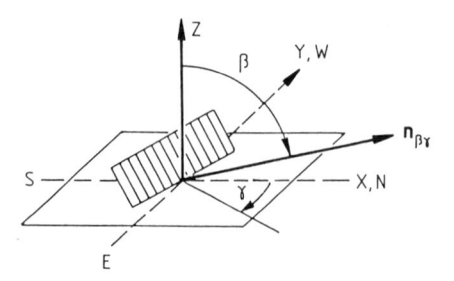

Fig. 2.7. Surface azimuth angle γ. Definition of surface azimuth angle γ. The direction vector of the receiver plane is $\mathbf{n}_{\beta\gamma}$. Rotation γ from north towards east is counted positive.

Finally, the scalar product of $\mathbf{n}_s$ and $\mathbf{n}_{\beta\gamma}$ yields the angle Θ of solar beam incidence on the tilted plane

$$\begin{aligned}\cos\Theta = \quad & - \sin\beta\cos\gamma\cos\delta\sin\phi\cos\omega + \cos\beta\cos\delta\cos\phi\cos\omega \\ & - \sin\beta\sin\gamma\cos\delta\sin\omega + \cos\beta\sin\delta\sin\phi + \sin\beta\cos\gamma\sin\delta\cos\phi.\end{aligned} \quad (2.36)$$

The equation is particularly useful if the tilt is fixed or varying in time by an independent change of β and γ. Other axes for rotating the receiver plane, e.g. parallel to the Earth's axis, cause the β and γ to be mutually linked. Various procedures are available for keeping, e.g. $\mathbf{n}_{\beta\gamma} = \mathbf{n}_s$: two axes motion of a focusing paraboloid. In central receiver systems the image of the Sun must be kept at a fixed position: heliostatic motion of mirrors, $\mathbf{n}_s \times \mathbf{n}_{\beta\gamma} = \mathbf{n}_{\beta\gamma} \times \mathbf{n}_{he}$. Here, $\mathbf{n}_{he}$ is the direction vector from a heliostat mirror plane to the fixed receiver aperture.

2.2.3 Terrestrial Solar Radiation

The extinction of solar radiation passing through the atmosphere depends on a variety of processes. These are taken to be independent of each other [28]. Therefore, a separate transmission coefficient τ_i can be assigned to any particular extinction process

$$E_\perp = E_{sc}\tau_{Ra}\tau_{O3}\tau_{Ga}\tau_{Wa}\tau_{Ae}\tau_{Ci}. \quad (2.37)$$

$E_\perp$ is the irradiance (beam flux density) arriving terrestrially at the receiver plane, which is assumed to be oriented towards the Sun. E_{sc} (solar constant) is the extraterrestrial input. Thick clouds, which are another important perturbance of solar beam irradiance, are considered in the next Sect. 2.2.4.

The meaning of the τ-indices is

Ra : Rayleigh scattering by molecules of the air
O3 : absorption by ozone
Ga : absorption by uniformly mixed gases (in particular CO_2 and O_2)
Wa : absorption by water vapor
Ae : extinction by aerosol particles
Ci : extinction by high clouds of cirrus type.

Scattering and absorption are strongly wavelength dependent, with the exception of cirrus scattering. Rayleigh scattering follows an $1/\lambda^4$-law, aerosol scattering an approximate $1/\lambda$-dependence (Mie scattering). Absorption occurs in several broad wavelength bands, which are characteristic of the absorbing species.

One is often merely concerned with the total radiative flux. Then the adequate transmission factors are spectrally integrated factors and functions of relative air mass AM and concentrations only. The following formulae are best fits by adjusted empirical constants to extended calculations and measurements.

The transmission factor for Rayleigh scattering is

$$\tau_{Ra} = \exp-\left[0.0903(1.0 + AM_* - AM_*^{1.01})AM_*^{0.84}\right]. \quad (2.38)$$

The relative air mass AM_* applies to the atmospheric pressure p at the receiver location, whereas AM refers to $p_o = 1,013$ hPa

$$AM_* = \frac{p}{p_o} \cdot AM. \quad (2.39)$$

Absorption by ozone follows a transmission factor

$$\tau_{O3} = 1 - \frac{0.1611 U_{O3}}{(1.0 + 139.48 U_{O3})^{0.3035}}. \tag{2.40}$$

The parameter U_{O3} is related to the vertical thickness d_{O3} of the ozone layer (in cm at normal temperature and pressure)

$$U_{O3} = d_{O3} AM. \tag{2.41}$$

A typical value for the atmosphere in midlatitudes is $d_{O3} = 0.32$ cm of ozone. A variation in ozone content in the usual magnitude observed will change the overall transmission only slightly.

The ordinary and uniformly mixed gases (CO_2, O_2, N_2) of the atmosphere reduce the transmission of solar radiation by

$$\tau_{Ga} = \exp-\left(0.0127 AM_*^{0.26}\right). \tag{2.42}$$

Water vapor lowers the transmission by

$$\tau_{Wa} = 1 - \frac{2.496 U_{Wa}}{6.385 U_{Wa} + (1.0 + 79.03 U_{Wa})^{0.683}}. \tag{2.43}$$

The parameter $U_{Wa} = d_{Wa} AM_*$ accounts for the amount of water vapor in a vertical column at the location of the receiver. d_{Wa} is in units of g·cm^{-2}. According to Tuller [54] approximate values are: in winter $d_{Wa} = 1.3$ gcm^{-2}; in summer $d_{Wa} = 3.0$ gcm^{-2}.

To evaluate the transmission for aerosol particles a measure of their local concentration is required. Spectral optical depth measurements in narrow bandwidths around wavelengths 0.38 and 0.5 μm are data which become more and more available from the meteorology network. Otherwise the local horizontal visibility VIS (in km) can be used, which is to a certain extent correlated with the aerosol transmission factor

$$\tau_{Ae} = \exp\left[AM_*^{0.9} \ln\left(0.97 - \frac{1.265}{VIS^{0.66}}\right)\right]. \tag{2.44}$$

The sixth and last transmission factor takes account of the extinction by cirrus. Employing an optical thickness parameter d_{Ci} (see Sect. 2.2.4), we obtain

$$\tau_{Ci} = \exp-(AM \cdot d_{Ci}). \tag{2.45}$$

The range of d_{Ci} is 0.01 for thin to 1.0 for thick cirrus coverage.

The quoted transmission factors (except the cirrus data) were taken from Bird and Hulstrom [6]. Originally, the authors used a solar constant and a spectrum different from the present World Radiation Center recommendation [20,21]. Iqbal [28] has shown that then the overall correction factor for the product of the six transmission factors is 0.9946. Considering in practice the approximations on ozone, water vapor, aerosol and cirrus contents such a small correction can be disregarded here.

A numerical example shows the order of deviation of the various transmission factors from 1. Values used are: $AM = 2$ (corresponding to $A = 30°$ solar elevation angle), standard pressure, 100 km visibility, 2 gcm^{-2} water content, a cirrus of $d_{Ci} = 0.1$. Insertion into the transmission factors yields

$\tau_{Ra} = 0.85$ for Rayleigh scattering
$\tau_{O3} = 0.97$ for absorption by ozone
$\tau_{Ga} = 0.99$ for absorption in CO_2, O_2, N_2
$\tau_{Wa} = 0.87$ for absorption by water vapor
$\tau_{Ae} = 0.84$ for extinction by aerosol particles
$\tau_{Ci} = 0.82$ for extinction through cirrus.

The overall transmission comes down to $\tau = 0.49$. The terrestrial solar flux density of the given composition is $E_\perp = E_{sc}\tau = 1,367$ Wm$^{-2} \cdot 0.49 = 669$ Wm^{-2}.

2.2.4 Beam Radiation and Clouds [2]

Knowledge about solar irradiance on a specific site is of paramount concern to solar power plant designers. For direct beam dependent installations the use of the time resolved data of the CFLOS World Atlas, referred to presently, provides substantial first information. Quantities derived from such large scale data do not include local phenomena, e.g. orographic clouds, fog. Therefore, prior to the final decision on location of a solar power plant, long term local meteorological observation and measurement remain indispensable.

There is a growing collection of regional and local daily direct and global insolation data, for instance the continued publications in *Solar Energy* and the specific reports on various present solar power plants, e.g. Barstow, Tabernas, Targasonne. In the following, basic rather than specific meteorology of solar radiation input is presented.

Clouds with an optical thickness > 3 scatter so efficiently that the direct solar beam disappears. *Cloudfree Lines Of Sight Probabilities*, CFLOS [38] indicate to what percentage for a given time and a given elevation angle the sky is cloudfree. CFLOS can be derived from usual meteorological information on cloudiness [39] (which is the fraction of cloud coverage of the sky in multiples of 1/8 or 1/10) and the type of cloud. Worldwide CFLOS data are available for monthly averages in January, April, July and October; for four values of local standard time: 00.00, 06.00, 12.00, 18.00; for elevation angles 10°, 30° and 90°.

Figure 2.8 is an example of such data. With CFLOS local probabilities for beam radiation and their daily course can be calculated [32]; a plot for four seasons for Almería (Spain) is shown in Fig. 2.9. The plot shows a steep decrease of probability for CFLOS towards sunrise and sunset and during winter. It can be concluded that in Almería direct radiation will be available in summer to roughly 80%, in winter for about 60%.

Insolation values averaged over, e.g. a monthly period, are only useful for concentrating devices, if they are close to maximum possible insolation. Cloudfree (direct beam insolation) and cloudy periods (prevailing diffuse radiation) average to a mean irradiance; the important temporal structure of direct/diffuse availability is lost. Thus, for the assesment of solar power plant sites short interval recordings of sunshine, direct and diffuse radiation among other meteorological data is required. The relevant measuring instruments are presented in an Appendix to this chapter.

Cirrus clouds are most frequent compared to other cloud types [37]. Thick cirrus, with an optical depth of $d_{Ci} \geq 2$, resembles a water cloud. The extinction by thin cirrus is incorporated in the calculations of Sect. 2.2.3. Consequently their optical depth distribution is required. Such data are not available from standard meteorological observations. Direct lidar measurements are rare. Three classes of optical depth, $2 \geq d_{Ci}^{(1)} > 0.2 \geq d_{Ci}^{(2)} > 0.02 \geq d_{Ci}^{(3)} > 0$ have been established, for which the probabilities can be correlated to available aircraft and satellite observations [32]. The classes are: (1) thin cirrus, (2) subvisible cirrus and (3) cirrus detectable in long path transmittance only [59]. As an example, cirrus data for Almería are presented in Fig. 2.10.

2.2.5 Diffuse and Global Radiation [3]

Part of the radiation scattered from the solar beam on its path through the atmosphere reaches the surface as *diffuse irradiance*, E_d. The global irradiance ^SE, measured on a horizontal surface, is the diffuse E_d and direct (beam) radiation, E_b,

$$^SE = E_b + E_d \quad \text{with} \quad E_b = E_\perp \sin A \tag{2.46}$$

---

[2] By Peter Köpke, Meteorologisches Institut der Ludwig-Maximilians-Universität München
[3] By Peter Köpke, Meteorologisches Institut der Ludwig-Maximilians-Universität München

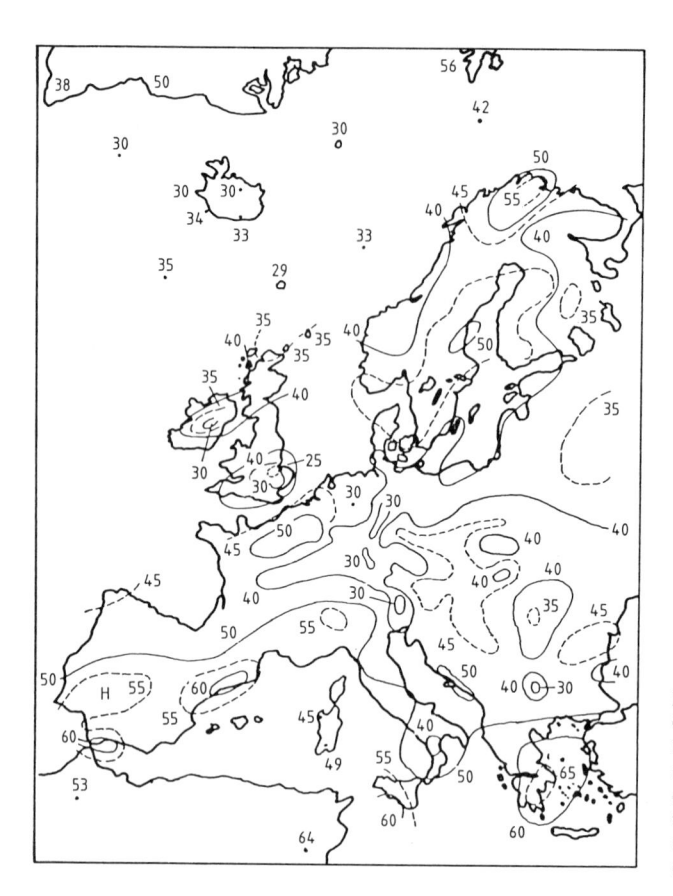

Fig. 2.8. Cloudfree lines of sight, CFLOS. CFLOS probabilities for Europe in April at noon, LST = 12.00–14.00. The numbers indicate for locations and isolines the percentage of cloudfree sight at any azimuth but 30° elevation angle.

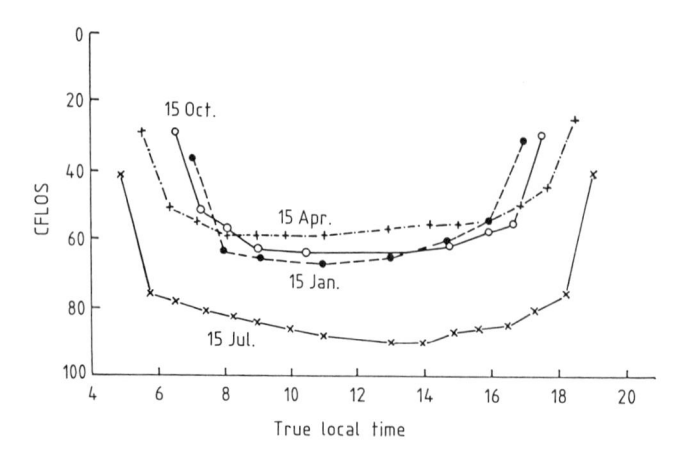

Fig. 2.9. Tabernas CFLOS probabilities. Probability in % of the occurrence of thick clouds for the line of sight between Sun and ground at Tabernas (Plataforma Solar de Almería) on the 15th of January, April, July, and October.

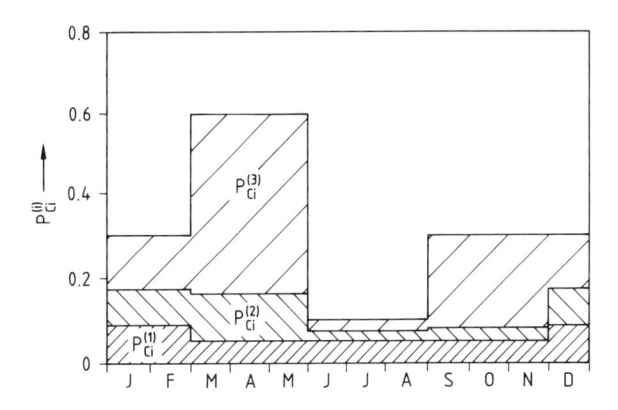

Fig. 2.10. Tabernas cirrus occurrence. Seasonal probabilities $P_{ci}^{(i)}$ of the occurrence of cirrus catalogued in three classes i of optical depth for Almería, Spain.

In solar energy utilization either direct E_b or global irradiance $^S E$ is the desired quantity, but in the following emphasis will be laid also on the diffuse component E_d, because of the specific difference between E_b and E_d.

Bird and Hulstrom [5] treated diffuse irradiance as a sum of contributions from scattering by molecules (Rayleigh scattering), $E_d^{(\text{Ra})}$, by aerosols, $E_d^{(\text{Ae})}$, and from multiple scattering between surface and atmosphere, $E_d^{(\text{MS})}$.

$$E_d = E_d^{(\text{Ra})} + E_d^{(\text{Ae})} + E_d^{(\text{Ci})} + E_d^{(\text{MS})} \qquad (2.47)$$

Note that in this equation a contribution by cirrus, $E_d^{(\text{Ci})}$, has been added. The source for the scattering processes is the horizontal irradiance from solar radiation not absorbed on its path through the atmosphere, E_{NA},

$$E_{\text{NA}} = 0.786 E_{sc} \cos Z \cdot \tau_{\text{O3}} \cdot \tau_{\text{Ga}} \cdot \tau_{\text{Wa}} \cdot \tau_{\text{AAe}}. \qquad (2.48)$$

τ_{AAe} is the transmittance due to aerosol absorption only

$$\tau_{\text{AAe}} = 1 - (1 - \omega_o)(1 - \tau_{\text{Ae}})(1 - AM_* + AM_*^{1.06}) \qquad (2.49)$$

The single scattering albedo of average continental aerosol is $\omega_o = 0.88$ [3]; other aerosol types have values between 0.6 and 1.0.

$$\begin{aligned} E_d^{(\text{Ra})} &= E_{\text{NA}}(1 - \tau_{\text{Ra}})/2(1 - AM_* + AM_*^{1.02}) \\ E_d^{(\text{Ae})} &= E_{\text{NA}}(1 - \tau_{\text{Ae}}/\tau_{\text{AAe}})F_{\text{Ae}} \\ E_d^{(\text{Ci})} &= E_{\text{NA}}(1 - \tau_{\text{Ci}})F_{\text{Ci}}. \end{aligned} \qquad (2.50)$$

The part of radiation scattered towards the surface is given by a factor F, which depends on the scattering asymmetry of the particular scatterer [49].

$$F_{\text{Ae,Ci}} = 0.5 + 0.375 \cos Z \cdot g_{\text{Ae,Ci}}. \qquad (2.51)$$

$g_{\text{Ae}} = 0.63$ for average continental aerosol [3] and $g_{\text{Ci}} = 0.986$ [5].

Both the direct and the diffuse irradiance are reflected at the surface and become backscattered again from the atmosphere. Values of the surface albedo ρ_s can be taken for the actual conditions; examples are given in Tab. 2.4. For the so-called atmospheric albedo, ρ_{At}, the backscatterance $(1 - F_{\text{Ae,Ci}})$ is used, calculated for a fixed zenith angle of $Z = 60°$, which is typical for diffuse backward scattering [60].

Table 2.4. Surface albedo of various surfaces in the solar spectral range (average solar elevation).

Snow, fresh	0.75–0.95	Grass	0.2–0.3
Snow, old	0.4–0.6	Green forest	0.1–0.2
Desert sand	0.3–0.4	Conifereous forest	0.1–0.15
Soil, dry	0.15–0.3	Concrete	0.25–0.35
Soil, wet	0.1–0.2	Asphalt	0.1–0.2

$$\rho_{At} = 0.0685 + (0.5 - 0.189 g_{Ae})(1 - \tau_{Ae}/\tau_{AAe}) + (0.5 - 0.189 g_{Ci})(1 - \tau_{Ci}). \tag{2.52}$$

Multiple reflections between surface and atmosphere produce

$$E_d^{(MS)} = (E_b + E_d^{(Ra)} + E_d^{(Ae)} + E_d^{(Ci)}) \frac{\rho_s \rho_{At}}{1 - \rho_s \rho_{At}}. \tag{2.53}$$

If the global irradiance $^S E$ is the desired quantity, no separate calculation of $E_d^{(MS)}$ is needed, as

$$^S E = (E_b + E_d^{(Ra)} + E_d^{(Ae)} + E_d^{(Ci)}) \frac{1}{1 - \rho_s \rho_{At}}. \tag{2.54}$$

With increasing turbidity the amount of downward scattered radiation increases; the direct beam irradiance decreases. For both E_b and E_d the absolute values decrease with decreasing solar elevation, although less pronounced for diffuse irradiance. For a solar elevation less than about 10°, both parts contribute to global radiation $^S E$ with approximately the same strength. For high Sun the direct radiation is up to 10 times the diffuse radiation in a clear atmosphere, but only about 3 times in a turbid atmosphere.

These numbers are valid for horizontal surfaces. If a receiver plane is tilted towards the Sun, the contribution of the beam radiation E_b increases, whereas the diffuse irradiance E_d decreases. However, the receiver then gains part of its irradiance from the foreground, which increases the albedo influence.

Assume the diffuse solar radiation to be homogeneously distributed over the sky; it renders a diffuse irradiance E_d on a horizontal plane. A collector plane with tilt angle β towards the horizon (see Fig. 2.5), receives only part of this diffuse irradiance because it sees only part of the luminous hemisphere

$$^{sky}E_d(\beta) = E_d \frac{1 + \cos \beta}{2} = E_d \cos^2 \frac{\beta}{2}. \tag{2.55}$$

However, the tilted collector gains a field of view towards its foreground. From there, due to ground albedo ρ_s (see Tab. 2.4), part of diffuse scattered global radiation, $\rho_s {}^S E$, can reach the collector plane

$$^{ground}E_d(\beta) = \rho_s {}^S E \frac{1 - \cos \beta}{2} = \rho_s {}^S E \sin^2 \frac{\beta}{2}. \tag{2.56}$$

In total a (β, γ)-tilted collector plane receives

$$E_{\beta\gamma} = E_b \frac{\cos \Theta}{\cos A} + E_d \cos^2 \frac{\beta}{2} + \rho_s {}^S E \sin^2 \frac{\beta}{2}. \tag{2.57}$$

The first term is the direct solar part with $\cos \Theta = \mathbf{n}_{\beta\gamma} \cdot \mathbf{n}_s$ according (2.36) and $\cos A = \sin Z$ for the elevation angle A or the zenith distance Z according (2.21). The three horizontal irradiances $^S E, E_b, E_d$ in (2.57) are not independent: $^S E = E_b + E_d$.

Contrary to the previous assumption it should be noted that diffuse radiation is often highly anisotropic. Then there is an angular range of preferential incidence which transforms geometrically in the same way as direct radiation. For a clear sky the darkest point is at about $A+90°$, where A is the solar elevation angle. An annular region around the Sun appears radiantly bright (the so-called *aureole*); the horizon is lightened. Figure 2.11 shows an example of spectrally measured radiance at 0.56 μm. Consequently, equations which are used for the diffuse radiation on a tilted plane should distinguish between a circumsolar contribution and a diffuse contribution from the rest of the sky [30].

In applications which rely on the direct beam irradiance the aureole radiation can also be significant. The aureole intensity depends, in addition to solar elevation and optical depth, on the size distribution of the scattering particles. The degree of forward scattering increases with particle size and is especially strong for cirrus clouds.

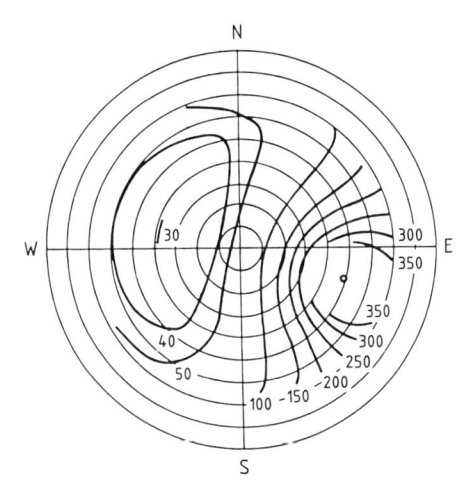

Fig. 2.11. Hemispheric radiance. Radiance distribution measured for a cloudless but hazy sky at $\lambda = 0.561$ μm (in units of $Wm^{-2}\mu m^{-1}sr^{-1}$). Horizontal visibility is 6 km. The right hand side spot is the temporary position of the Sun, here at $A = 40°$. The outer circle corresponds to the horizon, $A = 0°$, the center to $A = 90°$.

2.2.6 Spectral Direct and Diffuse Radiation [4]

In various solar energy applications the efficiency is sensitive to the spectral distribution of the radiation. This is particularly true for optically selective coatings, in photovoltaics and photochemistry. As mentioned in Sect. 2.2.3, in the solar spectral range all atmospheric components absorb and scatter wavelength dependently. This is also valid for the absorption coefficient of ice [56], but has not been found in the extinction cross section of ice crystals, typical for cirrus clouds [37].

The spectral transmission for any absorber is of the form $\tau(\lambda) = \exp(-AM \cdot d(\lambda))$. Thus, the spectral irradiance depends on both, the optical depth $d(\lambda)$ of the different atmospheric components and on the relative air mass AM. Aerosol extinction dominates in the spectral region near maximum of solar E_λ.

The effect of varying solar elevation on beam irradiance is shown in Fig. 2.12. It is strongest in the region of the maximum of the solar E_λ spectrum, because there the optical depth is roughly around 0.5 and, consequently, multiplication with AM results in strong variations of the transmission. With increasing air mass, as with increasing aerosol content, the spectrum shifts to longer wavelengths, an effect well known from red sunset. The diffuse spectral irradiance is also reduced by absorption, but enhanced by scattering. Since the scattering coefficients of gas molecules and aerosols increase with decreasing wavelength, the spectrum of diffuse radiation is shifted to shorter wavelengths, compared to the direct beam. The sky appears blue. As shown in Fig. 2.13, the effect becomes stronger the lower the aerosol content.

---
[4] By Peter Köpke, Meteorologisches Institut der Ludwig-Maximilians-Universität München

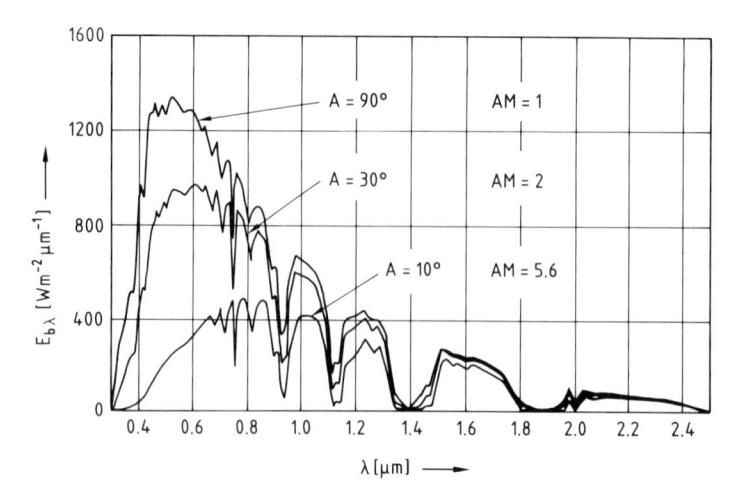

Fig. 2.12. Spectral beam irradiance. Spectral beam irradiance $E_{b\lambda}$ for various solar elevation angles A. $d_{O3} = 0.32$ cm, $d_{Wa} = 2$ g·cm^{-2}. Visibility is 25 km.

The influence of surface albedo ρ_s on diffuse and, therefore, on global radiation increases with turbidity and surface reflectance. In Fig. 2.14 reflectances are presented for different surface types [8]. The spectral reflectances depend on many variables, e.g. water content of soil, growing situation, surface coverage by plants, ice, dust on snow, even on solar elevation. Thus, the values in Fig. 2.14 are merely examples. On the other hand, radiation which contributes to the diffuse part of global irradiance is reflected from an area of several kilometers in diameter, depending on turbidity and position of the Sun. Hence for global irradiance an albedo averaged over the surroundings of the solar plant has to be taken into account.

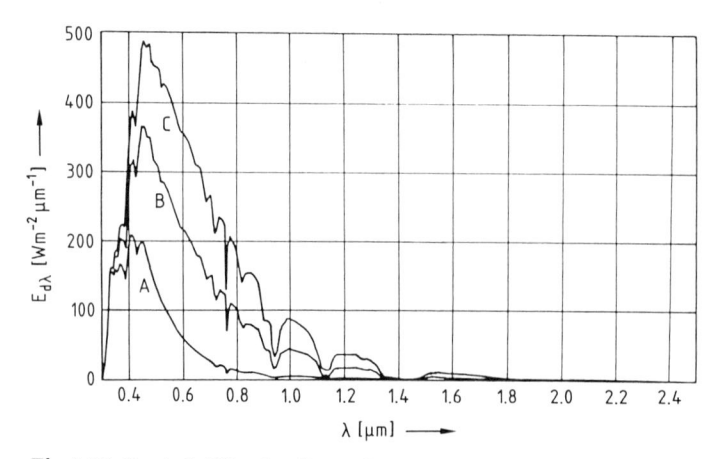

Fig. 2.13. Spectral diffuse irradiance. Spectral diffuse irradiance $E_{d\lambda}$ for various aerosol contents, given as visibility (A: 310 km, B: 25 km, C: 11 km). $d_{O3} = 0.32$ cm, $d_{Wa} = 2$ g · cm^{-2}, $\omega_o = 0.95, \rho_s = 0.2, A = 30°$.

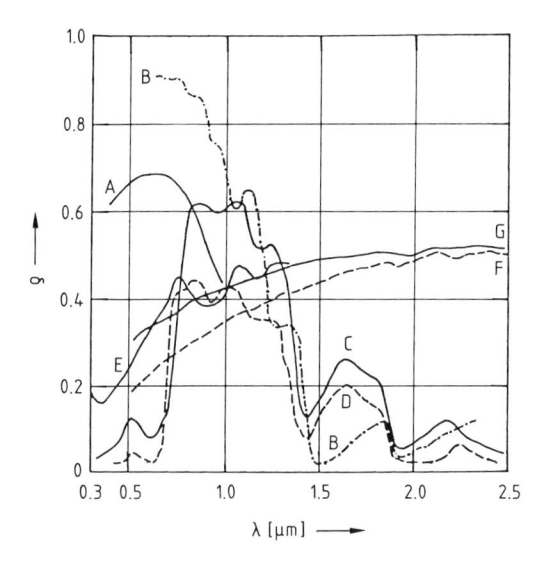

Fig. 2.14. Spectral reflectances. Spectral reflectances of various surfaces: A: wet snow; B: typical snow; C: Ponderosa pine needles; D: Kentucky blue grass; E: concrete; F: dry sandy soil; G: beach sand.

2.3 Thermodynamic Quality of Solar Radiation

In solar engineering the incident radiation is converted to useful energy fluxes: heat, motive or electric power, chemical energy carriers. The question is then about the quality of solar radiation [51]. In thermodynamics, a conventional measure of energy quality is the maximum work (termed *exergy*) W^* obtainable in a converter from solar energy input ^SE, or

$$\eta^* = \frac{W^*}{^SE}. \qquad (2.58)$$

The integral global irradiance $^SE = \int {^SE_\lambda} d\lambda$ turns out not to be a sufficient specification for η^*. It can be shown that for equal spectral flux densities E_λ, more work is obtainable from shortwave than from longwave radiation.

2.3.1 Measure of Quality

The situation is similar in the assessment of the thermodynamic quality of heat. Imagine two portions of water containing equal amounts of heat but of different temperature. A perfect heat engine delivers more work with the hot than with the colder water. The reason is that there is no pure energy flux: energy fluxes are always coupled to fluxes of other quantities which in this sense can be envisaged as carrying the energy. Therefore, in a continuously operated energy converter a balance of in/out energy fluxes *and* of the energy carrier fluxes must hold.

In Fig. 2.15a the converter (a heat engine) receives heat $E = Q$ of absolute temperature T. Coupled to E flows an entropy flux $S = Q/T$. The converter delivers work W, e.g. electric power, which is carried by an in/out flux of electric charges thereby being increased in potential, but also entropy due to irreversibilities ^iS. The balances for steady state operation are

$$\begin{aligned} \text{Energy} \qquad Q &= W + Q' \\ \text{Entropy} \qquad Q/T + {^iS} &= Q'/T' \end{aligned}$$

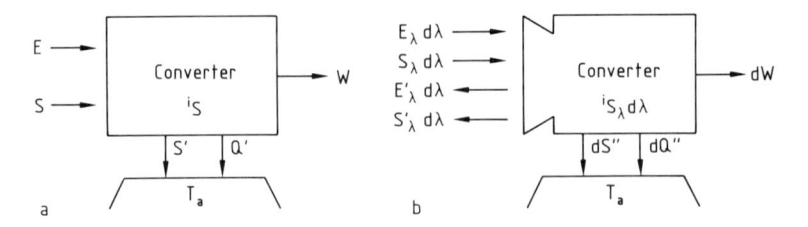

Fig. 2.15. Converter systems.
(a) Energy and entropy fluxes in a converter. The internal source term ^iS takes care of entropy production by irreversibilities. W is the work delivered in the conversion process from the entering energy flux E. The outflow of waste heat Q' is a consequence of draining surplus entropy $S' = S + {}^iS$ from the converter.
(b) Converter for incident radiation $E_\lambda d\lambda$ to work dW operated at ambient temperature T_a. Its entrance maintains an areal aperture A with angular aperture Ω. The fluxes are given as flux densities and need to be multiplied by the area A. $^iS_\lambda$ takes care of entropy production by irreversibilities.

A heat flow Q' (waste heat) is required to accomplish the entropy balance: Q' removes entropy $S' = Q'/T'$ at (absolute) temperature T'.

Elimination of Q' from both equations yields for the *reversible* case of $^iS = 0$ the familiar Carnot relation

$$\eta^* = \frac{W}{Q} = 1 - \frac{T'}{T}. \tag{2.59}$$

The best choice of T', the temperature of the waste heat flow, is obvious: it should be a lowest possible temperature to render a highest useful quality η^* of the input heat. However, the only practicable low temperature of T' is T_a, the ambient temperature. Any lower temperature would require auxiliary power expenditure for producing and keeping $T' < T_a$.

A high temperature T compared to the ambient T_a allows almost complete conversion of heat Q to work W. A low temperature heat source, e.g. of constant $T = 77°C = 350$ K and a $T_a = 300$ K of the entropy sink can at best produce power at the small efficiency of $\eta^* = 1 - 300/350 = 0.14$.

2.3.2 Quality of Radiation

The previous model example can be applied for evaluating the thermodynamic quality of a radiation flux. In Fig. 2.15b the appropriate scheme of the converter is shown. It differs in details from the converter of Fig. 2.15a. The entrance aperture which receives radiation $E_\lambda d\lambda$ is at the same time and inevitably an exit aperture, releasing radiation $E'_\lambda d\lambda$. However, the quality of E'_λ should be made identical with $^aE_\lambda$ of ambient radiation. If this is the case the lost radiation $E'_\lambda d\lambda$ cannot produce additional work. The balances for a spectral component of band width $d\lambda$ are

of energy flow $\qquad\qquad E_\lambda d\lambda = E'_\lambda d\lambda + dW + dQ''$
of entropy flow $\qquad S_\lambda d\lambda + {}^iS_\lambda d\lambda = S'_\lambda d\lambda + dQ''/T_a$

A waste heat flow dQ'' is necessary to remove surplus entropy. Elimination of dQ'' from both equations and assuming a perfect converter ($^iS = 0$), the maximum spectral power output $W_\lambda = dW/d\lambda$ of spectral irradiance E_λ for an ambient temperature T_a is obtained

$$W_\lambda = (E_\lambda - E'_\lambda) - T_a(S_\lambda - S'_\lambda). \tag{2.60}$$

An arbitrary spectral irradiance $E_\lambda(\lambda)$ from within a solid angle cone of half angle θ can be imagined to originate from a black body Planck emitter of adjusted temperature T so that

$$E_\lambda(\lambda) = \pi \sin^2 \theta \, L_\lambda(\lambda, T). \tag{2.61}$$

2.3 Thermodynamic Quality of Solar Radiation

$E_\lambda(\lambda)$ can be a mixture of solar radiation and any other radiation from the environment, e.g. from the sky. By solving for T the *equivalent black body* or *radiance temperature* of the incident E_λ is obtained

$$T(\lambda) = \frac{hc}{\lambda k} / \ln\left(1 + \frac{2\pi hc^2 \sin^2\theta}{\lambda^5 E_\lambda(\lambda)}\right). \tag{2.62}$$

In principle $T(\lambda)$ could be measured experimentally by using a thermometer sensitive in half angle cone θ with spectral absorptance $\alpha(\lambda) = 1$ at the given λ in bandwidth dλ and $\alpha(\lambda) = 0$ beyond dλ. The equilibrium temperature obtained is $T(\lambda)$.

In Fig. 2.16 the spectral components $E_\lambda(\lambda)$ and the corresponding radiance temperature $T(\lambda)$ are presented in a histogram of an $AM1$-spectrum. Applying radiance temperatures the difference $E_\lambda - E'_\lambda$ can be written in terms of Planck's radiances

$$E_\lambda - E'_\lambda = \pi \sin^2\theta \left(L_\lambda(T) - L_\lambda(T')\right). \tag{2.63}$$

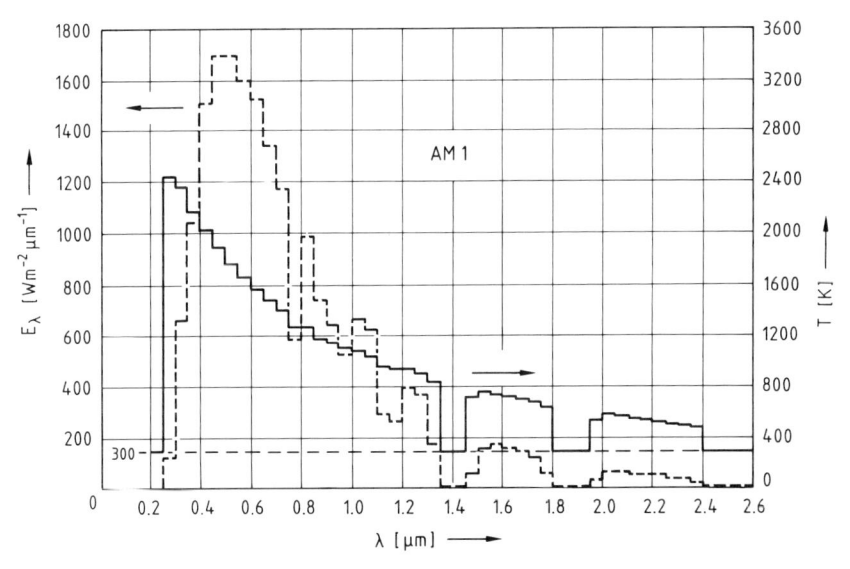

Fig. 2.16. *AM*1-Histograms. Histograms of the *AM*1 spectral irradiance E_λ and the corresponding radiance temperature $T(\lambda)$.

In the following a shorthand notation of $L_\lambda(T)$ is useful

$$L_\lambda(T) = \frac{2hc^2}{\lambda^5} \cdot X \quad \text{with} \quad X = \frac{1}{exp\frac{hc}{\lambda kT} - 1}. \tag{2.64}$$

The entropy flux density S_λ is related to the energy flux density by [35]

$$\mathrm{d}S_\lambda = \frac{\mathrm{d}E_\lambda}{T} = \pi \sin^2\theta \frac{\partial L_\lambda}{\partial T} \frac{\mathrm{d}T}{T} \tag{2.65}$$

By integration over the radiance temperature range T to T' for bringing the incident irradiance E_λ in quality down to ambient irradiance E'_λ it follows [35]

$$S_\lambda(T) - S_\lambda(T') = \int_{T'}^{T} \mathrm{d}S_\lambda = \pi \sin^2\theta \left(K_\lambda(T) - K_\lambda(T')\right). \tag{2.66}$$

The shorthand notation used is

$$K_\lambda = \frac{2kc}{\lambda^4} \cdot Y \quad \text{with} \quad Y = (1+X)\ln(1+X) - X\ln X. \tag{2.67}$$

The work W_λ obtainable with a perfect converter operating at ambient temperature $T' = T_a$ from an arbitrary spectral irradiance E_λ is,

$$W_\lambda = \pi \sin^2 \theta \left(L_\lambda(T) - L_\lambda(T_a) - T_a \Big(K_\lambda(T) - K_\lambda(T_a) \Big) \right). \tag{2.68}$$

More important than W_λ is the *integral work* $W = \int_0^\infty W_\lambda d\lambda$ from a solar spectrum. Using the formulae, W can be computed for any spectrum of E_λ.

2.3.3 Standard Spectra

In comparing the efficiency of solar conversion processes it is of advantage to *define* a smooth and simple *standard spectrum*. It should consist of a mixture of the two main contributions to the irradiance of a converter aperture: the radiance $L_\lambda(\lambda, T_S)$ of the Sun with $T_S = 5,777$ K and the radiance $L_\lambda(\lambda, T_a)$ of the surroundings with $T_a = 300$ K at sea level (radiance contribution from the terrestrial atmosphere) or $T_a = 3$ K at an extraterrestrial position (contribution from the thermal cosmic background radiation)

$$E_\lambda(\lambda) = \pi \sin^2 \theta \left[f L_\lambda(\lambda, T_S) + (1-f) L_\lambda(\lambda, T_a) \right]. \tag{2.69}$$

θ is the half angle of the cone, in which the irradiance enters the converter, see for illustration Fig. 2.19a (Sect. 2.4). The parameter f is a measure of the mixing ratio of radiation from both sources $L_\lambda(\lambda, T_S)$ and $L_\lambda(\lambda, T_a)$. Since $T_S \gg T_a$, the mixing can be regarded as a dilution of 'hot' solar radiation with 'cold' ambient radiation.

In particular $f = \tau \sin^2 \theta_S / \sin^2 \theta$, where θ_S is the half angle of the cone in which the solar component enters, in contrast to $\theta (\geq \theta_S)$, the half angle of the cone in which all of the radiation, solar and ambient, enters. τ is the usually wavelength dependent transmittance of the radiation through the atmosphere. f is then variable with the solar geometric concentration ratio $C = 46,200 \sin^2 \theta_S$, as explained in Sect. 2.4, (2.97)

$$f = \tau \frac{C}{46,200} \frac{1}{\sin^2 \theta} \leq 1. \tag{2.70}$$

This is true, if the atmospheric transmittance τ is dominated by *absorption*. However, if τ is dominated by atmospheric *scattering*, then solar radiation is to a high degree present as diffuse solar radiation beyond θ_S of the direct beam. In the limit of strong scattering (fog, cloud cover), solar radiation spreads evenly over the hemisphere. This is equivalent to $C = 1$ and $\theta = 90°$ or

$$f = \tau \frac{1}{46,200} = \tau f_s. \tag{2.71}$$

An aperture fully open to the hemisphere, $\theta = 90°$, corresponds to a flat plate receiver. The standard spectrum irradiance is in this case

$$E_\lambda(\lambda) = f\pi L_\lambda(\lambda, T_S) + (1-f)\pi L_\lambda(\lambda, T_a). \tag{2.72}$$

- A limiting case is $\tau = 1$: it corresponds to an extraterrestrial standard spectrum. If in particular $f = 1$, i.e. $C = 46,200$, then the irradiance E_λ is solely solar and identical with the radiosity M_λ of the photosphere of the Sun

$$^S E_\lambda(\lambda) = \pi L_\lambda(\lambda, T_S). \tag{2.73}$$

The integral over the spectrum $^S E_\lambda(\lambda)$ is $M_S = \sigma T_S^4 = 63.2$ MWm^{-2}.

2.3 Thermodynamic Quality of Solar Radiation

- An interesting case is $f = 1/46,200 = f_s$ (see Sect. 2.2), i.e. $C = 1$, for the extraterrestrial spectrum, $\tau = 1$

$$E_\lambda(\lambda) = f_s \pi L_\lambda(\lambda, T_S) + (1 - f_s)\pi L_\lambda(\lambda, T_a). \tag{2.74}$$

Because of $f_s \pi L_\lambda(\lambda, T_S) \gg \pi L_\lambda(\lambda, 3\,\text{K})$ the spectrum is again entirely dominated by the solar component. The integral $E = \int_0^\infty E_\lambda d\lambda$ produces the solar constant $E_{sc} = 1,367\,\text{Wm}^{-2}$.

- Another limiting case is $f = 0$. Only ambient radiation $^aE_\lambda$ is in the spectrum E_λ

$$^aE_\lambda(\lambda) = \pi L_\lambda(\lambda, T_a). \tag{2.75}$$

The integral over the spectrum is $^aE = \sigma T_a^4 = 459\,\text{Wm}^{-2}$ ($T_a = 300\,\text{K}$) and $4.59 \cdot 10^{-6}\,\text{Wm}^{-2}$ ($T_a = 3\,\text{K}$), respectively. The numbers show that with $f > 10^{-5}$ for $T_a = 300\,\text{K}$ or $f > 10^{-13}$ for $T_a = 3\,\text{K}$, the standard spectrum irradiance is dominated by solar radiation

$$E_\lambda(\lambda) \approx f\pi L_\lambda(\lambda, T_S). \tag{2.76}$$

- The integral value over the entire wavelength region of the standard spectrum is, assuming f to be wavelength independent,

$$E = {}^SE + {}^aE = f\sigma T_S^4 + (1-f)\sigma T_a^4. \tag{2.77}$$

In the following the parameter of the standard spectrum f is frequently assumed to be wavelength independent. The f-dependent radiance temperatures of the standard spectrum with $T_a = 300\,\text{K}$ are plotted in Fig. 2.17. The shortwave region remains at high equivalent temperatures even with strong dilution (small f-values). This observation is important for optically selective processes in absorption of solar radiation.

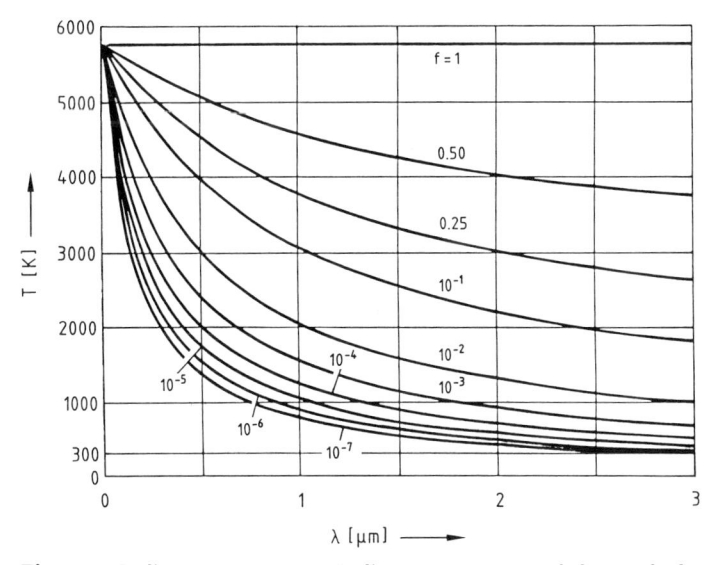

Fig. 2.17. Radiance temperatures. Radiance temperatures of the standard spectrum irradiances ($T_S = 5,777\,\text{K}$, $T_a = 300\,\text{K}$). The parameter is the dilution factor f, which is taken to be independent of wavelength.

2.3.4 Quality of Solar Irradiance

Employing standard spectra the integral quality

$$\eta^* = \frac{W^*}{{}^S E} = \frac{\int_0^\infty W_\lambda^* d\lambda}{\int_0^\infty {}^S E_\lambda d\lambda} = \frac{W^*(f)}{f\sigma T_S^4} \qquad (2.78)$$

of solar irradiance ${}^S E$ can be calculated. The convention is to relate the maximum obtainable power W^* to the solar part ${}^S E$ of the spectrum only. Figure 2.18 is the final result of the (numerical) evaluation of the thermodynamic quality $\eta^*(f)$ of solar radiation.

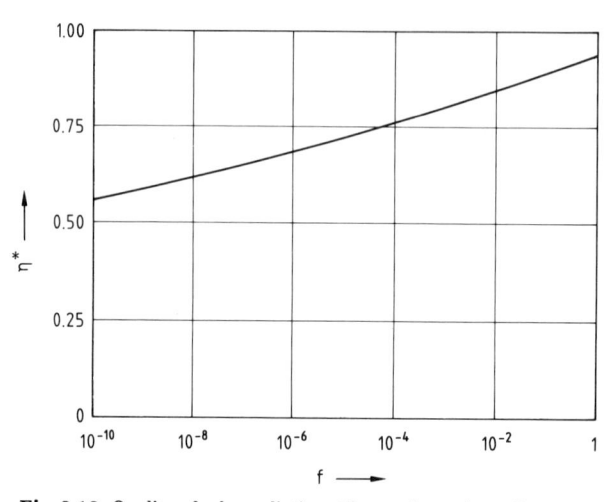

Fig. 2.18. Quality of solar radiation. Thermodynamic quality $\eta^* = W^*/{}^S E$ of solar radiation with a spectral distribution according to the standard spectrum ($T_S = 5,777$ K, $T_a = 300$ K). The parameter is the dilution factor f, which is taken to be independent of wavelength.

In Tab. 2.5 values of E (total irradiance), ${}^S E = f\sigma T_S^4$ (solar part of E), W^* (maximum power output from a perfect converter of E operating at $T_a = 300$ K) and the thermodynamic conversion ratio (exergy yield) $\eta^* = W^*/{}^S E$ are listed depending upon dilution factor f. The limiting case $f = 1$, where the radiance temperature is wavelength independent, $T(\lambda) = T_S$, is easily calculated. It follows $dS = (\partial \sigma T^4/\partial T)dT/T = 4\sigma T^2 dT$ and, therefore, $S - S' = 4\sigma \int_{T_a}^{T_S} T^2 dT = (4/3)\sigma(T_S^3 - T_a^3)$. Since $E - E' = \sigma(T_S^4 - T_a^4)$ the result of (2.60) is

$$W^* = \sigma T_S^4 \left(1 - \frac{4}{3}\frac{T_a}{T_S} + \frac{1}{3}\frac{T_a^4}{T_S^4}\right). \qquad (2.79)$$

The terms in the brackets represent the quality measure $\eta^* = W^*/\sigma T_S^4$. Putting $T_S = 5,777$ K and $T_a = 300$ K the maximum quality value of terrestrially available standard solar irradiance is

$$\eta^*_{(f=1)} = 0.93. \qquad (2.80)$$

Only 7% of (solar) energy in the incident flux is inevitably lost due to the requirement to drain, via waste heat, the entropy income which accompanies the radiant flux. An analytical approximation of $\eta^*(f)$ with $f < 1$ for standard spectra is possible, if f is again taken to be independent of λ [44]

$$\eta^*(f) \approx 1 - \frac{4}{3}\frac{T_a}{T}(1 - 0.28 \ln f). \qquad (2.81)$$

Table 2.5. E, ^SE, W^* and η^* of standard spectra. Total integral irradiance E and its solar part ^SE of the standard spectrum with $T_S = 5,777$ K and $T_a = 300$ K. W^* is the work obtainable in a perfect converter which operates at $T_a = 300$ K. The quality of the radiation, $\eta^* = W^*/^SE$, refers to the solar part of the spectrum. The dilution factor f is assumed to be wavelength independent.

f	E Wm^{-2}	^SE Wm^{-2}	W^* Wm^{-2}	η^*
1	$63.16 \cdot 10^6$	$63.16 \cdot 10^6$	$58.78 \cdot 10^6$	0.931
10^{-1}	$63.16 \cdot 10^5$	$63.16 \cdot 10^5$	$56.12 \cdot 10^5$	0.889
10^{-2}	$63.20 \cdot 10^4$	$63.16 \cdot 10^4$	$53.37 \cdot 10^4$	0.845
10^{-3}	$63.62 \cdot 10^3$	$63.16 \cdot 10^3$	$50.64 \cdot 10^3$	0.802
10^{-4}	$76.75 \cdot 10^2$	$63.16 \cdot 10^2$	$47.98 \cdot 10^2$	0.760
10^{-5}	$10.91 \cdot 10^2$	$63.16 \cdot 10^1$	$45.49 \cdot 10^1$	0.720
10^{-6}	$52.25 \cdot 10^1$	63.16	43.30	0.686
10^{-7}	$46.56 \cdot 10^1$	$63.16 \cdot 10^{-1}$	$41.31 \cdot 10^{-1}$	0.654
10^{-8}	$45.99 \cdot 10^1$	$63.16 \cdot 10^{-2}$	$39.07 \cdot 10^{-2}$	0.619
10^{-9}	$45.94 \cdot 10^1$	$63.16 \cdot 10^{-3}$	$36.76 \cdot 10^{-3}$	0.582
10^{-10}	$45.93 \cdot 10^1$	$63.16 \cdot 10^{-4}$	$34.79 \cdot 10^{-4}$	0.551

The approximation is accurate within –3% for $f > 1 \cdot 10^{-6}$ and $T_S = 5,777$ K, $T_a = 300$ K, which is sufficient for practical use.

Examples. With $f = f_s = 2.165 \cdot 10^{-5}$ the irradiance is $E \approx {}^SE = E_{sc} = 1,367$ Wm^{-2}. The quality is $\eta^* = 0.72$ for a converter ambiance at $T_a = 300$ K: only 984 Wm^{-2} are available as work W and 383 Wm^{-2} are waste heat for removing surplus entropy.

The numbers show that the entropy content of solar radiation increases with increasing dilution of the solar contribution to the total irradiance E. It is therefore important to keep the entropy content of the radiation as low as possible. This can be accomplished with optical concentrators and selective absorbers.

2.4 Concentration of Radiation

Concentration[5] aims at an increase of flux density E of a radiant flux Φ. Assuming conservation of radiant energy in the concentrating device or *concentrator* with input Φ and output Φ' the relations hold

$$\Phi = \Phi' \quad \text{or} \quad A \cdot E = A' \cdot E'. \tag{2.82}$$

A and A' are the geometric entrance and exit aperture areas of the concentrator; E and E' are the corresponding flux densities, averaged over A and A', respectively.

The ratio E'/E determines the flux density concentration ratio (or simply concentration) C. Because of the assumed conservation of radiant flux, it is identical with the (geometric) ratio A/A' of the aperture areas

$$C = \frac{E'}{E} = \frac{A}{A'}. \tag{2.83}$$

---

[5] For a detailed exposition of concentration of radiation see Chap. 3

Radiation within angular aperture θ (the half angle of a cone) enters the concentrator's areal aperture (see Fig. 2.19). It corresponds to a (spectral) radiant flux

$$\Phi_\nu(\nu)d\nu = A\, E_\nu(\nu)d\nu = A\pi \sin^2\theta\, L_\nu(\nu,T)d\nu. \tag{2.84}$$

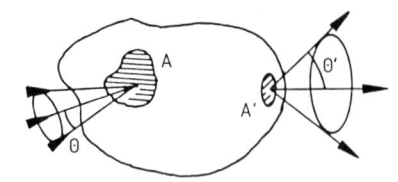

Fig. 2.19. Geometric quantities of a concentrator. A and θ are areal and angular apertures for the entering radiant flux. The effect of concentration is to have a reduced exit aperture area A'; to it belongs the angular exit aperture θ'.

In this equation the arbitrary spectral flux density E_ν has been equated to spectral irradiance of the same frequency ν of a black body of the correct radiance temperature $T(\nu)$ (see Sect. 2.3.2). For practical reasons, which will become obvious presently, frequency ν is used, not wavelength λ, to indicate spectral components, $E_\nu = dE(\nu)/d\nu$ and

$$L_\nu(\nu,T) = \frac{2h\nu^3}{c^2}\frac{1}{\exp(h\nu/kT)-1} = \frac{d^3\Phi}{dA\cos\theta d\Omega d\nu}. \tag{2.85}$$

The same energy flux $\Phi_\nu d\nu$ has in steady state to leave an *ideal* (conserving radiant flux) and *passive* (conserving frequency) concentrator

$$\Phi'_\nu d\nu' = A'\pi \sin^2\theta'\, L_\nu(\nu',T')d\nu' \equiv \Phi_\nu d\nu. \tag{2.86}$$

The ideal and passive concentrator conserves not only radiant energy but also entropy flux. The corresponding relations are (see 2.3.2)

$$A\cdot S_\nu d\nu = A\pi \sin^2\theta\, K_\nu(\nu,T)d\nu \quad \text{and} \quad A'\cdot S'_\nu d\nu' = A'\pi \sin^2\theta'\, K_\nu(\nu',T')d\nu' \equiv A\cdot S_\nu d\nu \tag{2.87}$$

with the previous definitions of K and Y in (2.67). Dividing the energy and entropy flux equations leaves

$$L_\nu(\nu,T)/K_\nu(\nu,T) = L_\nu(\nu',T')/K_\nu(\nu',T'). \tag{2.88}$$

In a passive concentrator the frequency ν is conserved: $\nu' = \nu$. Hence it follows from (2.88) that the concentrator is also conserving the radiance temperature, $T' = T$. Hence, $L_\nu(\nu,T) = L_\nu(\nu',T')$ and $K_\nu(\nu,T) = K_\nu(\nu',T')$. The consequence of the balance equation $\Phi'_\nu = \Phi_\nu$ (or $S'_\nu = S_\nu$) is a remarkable 'geometric' conservation law

$$A\sin^2\theta = A'\sin^2\theta' = \text{const.} \tag{2.89}$$

It says that in an ideal and passive concentrator the entity $A\cdot \sin^2\theta$, termed *throughput* or *Etendue*, remains preserved.

The definition of concentration ratio C can now be extended to a ratio in terms of angular apertures θ and θ'

$$C = \frac{A}{A'} = \frac{\sin^2\theta'}{\sin^2\theta}. \tag{2.90}$$

So far it has been assumed that concentrator entrance and exit happen to be in media of equal index of refraction, $n = n'$. If this is not the case [24], the phase velocity c in Planck's

2.4 Concentration of Radiation

radiance L_ν has to be substituted by c/n

$$C = \frac{A}{A'} = \frac{n'^2 \sin^2 \theta'}{n^2 \sin^2 \theta}. \tag{2.91}$$

Because of $\lambda/\lambda' = n'/n$ the wavelength of a spectral component is different for both media, whereas the frequency $\nu = \nu'$ remains unaffected. This is the reason for using ν rather than λ in the formulae of this section. Note that the Stefan-Boltzmann constant changes to $\sigma = n^2 5.67051 \cdot 10^{-8}$ Wm^{-2}K^{-4}.

In going through the derivation of Etendue again but this time with differential quantities of area dA and solid angle dΩ, the correct relation is

$$\int_{A,\Omega} n^2 \mathrm{d}A \cos\theta \mathrm{d}\Omega = \mathrm{const.} \tag{2.92}$$

Concentrators with radiation in a conical solid angle incident upon the entrance area A (so-called *point focusing concentrators*) produce an Etendue of $n^2 A \sin^2\theta$. Concentrators with a line aperture (so-called *line focusing concentrators*) of width B produce an Etendue of $nB\sin\theta$. The latter applies to troughs. The trough concentration factor, therefore, is for the simplest geometry,

$$C = \frac{A}{A'} = \frac{n'\sin\theta'}{n\sin\theta}. \tag{2.93}$$

The maximum concentration ratio is obtained with exit angular aperture $\theta' = 90°$: the radiation emerges from the concentrator in a hemispherical angular distribution

$$C^{\max} = \frac{n'^2}{n^2 \sin^2\theta} \quad \text{and} \quad C^{\max} = \frac{n'}{n\sin\theta}, \quad \text{respectively.} \tag{2.94}$$

The minimum solar aperture angle, $\theta_S^{\min} = 0.267°$, can be calculated from $\sin\theta_S^{\min} = R_S/D_{ES} = 4.65 \cdot 10^{-3}$. The theoretical maximum concentration ratio is for the point or line focusing concentrator, respectively, with $\theta = \theta_S^{\min}$

$$C_S^{\max} = n'^2 \left(\frac{D_{ES}}{R_S}\right)^2 = \frac{n'^2}{2.165 \cdot 10^{-5}} = n'^2 \cdot 46,200 \tag{2.95}$$

$$C_S^{\max} = n'\left(\frac{D_{ES}}{R_S}\right) = \frac{n'}{4.65 \cdot 10^{-3}} = n' \cdot 215. \tag{2.96}$$

Concentrators are non-imaging systems: they are not built to produce an undistorted point by point corresponding image of the Sun's disk [23,45,57]. This is particularly obvious for high concentration ratios, where the exit rays spread over a cone of 90° angular aperture.

The geometric contribution to the dilution factor f and the geometric solar concentration ratio $C = A/A' = n'^2 \sin^2\theta'/n^2 \sin^2\theta_S^{\min}$ of a point concentrator are because of $\theta' \equiv \theta_S$ related accordingly (θ being the entrance angular aperture of the receiver, where n' is the index of refraction)

$$f = \frac{\sin^2\theta_S}{\sin^2\theta} = C \frac{n^2 \sin^2\theta_S^{\min}}{n'^2 \sin^2\theta} = C \frac{2.165 \cdot 10^{-5}}{n'^2 \sin^2\theta}. \tag{2.97}$$

$\theta_S^{\min}$ is astronomically determined, 0.267°; θ_S and θ are illustrated in Fig. 2.20. In the case of a flat receiver, i.e. a concentrator with an exit aperture of 90°, it follows that

$$f = C \frac{1}{n'^2} 2.165 \cdot 10^{-5} = \frac{C}{46,200} \frac{1}{n'^2}. \tag{2.98}$$

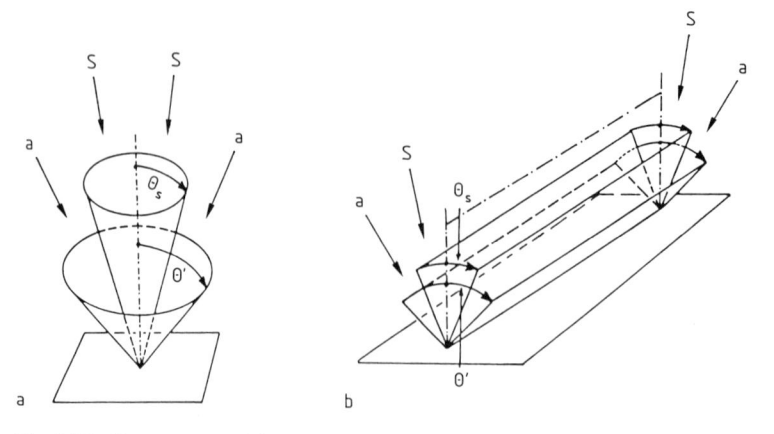

Fig. 2.20. Concentrators. (a) Point concentrator. Direct solar radiation S enters the angular aperture θ_S, whereas in the interstice $\theta' - \theta_S \geq 0$ ambient radiation a reaches the receiver. (b) Line concentrator. S, a and θ_S, θ' are similar to case (a).

2.5 Conversion to Heat

Conversion of radiation to process heat is a highly developed technique in solar engineering [16,53]. A prominent part of this book on solar power plants is devoted to realized installations, practical experiences and future prospects of solar process heat production. In the following, fundamentals of the conversion process are presented and exemplified.

The relation between flux density E of total (including solar and ambient) radiation incident upon a collector and process heat Q of temperature T produced per unit absorber area is

$$Q = (\alpha\tau)FE + (\alpha\epsilon)F\sigma T_C^4 - (\epsilon\bar{\rho})F\sigma T^4 - U_L F(T - T_a). \tag{2.99}$$

This equation is an energy balance of gains and losses of the receiver (the collector). It is instructive to derive the relation, which provides the occasion for defining the various quantities involved.

The incident flux density $E = \int_0^\infty E_\lambda d\lambda$ is absorbed with an efficiency $(\alpha\tau)$. The absorptance α_A of the absorber is a materials property for transforming radiation into heat. Before reaching the absorber, the radiation has frequently to pass through a cover, which transmits the radiation with transmittance τ_C; the fraction ρ_C is assumed to be backscattered and the fraction α_C to be absorbed by the cover. The symbol $(\alpha\tau)$ does not represent simply the product $\alpha_A \cdot \tau_C$, but because of multiple reflections between absorber and cover

$$(\alpha\tau) = \frac{\alpha_A \cdot \tau_C}{1 - \rho_C(1 - \alpha_A)} = \frac{\alpha_A \cdot \tau_C}{1 - \rho_C \rho_A}. \tag{2.100}$$

The second input term involves emission from the (hot) cover of temperature T_C towards the absorber; the factor $(\alpha\epsilon)$ is in structure analogous to $(\alpha\tau)$ except for the exchange of τ_C by ϵ_C ($\hat{=} \alpha_C$). The cover becomes heated by absorption of radiation from incident E_λ, from emission of the absorber and by internal convection; the cover loses heat into the ambiance. Note that the first parameter in $(\alpha\tau)$, $(\alpha\epsilon)$ and $(\epsilon\bar{\rho})$ always refers to the absorber, the second to the cover window.

2.5 Conversion to Heat

Note that $\alpha_C + \rho_C + \tau_C = 1$, $\alpha_A + \rho_A = 1$ and, according to Kirchhoff's law, $\alpha(\lambda) = \epsilon(\lambda)$. $\alpha_A, \epsilon_C, \rho_C, \tau_C$ are in general wavelength dependent. Therefore, $(\alpha\tau)$ is an average over the spectral components $E_\lambda(\lambda)$ in the incident flux density E while $(\alpha\epsilon)$ is an average over the black body emission spectrum at T_C

$$(\alpha\tau) = \frac{\int_0^\infty (\alpha\tau) E_\lambda(\lambda) d\lambda}{\int_0^\infty E_\lambda d\lambda} \quad \text{and} \quad (\alpha\epsilon) = \frac{\int_0^\infty (\alpha\epsilon) \pi L_\lambda(\lambda, T_C) d\lambda}{\sigma T_C^4}. \tag{2.101}$$

The loss terms are of two different kinds:
1. The absorber of temperature T_A radiates thermally through the aperture of the receiver or collector

$$(\epsilon\bar{\rho})\sigma T_A^4. \tag{2.102}$$

The radiation losses depend on emissivity ϵ_A ($\hat{=} \alpha_A$) of the absorber and on reflectivity ρ_C for the thermal radiation at the cover. In the shorthand symbol $(\epsilon\bar{\rho})$ the $\bar{\rho}$ indicates $(1 - \rho_C)$ or, equivalently, $(\alpha_C + \tau_C)$

$$(\epsilon\bar{\rho}) = \frac{\epsilon_A \cdot (1 - \rho_C)}{1 - \rho_C(1 - \alpha_A)} = \frac{\epsilon_A \cdot (\alpha_C + \tau_C)}{1 - \rho_C \rho_A}. \tag{2.103}$$

$(\epsilon\bar{\rho})$ averages over the spectral components of Planck's (hemispherical) radiosity $\pi L_\lambda(\lambda, T_A)$ of black body emission

$$(\epsilon\bar{\rho}) = \frac{\int_0^\infty (\epsilon\bar{\rho}) \pi L_\lambda(\lambda, T_A) d\lambda}{\sigma T_A^4} \quad \text{since} \quad \sigma T_A^4 = \int_0^\infty \pi L_\lambda(\lambda, T_A) d\lambda. \tag{2.104}$$

2. The convective (or conductive) thermal loss of the absorber to the ambiance at temperature T_a is

$$U_L(T_A - T_a). \tag{2.105}$$

The heat loss coefficient U_L depends on temperature and on forced convection by wind. U_L can usually be made small by thermal insulation.

The fluid temperature T and the absorber temperature T_A are not identical. A heat transfer resistance $1/U_{AF}$ between the absorber surface and the fluid main stream causes a temperature drop $T_A - T$. In steady state the energy fluxes are (see Fig. 2.21)

$$Q = U_{AF}(T_A - T) \text{ and } Q + U_L(T_A - T_a) + (\epsilon\bar{\rho})\sigma T_A^4 = (\alpha\tau)E + (\alpha\epsilon)\sigma T_C^4. \tag{2.106}$$

For good performance U_{AF} should be made as great as possible but U_L as small as possible. The first equation in (2.106) yields $T_A^4 = (T + Q/U_{AF})^4$, which can be expanded to $T^4 + Q\,4T^3/U_{AF}$, recalling that $Q/U_{AF} = T_A - T \ll T$. Then combining both equations of (2.106) by eliminating the absorber temperature T_A and solving for Q the initial equation (2.99) of this section is obtained. F (termed *fin factor*) is the abbreviation for

$$\frac{1}{F} = 1 + \frac{U_L}{U_{AF}} + \frac{4\sigma(\epsilon\bar{\rho})T^3}{U_{AF}}. \tag{2.107}$$

F is close to 1 for a well designed receiver or collector.

The incident flux density E is composed of solar radiation ^SE and radiation from the ambiance ^aE. This mixture constitutes the total radiation. The solar part ^SE of global irradiance can be divided into a direct radiation (beam radiation) ^SE_b and a diffuse radiation ^SE_d

$$E = {}^SE + {}^aE = {}^SE_b + {}^SE_d + {}^aE. \tag{2.108}$$

The thermal efficiency of a collector for conversion of radiation into heat is usually defined by

$$\eta = \frac{Q}{{}^SE} \quad \text{with} \quad {}^SE \text{ the } global \text{ irradiance.} \tag{2.109}$$

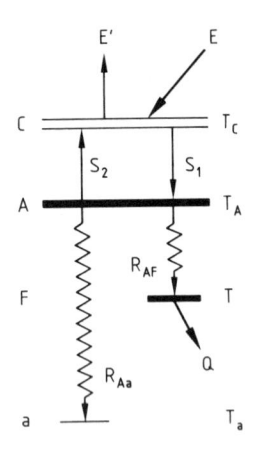

Fig. 2.21. Flow sheet of collector radiation and heat fluxes. The total irradiance of the collector is E, the lost back radiation is E'. Q is usable process heat. Heat flow resistances are $R_{AF} = 1/U_{AF}$ (between absorber and fluid) and $R_{Aa} = 1/U_L$ (between absorber and ambiance). The absorber with absorptivity α_A is covered with glass of transmissivity τ_C, refectivity ρ_C and emissivity (absorptivity) ϵ_C. The absorber (A) temperature is T_A, the cover (C) temperature T_C, the ambient (a) temperature T_a, the fluid (F) temperature T. S_1 and S_2 are collector internal radiation flux densities.

In case concentrated radiation is used the thermal efficiency of a receiver is, by convention, defined as

$$\eta = \frac{Q}{{}^S E_b} \quad \text{with} \quad {}^S E_b \text{ the } \textit{direct} \text{ solar irradiance,} \tag{2.110}$$

omitting the diffuse solar part in the incident global radiation. This is reasonable, because otherwise the thermal efficiency would become strongly dependent on the ratio ${}^S E_b/{}^S E$, i.e. on meteorology parameters.

Introducing the concentration ratio $C = E/{}^S E_b$, the thermal efficiencies of process heat production by absorption of solar radiation can be written

$$\eta = (\alpha\tau)F + (\alpha\epsilon)F \frac{\sigma T_C^4}{{}^S E} - (\epsilon\bar{\rho})F \frac{\sigma T^4}{{}^S E} - U_L F \frac{(T - T_a)}{{}^S E} \quad \text{and} \tag{2.111}$$

$$\eta = (\alpha\tau)F + (\alpha\epsilon)F \frac{\sigma T_C^4}{C \, {}^S E_b} - (\epsilon\bar{\rho})F \frac{\sigma T^4}{C \, {}^S E_b} - U_L F \frac{(T - T_a)}{C \, {}^S E_b}. \tag{2.112}$$

In both cases contributions of order ${}^a E/{}^S E$ have been neglected.

2.5.1 Process Heat and Concentrated Radiation

First assume the radiation loss term to dominate over the convective loss term

$$\eta = (\alpha\tau)F + (\alpha\epsilon)F \frac{\sigma T_C^4}{C \, {}^S E_b} - (\epsilon\bar{\rho})F \frac{\sigma T^4}{C \, {}^S E_b}. \tag{2.113}$$

The process heat temperature T for a given thermal efficiency is

$$T = \left(C \, {}^S E_b \frac{(\alpha\tau)F - \eta}{(\epsilon\bar{\rho})F \sigma} + \frac{(\alpha\epsilon)}{(\epsilon\bar{\rho})} T_C^4 \right)^{1/4}. \tag{2.114}$$

2.5 Conversion to Heat

As an example, consider the conditions stated in Fig. 2.22 but excluding convection losses. For $\eta = 0$ one obtains with $C = 1$ a maximum fluid temperature $T_{\eta=0,C=1} = 345$ K. For $C = 10; 100; 1,000; 10,000$ the corresponding maximum temperatures $T_{\eta=0,C}$ are 613 K; 1,090 K; 1,938 K; 3,446 K.

Then a term for convective loss ($U_L F = 20$ Wm^{-2}K^{-1}, $T_a = 300$ K) is added (thin line curves in Fig. 2.22). For $C = 1$ the stagnation temperature ($\eta = 0$) is only 308 K. Note the almost linear graph of $\eta(T)$ for low concentration (the convection dominates); the graph remains practically unperturbed by U_L at high concentration ratios. This proves the importance of high concentration ratios even in cases of low process heat temperatures.

An apparent measure of irreversibilities (i.e. of entropy production) involved in the conversion process of solar radiation to heat by absorption is the *work* obtainable from the heat flux. Imagine the collector being coupled to a perfect (isothermal) heat engine operating between the temperature level T of the collector output Q and the ambient T_a. By means of fluid rate variation the temperature T can be adjusted to produce a maximum of work: $\eta(T)$ decreases with increasing T (see Fig. 2.22), but the Carnot conversion factor increases with T. The result with $(\alpha\tau) = 1$ and $(\epsilon\bar{\rho}) = 1$, which is a *perfect black absorber with a perfect transparent cover*, is plotted and marked 'black' in Fig. 2.23. Incident radiation is according to standard solar spectra, varying the dilution factor f. For comparison the maximum work obtainable from solar radiation (see Sect. 2.3.4, Fig. 2.18) is included in the figure, marked 'solar maximum'. An important observation is that the degree of quality of conversion to heat by absorption processes increases with concentration ratio $C = 2.165 \cdot 10^{-5}/f$ of the radiation. For $C = 1$ or $f = 2.165 \cdot 10^{-5}$ if extinction in the atmosphere is neglected, the achieved quality η^* is a low 10% compared to a possible 72% of the incident solar (standard) spectrum.

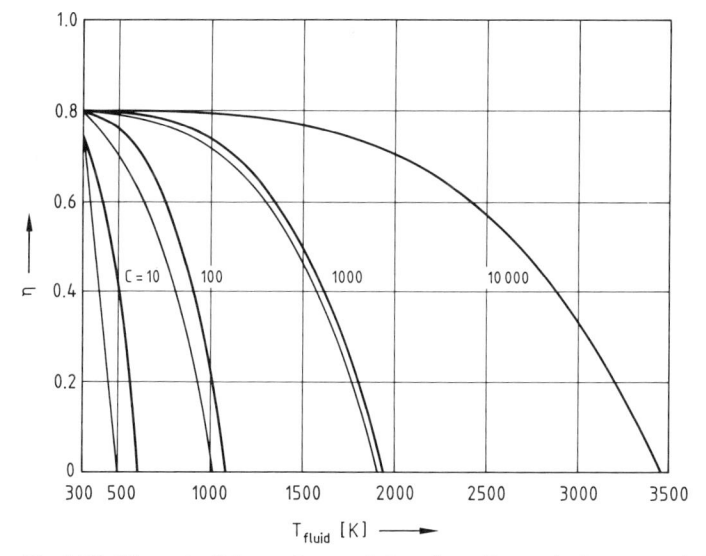

Fig. 2.22. Thermal efficiency characteristics of a collector. $(\alpha\tau)F = 0.8, (\epsilon\bar{\rho})F = 0.8$ and $(\alpha\epsilon)F = 0$. 800 Wm^{-2} irradiance: thick solid lines. Including a convective loss of $U_L F = 20$ Wm^{-2}K^{-1} and $T_a = 300$ K produces the thin solid lines. The geometric concentration ratio C of incident solar radiation $^S E_b$ has been varied in decimal steps.

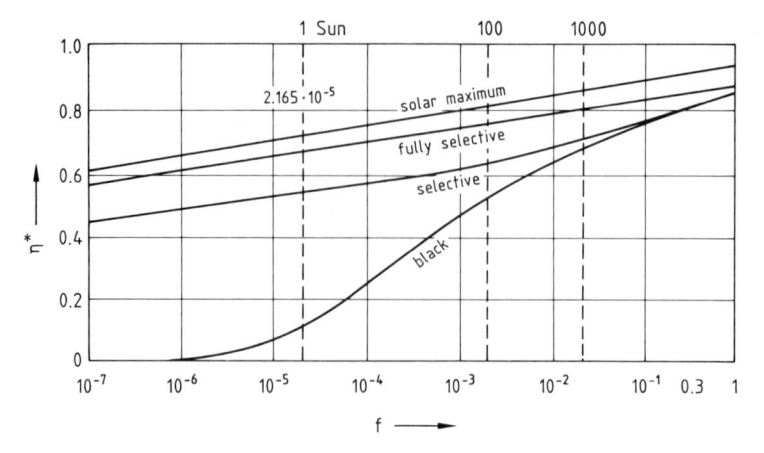

Fig. 2.23. Thermodynamic quality of conversion of radiation to heat by absorption. Quality measure $\eta^* = W^*/{}^S\!E$ of process heat produced in absorption from standard spectrum radiation. 'solar maximum' refers to reversible conversion of the incident radiation, 'black' to a perfect black collector ($\alpha_A = 1$, $\tau_C = 1$), 'selective' to a perfect optically selective collector, 'fully selective' to the limiting case of an infinite number of separate selective collectors each tuned in its infinitesimal spectral bandwidth to optimum delivery of work from a coupled perfect heat engine.

2.5.2 Selective Absorption-Transmission

Looking at the expression for collector efficiency (2.99), it becomes apparent that it is possible to produce medium temperature process heat without concentration of radiation by making the emission-reflection product $(\epsilon\bar{\rho})$ as small as possible. This mode of procedure seems to be equivalent to applying concentration, since $(\epsilon\bar{\rho})/C$ is decisive in the radiation loss term (see Sect. 2.5.1). However, there are two important differences. First, ${}^S\!E$ is the *global* irradiance comprising both the direct and diffuse solar insolation, which is an advantage over concentration. Second, as a disadvantage over concentration, the convective loss term $U_L F$ remains unaffected (i.e. is not at the same time reduced) by lowering $(\epsilon\bar{\rho})$; hence, convective losses may finally negate the advantages of further lowering $(\epsilon\bar{\rho})$.

Technically, a small $(\epsilon\bar{\rho})$ is possible by proper choice of materials [4,31]. However, $(\epsilon\bar{\rho})$ cannot be allowed to be zero (or close to zero) over all wavelengths. Kirchhoff's law, $\alpha(\lambda) = \epsilon(\lambda)$, implies in this case $(\alpha\tau) = 0$: the collector would not absorb any radiation.

The possibility exists, however, to make $\alpha_A(\lambda)$ and $\tau_C(\lambda)$ strongly wavelength dependent [7,18,26,33,34,40]. In the ideal case, both $\alpha_A(\lambda)$ and $\tau_C(\lambda)$ should be 1 in the (short optical wavelength) region of incident solar radiation; $\epsilon_A(\lambda)$ and $\rho_C(\lambda)$ should be 0 in the long wavelength regime of thermal emission of the hot absorber. Wien's displacement law of the wavelength $\lambda_{\max}$ of maximum radiance of a thermal emitter of temperature T ensures that such a requirement is feasible:

$$\lambda_{\max} T = 2,898 \quad \text{in} \quad \mu\text{m K}. \tag{2.115}$$

The Sun's temperature $T_S = 5,777$ K centers the maximum of insolation at $\lambda_{\max} = 0.5\ \mu$m, whereas a hot absorber of $T_A = 1,000$ K has its emission maximum at a distant $\lambda_{\max} = 2.9\ \mu$m.

The question is, where to locate the transition from $(\alpha\tau) = 1$ (which implies $(\alpha\epsilon) = 0$) to $(\epsilon\bar{\rho}) = 0$ for best performance of the collector. For this purpose the collector equation is

2.5 Conversion to Heat

rewritten, replacing the spectral averages $(\alpha\tau)$ and $(\epsilon\bar{\rho})$ by the wavelength dependent input and loss terms and neglecting convective losses,

$$Q = \int_0^\infty (\alpha\tau) E_\lambda(\lambda) \mathrm{d}\lambda - \int_0^\infty (\epsilon\bar{\rho}) \pi L_\lambda(\lambda, T_A) \mathrm{d}\lambda. \tag{2.116}$$

T_A is the absorber temperature. Of interest are the wavelength regions where Q is positive

$$E_\lambda(\lambda) > \pi L_\lambda(\lambda, T_A). \tag{2.117}$$

It is again instructive (see Sect. 2.3.2) to substitute any spectral component $E_\lambda(\lambda)$ by an equivalent black body radiation component $\pi L_\lambda(\lambda, T)$

$$E_\lambda(\lambda) = \pi L_\lambda(\lambda, T). \tag{2.118}$$

The actual $E_\lambda(\lambda)$ can originate from a mixture of various radiation sources: the Sun, the sky, the ambiance, artificial radiation sources shining onto the (flat) absorber. The equivalent spectral black body radiation defines by (2.118) the *radiance temperature*

$$T(\lambda) = \frac{hc/k\lambda}{\ln(1 + 2\pi hc^2/E_\lambda \lambda^5)}. \tag{2.119}$$

The condition (2.117)

$$E_\lambda(\lambda) = \pi L_\lambda(\lambda, T) > \pi L_\lambda(\lambda, T_A) \tag{2.120}$$

appears to be synonymous to

$$T(\lambda) > T_A \quad \text{with maximum of absorption required and} \tag{2.121}$$
$$T(\lambda) < T_A \quad \text{with minimum of emission required.} \tag{2.122}$$

Only if the spectral radiance temperature $T(\lambda)$ at wavelength λ is greater than T_A does heat flow from radiation to the absorber. For that wavelength it is of advantage for the yield of process heat if $(\alpha\tau)$ is close to 1. Otherwise, if the radiance temperature $T(\lambda)$ of incident radiation is lower than T_A, heat flows from the absorber into radiation. Thus, for that wavelength it is of advantage if $(\epsilon\bar{\rho})$ is close to zero. The concept of radiance temperature can be used to find the wavelength regimes with optimum $(\alpha\tau) = 1$ and $(\epsilon\bar{\rho}) = 0$. First the available spectrum $E_\lambda(\lambda)$ is transformed into a spectrum $T(\lambda)$ of radiance temperature. In Fig. 2.24 this has been accomplished for an $AM1$ spectrum. The required process heat temperature T_A (here 600 K is assumed) is marked in Fig. 2.24 as a horizontal line. The regions with $T(\lambda) > T_A$ are the regions where the collector should be 'black'. In the regions with $T(\lambda) < T_A$ the collector should be transparent, reflective or 'white'. Such a wavelength dependent characteristic of a collector is termed *selective absorption-transmission*.

2.5.3 Yield of Process Heat

The yield of process heat of temperature T_A of a collector depends on the combined behavior of absorptance (emittance) $\alpha_A(\lambda)$ of the absorber and reflectance $\rho_C(\lambda)$ of the cover. For estimating the achievable thermal efficiencies a simple model collector is considered. Its $(\alpha\tau)$ is assumed to be constant (and high) for all wavelengths which fulfil $T(\lambda) > T_A$; for all wavelengths where $T(\lambda) < T_A$ it is assumed that $(\epsilon\bar{\rho})$ is constant (and small); finally it is assumed that there is no absorption in the cover, $(\alpha\epsilon) = 0$. In particular, in the standard solar spectrum $E_\lambda(\lambda) = f \pi L_\lambda(\lambda, T_S) + (1-f) \pi L_\lambda(\lambda, T_a)$, the factor f is taken to be $1.3 \cdot 10^{-5}$,

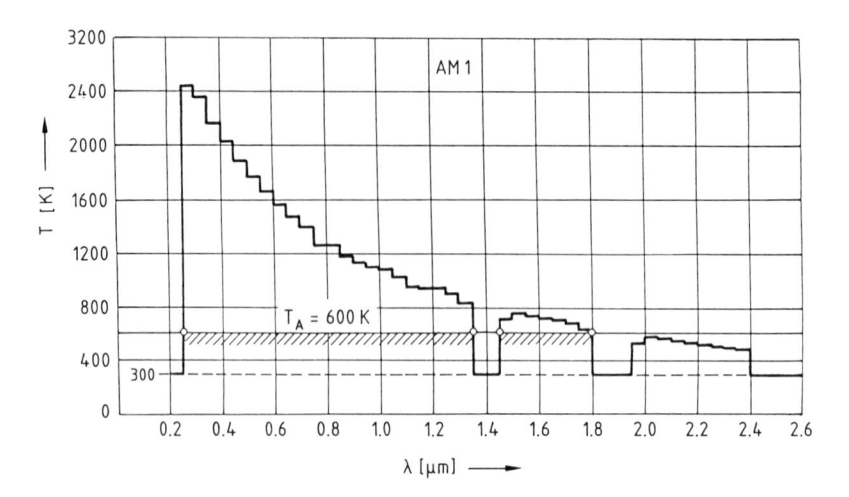

Fig. 2.24. Radiance temperature $AM1$ spectrum. For illustration a process heat temperature $T_A = 600$ K has been assumed. In the shaded spectral region $(\alpha\tau)$ should be made 1; in the other regions $(\epsilon\bar{\rho})$ should be made 0.

a value which can be regarded as typical for terrestrial $AM2$ insolation. The integral solar part of this spectrum is $^SE = f\sigma T_S^4 = 821$ Wm^{-2}.

An absorber temperature of 1,000 K is assumed. In Fig. 2.17 of Sect. 2.3.3 showing radiance temperatures $T(\lambda)$ for standard spectra the cut-off wavelength is found to be $\lambda_o = 1.065$ μm for $f = 1.3 \cdot 10^{-5}$.

Hence, for all $\lambda < \lambda_o$ the model collector is assumed to be strongly absorptive with constant $(\alpha\tau)$; for $\lambda > \lambda_o$ the collector is weakly emissive with constant $(\epsilon\bar{\rho})$.

The collector equation then becomes

$$Q = (\alpha\tau)\int_0^{\lambda_o} E_\lambda(\lambda)d\lambda + (\epsilon\bar{\rho})\int_{\lambda_o}^{\infty} E_\lambda(\lambda)d\lambda \qquad (2.123)$$
$$- (\alpha\tau)\int_0^{\lambda_o} \pi L_\lambda(\lambda, T_A)d\lambda - (\epsilon\bar{\rho})\int_{\lambda_o}^{\infty} \pi L_\lambda(\lambda, T_A)d\lambda - U_L(T_A - T_a).$$

The thermal efficiency becomes

$$\eta = \frac{Q}{^SE} = \{(\alpha\tau)F_1 + (\epsilon\bar{\rho})(1 - F_1)\}\frac{E}{^SE} \qquad (2.124)$$
$$- \{(\alpha\tau)F_2 + (\epsilon\bar{\rho})(1 - F_2)\}\frac{\sigma T_A^4}{^SE} - U_L\frac{(T_A - T_a)}{^SE}.$$

The abbreviations are, since $(\alpha\tau)$ and $(\epsilon\bar{\rho})$ assumed to be locally constant

$$F_1(\lambda_o) = \frac{\int_0^{\lambda_o} E_\lambda(\lambda)d\lambda}{E} \qquad E = \int_0^{\infty} E_\lambda(\lambda)d\lambda \qquad (2.125)$$
$$F_2(\lambda_o, T_A) = \frac{\int_0^{\lambda_o} \pi L_\lambda(\lambda, T_A)d\lambda}{\sigma T_A^4} \quad \text{with} \quad \sigma T_A^4 = \int_0^{\infty} \pi L_\lambda(\lambda, T_A)d\lambda. \qquad (2.126)$$

^SE is the solar part of the total incident radiation E. With a standard spectrum and $f = 1.3 \cdot 10^{-5}$ it follows $E = f\sigma T_S^4 + (1-f)\sigma T_a^4 = 1,280$ Wm^{-2}. From the graph Fig. 2.25 $F_1(\lambda)$ for this standard spectrum at $\lambda = \lambda_o = 1.065$ μm a value $F_1 = 0.484$ is found.

2.5 Conversion to Heat

In Fig. 2.26 various diagrams are plotted to demonstrate the influences of the collector parameters $(\alpha\tau)$, $(\epsilon\bar{\rho})$ and U_L. Note in the plots the sensitivity of the thermal efficiency η to $(\epsilon\bar{\rho})$ and to the convective heat loss coefficient U_L. The consequence is that $(\epsilon\bar{\rho})$ should be made as small as possible, irrespective of the operating temperature of the collector. For such collectors with a geometric concentration ratio of unity the influence of U_L on the shape of the characteristic diagram is strong. That is even true for the small value of $U_L = 0.1$ Wm^{-2}K^{-1}.

The main practical problem remains to manufacture absorbers with low $\epsilon_A(\lambda)$ and covers with high $\rho_C(\lambda)$ in the thermal reradiation region $\lambda > \lambda_o$. At present stable and high temperature resistant $(\epsilon\bar{\rho})$ smaller than 0.1 are hardly available [7,27,34], whereas an $(\epsilon\bar{\rho})$ of an order of 0.01 would be desirable. In this process it is important that $(\alpha\tau)$ not be sacrificed in the range $0 < \lambda < \lambda_0$ as the first integral in (2.124) is typically much greater than the fourth, particulary if the concentration ratio is high.

The absorber temperature T_A is related to the temperature of the heat transfer fluid T by $Q = U_{AF}(T_A - T)$ or $T = T_A - \eta^S E/U_{AF}$. The heat transfer U_{AF} between absorber and fluid should be as good as possible to avoid loss of heat quality. For example, with $U_{AF} = 200$ Wm^{-2}K^{-1} (which is typical for turbulent fluid flow) and $^S E = 800$ Wm^{-2} insolation at most 4 K in process heat temperature is lost. Selective collector systems for medium temperature process heat production are attractive for several reasons: utilization of both direct and diffuse insolation, no need for tracking the Sun, no need for frequent cleaning of the collector covers.

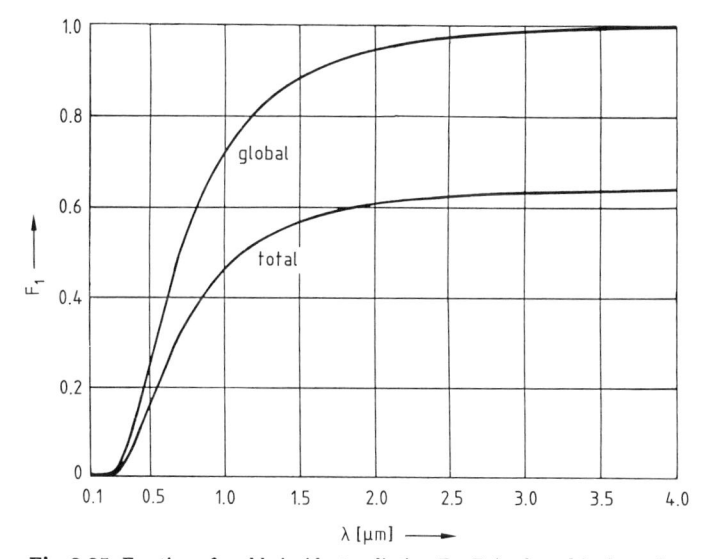

Fig. 2.25. Fraction of usable incident radiation F_1. F_1 is plotted in dependence on wavelength for a standard spectrum with $f = 1.3 \cdot 10^{-5}$ (no concentration of incident radiation), $T_S = 5,777$ K and $T_a = 300$ K. The lower curve refers to E being the sum of solar and ambient radiation; the upper curve refers to global (solar) radiation only.

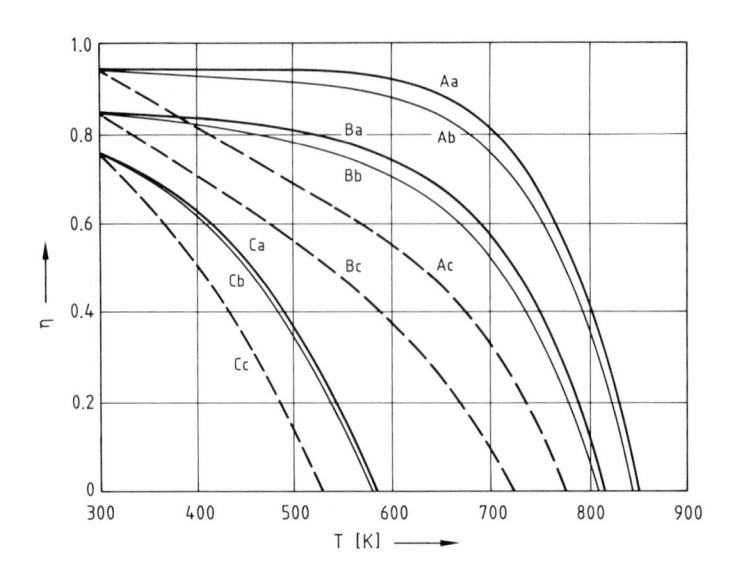

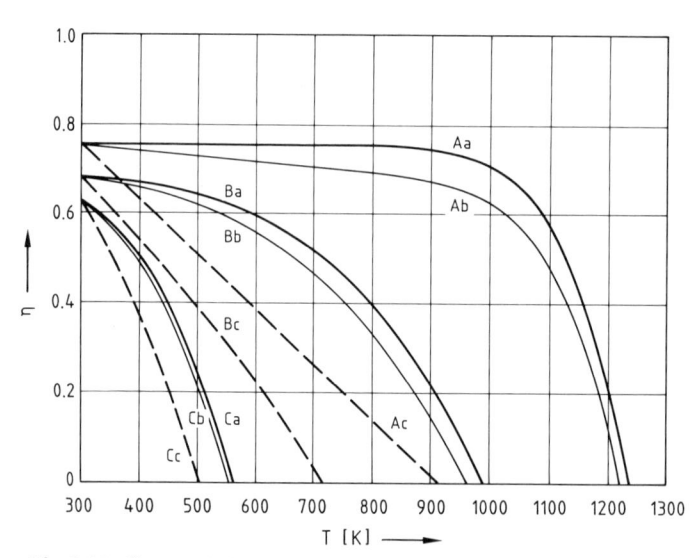

Fig. 2.26. Characteristic curves of collector performance.
Standard spectrum with $f = 1.3 \cdot 10^{-5}$, $T_S = 5,777$ K and $T_a = 300$ K. In both cases the following combinations of $(\alpha\tau)$, $(\epsilon\bar{\rho})$ (with $(\alpha\epsilon) = 0$) and U_L were used

A: $(\alpha\tau) = 1$; $(\epsilon\bar{\rho}) = 0$ B: $(\alpha\tau) = 0.9$; $(\epsilon\bar{\rho}) = 0.01$ C: $(\alpha\tau) = 0.8$; $(\epsilon\bar{\rho}) = 0.1$
a: U_L (Wm^{-2}K^{-1}) = 0   b: U_L (Wm^{-2}K^{-1}) = 0.1   c: U_L (Wm^{-2}K^{-1}) = 1

Design temperature 600 K: cut-off wavelength $\lambda_o = 1.970$ µm; $F_1 = 0.605$; $^S E = 821$ Wm^{-2}
Design temperature 1,000 K: cut-off wavelength $\lambda_o = 1.065$ µm; $F_1 = 0.484$; $^S E = 821$ Wm^{-2}.

2.5.4 Simultaneous Concentration and Selective Absorption

At low concentration ratios of about $C = 10$ good collector efficiency is limited to temperatures of about 450 K (see Sect. 2.5.1). With selective absorbers the yield at high temperatures depends sensitively on $(\epsilon\bar{\rho})$ and U_L. The combination of both selectivity and concentration has the advantage of reduced loss factors $(\epsilon\bar{\rho})/C$ and U_L/C. The disadvantage is that the acceptance of the diffuse insolation component is reduced to $1/C$. In Fig. 2.27, the situation with present technology of $(\epsilon\bar{\rho}) = 0.1, U_L = 1$ Wm^{-2}K^{-1} and $C = 10$ is shown. It should be compared to the separate cases of $C = 10$ in Sect. 2.5.1 and to $(\epsilon\bar{\rho}) = 0.1, U_L = 1$ Wm^{-2}K^{-1} in Sect. 2.5.2.

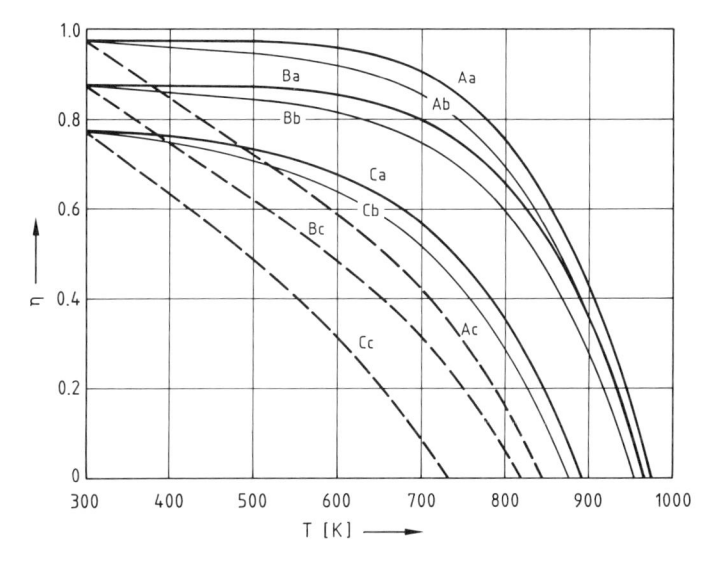

Fig. 2.27. Characteristic collector curves: concentration ratio $C = 10$.
Graphs of thermal efficiency as function of absorber temperature for a standard spectrum with $f = 1.3 \cdot 10^{-5}, T_S = 5,777$ K and $T_a = 300$ K. Design temperature 600 K, geometric concentration $C = 10$: cut-off wavelength $\lambda_o = 2.535$ μm; $F_1 = 0.9733$; $^SE' = 8,210$ Wm^{-2}.
The following combinations of $(\alpha\tau)$, $(\epsilon\bar{\rho})$ (with $(\alpha\epsilon) = 0$) and U_L are plotted
    A: $(\alpha\tau) = 1$; $(\epsilon\bar{\rho}) = 0$ B: $(\alpha\tau) = 0.9$; $(\epsilon\bar{\rho}) = 0.01$ C: $(\alpha\tau) = 0.8$; $(\epsilon\bar{\rho}) = 0.1$
    a: U_L (Wm^{-2}K^{-1}) = 0   b: U_L (Wm^{-2}K^{-1}) = 0.1   c: U_L (Wm^{-2}K^{-1}) = 1

Concentration increases λ_o and thereby the fraction F_1 of usable incident solar flux. Once λ_o shifts beyond 2.4 μm (the natural cut-off of the terrestrial solar spectrum), and once the absorber temperature is higher than 1,500 K (which locates λ_{max} at 2 μm), selective absorption ceases to be important for improving (terrestrial) collector systems.

The quality of the conversion of solar radiation to heat by optically selective absorbers can be calculated, analogous to the procedure presented in the last paragraph of Sect. 2.5.1. The result is illustrated in Fig. 2.23 with the curve marked 'selective'. The drastic quality improvement is particularly obvious for low concentration ratios.

Finally, further improvement is possible by separation of the solar spectrum into two or more wavelength bands and adjustment of separate selective collectors to each band. The result for the theoretically limiting case of an *infinite* number of such separate optically selective collectors is plotted in Fig. 2.23 with the curve marked 'fully selective'.

2.6 Conversion of Radiation to Electrical Energy

Large scale generation of electric power is carried out in solar heat-driven thermal power plants, employing the so-called *Carnot conversion*, see Chap. 6. However, a solar specific process exists, the *inner photoeffect*, e.g. in semiconductors such as silicon. It can be employed in an elegant *photovoltaic conversion* of solar radiation to electric energy, see Chap. 7. By photoelectrochemical effects a third path of conversion is available. It is in a status of research and development and so far not available for power production.

In this section the thermodynamic basis of photovoltaics is examined. Ideal photocells are defined and their maximum conversion yield for solar radiation is calculated.

2.6.1 Photoionization

Absorbed radiation releases electrons from the covalent bonds which represent the chemical binding of the atoms in semiconductors. This ionization process produces mobile charge carriers of both negative and positive sign in a semiconductor lattice: free negative electrons and free positive holes (defect electrons). The atomic (covalent) bonds consist of electron pairs. In silicon (Si), which is a 4th column element in the periodic table, adjacent atoms contribute one electron each to the pair bond. For a III–V semiconductor, e.g. gallium arsenide (GaAs), the electron pair is provided by the atoms of the 5th column; the 1:1 mixture of III and V column elements behaves as if the average lattice atom contributes again four valence electrons. The same arguments apply to II–VI semiconductors such as cadmium sulfide, CdS. At room temperature, non-irradiated semiconductors are electric insulators.

In the photoionization process radiation appears absorbed as *photons*, i.e. discrete amounts (quanta) of energy per electron-hole pair produced. Therefore, the incoming radiation is considered to consist of a flux of photons, each of energy $h\nu$. Here, ν is the frequency of a spectral component of the radiation and h Planck's constant, $h = 6.6260755 \cdot 10^{-34}$ Ws^2. The number density N_ν of photons $h\nu$ equivalent to a spectral flux density E_ν of electro-magnetic radiation is

$$N_\nu = \frac{\mathrm{d}N}{\mathrm{d}\nu} = \frac{E_\nu}{h\nu}. \tag{2.127}$$

The ionization process requires a minimum photon energy $h\nu_g$, which is characteristic of the bond strength in the semiconductor. The energy $E_g = h\nu_g$ is the *gap energy* of the semiconductor material. Table 2.6 shows numerical values of E_g (in eV) for several semiconductors and the related (vacuum) wavelength $\lambda_g = hc/E_g$ (in μm). The type of transition (d for direct, i for indirect) is a specific materials property. It indicates the mechanism of the photon-lattice interaction. An indirect transition requires a phonon (a quantum of vibrational energy of the lattice) to match energy and momentum conservation. The direct photoionization process is independent of phonons and manifests itself in a much greater absorption coefficient of d- than i-semiconductors. The spectral part $\int_0^{\nu_g} E_\nu \mathrm{d}\nu$ or $\int_{\lambda_g}^\infty E_\lambda \mathrm{d}\lambda$ is inactive in photoionization; the semiconductor is transparent for radiation of wavelengths longer than λ_g. The surplus energy $h\nu - h\nu_g$ of a photon with $h\nu > E_g$ appears primarily as kinetic energy (motional energy) of the produced pair of charge carriers. By collision interaction this kinetic energy is quickly transferred to the semiconductor lattice, causing heating of the material. The fraction of solar energy input actually used for charge carrier generation, E_p, is therefore the number N_p of photons with $h\nu > E_g$ in the spectrum E_ν times the ionization energy E_g

$$E_p = E_g N_p = E_g \int_{\nu_g}^\infty \frac{E_\nu \mathrm{d}\nu}{h\nu}. \tag{2.128}$$

Table 2.6. Collection of semiconductor data. d is direct, i is indirect transition in the photoionization. E_g is the energy for photoionization of the material at 300 K; the corresponding vacuum wavelength of photons is $\lambda_g = hc/E_g$.

Semiconductor	Symbol	Transition	E_g eV	λ_g μm
Silicon, crystalline	Si	i	1.12	1.11
Silicon, amorphous	Si	d	1.65	0.75
Germanium	Ge	i	0.66	1.88
Gallium phosphide	GaP	i	2.26	0.55
Gallium arsenide	GaAs	d	1.42	0.87
Gallium antimonide	GaSb	d	0.72	1.59
Indium phosphide	InP	d	1.35	0.92
Indium arsenide	InAs	d	0.36	3.44
Indium antimonide	InSb	d	0.17	7.30
Zinc oxide	ZnO	d	3.35	0.37
Zinc sulfide	ZnS	d	3.68	0.34
Cadmium sulfide	CdS	d	2.42	0.51
Cadmium selenide	CdSe	d	1.70	0.73
Lead sulfide	PbS	d	0.41	3.02
Lead selenide	PbSe	d	0.27	4.59

The energy used for ionization in relation to the incident solar energy is

$$\frac{E_p}{{}^S\!E} = E_g \frac{\int_{\nu_g}^{\infty}(E_\nu d\nu/h\nu)}{\int_0^{\infty} {}^S\!E_\nu d\nu}. \tag{2.129}$$

For illustration of the numerical order of this fraction, the solar part ${}^S\!E_\nu = f\pi L_\nu(\nu, T_S)$ of the standard spectrum with $f = 2.165 \cdot 10^{-5}$, i.e. extraterrestrial irradiance, and $T_S = 5,777$ K is used. The integral solar irradiance ${}^S\!E = \int_0^{\infty} {}^S\!E_\nu d\nu$ is in this case 1,367 Wm^{-2}. The (ideal) gap energy of the semiconductor material is taken to be $E_g = 1.12$ eV. Then $E_p/{}^S\!E = 529$ Wm^{-2}/1,367 Wm^{-2} = 38.7%. The fraction of solar energy which escapes absorption is $\int_0^{\nu_g} {}^S\!E_\nu d\nu/{}^S\!E = 474$ Wm^{-2}/1,367 Wm^{-2} = 34.7%. The remaining fraction of 26.6% heats the semiconductor.

2.6.2 Photovoltaics

The photoeffect of charge carrier generation in a semiconductor can be employed for delivering electric power. What is needed is an internal electric field inside of a slab of semiconductor material. It drives the photogenerated negative charge carriers to one side, the positive carriers to the other side (see Fig. 2.28). The charge separation builds up a difference of electric potential, U_e, over the two sides of the slab. By electrodes placed on either side, an electric current i can be drawn through an external load. Electric power $P_e = i\,U_e$ is delivered to the load, where U_e is the voltage drop across that load. The required internal field can be produced by, e.g. introducing on both sides of the semiconductor slab different foreign atoms. This process of *doping* requires foreign atoms with more than four valence electrons at the negative side; these are atoms of the elements in the 5th column of the periodic table: phosphorus, arsenic, antimony. At the positive side, doping atoms with less than three valence electrons are used; they are atoms of the elements in the 3rd column: boron, aluminium, gallium, indium.

5th column doping atoms provide surplus electrons; these behave at room temperature as free negative charge carriers over a background of fixed positive charges:

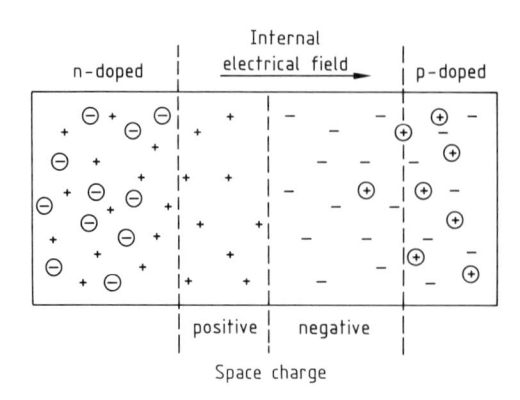

Fig. 2.28. n-p-doped semiconductor. The n-p-junction generates a region of space charges which provides the internal electric field for charge carrier separation.

As $\to$ As$^+_{\text{fixed}}$ + e_{free}. Silicon, doped with arsenic is, therefore, an electric conductor with negative carriers (n-type silicon). 3rd column doping atoms are lacking an electron for forming the covalent electron pair bond. They provide positive holes (defect electrons), which at room temperature are mobile along the valence bonds of the semiconductor lattice. The immobile boron atoms cause a negative background: B $\to$ B$^-_{\text{fixed}}$ + e^+_{free}. Silicon doped with boron is a p-type conductor. The conductivity is proportional to the concentration of the dopants, at least at low concentration levels.

The junction of n- and p-type silicon produces the internal electric field. The mobile carriers e_{free} of n-doped Si diffuse into the p-doped region; there they neutralize defect electrons. Defect electrons e^+_{free} of the p-doped Si diffuse into the n-doped region; there they combine with the free electrons. The overall effect is the formation of space charges, positive in n-Si and negative in the adjacent p-Si. The electric field strength over the p-n junction increases with continued diffusion of carriers and their recombination. Finally it can become sufficiently strong to stop further interdiffusion because its polarity counteracts the motion of the charge carriers. The p-n junction exhibits, because of the internal field, polar directive properties: the p-n combination produces an electric diode. Electrons supplied to the n-Si side from an external voltage source can freely flow through the junction. There is no such current through the p-n junction if the electrons are supplied to the p-Si side. Figure 2.29 shows the diagram of a diode with its distinct unidirectional current-voltage (i-V) characteristics. Irradiation of a semiconductor diode produces a shift of the i-V curve: the diode becomes an electrically active element, Fig. 2.32. On account of the intelligent structure of the p-n junction, the photogenerated charge carriers become separated in the internal field. An external voltage across the p-n junction shows up. This *photovoltaic effect* can deliver electric power, originating from incident radiation, to an external load.

E_p as calculated above is, however, not entirely available as electric energy output of a photovoltaic cell (*solar cell*) [50]. Photogenerated charge carriers tend to recombine in the semiconductor, which reduces the usable photovoltaic current. Thermodynamically, a back reaction by recombination must exist to limit the charge carrier concentration. Qualitatively, this is particularly obvious in open circuit operation of a photovoltaic cell: without recombination the charge carrier concentration would increase beyond limits.

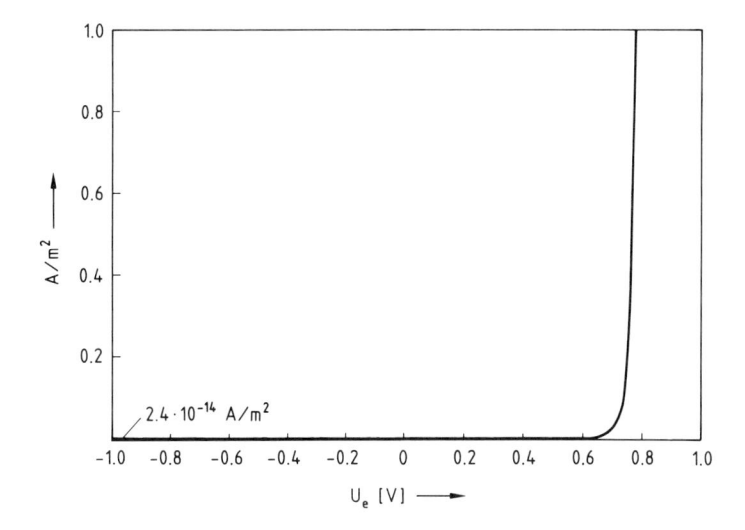

2.6 Conversion of Radiation to Electrical Energy 57

Fig. 2.29. Current-voltage or *I-V* diagram of a *p-n* junction diode. The numerical values refer to an ideal diode with $E_g = 1.12$ eV at cell temperature $T_c = 300$ K.

2.6.3 Ideal Photocell

An ideal photocell (photodiode) operates reversibly for a particular flux density of a given monochromatic radiation. Imagine such a cell to be contained in a cavity with perfectly reflective walls (see Fig. 2.30). It is thermally coupled to a heat reservoir to remain at constant temperature T_c. The cell emits and absorbs radiation. It fills the cavity with thermal radiation of Planck's spectral distribution $L_\nu(\nu, T_c)$, but with a cut-off at $\nu = \nu_g$: the semiconductor material is a selective emitter, due to the fact that it does not absorb for $\nu \leq \nu_g$. If a constant external voltage U_e is applied to the ideal cell, i.e. a cell without radiationless transitions or ohmic (shunt) resistances, a short transient current $i(t)$ appears: the radiation energy in the cavity changes as the electric field across the *p-n* junction enhances or depresses the luminescent recombination of the charge carriers. The new spectral distribution, which is

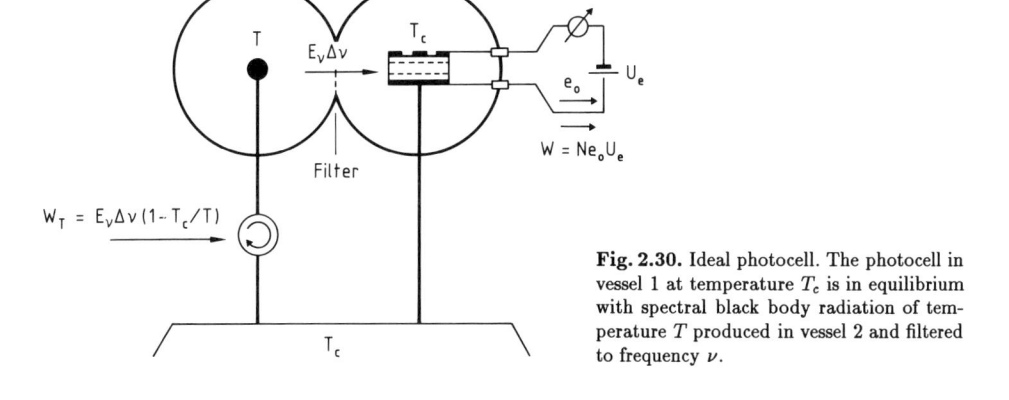

Fig. 2.30. Ideal photocell. The photocell in vessel 1 at temperature T_c is in equilibrium with spectral black body radiation of temperature T produced in vessel 2 and filtered to frequency ν.

dependent on U_e and is termed *photodiode luminescence radiation*, can easily be derived. To this end the cavity is coupled with a second similar cavity containing a black body emitter of adjustable temperature T. An optical filter transparent for radiation of a frequency ν with band width $\Delta\nu$ separates the two cavities. T is adjusted so that for given voltage U_e no electric current is flowing through the photodiode: then the photon flux densities across the filter are the same in both directions, from 1 to 2 and from 2 to 1.

A steady state net radiation flux $E_\nu \Delta\nu = \pi L_\nu(\nu, T)\Delta\nu$ is assumed passing through the filter per unit area from 2 to 1. It does not change the equilibrium in the system

- if an equivalent heat flux $\delta Q = E_\nu \Delta\nu$ is supplied to the black body emitter in 2. This can be achieved by using a heat pump, which brings the required heat with (mechanical) power $\delta W_T = \delta Q \left(1 - T_c/T\right)$ reversibly from an external reservoir at T_c up to T;
- if the photocell absorbs the radiation $E_\nu \Delta\nu$. Then it produces electric power δW and heat, which is removed at T_c.

$$\delta W = \delta i \cdot U_e = \left(\frac{E_\nu \Delta\nu}{h\nu}\right) e_o \cdot U_e. \tag{2.130}$$

e_o is the elementary charge of an electron, $e_o = 1.602177 \cdot 10^{-19}$ As [11].

The process is reversible, the direction of the photon flux can be reversed. The device Fig. 2.30 behaves as an ideal electric power generator: in 2 mechanical power δW_T is supplied and the equivalent electric power δW is delivered from 1. The equivalence $\delta W_T = \delta W$ yields the relationship

$$e_o U_e = h\nu \left(1 - \frac{T_c}{T}\right) \quad \text{or} \quad \frac{h\nu}{kT} = \frac{h\nu - e_o U_e}{kT_c}. \tag{2.131}$$

The result for $h\nu/kT$ is inserted into Planck's equation. This produces the spectral U_e-dependent distribution of luminescence radiation in equilibrium with an ideal photodiode of cell temperature T_c [13]

$$Lu_\nu(\nu, U_e, T_c) = \frac{2h\nu^3}{c^2} \left[\exp\left(\frac{h\nu - e_o U_e}{kT_c}\right) - 1\right]^{-1}. \tag{2.132}$$

For $U_e = 0$ Planck's spectral distribution occurs. For $U_e > 0$ a steep vertical asymptote located at $h\nu = e_o U_e$ appears, which drastically changes Planck's distribution.

2.6.4 Ideal Solar Cell Equation

Irradiation of a semiconductor generates a current density

$$i_G = e_o \int_{\nu_g}^{\infty} \frac{E_\nu d\nu}{h\nu}. \tag{2.133}$$

Recombination of charge carriers causes a loss current density

$$i_D = -e_o \int_{\nu_g}^{\infty} \frac{\pi Lu_\nu(\nu, U_e, T_c) d\nu}{h\nu}. \tag{2.134}$$

Here the voltage dependent luminescent radiance $Lu_\nu(\nu, U_e, T_c)$ of a perfect diode (Sect. 2.6.3) has been used. The radiation is assumed to escape hemispherically from the diode. Figure 2.31 shows the equivalent circuit diagram of a photocell with load R_L. The balance of the currents is

$$i_G + i_D + i = 0 \tag{2.135}$$

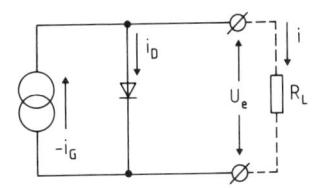

2.6 Conversion of Radiation to Electrical Energy

Fig. 2.31. Equivalent ideal photocell circuit. A current $-i_G$ is generated by irradiation; part of it passes through the diode, i_D. The current through the load R_L is i, producing the voltage drop $U_e = R_L i$.

and the useful electric power is

$$P_e = iU_e = -(i_G + i_D)U_e = -\left(e_o \int_{\nu_g}^\infty \frac{E_\nu d\nu}{h\nu} - e_o \int_{\nu_g}^\infty \frac{\pi L u_\nu(\nu, U_e, T_c) d\nu}{h\nu}\right) U_e. \quad (2.136)$$

U_e across the load depends on the load resistance R_L. In an open circuit, $R_L = \infty$ and $i = 0$: the voltage U_e assumes its maximum value. In absence of solar radiation, E_ν is not zero but still contains the black body ambient component $\pi L_\nu(\nu, T_a)$. Then the open circuit voltage U_{oc} of the photocell, calculated from $i_G + i_D = 0$, is of course zero, provided the cell is also at ambient temperature, $T_c = T_a$. Such a 'non-irradiated' photocell is assumed to be connected to an external current source of voltage U_e and the corresponding 'dark current' i measured is

$$i = -e_o \int_{\nu_g}^\infty \frac{\pi L_\nu(\nu, T_a) d\nu}{h\nu} + e_o \int_{\nu_g}^\infty \frac{\pi L u_\nu(\nu, U_e, T_a) d\nu}{h\nu}. \quad (2.137)$$

A plot of the i-V characteristic of the dark ideal photodiode is shown in Fig. 2.29. For negative bias the current saturates at the level of the charge carrier generation

$$i_{U \to -\infty} = -e_o \int_{\nu_g}^\infty \frac{\pi L_\nu(\nu, T_a) d\nu}{h\nu}. \quad (2.138)$$

The present diode equation can be approximated by a simpler expression. Usually $E_g > kT_a$ and also $E_g - e_o U_e > kT_a$. Then $\exp(h\nu/kT_a) \gg 1$ and $\exp[(h\nu - e_o U_e)/kT_a] \gg 1$: the -1 in the denominator of $L_\nu(\nu, T_a)$ and $Lu_\nu(\nu, U_e, T_a)$ can be neglected, which allows solution of the integrals in i_S and i_D. The result is

$$i_a = e_o \frac{2\pi}{c^2} \left(\frac{kT_a}{h}\right)^3 (2 + 2x_g + x_g^2) \exp(-x_g) \quad \text{and} \quad i_D = -i_a \exp\frac{e_o U_e}{kT_a}. \quad (2.139)$$

i_G has been renamed to i_a to point to its sole dependence on the (dark) ambient radiation; $x_g = h\nu_g/kT_c$. The well-known diode equation follows

$$i = -(i_a + i_D) = i_a \left(\exp\frac{e_o U_e}{kT_a} - 1\right). \quad (2.140)$$

It can be directly derived from simple approximate arguments for the electric behavior of p-n junctions.

A numerical example demonstrates the relevant orders of magnitude usually encountered in p-n diodes. Assume an ideal semiconductor material with $E_g = 1.12$ eV at $T_a = 300$ K. Then $\exp(E_g/kT_a) = 6.5 \cdot 10^{18} \gg 1$; with $U_e = 1$ V one obtains $\exp(E_g - e_o U_e)/kT_a = 104 \gg 1$ and calculates $i_a = 8.2 \cdot 10^{-13}$ Am^{-2}. The diode current rises steeply but not before $U > 0.7$ V, where it attains a value of about 1 Am^{-2}. At $U = 1$ V the current density is already up to $5 \cdot 10^4$ Am^{-2}.

2.6.5 Parameters of Solar Cells

With total radiation $E_\nu = {}^S E_\nu + {}^a E_\nu$ the cell receives a photon flux density $\int_{\nu_g}^\infty (E_\nu d\nu/h\nu)$ which is often orders of magnitude greater than $\int_{\nu_g}^\infty L_\nu(\nu, T_a) d\nu/h\nu$. Consequently, the generated current density i_G is much higher than i_a in the dark cell. It helps to divide the incident spectrum E_ν into two parts: the true solar contribution $^SE_\nu = E_\nu - {}^aE_\nu$ and the ambient thermal radiation $^aE_\nu$. In case of a standard spectrum

$$^SE_\nu = f\pi L_\nu(\nu, T_S)d\nu \quad \text{and} \quad {}^aE_\nu = (1-f)\pi L_\nu(\nu, T_a)d\nu. \tag{2.141}$$

Accordingly, consider the total generated photocurrent i_G to be the sum of a solar part and an ambient part: $i = i_S + i_a$. Employing the diode equation, one obtains

$$i = -(i_G + i_D) = -(i_S + i_a + i_D) = -i_S - i_a + i_c \exp\frac{e_o U_e}{kT_c} \tag{2.142}$$

and

$$i_S = e_o \int_{\nu_g}^\infty \frac{^SE_\nu d\nu}{h\nu} \quad i_a = e_o \int_{\nu_g}^\infty \frac{^aE_\nu d\nu}{h\nu} \quad i_c = e_o \int_{\nu_g}^\infty \frac{\pi L_\nu(\nu, T_c)d\nu}{h\nu}. \tag{2.143}$$

For $T_c = T_a$ and the modest approximation that $i_c = i_a$ the better known form of the solar cell equation results [12,19]

$$i = -i_S + i_a(\exp\frac{e_o U_e}{kT_a} - 1). \tag{2.144}$$

Note that the effect of solar irradiation is to shift the i-V dark diode characteristic along the current axis. This is illustrated in Fig. 2.32 for various irradiation levels. Four parameters of the photocell appear which provide independent information about its activity.

Short Circuit Current i_{sc}.
It is the point in Fig. 2.32, where $U_e = 0$. This yields

$$i_{sc} = -i_S - i_a + i_c \tag{2.145}$$

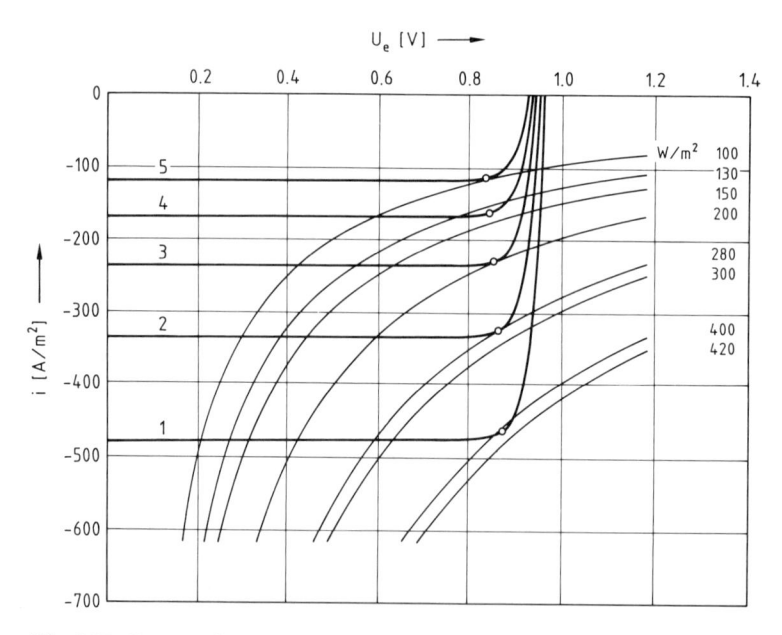

Fig. 2.32. Current-voltage diagram of a photodiode. Gap energy $E_g = 1.2$ eV, cell temperature $T_c = 300$ K. The numbers 1 to 5 refer to irradiances of standard spectra ($T_S = 5,777$ K, $T_a = 300$ K) with $f_1 = 2.10 \cdot 10^{-5}$; $f_2 = 1.48 \cdot 10^{-5}$; $f_3 = 1.05 \cdot 10^{-5}$; $f_4 = 7.42 \cdot 10^{-6}$; $f_5 = 5.25 \cdot 10^{-6}$. The maximum power points (open circles) correspond to efficiencies of 30.38, 30.07, 29.75, 29.44 and 29.13%, respectively.

2.6 Conversion of Radiation to Electrical Energy

or, fully written for an ideal photocell

$$i_{sc} = -e_o \int_{\nu_g}^{\infty} \frac{E_\nu d\nu}{h\nu} + e_o \int_{\nu_g}^{\infty} \frac{\pi L_\nu(\nu, T_c) d\nu}{h\nu}. \tag{2.146}$$

E_ν stands for a spectral component of the total incident radiation. With the standard spectrum and cell temperature $T_c = T_a$ the short circuit current density is strictly proportional to the dilution factor f

$$i_{sc} = -f e_o \int_{\nu_g}^{\infty} \frac{\pi [L_\nu(\nu, T_S) - L_\nu(\nu, T_a)] d\nu}{h\nu}. \tag{2.147}$$

As $T_S \gg T_a$, the integral is dominated by the solar radiance $L_\nu(\nu, T_S)$. Then i_{sc} is a direct measure of the solar part of incident radiation. In Tab. 2.7 various numerical values for i_{sc} at $f = 1$ are listed for the ideal solar cell. Other f values than 1 produce $i_{sc} = f \cdot i_{sc(f=1)}$.

Table 2.7. Short circuit current densities i_{sc} of an ideal solar cell. Numerical values for $E_\nu = \pi L_\nu (\nu, T_S = 5,777K)$. i_{sc} in 10^5 Am^{-2}, E_g in eV.

E_g	0.5	1	1.5	2	2.5	3
i_{sc}	−400	−276	−167	−92.7	−48.1	−23.8

Open Circuit Voltage U_{oc}
Here $i = 0$. From $0 = -i_S - i_a + i_c \exp(e_o U_{oc}/kT_c)$ the open circuit voltage is found to be

$$U_{oc} = \frac{kT_c}{e_o} \ln \frac{i_G}{i_c}. \tag{2.148}$$

There is a remarkable relation between U_{oc} and the gap voltage $U_g = E_g/e_o$ of the semiconductor. First substitute any arbitrary incident spectrum by a Planck spectrum of temperature T^*, which produces the same photon flux density

$$\int_{\nu_g}^{\infty} \frac{E_\nu d\nu}{h\nu} = \int_{\nu_g}^{\infty} \frac{\pi L_\nu(\nu, T^*) d\nu}{h\nu}. \tag{2.149}$$

By its definition T^* is the effective *photon radiance temperature* of the incident radiation. It follows

$$\frac{i_G}{i_c} = \frac{\int_{\nu_g}^{\infty} \pi L_\nu(\nu, T^*)(d\nu/h\nu)}{\int_{\nu_g}^{\infty} \pi L_\nu(\nu, T_c)(d\nu/h\nu)}. \tag{2.150}$$

The ratio of integrals can be approximated by

$$\left(\frac{T^*}{T_c}\right)^3 \frac{x_g^{*2} + 2x_g^* + 2}{x_g^{c2} + 2x_g^c + 2} \exp\left[\frac{-E_g}{k}\left(\frac{1}{T^*} - \frac{1}{T_c}\right)\right]. \tag{2.151}$$

The abbreviations are $x_g^* = E_g/kT^*$ and $x_g^c = E_g/kT_c$. A numerical example shows the orders of magnitude involved. For $T_c = 300$ K and $E_g = 1$ eV and because T^* remains in our case $\leq 5,777$ K it follows $i_G/i_c \geq 10 \exp[-(E_g/k)(1/T^* - 1/T_c)]$. Inserting it into (2.148) for U_{oc},

$$e_o U_{oc} \geq E_g \left(1 - \frac{T_c}{T^*}\right). \tag{2.152}$$

The open voltage U_{oc} appears to be proportional to the gap voltage E_g/e_o with a Carnot type of reduction factor $(1 - T_c/T^*)$. The approximation is accurate within −5% of the exact value of U_{oc}. For several dilution factors f (or, equivalently, when extinction is disregarded, for concentration ratios $C = f \cdot 46,200$) photon radiance temperatures T^* are calculated using the standard spectrum (see Tab. 2.8). Terrestrial solar radiation corresponds typically to $f = 10^{-5}$. For crystalline silicon ($E_g = 1.12$ eV) the photon radiance temperature would be 1,130 K. This produces an estimated $U_{oc} = 0.82$ V for the ideal solar cell.

Table 2.8. Photon radiance temperatures T^* in Kelvin in dependence on dilution factor f and gap energy E_g. Standard spectrum with $T_S = 5,777$ K and $T_a = T_c = 300$ K.

f	E_g (eV)					
	0.5	1	1.5	2	2.5	3
1	5,777	5,777	5,777	5,777	5,777	5,777
10^{-1}	3,001	3,410	3,741	4,000	4,204	4,368
10^{-2}	1,757	2,272	2,677	3,001	3,265	3,483
10^{-3}	1,162	1,659	2,055	2,380	2,652	2,883
10^{-4}	845	1,292	1,656	1,963	2,225	2,453
10^{-5}	655	1,051	1,381	1,665	1,913	2,131
10^{-6}	532	883	1,182	1,443	1,675	1,882
10^{-7}	446	760	1,032	1,272	1,488	1,683
10^{-8}	384	666	914	1,137	1,338	1,522

Maximum Power Point P_{mpp}

The electric power output from the solar cell is

$$P_e = U_e\, i = -U_e \left[i_S - i_a \left(\exp \frac{e_o U_e}{kT_a} - 1 \right) \right]. \tag{2.153}$$

Obviously, P_e is a function of U_e: it assumes zero for both $U_e = 0$ (short circuit operation) and $U_e = U_{oc}$ (open circuit operation). In between the output power attains a maximum at an U_{mpp}. The condition for the maximum is $dP_e/dU_e = 0$ in the range $0 < U_e < U_{oc}$. The solution can be found numerically. Table 2.9 presents maximum power densities (units Wm^{-2}) for an ideal solar cell, irradiated with solar standard spectrum radiation.

Table 2.9. Maximum electric power density P_{mpp} in units Wm^{-2} of an ideal solar cell. Solar irradiation according to standard spectra with $T_S = 5,777$ K and $T_a = T_c = 300$ K and various dilution factors f; E_g is the gap energy of semiconductor material.

f	E_g (eV)					
	0.5	1	1.5	2	2.5	3
1	$1.78 \cdot 10^7$	$2.48 \cdot 10^7$	$2.25 \cdot 10^7$	$1.65 \cdot 10^7$	$1.06 \cdot 10^7$	$6.27 \cdot 10^6$
10^{-1}	$1.72 \cdot 10^6$	$2.40 \cdot 10^6$	$2.18 \cdot 10^6$	$1.61 \cdot 10^6$	$1.04 \cdot 10^6$	$6.14 \cdot 10^5$
10^{-2}	$1.47 \cdot 10^5$	$2.24 \cdot 10^5$	$2.09 \cdot 10^5$	$1.56 \cdot 10^5$	$1.01 \cdot 10^5$	$6.01 \cdot 10^4$
10^{-3}	$1.27 \cdot 10^4$	$2.09 \cdot 10^4$	$2.00 \cdot 10^4$	$1.51 \cdot 10^4$	$9.86 \cdot 10^3$	$5.89 \cdot 10^3$
10^{-4}	$1.06 \cdot 10^3$	$1.93 \cdot 10^3$	$1.91 \cdot 10^3$	$1.46 \cdot 10^3$	$9.61 \cdot 10^2$	$5.76 \cdot 10^2$
10^{-5}	$8.46 \cdot 10^1$	$1.77 \cdot 10^2$	$1.82 \cdot 10^2$	$1.41 \cdot 10^2$	$9.35 \cdot 10^1$	$5.63 \cdot 10^1$
10^{-6}	$6.34 \cdot 10^0$	$1.61 \cdot 10^1$	$1.72 \cdot 10^1$	$1.35 \cdot 10^1$	$9.09 \cdot 10^0$	$5.50 \cdot 10^0$
10^{-7}	$4.28 \cdot 10^{-1}$	$1.45 \cdot 10^0$	$1.63 \cdot 10^0$	$1.31 \cdot 10^0$	$8.83 \cdot 10^{-1}$	$5.38 \cdot 10^{-1}$
10^{-8}	$2.44 \cdot 10^{-2}$	$1.30 \cdot 10^{-1}$	$1.54 \cdot 10^{-1}$	$1.26 \cdot 10^{-1}$	$8.57 \cdot 10^{-2}$	$5.25 \cdot 10^{-2}$

A simple and general graphical construction which delivers directly the maximum power point values $U_{mpp}, i_{mpp}, P_{mpp}$ is shown in Fig. 2.32: a hyperbola $U_e\, i = P_e$ is drawn which just touches the given i-V characteristic. The point of contact is the maximum power point, $P_e = P_{mpp}$.

2.6 Conversion of Radiation to Electrical Energy

The information obtained with P_{mpp} is new and not contained in U_{oc} and i_{sc}. It is used for deriving two other cell parameters

- the *cell efficiency* $\eta^* = P_{mpp}/{}^S E$ with ${}^S E$ the solar part of the incident radiation. In terms of the standard solar spectrum, ${}^S E = f\,\sigma T_S^4$. Obviously, η^* measures the conversion efficiency of the solar energy flux to electric power via a photocell;
- the *fill factor* $FF = P_{mpp}/(U_{oc}\,i_{sc})$. It measures the quality of a solar cell in terms of a hypothetical (thermodynamically never attainable) maximum power $P_e = U_{oc}\,i_{sc}$. Fill factors of the ideal photocell, irradiated with standard solar radiation are listed in Tab. 2.10.

Table 2.10. Fill factors FF. $FF = P_{mpp}/(U_{oc}\,i_{sc})$ of an ideal solar cell of given band gap E_g with variation of the dilution factor f (equivalent to concentration $C = f \cdot 46{,}200$ when extinction of direct solar radiation in the atmosphere is disregarded). Standard solar spectrum with $T_S = 5{,}777$ K and $T_a = T_c = 300$ K.

f	E_g (eV)					
	0.5	1	1.5	2	2.5	3
1	0.896	0.897	0.900	0.900	0.901	0.899
10^{-1}	0.861	0.885	0.899	0.900	0.900	0.899
10^{-2}	0.782	0.871	0.897	0.899	0.900	0.899
10^{-3}	0.782	0.869	0.896	0.899	0.900	0.899
10^{-4}	0.749	0.862	0.895	0.899	0.900	0.899
10^{-5}	0.718	0.853	0.893	0.898	0.900	0.899
10^{-6}	0.695	0.844	0.891	0.898	0.900	0.899
10^{-7}	0.609	0.833	0.889	0.898	0.900	0.899
10^{-8}	0.525	0.821	0.885	0.898	0.900	0.899

2.6.6 Maximum Photovoltaic Efficiencies

For given spectrum E_ν and fixed cell temperature T_c there are two free parameters left for optimizing the output of electrical power P_e: the voltage U_e (or, alternatively, the current density i) and the semiconductor material, represented by the gap energy E_g. Consequently, the two necessary conditions for maximum output P_e are

$$\frac{\partial P_e}{\partial U_e} = 0 \quad \text{for constant} \quad E_g \quad \text{and} \quad \frac{\partial P_e}{\partial E_g} = 0 \quad \text{for constant} \quad U_e. \tag{2.154}$$

The first is synonymous with tuning the cell to the maximum power point. The latter produces from (2.136) the relationship

$$0 = E_\nu(\nu_g) - \pi L u_\nu(\nu_g, U_e, T_c) \tag{2.155}$$

and leads to a new measure of the quality of radiation in excitation processes. If $E_\nu(\nu) > \pi L u_\nu(\nu, U_e, T_c)$, then that particular spectral component is able to deliver electrical power to a load operated at voltage U_e. If the opposite inequality holds, then the cell at voltage U_e would lose more by luminescence than receive power by irradiation.

This can be looked upon from another but physically significant point of view. First the given energy flux density spectrum E_ν is transformed into an equivalent spectrum of *chemical potential* $\mu = e_o U_e$,

$$E_\nu = \pi L u_\nu(\nu, T, \mu). \tag{2.156}$$

Solving explicitly for μ

$$\mu = h\nu - kT \ln\left(1 + \frac{2\pi\nu^2}{c^2 E_\nu}\right). \tag{2.157}$$

Note that there is a relation between μ and the assumed temperature T. Instead of invoking a fixed T, a (variable) temperature $T(\nu)$ could always be found which reduces the chemical potential to zero. The result is

$$T(\nu) = \frac{h\nu}{k}\left[\ln\left(1 + \frac{2\pi\nu^2}{c^2 E_\nu}\right)\right]^{-1} \tag{2.158}$$

One recognizes $T(\nu)$ to be the *radiance temperature* of the spectral component E_ν, worked out in Sect. 2.3.2. For a constant $T = T_a$ it follows from (2.157) with (2.158)

$$\mu(\nu) = h\nu\left(1 - \frac{T_a}{T(\nu)}\right). \tag{2.159}$$

This is a simple relation between radiance temperature and chemical potential of a spectral component. The choice of a constant T in the stipulation of μ is reasonable, regarding the radiation to interact with a system of constant temperature T, e.g. the ambient temperature T_a. Examples for such systems are solar cells and (photo-)chemical reactors.

For the standard spectrum $E_\nu = f \pi L_\nu(\nu, T_S) + (1-f)\pi L_\nu(\nu, T_a)$ with several dilution factors f the chemical potential $\mu(\nu)$ is plotted in Fig. 2.33. The AM 1 energy spectrum has also been transformed into an AM 1 spectrum of chemical potential $\mu(\lambda)$ (Fig. 2.33). $\mu(\nu)$ or $\mu(\lambda)$ can be used for finding the very location ν_g (or λ_g, respectively), where $\mu(\nu) > e_o U_e$ changes to $\mu(\nu) < e_o U_e$. In the way indicated, a single definite E_g can be found in case of a standard spectrum. In a complex spectrum, however, such as AM 1, more than one location for ν_g can exist. The same complication is encountered in the discussion of optimum selective absorber surfaces (Sect. 2.5.2). For any given cell voltage U_e the corresponding ν_g can be inserted into

$$P_e(\nu_g, U_e) = -U_e\left(e_o \int_{\nu_g}^\infty E_\nu d\nu/h\nu - e_o \int_{\nu_g}^\infty \pi L u_\nu(\nu, U_e, T_a) d\nu/h\nu\right) \tag{2.160}$$

and the optimum efficiency $\eta^*(U_e) = P_e(U_e)/{}^S E$ can be calculated. In Fig. 2.33 $U_e = 1.5$ V has been chosen as an example. To $f = 1$ corresponds $E_g = h\nu_g = 1.58$ eV as the best gap energy; for $f = 10^{-10}$ the optimum gap energy is $E_g = 2.20$ eV. The efficiency for $f = 1$ is calculated to be approximately 35% and for $f = 10^{-10}$ to be approximately 15%.

The better strategy for a given $U = 1.5$ V would have been to use two solar cells connected in series, each cell operating at 0.75 V. Then at $f = 1$ the best band gap energy would be $E_g = 0.76$ eV; the efficiency $P_e(U_e)/{}^S E$ comes up to approximately 37%. The improvement is even more pronounced at $f = 10^{-10}$. Here the best E_g with $U_e = 0.75$ V is found to be 1.40 eV; the corresponding efficiency $P_e(U_e)/{}^S E$ is up to 23% compared to the 15% of a single cell. These examples show how important it is to optimize the lay-out of arrays of solar cells.

Finally, one can ask for the best combination of U_e and E_g to produce maximum electrical power. The two conditions $\partial P_e/\partial U_e = 0$ and $\partial P_e/\partial E_g = 0$ are sufficient to fix the two parameters U_e and E_g. Unfortunately, there is no easy analytical solution or graphical construction. Numerical methods are required. Results are shown in Fig. 2.34, where $\eta^* = P_{mpp}/{}^S E$ values for standard spectra are plotted. For $f = 1.6 \cdot 10^{-5}$ the maximum efficiency is $\eta^* = 0.30$; the corresponding best semiconductor is a material with $E_g = 1.25$ eV. Compared to the thermodynamic quality (exergy yield) related to $f = 1.6 \cdot 10^{-5}$ of the incident radiation, $\eta^* = 0.71$, the best exergy conversion efficiency of a single gap solar cell is $0.30/0.71 = 0.42$. This refers to a standard spectrum. Similar calculations can be performed with any spectrum E_ν.

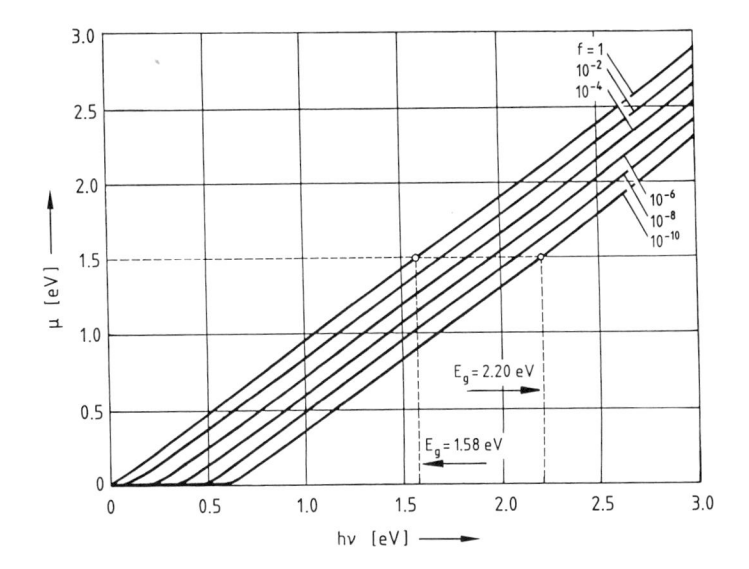

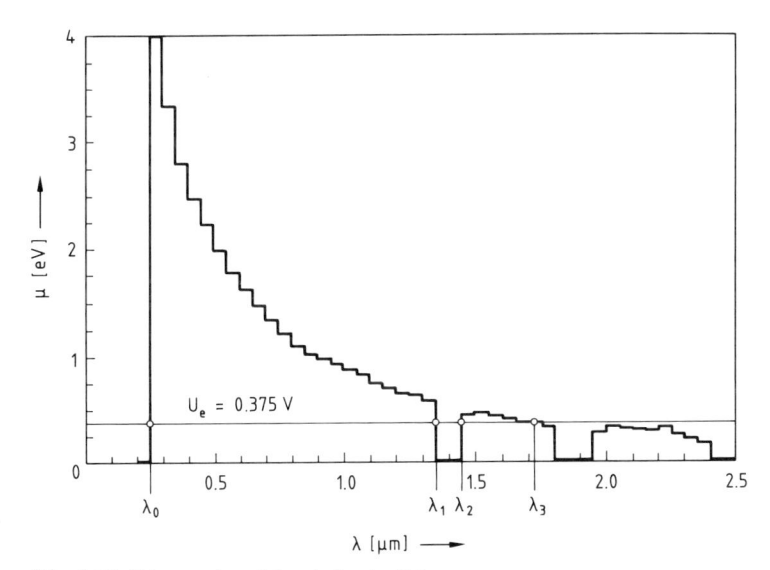

2.6 Conversion of Radiation to Electrical Energy 65

Fig. 2.33. Solar spectra of chemical potential.
Upper: standard spectra of chemical potential $\mu(\nu)$ with system temperature $T = 300$ K. The dilution factor f is varied over ten orders of magnitude; $T_S = 5,777$ K, $T_a = 300$ K.
Lower: AM 1 spectrum of chemical potential $\mu(\lambda)$ with system temperature $T = 300$ K.

For illustration, the calculation can be performed with the $AM1$ spectrum. Then with ambient temperature $T_a = 300$ K we find $\eta^* = 0.36$ for the best value $E_g = 1.4$ eV (i.e. $\lambda_g = 0.875\mu$m). Note that η^* is greater for AM 1 than for the standard spectrum $f = 1.6 \cdot 10^{-5}$, although both spectra are identical in $^SE = 1,000$ Wm^{-2}. However, AM 1 contains less of the ineffective long wavelength components E_ν.

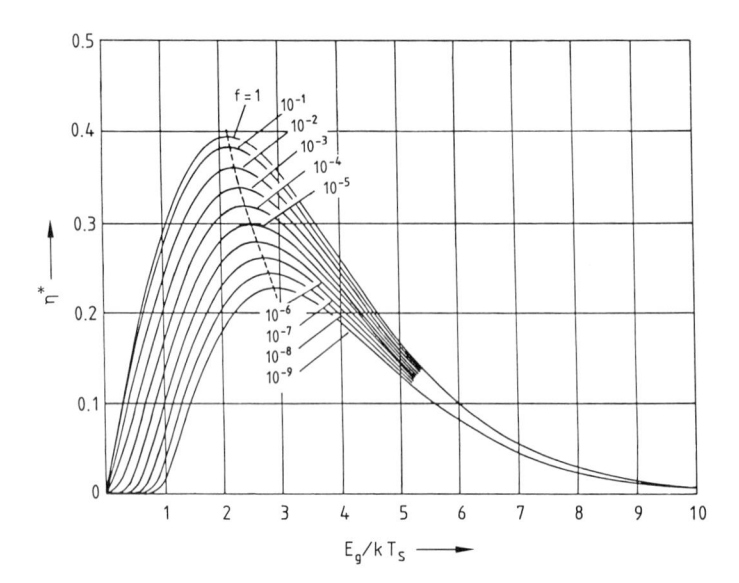

Fig. 2.34. Optimum efficiencies $\eta^* = P_e/{}^S\!E$ of ideal solar cells dependent on E_g cell temperature 300 K. Standard spectra with $T_S = 5,777$ K, $T_a = 300$ K are used with various dilution factors f. The abscissa is a general variable E_g/kT_S; for given Sun temperature $T_S = 5,777$ K one obtains $2.009 \cdot E_g$ with E_g in units of eV.

$\eta^*(f) = P_{mpp}/{}^S\!E$ has been plotted for standard spectra with several dilution factors f. In Fig. 2.35 the dependence of η^* on f is shown with the line marked 1× (single gap solar cell). The upper limit, the maximum exergy obtainable from radiation in thermodynamically reversible processes (see Sect. 2.3.3), is also shown, marked 'maximum solar', again referring to standard spectra with $T_S = 5,777$ K and $T_a = 300$ K.

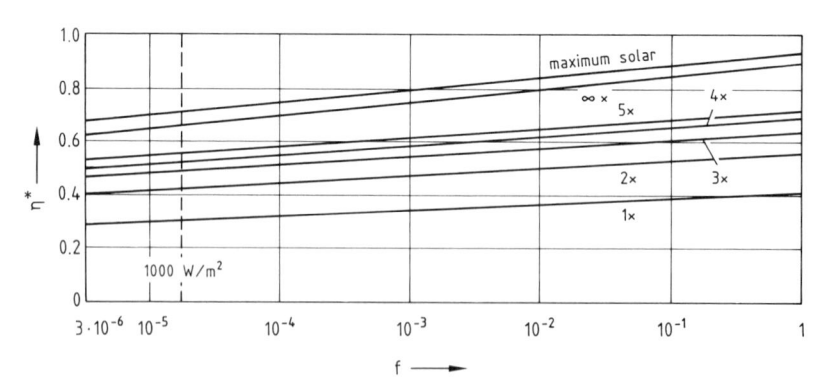

Fig. 2.35. Optimum efficiencies $\eta^* = P_e/{}^S\!E$ of ideal solar cells. Cell temperature 300 K. η^* in dependence on the dilution factor f of standard spectra with $T_S = 5,777$ K, $T_a = 300$ K. The upper line 'maximum solar' is the yield of work (electrical power) from standard spectra obtainable with a *perfect converter*. The other lines are optimum yields with *photovoltaics* from single gap solar cells (1×), two-stack tandem cells (2×), etc. up to a stack of an infinite number of cells, ∞×.

2.6.7 Spectral Matching of Solar Cell Devices

The photovoltaic yield $\eta^* = P_e/{}^S E$ is meager as seen from Fig. 2.35: at maximum 30% is obtained with an ideal photocell compared to the about 70% available for non-concentrated standard spectrum irradiation. Thus, thermodynamic quality (exergy) of radiation is degraded although an ideal photocell is employed. Recall that the term 'ideal' refers to reversible operation with irradiance of the proper luminescence spectrum, i.e. the spectrum corresponding to the gap energy E_g, cell voltage U_e and cell temperature T_c of the diode. In practice a compromise is necessary to adapt given solar radiation to a single band gap E_g. The consequence is that

- all the radiation of frequency below $\nu_g = E_g/h$ remains unused (either transmitted through or reflected by the solar cell);
- all the radiation beyond ν_g produces electron-hole pairs but consuming to that end per photon only the ionization energy E_g; the residual energy $h\nu - E_g$ heats the semiconductor material. Such heat of $T_c > T_a$ may be used in hybrid applications combined with the photoelectrical power. In the following, however, maximum electrical power output is requested and so the cell should be kept at ambient temperature.

Improvement of solar photovoltaic cell devices is possible by appropriate adjustment to the broad solar spectrum. This cannot be accomplished with a single gap cell, because it has already been tuned with the two adjustable parameters E_g, U_e to best performance. The natural path to follow is to divide the wide solar spectrum into spectral sections. Each section then works on a separate cell of appropriate E_g and U_{mpp}. There are several ways for dividing the spectrum of the incident radiation:

- by employing an *optically dispersive element* (prism, diffraction grating, diffraction hologram), which spatially separates the incoming radiation. In the fanned spectrum, two or more photocells are located and tuned to maximum output. Only direct but not diffuse solar radiation can produce spatially separated spectral images of the Sun;
- another way not hampered by the requirement of direct radiation is the *sandwich configuration*. Individual photocells of different E_g are stacked one upon the other. On top of the stack is located the cell with highest band gap, E_{g1}. It absorbs all the radiation with $\nu > \nu_{g1}$ but transmits radiation of $\nu < \nu_{g1}$. The cell down next in the stack with band gap $E_{g2} < E_{g1}$ absorbs radiation between ν_{g2} and ν_{g1}, etc. A sandwich of two cells (a two-stack tandem device) is the simplest configuration to be examined first in the following Sect. 2.6.8. The goal is to employ stacks of N individual cells: there are $2N$ free parameters E_{gi}, U_i to adjust to the spectrum for best performance;
- a further possibility for spatial spectral separation is by *fluorescence collectors* [25,46,47]. Fluorescent dyes are imbedded in flat plate clear glass or plastic. Short wavelength incident radiation excites the dye, which then emits its characteristic longer wavelength fluorescence. The difference in photon energy $h\nu_{\text{incident}} - h\nu_{\text{fluorescent}}$, the Stokes-shift, appears as heat in the glass. The direction of emission is independent of the incident solar radiation and isotropic in space. Consequently, there is a cone of directions of half angle $\gamma = \arcsin(1/n)$, within which the fluorescence radiation escapes (n is the index of refraction of the glass). The fraction of fluorescence radiation captured in the glass by total reflection, see Fig. 2.36, is $F = \sqrt{1 - n^{-2}}$.

For instance with $n = 1.5$ the cone half angle is $\gamma = 41.8°$ and $F = 0.75$. This radiation cannot escape except at the rim, where (small size) photocells are fixed. The fluorescence is, compared to the solar spectrum, a narrow band spectrum. The photocell band gap can be optimally adjusted to that spectrum. In principle high geometric flux concentrations (the ratio of

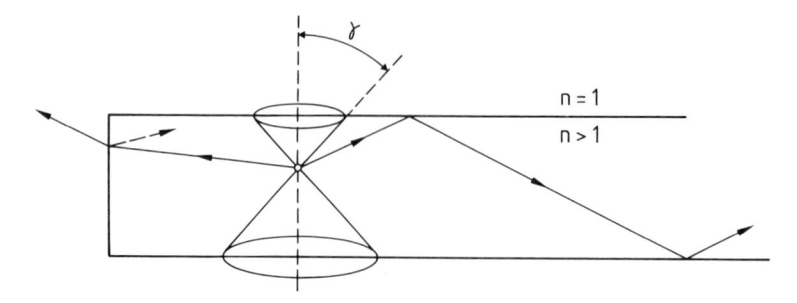

Fig. 2.36. Fluorescence collector for concentrating global solar radiation. The index of refraction n of the glass pane is greater than 1. The emission of fluorescence radiation is isotropical in space.

the flat area of the glass over its rim side area corrected for losses $1-F$ and quantum yields of fluorescence) are conceivable. There are, however, practical limits, e.g. by internal scattering and reabsorption processes. The glass must be kept at ambient temperature; otherwise the back reaction of a fluorescence photon with a thermal phonon will produce randomly scattered short wavelength photons again. Such a back reaction is thermodynamically required: a spectral photon flux density of radiance temperature higher than the Sun's temperature of 5,777 K could otherwise build up. It would contradict the second law of thermodynamics.

A stack of two or three fluorescence plates can be used, each tuned to consecutive spectral parts of solar radiation. It should be optimized to convert the exergy available better with the incident diffuse and direct radiation.

2.6.8 Tandem Solar Cells

The upper cell is an ideal photodiode with cut-off frequency ν_{g1}. The lower ideal cell has its cut-off at $\nu_{g2} < \nu_{g1}$, Fig. 2.37.

The individual power outputs are

$$P_{e1} = -e_o U_1 \left(\int_{\nu_{g1}}^{\infty} \frac{E_\nu d\nu}{h\nu} - \int_{\nu_{g1}}^{\infty} \frac{\pi L u_\nu(\nu, U_1, T_c) d\nu}{h\nu} \right) \qquad (2.161)$$

$$P_{e2} = -e_o U_2 \left(\int_{\nu_{g2}}^{\nu_{g1}} \frac{E_\nu d\nu}{h\nu} - \int_{\nu_{g2}}^{\nu_{g1}} \frac{\pi L u_\nu(\nu, U_2, T_c) d\nu}{h\nu} \right). \qquad (2.162)$$

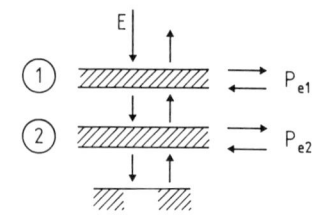

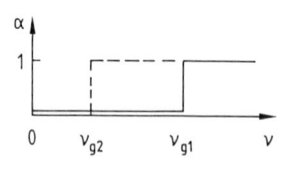

Fig. 2.37. Two-stack tandem. In cell 1 the short wave region of the solar spectrum, in cell 2 part of the longwave region which is transmitted through cell 1 is absorbed. Luminescence radiation from cell 1 (and cell 2) is assumed to be backreflected by a spectrally selective mirror at the bottom of cell 1 (and cell 2, respectively). Separate adjustments of E_g and U_e of the two cells is made to maximum output of the sum P_e of electrical power of the two cells.

2.6 Conversion of Radiation to Electrical Energy

The four free parameters are E_{g1}, E_{g2}, U_1 and U_2. The four necessary conditions for the maximum total electric power output $P_e = P_{e1} + P_{e2}$ are

$$\frac{\partial P_e}{\partial \nu_1} = 0; \quad \frac{\partial P_e}{\partial U_1} = 0; \quad \frac{\partial P_e}{\partial \nu_2} = 0; \quad \frac{\partial P_e}{\partial U_2} = 0. \quad (2.163)$$

The resulting equations are numerically soluble. In Fig. 2.35 efficiencies of tandem cells at maximum power point, $\eta^*(f) = P_{mpp}/{}^S E(f)$, for standard spectra and cell temperature $T_c = 300$ K are plotted. For $f = 2.165 \cdot 10^{-5}$ one obtains $\eta^* = 42\%$, which may be compared with 30% maximum efficiency of the best spectrally adjusted single cell. The curves marked 2×, 3×, 4× and 5× belong to two-stack, three-stack, four-stack and five-stack tandem cells, respectively. Included with ∞× is the limiting case of a *fully selective stack*, consisting of an infinite number of spectrally different photocells in the sandwich.

The individual cells of a stack are assumed to be galvanically isolated circuits, individually adjusted to maximum power output. This is impractical. The question is, is the maximum power output strongly reduced if the cells are connected in series. Then the device would have one common current loop and one voltage output, as if it were a single cell. In case of a two-stack tandem device, one free parameter of the four for best adjustment to the given spectrum is lost: E_{g1}, E_{g2} and $U_e \,(= U_1 + U_2)$ are the remaining parameters for tuning (see Fig. 2.38).

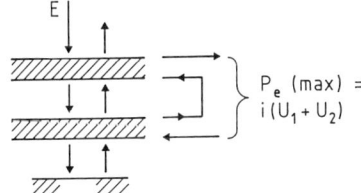

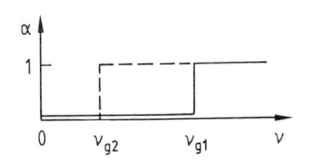

Fig. 2.38. Current coupled two-stack tandem. In cell 1 the shortwave region of the solar spectrum, in cell 2 part of the longwave region which is transmitted through cell 1 is absorbed. Luminescence radiation from cell 1 (or cell 2) is assumed to be backreflected by a spectrally selective mirror at the bottom of cell 1 (or cell 2, respectively). Separate adjustments of E_g of the two cells and of the common i to maximum output of total electrical power P_e of the tandem.

Numerical results for the standard spectrum with $f = 2.165 \cdot 10^{-5}$ and cell temperature $T_c = 300$ K are with two *galvanically separated* cells:

for and
$P_e(\text{max}) = 571$ Wm^{-2}
$P_{e1} = 352$ Wm^{-2} $\quad P_{e2} = 219$ Wm^{-2}
$E_1 = 1.79$ eV $\quad E_{g2} = 0.945$ eV
$U_1 = 1.41$ V $\quad U_2 = 0.623$ V
$i_1 = 250$ Am^{-2} $\quad i_2 = 352$ Am^{-2}.

These values can be compared to the data of a *current-coupled* two-stack tandem cell

for and
$P_e(\text{max}) = 567$ Wm^{-2}
$U_e = 1.87$ V
$i = 302$ Am^{-2}.
$P_{e1} = 381$ Wm^{-2} $\quad P_{e2} = 186$ Wm^{-2}
$E_1 = 1.63$ eV $\quad E_{g2} = 0.936$ eV
$U_1 = 1.26$ V $\quad U_2 = 0.614$ V

The difference is less than 1% in the two-stack tandem case.

A few observations are notable with multiple cells:

- the greatest improvement by adding another cell to the stack occurs in going from the single to a two-stack tandem cell;
- the spectrally fully selective solar cell device exhibits a performance identical to the fully selective thermal absorber. The first delivers electrical power directly, the latter is assumed to be coupled to separate ideal Carnot thermal power converters;
- a twofold infinite number of adjustable parameters are available in a fully selective solar cell. However, the electric power yield of this infinite-stack tandem does not fully coincide with the maximum yield level of the incident radiation (maximum solar in Fig. 2.35).

The infinite array emits luminescence radiation, which differs from ambient radiation. Therefore, radiation capable of doing work is lost. This cannot be avoided, because the direction of the luminescence emission is opposite to the direction of the incident rays. A gadget is needed which is not invariant to the ray direction, e.g. a circulator. Then outgoing and incoming radiation can be separated; the luminescence can be used for producing additional electrical power in a second fully selective stack, etc. Performing the calculations yields the result that theoretically the exergy of solar radiation can indeed be *completely* transferred to electrical power by means of the photovoltaic effect.

The following statement is indispensable [48]. A given multiple-stack tandem cell is a device adjusted with its bandgaps E_{gi} to a particular solar spectrum and to a certain cell temperature. If cell temperature, flux density or spectral distribution of the incident radiation vary, the fixed E_{gi} are generally no longer the best. This effect has already been encountered with single cells (see Fig. 2.34). However, in single cells there is a broad flat maximum of best performance. In galvanically coupled multi-stack tandem cells the sensitivity is particularly great to changes in the spectral distribution of incident radiation. It is a price to be paid for the increased yield of electrical power from such multiple cells.

2.7 Photochemical Conversion

The significance of photochemical processes is of two different sorts:

- absorption of radiation is used for activation or acceleration of a kind of chemical processes, which thermodynamically could also move ahead without the assistance of radiation. With photochemical activation a reaction can proceed with higher yield, and often at a lower temperature, along a specific reaction path than in the absence of radiation *(solar activation, solar chemicals)*
- absorption of radiation energy is used for the production of energy rich chemicals *(solar fuels)*.

The first case is concerned with kinetics of reactions in which the free enthalpy of reaction, ΔG, may be slightly positive or, for so-called downhill reactions, negative. Accumulation of energy from radiation in the products is of no concern. The second case, the production of solar fuels, is a field of reaction kinetics too, regarding reaction rates and reaction paths. However, the free enthalpy of reaction is strongly positive, $\Delta G > 0$. Then energy and entropy *input* from radiation determine the maximum possible yield of products.

Solar fuels are energy carriers. They excel in providing a long term stable and storable product of accumulated solar energy; they allow large scale transportability of converted solar energy with standard techniques and equipment such as tankers and pipelines, for e.g. liquid

2.7 Photochemical Conversion

or chemically fixed or gaseous hydrogen [58]. A solar fuel is a new path to manufacture, not a new brand of chemical compound.

In this section the thermodynamic basis of the production of solar fuels is examined. The cardinal relationship for maximum product yield is

$$W^* = \Delta G. \tag{2.164}$$

The left hand side term, $W^* = \eta^* {}^S\!E$, is the work (or, since the symbols mean flux densities $^S\!E$, i.e. fluxes per unit area of incident radiation, rather *power* density than work) obtainable from solar radiation (see Sect. 2.3.4). η^* refers to ambient temperature, $T_a = 300$ K.

The right hand side term is the free reaction enthalpy ΔG (again rather a rate referred to unit area of incident radiation). ΔG is the difference in Gibbs free enthalpy between final and initial products at reference pressure and reference temperature T, which is frequently T_a. For a reaction, e.g. A + B $\rightarrow$ C it follows $\Delta G = G_C - (G_A + G_B)$. On account of $\Delta G = \Delta H - T\Delta S$, the connection to the enthalpies and entropies of the reactants is $\Delta H = H_C - (H_A + H_B)$ and $\Delta S = S_C - (S_A + S_B)$.

The equation $W^* = \Delta G$ implies decoupling of the entropy fluxes of radiation and chemical turnover. The entropy difference between radiation input and output of reaction products is assumed to be delivered to (or extracted from) the environment.

The preceding analysis of means for obtaining W^* from radiation has shown that perfect *and* at the same time practical conversion techniques are not at hand. It seems unavoidable and involves considerable quality losses to use the imperfect conversion paths of producing first by absorption *process heat* (Sect. 2.5) or by photovoltaics *electrical power* (Sect. 2.6). In this sense 'solar chemistry' would be merely conventional chemical engineering of thermochemical processes (coupling with process heat) and of electrochemical processes. The so-called 'solarization' of conventional chemical processes then focuses on

- adaptation to fluctuations and transients of the energy supply from solar radiation. This brings up the importance of short- and long-term storage capacity particularly of heat and electric energy;
- utilization of high temperatures of 1,000 °C and more with the advantage of a clean heat input by radiation. This poses a materials problem of high temperature resistant chemical reactors.

The example of water decomposition $H_2O_l \rightarrow H_2 + \frac{1}{2}O_2$ can be used to illustrate yield data involved in the conversion paths. For $T_a = 300$ K and standard pressure of both hydrogen and oxygen and of the initial liquid water the free reaction enthalpy is $\Delta G = 237$ kJ per formula turnover, corresponding to a hydrogen gas volume of $9.45 \cdot 10^{-8}$ m^3/Joule. This can be used by applying (2.164) to calculate the maximum of hydrogen yield from solar radiation of a standard spectrum ($T_S = 5,777$ K, $T_a = 300$ K) with $f = 2.165 \cdot 10^{-5}$ ($^S\!E = 1,367$ Wm^{-2}), see Tab. 2.11.

However, there is a third conversion path of radiation, the *photochemical path* [9]. It is based upon direct absorption of radiation in a chemical system, causing chemical transformations. Natural photosynthesis is a conspicuous case in point. A photochemical process resembles photovoltaics in principle and in many details of its thermodynamics: a flow of chemicals of increased chemical potential is delivered, in contrast with a flow of electric charges (i.e. an electric current) of increased electrical potential.

2.7.1 Equation of Ideal Photochemical Processes

The maximum yield of photochemical processes can be found by using a technique analogous to the yield in photovoltaics. Both processes are based on electronic excitation by absorp-

Table 2.11. Maximum conversion yields of hydrogen from solar irradiance. Irradiance by a standard spectrum with $f = 2.165 \cdot 10^{-5}$, $T_S = 5,777$ K, $T_a = 300$ K. The electric energy from photovoltaics is assumed to be produced by ideal single gap cells.

Conversion process	Quality η^*	Hydrogen yield mol m^{-2}s^{-1}	Hydrogen yield m^{3}m^{-2}h^{-1}
Direct	0.72	$4.15 \cdot 10^{-3}$	0.335
Process heat	0.53	$3.10 \cdot 10^{-3}$	0.247
Photovoltaics	0.30	$1.67 \cdot 10^{-3}$	0.135

tion of radiation. There is a minimum excitation energy $E_g = h\nu_g$; radiation with (vacuum) wavelengths longer than $\lambda_g = hc/E_g$ is not absorbed. For shorter wavelengths continuum absorption can be assumed, because the corresponding vibrational and rotational excitation levels beyond the threshold E_g are usually closely spaced. Internal relaxation to the lowest vibration-rotation level of the electronically excited state is fast; the surplus energy $h\nu - h\nu_g$ heats the system.

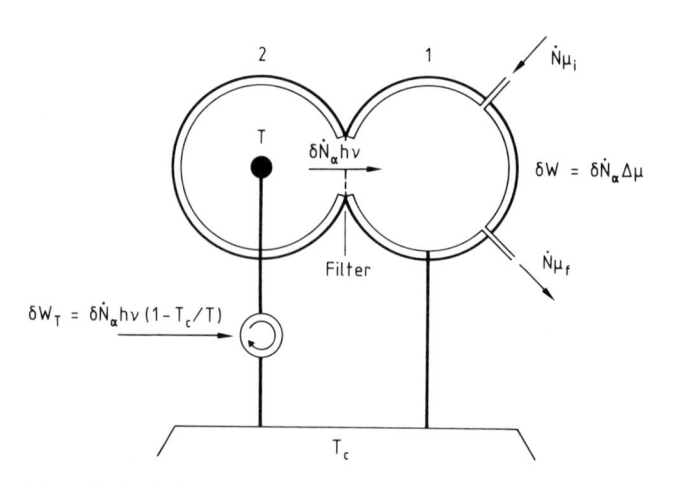

Fig. 2.39. Ideal photochemical reaction. Reaction vessel 1 with throughput of photochemically active reactants, kept at a system temperature T_c. The walls are perfect mirrors. Thermal black body emitter 2 at temperature T. Radiation is flowing through a spectral filter of band width $\Delta\nu$ at frequency ν. The corresponding heat loss of the emitter is supplied by an ideal heat pump working between T_c and T. The vessels 1 and 2 are in equilibrium for radiation of frequency ν.

In Fig. 2.39 a reactor vessel 1 is shown with walls made of perfect mirrors. The wall temperature and so the temperature of the reactants which are enclosed in the vessel is kept at a constant T_c by means of thermal coupling to an ambiance. Radiation of frequency $\nu \geq \nu_g$ communicates through an optical filter of bandwidth $\Delta\nu$ and area A between reactor 1 and a black body emitter held at temperature T. The filter is impermeable to matter. The chemical composition of the reactants in 1 is in equilibrium with the radiation density in 1. Reversible absorption and emission is assumed for calculating the maximum photochemical yield.

A slow and steady flux of reactants is passing through reactor 1. The incoming and outgoing products are measured in differences of particle fluxes $\delta \dot{N}_\alpha$; the index α indicates the

2.7 Photochemical Conversion

species which are involved in the photochemical reaction. The energy balance is

$$\delta \dot{N}_\alpha h\nu = A\,E_\nu(2)\Delta\nu - A\,E_\nu(1)\Delta\nu = A\pi L_\nu(\nu,T)\Delta\nu - A\,E_\nu(1)\Delta\nu. \tag{2.165}$$

The right hand side indicates the difference of radiation fluxes between $1 \to 2$ and $2 \to 1$. The black body emitter in 2 loses heat

$$\delta Q = \delta \dot{N}_\alpha\, h\nu \tag{2.166}$$

which is for steady state reversibly supplied by a heat pump. The power necessary for driving the heat pump is

$$\delta W_T = \delta Q \left(1 - \frac{T_c}{T}\right). \tag{2.167}$$

The external low temperature heat reservoir is assumed to be at T_c, the temperature of the reactants. The change in free enthalpy in the reaction vessel is

$$\delta W = \left(\frac{\partial \Delta G}{\partial N_\alpha}\right) \delta \dot{N}_\alpha = \Delta\mu\, \delta \dot{N}_\alpha. \tag{2.168}$$

$\Delta\mu$ is the difference in chemical potential between the final and initial products of the chemical reaction. Invoking reversibility requires $\delta W_T = \delta W$, independent of the direction of net flux. It leads to

$$\Delta\mu = h\nu \left(1 - \frac{T_c}{T}\right). \tag{2.169}$$

This can be used to substitute the black body temperature T of emitter 2 by a dependence on $\Delta\mu$ and system temperature T_c. Insertion into Planck's formula for hemispherical black body radiant flux density produces an expression for spectral radiation flux density in equilibrium with a reaction of change $\Delta\mu$ of chemical potential. In the ideal photochemical process the equilibrium luminescence radiation can then be looked upon to contribute with a photochemical potential $\mu = \Delta\mu$. It follows that

$$E_\nu(1) = \pi L u_\nu(\nu,\mu,T_c) = \frac{2\pi h\nu^3}{c^2}\,\frac{1}{\exp\left[(h\nu - \mu)/kT_c\right] - 1}. \tag{2.170}$$

The overall balance of photon processes in the photochemical reaction driven with an arbitrary spectrum $E_\nu(\nu)$ (which replaces the $\pi L_\nu(\nu,T)$ used in the model calculation) is

$$\dot{N}_\alpha = \int_{\nu_g}^\infty \frac{E_\nu}{h\nu}\,d\nu - \int_{\nu_g}^\infty \frac{\pi L u_\nu(\nu,\mu,T_c)}{h\nu}\,d\nu. \tag{2.171}$$

The equation corresponds closely to the ideal solar cell equation, (2.137), derived in Sect. 2.6.4.

In photovoltaics as in photochemistry the available spectral photon flux density $N_\nu = E_\nu/h\nu$ is of foremost interest. Energy and photon flux density spectra differ in shape, in particular in the position of the maximum and in the width of the distribution. This is illustrated in Fig. 2.40 for a standard spectrum.

Wien's displacement law for the spectral location of the maximum in *energy flux density spectra*

$$\lambda_{\max} T = 2{,}898\ \mu\mathrm{m\,K}\quad \text{in } E_\lambda(\lambda) \quad \text{or} \quad \frac{h\nu_{\max}}{kT} = 2.821 \quad \text{in } E_\nu(\nu)$$

can be formulated for *photon flux density spectra*

$$\lambda_{\max} T = 3{,}670\ \mu\mathrm{m\,K}\quad \text{in } N_\lambda(\lambda) \quad \text{or} \quad \frac{h\nu_{\max}}{kT} = 1.5937 \quad \text{in } N_\nu(\nu).$$

The shift of $\lambda_{\max}$ for E_λ towards longer wavelengths for N_λ and the corresponding shift towards lower photon energies is clearly perceivable.

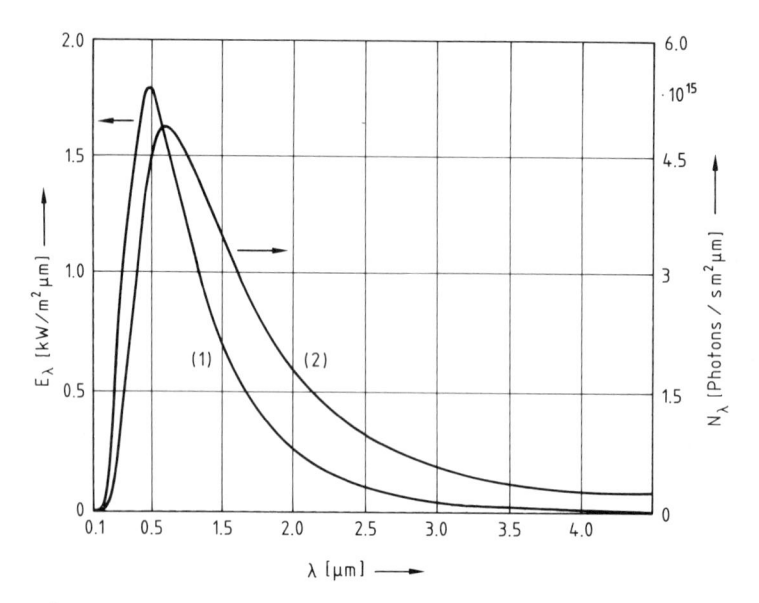

Fig. 2.40. Standard spectra E_λ and N_λ. (1) Energy flux densities E_λ and (2) photon flux densities N_λ. The dilution factor used is $f = 2.165 \cdot 10^{-5}$; $T_S = 5,777$ K and $T_a = 300$ K.

2.7.2 Maximum Yield in Photochemical Processes

The exergy yield in a photochemical process is frequently referred to the solar part of the spectrum

$$\eta^* = \frac{\Delta\mu \dot{N}_\alpha}{\int_0^\infty {}^s E_\nu d\nu}. \tag{2.172}$$

There are three adjustable reaction parameters, T_c, $\mu(=\Delta\mu)$ and ν_g. The temperature of the reaction vessel (of the reactants) should be as low as possible to suppress the luminescence losses (the aperture of incident radiation is inevitably an aperture of luminescence losses): T_c should be kept equal to ambient temperature T_a. The excitation energy $E_g = h\nu_g$ is system specific; hence E_g is variable with the particular choice of chemical reaction. Limited variations of $\Delta\mu$ for a given system of reactants are possible by changing pressures and concentrations.

The chemical potential μ of the *radiation* can be varied by optical concentration, $E'_\nu = C E_\nu$, as can be seen from (2.170). Solving for the chemical potential

$$\mu = h\nu - kT_a \ln\left(1 + \frac{2\pi h\nu^3}{c^2 E'_\nu}\right). \tag{2.173}$$

The case of standard spectra is presented in Fig. 2.33 of Sect. 2.6.6. The almost linear relationship of μ on the logarithm of the dilution factor $f = C/46,200$ can be approximated by

$$\mu = h\nu\left(1 - \frac{T_a}{T_S}\right) + kT_a \ln f \tag{2.174}$$

for $h\nu > -1.2\, kT_a \ln f$, whereas for $0 < h\nu < -kT_a \ln f$ the chemical potential μ remains of order kT_a only. Note that μ in addition to its dependence on $h\nu$ also changes with flux density E_ν, represented by the dilution factor f in (2.174).

The important consequence is that the photochemical activity of radiation of frequency ν not only depends on the photon energy $h\nu$ but also on its flux density E_ν and on the system temperature. This is equally true for the photovoltaic activity of incident radiation, as found in Sect. 2.6.6.

Pure longwave ambient radiation, for example, contains as a consequence of Planck's spectral distribution also high energy photons, $h\nu \gg h\nu_g$. However, with $E_\nu = \pi L_\nu(\nu, T_a)$ the chemical potential for *any* photon energy remains zero, see (2.173), provided the system is also at ambient temperature T_a.

In (2.171) with (2.172) the yield $\eta^*(f)$ can by variation of ν_g and μ be adjusted to a maximum value for any given f. It turns out that this maximum $\eta^*(f)$ is identical to the single gap solar cell $\eta^*(f)$, plotted in Fig. 2.35 of Sect. 2.6.7.

This is not surprising since both systems are essentially 'chemical'. There is, however, a practical difference. The cut-off wavelength λ_g for maximum exergy yield by a standard solar spectra is located in the near infrared region at about 1 μm, see Fig. 2.34 of Sect. 2.6.6. In semiconductor material of a solid state solar cell the high dielectric constant of order of 10 enables low bond breaking ionization energies E_g, which indeed cover this infrared region. In small molecule photochemistry, however, the high frequency dielectric constant is near 1 and E_g is in the blue or at most in the red region of the solar spectrum: a photochemical reaction system best matching to the solar spectrum can hardly be realized.

2.7.3 $h\nu$-, e_0V-, and kT- Reaction Paths

Finally, the question arises whether driving a chemical reaction with a combination of simultaneous radiation ($h\nu$-) and heat (kT-) or electric energy (e_0V-) inputs can have advantages over separate $h\nu$-, kT- or e_0V-dominated turnover [41,52].

The situation is familiar in electrochemical and thermochemical production of energy carriers, e.g. electrolysis of water [17].

$$\Delta G = \Delta H - T\Delta S > 0. \tag{2.175}$$

The free enthalpy of reaction ΔG has to be provided by electrical energy input. Keeping the reaction temperature T at an ambient T_a, the necessary (but quality zero) heat input $T_a \Delta S$ can be delivered from the ambiance. Increasing the reaction temperature, $T > T_a$, reduces ΔG and, therefore, the required electric energy for chemical turnover. Then, however, the necessary endothermic heat input $T\Delta S$ is high quality heat.

In the limit, the level of T can be chosen to accomplish a $\Delta G = 0$, i.e. $T = \Delta H/\Delta S$. The reaction moves on purely thermochemically; the heat input is identical with the enthalpy of reaction, ΔH, which is a measure of the change in bond energies in the course of the reaction.

For reversible processes there seems to be no preference for any choice of a more e_0V- or a more kT-dominated turnover: with perfect thermal power generators, perfect heat pumps or perfect heat transformers any mixture of e_0V- and kT-contributions could be supplied with the same expenditure of initial input of electrical energy or heat. In reality there are *practical* and *principle* differences.

Practically, if high temperature heat is available, a reaction temperature at the thermochemical limit $T = \Delta H/\Delta S$ is of advantage, because then no electric power is necessary, thus avoiding the usual exergy losses in thermal power conversion. In cases (e.g. direct thermochemical decomposition of water to hydrogen and oxygen at >3,000 K [22]), where this thermochemical limit of reaction temperature is too high, at least for reasons of materials properties, additional electrical energy input of ΔG helps to lower the temperature [17]

$$T \geq \frac{\Delta H - \Delta G}{\Delta S}. \tag{2.176}$$

If primarily electrical energy is available (e.g. from hydropower), a low temperature electrochemically dominated reaction has its technical merits.

Principle differences are perceptible in reaction kinetics. In an electrochemically dominated process the main product is related to the reaction path susceptible for charge carriers. In a thermochemically dominated process other reaction channels become possible and other main and side products may emerge.

Similar considerations are valid for the combination of $h\nu$- and kT-, or $h\nu$-, e_0V-, and kT- input. A high photon flux density with a radiance temperature exceeding the system temperature is, in its quality, comparable to electric energy and can cover free enthalpy of reaction. Increasing the temperature allows reduction of ΔG in the endothermic reaction at the price of absorption of heat of high quality.

The second aspect, the selection of the reaction path is also valid: the more $h\nu$- input is involved the better is the particular photochemical path in favor over thermochemically activated reaction channels. Any contribution of irradiance $E_\nu(\nu)$ of sufficient radiance temperature to ΔG will reduce the reaction temperature, $T = (\Delta H - \Delta G)/\Delta S$. This (solar) radiation unique effect has not yet been clearly observed; it is at present under investigation [41].

2.8 Appendix 1: Measurement of Solar Radiation

2.8.1 Introduction [6]

An overview is presented of standard instruments for solar radiation measurements; instruments receiving longwave and total radiation will be mentioned shortly. Commercially available instruments are mainly described. Sections of this appendix are taken from [2]. Several other reviews on solar radiation and its measurement are available, e.g. [43].

2.8.2 Basic Quantities and Instrumentation

In solar energy utilization important fluxes related to solar radiation and referred to a horizontal surface are:

- *Direct solar radiation* E_b, where $E_b = E_\perp \cdot \cos Z$, with $E_\perp$ being the direct solar irradiance at normal incidence and Z the zenith distance of the Sun's position
- *Diffuse solar radiation* E_d, which is solar radiation scattered or diffusely reflected in traversing the atmosphere
- *Global solar radiation* ^SE, the sum of the direct and the diffuse radiation: $^SE = E_b + E_d$
- *Reflected solar radiation* $E_r = \rho_s{}^SE$, with ρ_s being the hemispherical surface albedo. E_r is of importance for tilted mirrors and collectors.
- *Total radiation* E, the sum of (solar) global radiation and radiation of terrestrial and atmospheric origin: $E = {}^SE + {}^aE$.

For the measurement of $E_b, E_d, {}^SE, E$ the following instruments are used (the characterization follows [2]):

Pyrheliometer is an instrument for measuring the strength of direct solar irradiation $E_\perp$ at normal incidence. It usually employs a cone of $\approx 5°$ acceptance half angle.

Pyranometer is an instrument for the measurement of the solar irradiation received from the whole hemisphere. It is suitable for measuring global radiation ^SE, diffuse radiation E_d

---

[6] By Reinhold Busen, Institut für Physik der Atmosphäre, Deutsche Forschungsanstalt für Luft- und Raumfahrt (DLR), D-8031 Oberpfaffenhofen

(special shading attachement necessary), and reflected solar radiation E_r (instrument faced downward). Pyrheliometers and pyranometers are obtainable either for the total terrestrially available solar spectral range (0.29 to 3.0 µm) or for single broad or narrow spectral bands.

Pyrgeometer is an instrument for measuring longwave radiation received from the hemisphere. Most pyrgeometers eliminate the short wavelengths by means of filters and are then almost opaque up to $\lambda = 3.0$ µm. Their transmission is constant in the longwave region.

Pyrradiometer is an instrument for the measurement of both solar and terrestrial radiation (total radiation E) received from a hemisphere.

Sunshine Recorder is an instrument that records the hourly or daily totals of duration of sunshine accurately to the nearest tenth of an hour.

2.8.3 Detectors, Windows, Filters

Most detectors used in today's instruments transform the incoming solar radiation into an electrical signal. Through absorption by a blackcoated receiver surface the radiant energy is converted to heat. A thermopile measures the temperature increase. Conventional thermopiles (e.g. copper-constantan) and semiconductor thermopiles (e.g. Bi-Sn-Sb) are in use, each built of up to 50 single thermocouples. The passive thermocouple contacts are either inside the instrument and, therefore, shaded and held at the instrument's temperature, or they are connected to special white painted segments of the receiving surface. The black surface coatings used have 97–99% absorptivity. The main advantage of black absorbers is that the radiometers respond independently of wavelength within the spectral range they are intended to cover. For narrow-band measurements, where the radiation flux input is too small to produce an accurately detectable heating effect, photodetectors are used.

Most meteorological radiation instruments are covered with windows to protect the sensors against wind, humidity and dust. The windows can be simultaneously used as optical filters. For the terrestrially available solar spectral region normal glass offers nearly uniform transmissivity; in special cases quartz glass is used. Instruments having a 2π steradians solid angle of view (e.g. pyranometers) are usually covered with a hemispherical dome. The receiving surface is located near the center of the sphere.

In broad-band spectral measurements colored glasses are used as band pass filters for the spectral region to be transmitted (see Tab. 2.12). They are also available as hemispherical domes. Narrower spectral bands can be obtained by subtracting separate measurements made with different coloured glasses. Narrow-band measurements are obtained with sunphotometers, which consist of pyrheliometers equipped with photodetectors and narrow-band interference filters.

2.8.4 Description of Instruments

Pyrheliometers measure direct solar radiation. The receiving surfaces are perpendicular to the line joining Sun and receiver. Diaphragms ensure that only direct beam radiation and a narrow annulus of sky around the sun is detected. A sighting device is usually included in which a small spot of light falls upon a mark in the center of a target when the receiving surface is exactly normal to the direct solar beam.

For continuous recording of the direct solar radiation an equatorial mount is required. Care must be taken to ensure that windows and filters are kept clean and that no inside condensation occurs. Pyrheliometers are of different type:

- Ångström compensation pyrheliometer

- Silver disk pyrheliometer
- Linke-Feussner pyrheliometer
- Eppley normal incidence pyrheliometer (see Fig. 2.41).

Fig. 2.41. Eppley Pyrheliometer with suntracker. Normal Incidence Pyrheliometer, Model NIP, equipped with Schott glass filters; it is mounted on a Solar Tracker, Model ST-3. Manufactured by The Eppley Laboratory, Newport, R.I. The response time is $\approx$ 1 second for reaching an 1/e deviation from the final value. The sensitivity is approximately 8 μV/(Wm^{-2}). The instrument is about 20 cm long; its weight is 2.3 kg. The Solar Tracker accommodates up to three normal incidence pyrheliometers or other instruments to measure direct solar radiation (e.g. photometers).

All these instruments need calibration with an absolute pyrheliometer, using the Sun as source. Calibration should be repeated in intervals of about two to five years at a national or international radiation center.

The precision over one year of operation is about $\pm 0.5\%$ for the Ångström and the silver disk pyrheliometer. Most commonly used for continuous recording are the Linke-Feussner and the Eppley pyrheliometer; both are based on thermopile detectors. Their precision over one year is about $\pm 1\%$. Both instruments are available with filter wheels containing a quartz window and Schott special filters according to Tab. 2.12.

Table 2.12. Specifications of coloured glass filters after WHO.

	OG 530	RG 630	RG 700
Schott type glass			
Former designation	OG 1	RG 2	RG 8
Typical 50% cut-off wavelengths	526 ± 2 nm	630 ± 2 nm	702 ± 2 nm
	2,900 nm	2,900 nm	2,900 nm
Mean transmission (3 mm thickness)	0.92	0.92	0.92
Approximate temperature coefficient of shortwave cut-off (nmK^{-1})	0.12	0.17	0.18

Recently *sun photometry* was introduced. Sun photometers are used for air-pollution monitoring and measurements of spectral turbidity. Five center wavelengths are recommended by WMO: 368, 500, 675, 778, and 862 μm, each with 5 to 10 nm bandwidth.

Pyranometers measure the solar radiation received on a horizontal surface from within a solid angle of 2π steradians, referred to as global radiation (see Fig. 2.42). They are also used to measure solar radiation on surfaces inclined to the horizontal.

Inverted they can measure reflected solar radiation. The combination of one upward and one downward oriented pyranometer is called an *albedometer*. It is employed to determine the albedo and for direct measurement and recording of the shortwave radiation balance $^S E - E_r$. For measuring or recording *diffuse* sky radiation, the pyranometer can be shaded either by a small metal disk obstructing the Sun's direct beam (it is fixed to a power driven equatorial

Fig. 2.42. Sectored pyranometer. Pyranometer Model No. 8101, manufactured by Ph. Schenk, Vienna, Austria. The temperature difference between the black and white sectors of the receiver surface is measured by thermopiles. The response time for 95% of the final value is 20 seconds; the sensitivity is about 15 μV/(Wm^{-2}). The instrument is 90 mm high, 135 mm in diameter; its mass is approximately 1 kg.

mount) or by a shadow ring oriented towards the polar axis. The latter requires adjustments because of the variation of solar declination and corrections because of considerable screening of diffuse radiation by the shading ring [15,36]. Since radiation from a cloudless sky may be less than one-tenth of the global radiation, careful attention should be given to the sensitivity of the recording system.

For measuring reflected solar radiation the height above the surface should be 1 to 2 m, with the mounting device not greatly interfering with the field of view of the instrument. The resolution of global radiation measurements is, depending on the instrument type, ±1 to ±5 Wm^{-2}. Calibration drift has to be considered, especially for long term recording. The drift can be about ±1 to ±2% per year.

Calibration of pyranometers is performed in two different ways:

1. Comparison with a calibrated reference pyranometer, either outdoors with natural illumination or in the laboratory using artificial light sources;
2. Comparison between the direct beam irradiance $E_\perp$ at solar zenith distance Z, measured by a well calibrated pyrheliometer and the horizontal irradiance E_b, measured by the pyranometer. Suitable times should be selected with clear skies and steady radiation. E_b is obtained from the signal U_G (proportional to global irradiance) of the pyranometer and the signal U_D (proportional to diffuse irradiance) after shading the pyranometer with a disk held some distance away. Then

$$E_b = E_\perp \cos Z = k(U_G - U_D).$$

k is the requested calibration factor of the pyranometer. The solar zenith distance Z can be computed ([42] and (2.22) in Sect. 2.2.1). An average calibration factor is obtained from a series of measurements with various zenith distances.

Pyrradiometers. Total radiation includes short solar wavelengths (0.29 to 3.0 μm) and longer wavelengths of terrestrial and atmospheric origin (3.0 to 100 μm). The instruments used for this purpose are pyrradiometers; they measure either upward or downward radiation flux components or, in a tandem combination, the net difference between the two fluxes.

A difficulty in measuring total radiation is that there are no absorbers with constant sensitivity over the extended range of wavelengths concerned.

In pyrradiometers a variety of methods is employed for eliminating the convective heat losses by wind:

1. No protection provided: empirical formulae are used to correct for wind effects;
2. Compensation of the wind effect by electric heating;
3. Exceeding wind effects by artificial ventilation;
4. Elimination of wind effects by shielding the sensor.

In case 4 the receiving surfaces are covered with domes. The usual material is thin (0.1 mm) polyethylene film, which has been reported as having an integrated transmission of approximately 85% in the range of 0.3–100 μm. Because of the different absorption of the window in the longwave and shortwave region the calibration of the instrument slightly depends on the wavelength. This is in some instruments overcome by applying a strip or a small disk painted white with magnesium oxide (or similar material) on the sensing element. The white paint reflects shortwave radiation but absorbs longwave radiation: longwave radiation being attenuated by the dome finds a larger receiving area than short wavelength radiation [14].

Fig. 2.43. Sunshine recorder. Solar Energy Sensor, type SONIe 6.0081, manufactured by H.Siggelkow, Hamburg. A photocell senses radiation from narrow segments of the sky with a rotating slit diaphragm. 1 V corresponds to 100 Wm^{-2}. The spectral range of the instrument is 0.4 to 1.1 μm. The instrument can be operated between –30 °C and +50 °C. It is 120 × 140 mm wide and 210 mm high; its mass is approximately 1.2 kg.

Sunshine recorders provide information on whether the Sun is shining or not. The main types commonly used are the Campbell-Stokes recorder (a trace is burned in a chart by sunshine focused with a glass sphere) and the Foster sunshine switch (sunshine is indicated by an electric signal).

The latter instrument consists of a shaded and an unshaded photocell, both mounted inside a translucent tube. The output differential increases during periods of sunshine sufficiently to activate a relay which then is used to mark the sunshine duration on a recorder.

The threshold value used for discriminating between direct and diffuse solar radiation is usually set to 120 Wm^{-2} (WMO recommendation). Recently instruments became available which besides sunshine detection provide an approximate value of the direct solar radiation E_b (see Fig. 2.43). Then it is possible to record time intervals in which the solar flux density exceeds a specific threshold value, e.g. for starting solar power plant operation.

2.9 Appendix 2: Frequently Used Symbols

A	area, areal aperture, elevation angle	AM	relative air mass
c	vacuum speed of light	C	geometric concentration ratio
$CFLOS$	cloudfree lines of sight probabilities	D_{ES}	distance Earth-Sun
e_o	electrical elementary charge	E	energy (flux density), irradiance
E_{sc}	solar constant	E_λ, E_ν	spectral irradiances
$^sE, {}^aE$	solar, ambient irradiance	E_b, E_d	solar direct, diffuse irradiance
E_g	gap energy, ionization energy	EOT	equation of time
f	dilution factor, see Sect. 2.3.3	F	collector fin factor

G	Gibbs free enthalpy (flux density)	h	Planck's constant
H	enthalpy (flux density)	i	electric current density
I	radiant intensity	K	radiation entropy (flux density)
K_λ, K_ν	spectral radiation entropies	k	Boltzmann's constant
L	Planck's radiance	L_λ, L_ν	spectral Planck's radiances
Lu	luminescence radiance	Lu_λ, Lu_ν	spectral luminescence radiances
LST	local standard time	M	radiosity
n	index of refraction	p	gas (atmospheric air) pressure
P_e	electric power density	Q	heat (flux density)
R_S	radius of the Sun	S	entropy (flux density)
T	absolute (or Celsius) temperature	U_L	heat loss coefficient
U_e	voltage, electrical potential difference	VIS	local horizontal visibility
W	work (flux density)	Z	zenith distance
α	absorptance for radiation	β	surface tilt angle
γ	surface azimuthal angle	δ	(solar) declination
ϵ	emittance of radiation	η	efficiency
η^*	thermodynamic efficiency, exergy yield	ϕ	geographic latitude
Φ	radiant flux	λ	(vacuum) wavelength of radiation
μ	chemical potential	ν	frequency of radiation
ρ	reflectance for radiation	σ	Stefan-Boltzmann's constant
τ	transmittance for radiation	θ	angular aperture
Θ	angle of incidence	ω	hour angle
Ω	solid angle of radiation incidence	$(\alpha\tau)$	$\alpha\tau/[1-\rho(1-\alpha)]$
$(\alpha\epsilon)$	$\alpha\epsilon/[1-\rho(1-\alpha)]$	$(\epsilon\bar{\rho})$	$\epsilon(1-\rho)/[1-\rho(1-\alpha)]$

Bibliography

[1] *The Astronomical Almanac for the Year 1985.* London (UK): Her Majesty's Stationary Office: Washington/DC: U.S. Government Printing Office, 1985
[2] *Guide to Meteorological Instruments and Methods of Observation.* Volume Vol. 5 of *WMO No. 8, Fifth Edition*, WMO, 1983
[3] *A Preliminary Cloudless Standard Atmosphere for Radiation Computation.* Technical Repo.t WCP-112, WMO/TD-Nr. 24, WMO/WCP, 1986
[4] Agnihotri, O. P.; Gupta, B. K.: *Solar Selective Surfaces.* New York: John Wiley & Sons, 1981
[5] Bird, R. E.; Hulstrom, R. L.: *Applications of Monte Carlo Technique to Insolation Characterization and Prediction.* Technical Report SERI/RR-36-306, Solar Energy Research Institute, Golden/CO, 1979
[6] Bird, R. E.; Hulstrom, R. L.: Review, Evaluation and Improvement of Direct Irradiance Models. *Trans. ASME, J. Sol. Energy Engg.*, 103 (1981) 182–192
[7] Bogaerts, W. F.; Lampert, C. M.: Materials for Photothermal Solar Energy Conversion. *Materials Sciences*, 18 (1983) 2847–2875
[8] Bowker, D. E.; Davis, R. E.; Myrich, D. L.; Stacy, K.; Jones, W. T.: *Spectral Reflectances of Natural Targets for Use in Remote Sensing Studies.* NASA Ref. Publ., 1985
[9] Calzaferri, G.: Photochemical, Electrochemical and Thermochemical Transformation and Storage of Solar Energy. In *Proc. "Solar Energy '85" Summer School, Igls/Austria*, pp. 93–100, Köln (D): ASA/DLR, 1985
[10] Carnot, S.: *Reflections About the Motive Force from Heat* (in French). Paris (F), 1824
[11] Cohen, E. R.; Taylor, B. N.: *The 1986 Adjustments of the Fundamental Physical Constants.* Volume 63 of *CODATA Bulletin*, Oxford (UK): Pergamon Press, 1986
[12] Coutts, T. J.; Meakin, J. D. (Ed.): *Current Topics in Photovoltaics.* Orlando: Academic Press, 1985
[13] De Vos, A.; Pauwels, H.: On the Thermodynamic Limits of Photovoltaic Energy Conversion. *Applied Physics*, 25 (1981) 119–125
[14] Dehne, H.: *Development of an Evacuated Receiver for Line-Focus Solar Thermal Collectors.* Technical Report SAND86-7041, Sandia National Laboratories, Albuquerque/NM, 1986

[15] Drummond, A. J.: On the Measurement of Sky Radiation. *Archiv Met. Geophys. Bioklim.*, B7 (1956) 413–436
[16] Duffie, J. A.; Beckmann, W. A.: *Solar Engineering of Thermal Processes.* John Wiley & Sons, 1980
[17] Erdle, E.; Gross, J.; Meyringer, V.: Possibilities for Hydrogen Production by Combination of a Solar Thermal Central Receiver System and High-Temperature Electrolysis of Steam. In *Proc. 3rd Intl. Workshop on Solar Thermal Central Receiver Systems, Konstanz*, Becker, M. (Ed.), pp. 727–736, Berlin, Heidelberg, New York: Springer, 1986
[18] Fan, J. C. C.: Sputtered Films for Wavelength Selective Applications. *Thin Solid Films*, 80 (1981) 125–136
[19] Fonash, S. J.: *Solar Cell Device Physics.* Orlando: Academic Press, 1981
[20] Froehlich, C.; Brusa, R. W.: Solar Radiation and its Variation in Time. *Sol. Phys.*, 74 (1981) 209–215
[21] Froehlich, C.; Wehrli, C.: *Spectral Distribution of Solar Irradiance from 250 to 2500 nm.* Davos (CH): World Radiation Center, 1981
[22] Funk, J. E.; Knoche, F. K.: Hydrogen by Thermochemical Water Splitting. In *Proc. ISES Solar World Congress, Hamburg 1987*, Bloss, W.; Pfisterer, F. (Ed.), pp. 2895–2901, Oxford (UK): Pergamon Press, 1988
[23] Gleckman, P.; Winston, R.; O'Gallagher, J.: Concentration of Sunlight to Solar Surface Level Using Non-Imaging Optics. *Nature*, 339 (1989) 189–200
[24] Gleckmann, P.; Winston, R.; O'Gallagher, J.: Attaining the Maximum Solar Concentration. In *Proc. ISES Solar World Congress, Hamburg 1987*, pp. 2957–2960, Oxford (UK): Pergamon Press, 1988
[25] Goetzberger, A.; Greubel, W.: Solar Energy Conversion with Fluorescent Collectors. *Applied Physics*, 14 (1977) 123–139
[26] Hutchins, M. G.: Selective Thin Film Coatings for the Conversion of Solar Radiation. *Surface Technology*, 20 (1983) 301–320
[27] Inal, O. T.; Scherer, A.: Optimization and Microstructural Analysis of Electrochemically Deposited Selective Solar Absorber Coatings. *Materials Sciences*, 21 (1986) 729–736
[28] Iqbal, M.: *And Introduction to Solar Radiation.* New York: Academic Press, 1983
[29] Kasten, F.; Young, A. T.: Revised Optical Air Mass Tables and Approximation Formula. *Applied Optics*, 28 (1989) 4735–4738
[30] Klucher, T. M.: Evaluation of Models to Predict Insolation of Tilted Surfaces. *Solar Energy*, 23 (1979) 111–114
[31] Koltun, M. M.: *Selective Optical Surfaces for Solar Energy Converters.* New York: Allerton, 1981
[32] Köpke, P.; Quenzel, H.; Sizmann, R.: Yearly Yield of Solar CRS-Process Heat and Temperature of Reaction. In *Solar Thermal Energy Utilization*, Becker, M. (Ed.), pp. 3–96, Berlin, Heidelberg, New York: Springer, 1987
[33] Lampert, C. M.: Advanced Optical Materials for Energy Efficiency and Solar Conversion. *Solar & Wind Technology*, 4 (1987) 347–379
[34] Lampert, C. M.: Coatings for Photothermal Energy Collection. *Solar Energy Mat.*, 2 (1980) 1–17
[35] Landsberg, P. T.: *Thermodynamics.* New York: Interscience Publishers, 1961
[36] Le Baron, B. A.; Peterson, W. A.; Dirmhirn, I.: Corrections for Diffuse Irradiance Measured with Shadow Bands. *Solar Energy*, 25 (1980) 1–13
[37] Liou, K. N.: Influence of Cirrus Clouds on Weather and Climate Processes. *Monthly Weather Rev.*, 114 (1986) 1167–1199
[38] Lund, I. A.; Grantham, D. D.; Elam, C. B. B.: *Atlas of Cloud Free Lines of Sight Probabilities.* AF Surveys in Geographics No. 400, Part 4: Europe AF GL-TR-78-0276, 1978
[39] Lund, I. A.; Shanklin, M. D.: Universal Methods for Estimating Probabilities of Cloud-Free Lines-of-Sight Through the Atmosphere. *J. Appl. Meteorol.*, 12 (1973) 28–35
[40] Niklasson, G. A.; Granquist, C. G.: Surfaces for Selective Absorption of Solar Energy: An Annotated Bibliography 1955–1981. *Materials Sciences*, 18 (1982) 3475–3534
[41] Nix, G.; Sizmann, R.: High Temperature, High Flux Density Solar Chemistry. In *Solar Thermal Technology - Proc. 4th Intl. Symposium, Santa Fe/NM, 1988*, Gupta, B. P.; Traugott, W. H. (Ed.), p. 351, New York: Hemisphere Publ. Co., 1990
[42] Paltridge, G. W.; Platt, C. M. R.: *Radiative Processes in Meteorology and Climatology.* Volume Vol. 5, Amsterdam (NL): Elsevier Sci. Publ. Co., 1976
[43] Perrin de Brichambaut, C.; Lamboley, G.: *Solar Radiation and Its Measurement* (in French). Volume Cahiers A.F.E.D.E.S. No. 1, Paris (F): Editions Européennes: Thermique et Industrie, 1968 (edition complétée 1974)
[44] Press, W. H.: Theoretical Maximum for Energy from Direct and Diffuse Sunlight. *Nature*, 264 (1976) 734–735
[45] Rabl, A.: *Active Solar Collectors and Their Applications.* New York: Oxford University Press, 1985

[46] Ries, H.: Complete and Reversible Absorption of Radiation. *Applied Physics*, B 32 (1983) 153–156
[47] Ries, H.: Thermodynamic Limitations of the Concentration of Electromagnetic Radiaton. *J. Opt. Soc. Am.*, 72 (1982) 380–385
[48] Riordan, C. J.; Hulstrom, H. L.: *Summary of Studies that Examine the Effects of Spectral Solar Radiation Variations on PV Device Design and Performance.* Technical Report, Solar Energy Research Institute, Golden/CO, 1989
[49] Schmetz, J. *Tellus*, 36 A (1986) 417–166
[50] Shockley, W.; Queisser, H. J.: Detailed Balance Limit of Efficiency of p-n-Junction Solar Cells. *Applied Physics*, 32 (1961) 510–519
[51] Sizmann, R.: Process Heat Applications. In *Proc. "Solar Energy '85" Summer School, Igls/Austria*, pp. 93–100, Köln (D): ASA/DLR, 1985
[52] Sizmann, R.: Solarchemical Potential of Solar Radiation (in German). In *Proc. Solarchemisches Kolloquium*, Becker, M.; Funken, K. H. (Ed.), pp. 5–53, Heidelberg, Berlin, New York: Springer, 1989
[53] Stine, W. B.; Harrigan, R. W.: *Solar Energy Fundamentals and Design.* New York: John Wiley & Sons, 1985
[54] Tuller, S. E.: World Distribution of Mean Monthly and Annual Precipitable Water. *Month. Weath. Rev.*, 96 (1986) 786–797
[55] Waldmeier, M.: *Results and Problems of Solar Research* (in German). Leipzig (D): Akademische Verlagsanstalt, 1941
[56] Warren, S. G.: Optical Constants of Ice. *Applied Optics*, 23 (1984) 1206–1225
[57] Welford, W. T.; Winston, R.: *High Collection Non-Imaging Optics.* New York: Academic Press, 1989
[58] Winter, C.-J.; Nitsch, J. (Ed.): *Hydrogen as an Energy Carrier – Technologies, Systems, Economy.* Berlin, Heidelberg, New York: Springer, 1988
[59] Woodbury, G. E.; McCormick, M. P.: Zonal and Geographical Distribution of Cirrus Clouds Determined from SAGE Data. *J. Geophys. Res.*, 91 C2 (1986) 2772–2785
[60] Zdunkowski, W. G.; Welsch, R. M.; Korb, G.: An Investigation of the Structure of Typical Two-Stream-Methods for the Calculation of Solar Fluxes and Heating Rates in Clouds. *Contr. Phys. Atm.*, 53 (1980) 147–166

3 Concentrator Optics

L.L. Vant-Hull [1]

3.1 Introduction

In this chapter we will discuss the optical principles involved, and their application, in the design of solar power systems.

After a brief discussion of the solar design environment we will reiterate the required basic optics. In fact, little more than first year optics is required. However, the unique requirements to efficiently collect energy dispersed over square kilometers, and to do this at the least possible cost, requires a unique discipline.

In Sect. 3.3 we define this discipline as concentration optics, and present the primary features and mind set which distinguish it from the usual imaging optics. Following a discussion of the primary concepts involved in concentrator optics we will describe the limiting case, that is the ideal concentrator, as well as non-imaging concentrators of which the compound parabolic concentrator (CPC) is an example.

The primary optical element in many solar collectors is a parabolic mirror, and we will discuss such systems in some detail including parabolic mirrors with terminal concentrators, usually some sort of a non-imaging concentrator. Approximations to parabolic mirrors (or lenses) involving a multitude of facets are really Fresnel concentrators, which must still track the Sun. We will describe both monolithic trackers and variations in which the individual facets track separately.

The extreme case of individual Fresnel facets is the Central Receiver system, in which each of several hundred to several thousand facets may have an area of 50 m^2 to 200 m^2. Design considerations here are sufficiently different from the smaller concentrators that we will discuss their optics separately, along with the special techniques developed for their design and optimization.

A discussion of design issues and constraints unique to solar optics appears in Sect. 3.8. The constraints such as low initial and operating costs, predictable and limiting flux density distributions etc. control the optical design.

3.2 Basic Optics

1. The speed of light in a vacuum is constant at 3.00×10^8 m/s. When propagating through a medium of refractive index n ($n > 1$) the phase velocity is reduced to $v = c/n$. Typical values of n are unity for a vacuum, 1.0003 for air, 1.33 for water, 1.45–1.8 for glass of various

---

[1] Lorin L. Vant-Hull, Energy Laboratory, University of Houston, 4800 Calhoun Road, Houston, TX 77004, USA

3.2 Basic Optics

compositions and for plastics. Note that, depending upon the medium, n may be a function of frequency.

2. Snell's law follows from the principle of least time. At the interface between two media of refractive index n_1 and n_2, Snell's law provides that $n_1 \sin\theta_1 = n_2 \sin\theta_2$ where θ_1 is the angle of incidence (measured from the normal in medium 1) and similarly for θ_2. As we shall see later, at the interface the incident beam splits into a reflected beam and a transmitted beam, both subject to Snell's law (Fig. 3.1).

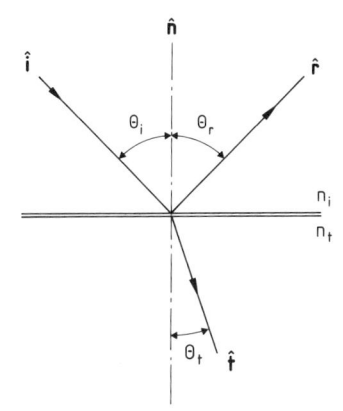

Fig. 3.1. Snell's law geometry for light incident on a medium of higher refractive index. The three vectors $\hat{\imath}$, $\hat{n}$, and $\hat{r}$ lie in, and define, the plane of incidence.

Reflection. Clearly the reflected beam remains in the first medium, so $n_r = n_i$. Consequently $\theta_r = \theta_i$ and the angle of incidence is equal to the angle of reflection. Note that there can be no chromatic aberration on reflection as the angles are independent of n_i and thus of wavelength. The angle of reflection is determined solely by the orientation of the local surface normal $\vec{n}$ with unit vector $\hat{n}$. In vector form:

$$\hat{r} = \hat{\imath} - 2(\hat{\imath}\cdot\hat{n})\hat{n}. \tag{3.1}$$

Refraction. A portion of the incident beam is transmitted through the interface, and, if the second medium is transparent (i.e. if n_t is real or has only a small imaginary (loss) component), this beam will continue to propagate into the second medium. If $n_t > n_i$, it will be bent toward the normal in accordance with Snell's law or the principle of least time. On emerging from a flat glass plate, the beam will reassume its original direction, having traversed a shorter path (lost less time) in the glass than if it had travelled in a straight line. If the second surface is not parallel to the first, the beam will be deflected, as by a prism.

Lenses are seldom used in solar power systems as their volume and weight become excessive, even for relatively small scale systems. Fresnel lenses (and reflectors) have been used, particularly as facetted concentrators for photovoltaic systems, but also in designs for mid-size dishes and troughs.

Due to dispersion, the index of refraction is somewhat dependent on wavelength; the resulting chromatic aberrations are not usually of concern in solar designs. At worst the ultraviolet, solar (0.3–1 μm), and IR bands may be treated separately. Of course for a mirror there is no chromatic aberration.

3. Further physical principles govern the fraction of a beam of electromagnetic radiation (light) propagating into, and through, a medium of index of refraction $n = n_t/n_i$. Maxwell's equations impose the boundary condition that both electric and magnetic fields be continuous at the boundary. Remembering that $E = vB = (c/n)B$, one can obtain the Fresnel relations for both the transverse electric (TE) mode (electric field direction perpendicular to the plane of incidence) and for the orthogonal transverse magnetic (TM) mode.

The reflectance ρ and transmittance τ of a beam of light traversing the interface between two dielectric media is a complicated function of the angle of incidence. Its derivation involves Maxwell's boundary conditions on electric and magnetic fields, conservation of energy, etc. The reflection and transmission coefficient for the transverse electric and transverse magnetic modes can be written as:

$$\text{TE:} \quad r_{\text{TE}} = -\frac{\sin(\theta_i - \theta_t)}{\sin(\theta_i + \theta_t)} \quad ; \quad t_{\text{TE}} = r_{\text{TE}} + 1 \tag{3.2}$$

$$\text{TM:} \quad r_{\text{TM}} = \frac{\tan(\theta_i - \theta_t)}{\tan(\theta_i + \theta_t)} \quad ; \quad t_{\text{TM}} = \frac{r_{\text{TM}} + 1}{n_t/n_i} \tag{3.3}$$

where θ_t is the angle of the refracted beam, and $n_i \sin \theta_i = n_t \sin \theta_t$.

Fortunately, at near normal angles of incidence ($\theta < 30°$), the result may be approximated by

$$\rho \cong r_{\text{TE}}^2 \cong r_{\text{TM}}^2 \cong [(n_i - n_t)/(n_i + n_t)]^2 \tag{3.4}$$

where ρ is the reflectance (or relative reflected intensity) and r is the reflection coefficient (or the relative amplitude of the reflected electric field). To ensure energy conservation, the transmittance τ is given by:

$$\tau = 1 - \rho = 4 n_i n_t / (n_i + n_t)^2. \tag{3.5}$$

The reflectance remains remarkably constant at angles of incidence up to about 20° (1/3 of Brewster's angle) for each component of polarized light, or to about 30° (1/2 of Brewster's angle) for unpolarized light[2].

4. Electromagnetic radiation (light) propagating through any medium (other than a vacuum) interacts with the medium. Fluctuations in density lead to Rayleigh scattering, proportional to the fourth power of the frequency; while small inclusions ($d < \lambda$) lead to the more complex Mie scattering (both are discussed in Chap. 2). Absorption (by atmospheric gasses or by impurities such as iron ions in a glass) tends to affect specific wavelengths or bands. These complications mean that any detailed discussion of losses requires an integration over the solar spectrum, where the loss factors at each frequency from all mechanisms are first combined over the entire path, and weighted by the appropriate incident solar spectrum, again, as described for the atmosphere in Chap. 2. In most cases, subject to sufficiently stringent limitations (such as small fractional losses, or specific components of the radiation (e.g. visible, far IR), simple empirical relations can be developed to describe the loss. The simplest, of course, is that the loss is proportional to the radiant flux with an appropriate optical depth t, leading to standard exponential decay:

$$\begin{aligned} I &= I_0 \exp(-x/t) \\ &\approx I_0(1 - x/t) \text{ for } x \ll t. \end{aligned}$$

If saturation occurs for a portion of the spectrum, the apparent optical depth will increase as x increases, resulting in curvature in a plot of $\ln(I/I_0)$ vs. x/t. This can be represented by an exponent s on x/t, thus:

$$I = I_0 \exp(-x/t)^s. \tag{3.6}$$

---

[2] At Brewster's angle the TM mode is totally reflected, and $\theta = \tan^{-1} n$, where $n = n_i/n_t$ for the incident beam and n_t/n_i for the transmitted beam.

Values of s from 0.56 to 0.92 have been used to fit atmospheric transmission data. Alternatively, a sum of several exponentials with different t can be used, each representing the behaviour in a specific spectral region, with the prefactors adding to I_0.

5. The focal conditions for solar power concentrators follow from Snell's law. However most solar collectors are designed to gather light effectively and thus have aperture diameters several times their focal length. As a consequence the usual thin lens and paraxial ray approximations are grossly invalid. Fortunately the Sun presents an object which is essentially at infinity. Thus the usual p, q, f (or a, b, f) focal relations are not necessary, the image will be formed at the focus. Although the solar literature is rife with multi-element optical systems, these are generally inefficient and impractically expensive. Only terminal concentrators, which act in the focal zone to compress the primary image, show any promise as second optical elements.

While ordinary lenses are not used in solar concentrators because of their weight and cost, Fresnel lenses are. They can be visualized as a lens from which superfluous plane sheets of dielectric have been removed to form a stepped refracting surface (like the stage lens in an overhead projector). Image quality (for solar purposes) suffers hardly at all, and the thickness of even a very large lens can be reduced to a few mm. The standard lens makers formula (for a thin lens comprised of surfaces having radii of curvature of R_1, R_2 respectively) is:

$$\frac{1}{f} = \frac{(n_2 - n_1)}{n_1} \left(\frac{1}{R_1} - \frac{1}{R_2} \right) \tag{3.7}$$

(with quantities to the right of the thin lens vertex taken as positive, to the left negative, with light propagating from left to right). Thin lens theory also shows that spherical aberration and coma (an off-axis aberration) are both reduced if R_1 and R_2 are negative, that is, if the Fresnel lens is *domed* over the receiver (focal point). The lenses of interest typically have short focal length f compared to the lens diameter D. For such small f/D lenses, very detailed calculations are required to design the lens facets so all nearly paraxial rays reach the receiver. This includes shaping the facets so the *steps* do not interfere with the refracted light, but strong protective enclosures can be generated incorporating the focusing element and having $f/D < 1$.

A parabola is the locus of points which, for parallel incident light, satisfies the requirement that the distance to a given point (called the focus) plus the perpendicular distance to a line is a constant. Consequently, by the principle of least (equal) time, parallel light (sunlight) will be redirected toward the focus. The equation is $y = x^2/4f$ with the focal point at $(x, y) = (0, f)$. If the rim angle is 90°, then $(X_{\text{rim}}, Y_{\text{rim}}) = (2f, f)$ and the collector has an f/D of 1/4, compared to a fast lens with $f/D \approx 2$.

3.3 Concentration Optics

3.3.1 Concepts of Concentrator Optics

Solar thermal power plants generally concentrate the direct beam solar radiation, incidentally rejecting a majority of the diffuse radiation. Concentration by a factor C increases the incident flux density at the solar receiver to the extent that radiation and convection losses are acceptable even at elevated receiver temperatures. Even photovoltaic power systems may use concentration, but here the objective is to reduce the required area of the expensive photovoltaic element by a factor of C. In any case, there are several points which must be made about the optical systems employed so the design elements involved can be appreciated.

1. All the basic laws of optics apply without modification.
2. The objectives are completely different (to concentrate energy cheaply rather than to form a precise image), consequently different approximations are valid, resulting in sometimes surprising configurations.

3. Most concentrating systems of interest for power applications have acceptance angles[3] of 5 mrad to perhaps 50 mrad. Consequently only the direct solar and near circumsolar(beam) radiation is of use or interest in solar design, and the collectors must track the Sun in some way.
4. Collectors must be very large (compared to a camera lens) in order to intercept significant power: of order 2 m^2 per thermal kilowatt or 2 km^2 per thermal gigawatt.
5. Interference and diffraction effects play no role in the optics of solar concentrators. Rather, the object size of $\approx 1/4°$, corresponding to the solar limb angle, plays a dominant role[4]. On the other hand, interference filters and multilayer antireflective coatings may prove useful in some cases.
6. Thermodynamic limits may play a role. Thus considerations of *Etendue* (throughput) vs. concentration, the surface temperature of the Sun, beam divergence, etc. replace second order aberrations, phase coherence, and interference phenomena in the designer's lexicon.

The term Concentrator Optics has been coined to describe the appropriate mind set involved in such analysis. For a more extensive discussion of concentrating systems see the literature, such as [33] or [14].

3.3.2 Solar and Circumsolar Brightness Distribution – Sunshape

Image formation and magnification are important concepts in solar thermal power, but not in quite the same context as in conventional optical design. The Sun presents an object which is effectively at infinity, but has an angular radius of $\alpha = 4.653$ mrad or 16 minutes of arc. Consequently an ideal image formed at a distance of 1000 m would have a radius of 4.7 m (typical of central receiver systems) or at 10.0 m would have a radius of 47 mm (typical of distributed receiver systems such as in parabolic troughs).

Several representations of the solar irradiance are shown in Tab. 3.1. A simple Lambertian radiator would provide a uniform distribution of radiance across the solar disc. Then $E = \int L d\Omega = \pi \alpha^2 L_0$ requiring $L_0 = 13.23$ MW/m^2 to provide a terrestrial irradiance of 900 W/m^2 from the 4.653 mrad solar disc.

Scattering and absorption in the solar photosphere modify the uniform distribution of radiance expected from a black body radiator, so the simple 'flat top' distribution is frequently replaced by a more realistic 'limb darkened' distribution. Although UV radiation is more strongly attenuated in the photosphere than is IR, after integrating over wavelength an empirical fit to the data results in a radiance distribution given by [1]

$$L/L_0 = 0.36 + 0.84\sqrt{1 - \xi^2/\alpha^2} - 0.20(1 - \xi^2/\alpha^2)$$

where L_0 is the central radiance (at $\xi = 0$), $\xi = r/D_{ES} \leq \alpha$ and $\alpha = R/D_{ES} = 4.653$ mrad. (R is the solar radius, D_{ES} the Earth-Sun distance). A simpler fit to the data providing nearly as good agreement is

$$L/L_0 = 1 - 0.5138\, \xi^4/\alpha^4 \tag{3.8}$$

and the parameters of this distribution are also shown in Tab. 3.1.

The extraterrestrial irradiance is modified upon traversing the earth's atmosphere due to absorption, and to multiple small angle scattering events which produce the solar aureole.

---

[3] Acceptance angle is the angular range over which essentially all incident rays are collected, i.e. the allowed source size

[4] At $\lambda = 1$ µm and $\theta = 4.65$ milliradian (mrad) $\approx 1/4°$, the diffraction limit corresponds to a collector diameter $\approx 1.2\lambda/\sin\theta \approx 0.2$ mm. Such small dimensions are only approached in the facets of Fresnel lenses for photovoltaic systems.

3.3 Concentration Optics

Table 3.1. Representative solar distribution functions and Gaussian (G) approximations to them. All distributions are normalized so $E = \int L d\Omega = 900$ W/m^2.

Distribution function	L_0 MW/m^{2}sr	Width of distribution σ	mrad
Uniform radiance	13.23	α^*	4.653
• G fit to central radiance	13.23	$\alpha/\sqrt{2}$	3.290
• G least squares fit	18.34	$\alpha/\sqrt{2\ln 4}$	2.794
Limb Darkened Sun	15.96	α^*	4.653
• G fit to central radiance	15.96	$\sqrt{1-k/3}\alpha/\sqrt{2}$	2.995
• G least squares fit	18.67	$0.8419\alpha/\sqrt{2}$	2.770
Circumsolar Data	17.26	$\alpha^\dagger$	4.4–4.9
• G fit to central radiance	$17.87(1-\gamma)$	$0.619\,\alpha$	2.880

† Including instrumental resolution, the transition from solar disc to aureole is ≈ 0.5 mrad wide centered near α
* $L = 0$ beyond α, the solar limb. Also, $k = 0.5138$

Particularily on hazy days, and on days with high cirrus clouds, the region of sky within a few degrees of the Sun is very bright due to light scattered out of the solar disc, and the rim of the disc is further darkened. The distribution of brightness from the center of the Sun, across the solar limb, and out to 3.2° was measured during the period 1977–79 at several locations [18] using specially designed high resolution circumsolar telescopes. A rather detailed analysis of the resulting 'sunshape' (as the brightness distribution is called) has been carried out, allowing estimation of the loss in concentration resulting from the aureole [5]. (Energy in the aureole, out to 2.6°, is included in normal incidence pyrheliometer measurements but does not contribute to the central intensity and may not be intercepted by a receiver sized to capture light from the solar disc.) A 'universal sunshape' which is a simple average over the reported data is proposed as representative.

The simplest fit to this sunshape requires solar and circumsolar regions. Thus

$$L = L_s[1 - (0.5051\xi/\alpha)^2 - (0.9499\xi/\alpha)^8], \qquad \xi \leq 4.653 \text{ mrad} \qquad (3.9)$$

$$E_s = \int_{\text{Sun}} L d\Omega = 0.7402 L_s \pi \alpha^2 = 50.35 \cdot 10^{-6} L_s$$

and

$$L = L_a(\xi/\alpha)^{-2}, \qquad 4.653 \text{ mrad} < \xi < 55.85 \text{ mrad} \qquad (3.10)$$

$$E_a = \int_{\text{aureole}} L d\Omega = L_a \pi \alpha^2 \cdot 2\ln 12 = 338 \cdot 10^{-6} L_a.$$

For the *standard circumsolar scan* $L_s = 13.639 \cdot 10^6$ W/m^{2}sr and $L_a = 72,200$ W/m^{2}sr, giving $E = E_s + E_a = 686.7 + 24.4 = 711.1$ W/m^2, in good agreement with the tabulated scans. Values scaled to 900 W/m^2 are given in Tab. 3.1. The circumsolar ratio, $\gamma = E_a/E$ varies considerably with atmospheric conditions, but for good solar sites the monthly average value rarely exceeds 5% for an operating threshold of 300 W/m^2. L_a and L_s should be adjusted to give observed or desired value of E and γ, thus $L_a = E\gamma/(338 \cdot 10^{-6} \text{ sr})$ and $L_s = E(1-\gamma)/(50.35 \cdot 10^{-6} \text{ sr})$.

The bivariate Gaussian is frequently used to approximate the sunshape as it is well behaved everywhere and falls rapidly to zero for angles greater than two sigma. In addition, it is easy to approximate the effects of random errors of the concentrator by simply adding the standard deviations in quadrature, i.e. $\sigma^2 = \sum \sigma_i^2$, to obtain the resulting distribution function. For these reasons we also show in Tab. 3.1 the parameters of two Gaussians for several of the sunshapes discussed above; one with equal central irradiance, and one resulting from a least squares fit; all providing 900 W/m². Note however, that the least squares fit will significantly overpredict the peak fluxes. On the other hand, the least squares fit to the uniform Sun is probably a more realistic representation of the sunshape than is the uniform Sun model, itself.

3.3.3 The Degraded Sun

These images formed by any concentrator will, of course, be degraded by any lack of perfection in the optical system. However, as we are interested in concentration, rather than image quality, 4.65 mrad provides a useful scale against which to compare optical imperfections. Thus, mirror errors which deflect the reflected beam substantially less than 4.65 mrad are of little importance, while deviations greater than ≈ 1.5 mrad will contribute to reductions in concentration and increases in spillage (radiation bypassing the receiver).

The actual *image* formed by a solar concentrator can be thought of as a summation of solar images, each distorted and displaced by local imperfections in the concentrator, and if a time-average value is sought, by variations from perfect tracking. If the concentrator can be characterized by a single error function, then it is convenient to define a *degraded Sun* which would combine with a perfect concentrator to produce an equivalent image to that of the true Sun and the actual imperfect concentrator. A convolution of the actual brightness distribution of the Sun S with the error function for the concentrator G produces the desired result

$$D(x,y) = G * S = \int_{-\infty}^{\infty} \int_{-\infty}^{\infty} G(x - x_1, y - y_1) S(x_1, y_1) \, dx_1 dy_1 \qquad (3.11)$$

where (x, y) is a point on the *image plane*.

It is convenient to express both G and S in terms of weighted sums over the two dimensional moments of their respective distribution functions. These moments are calculated via

$$\mu_{i,j}^S = \int_{-\infty}^{\infty} \int_{-\infty}^{\infty} S(x,y) \, x^i y^j \, dx dy \qquad (3.12)$$

and similarly for $\mu_{i,j}^G$.

In the same way, the moments of the convolution of S and G can be written:

$$\begin{aligned}
\mu_{m,n}^D &= \int_{-\infty}^{\infty} \int_{-\infty}^{\infty} x^m y^n \, D(x,y) \, dx dy \\
&= \int_{-\infty}^{\infty} \int_{-\infty}^{\infty} x^m y^n \, (G * S) \, dx dy \\
&= \int_{-\infty}^{\infty} \int_{-\infty}^{\infty} x^m y^n \int_{-\infty}^{\infty} \int_{-\infty}^{\infty} G(x - x_1, y - y_1) S(x_1, y_1) \, dx_1 dy_1 \, dx dy
\end{aligned}$$

3.3 Concentration Optics

$$\mu_{m,n}^D = \sum_{i=0}^{m}\sum_{j=0}^{n} \binom{m}{i}\binom{n}{j} \mu_{m-i,n-j}^G \mu_{i,j}^S \qquad (3.13)$$

where the substitution $X = x - x_1$ has been used and where $\binom{m}{i}$ represents the i-th coefficient in the binomial expansion of $x^m = (X + x_i)^m$, performed prior to the intergration over X or x_i.

Calculations are greatly simplified [26] by using normalized functions, carrying out calculations in a coordinate system centered at the point which makes $\mu_{1,1} = 0$ for the combined functions, and standardizing the coordinates by dividing with the second order moments. Under these conditions the moments of a Gaussian distribution ($\mu_{i,j}^G$), and of several representative solar distribution functions ($\mu_{i,j}^S$) are given in Tab. 3.2. The Flat Top distribution is characteristic of a constant probability function or of a uniform brightness solar disc. The Limb Darkened Sun (LDS) accounts for absorption in the solar photosphere, while the Lawrence Berkeley Laboratories (LBL) measured sunshape also includes effects of atmospheric scattering (see Sect. 3.3.2).

Table 3.2. Moments of several solar distribution functions

	Gaussian	Flat top	LDS[d]	LBL (Solar + Circumsolar)[e]	
$E^{(a)}$	$2\pi\sigma^2 L_0$	$\pi\alpha^2 L_0$	$0.8287\pi\alpha^2 L_0$	$0.7402\pi\alpha^2 L_s$	$+\pi\alpha^2 L_a \cdot 2\ln 12^{(f)}$
$\mu_{00}^{(b)}$	1	1	1	$1-\gamma$	$+\gamma$
$\mu_{20}^{(c)}$	σ^2	ρ^2	ρ^2	$(1-\gamma)\rho_s^2$	$+\gamma\rho_a^2$
	$\sigma = \alpha/\sqrt{2}$	$\rho = 0.5\alpha$	$\rho = 0.4735\alpha$	$\rho_s = 0.4537\alpha$	$\rho_a = 3.793\alpha$
μ_{40}	$3\sigma^4$	$3(0.6667\rho^4)$	$3(0.6917\rho^4)$	$3(1-\gamma)(0.6977\rho_s^4)$	$3\gamma(1.260\rho_a^4)$
μ_{22}	σ^4	$0.6667\rho^4$	$0.6917\rho^4$	$(1-\gamma)(0.6977\rho_s^4)$	$\gamma(1.260\rho_a^4)$
μ_{60}	$15\sigma^6$	$5\rho^6$	$5(1.1003\rho^6)$	$5(1-\gamma)(1.126\rho_s^6)$	$5\gamma(4.204\rho_a^6)$
μ_{42}	$3\sigma^6$	ρ^6	$1.1003\rho^6$	$(1-\gamma)(1.126\rho_s^6)$	$\gamma(4.204\rho_a^6)$

[a] Energy in solar beam $= \int L d\Omega$.
[b] Normalization by E reduces μ_{00} to unity.
[c] All odd moments of the Sun are zero by symmetry, also $\mu_{0i} = \mu_{i0}$.
[d] Limb Darkened Sun; the simpler fit, $L/L_0 = 1 - 0.5138\xi^4/\alpha^4$ is used. Coefficients for the more complicated expression differ by less than 1%.
[e] Lawrence Berkeley Laboratory (LBL) 'standard sunshape'. Energy and moments are sum of disc and aureole contributions. The circumsolar ratio γ is 0.0343 for the standard LBL scan.
[f] The LBL circumsolar data extends to $\xi = 12\alpha$.

The two-dimensional Gaussian function is particularly easy to work with (defined and well behaved over all space, and normalized to unit volume) and is a suitable approximation for many error functions. Hence we will choose it as the basis function for an orthogonal expansion of D. Then the appropriate orthogonal polynomials are the Hermite polynomials.

Using the orthogonality properties of the Hermite polynomials, a few of which are tabulated in Tab. 3.3 in our standardized coordinate system, we can determine the coefficients in an orthogonal expansion of D in our selected coordinate system [26], and these are also tabulated in Tab. 3.3. Finally, we can write the expression for the degraded Sun as:

$$D(x,y) = \frac{1}{2\pi\sigma_x\sigma_y}\exp\left\{-\frac{1}{2}\left(\frac{x^2}{\sigma_x^2}+\frac{y^2}{\sigma_y^2}\right)\right\}\cdot\sum_{i,j=1}^{\infty} C_{ij} H_i\left(\frac{x}{\sigma_x}\right) H_j\left(\frac{y}{\sigma_y}\right) \qquad (3.14)$$

where the first term in the summation is unity (returning a two dimensional Gaussian distribution), and the second term involves $(x/\sigma_x)^4$ or $(y/\sigma_y)^4$ in these standardized coordinates. Note however, that if both G and S were true Gaussians, substitution of the expression for μ from Tab. 3.2 reveals that the coefficients of the 4th order (and all higher) terms become exactly zero, reflecting exact correspondence between the basis function and the convolved functions.

The equation above for the degraded Sun $D(x,y)$, when convolved with a function $M(x,y)$ which gives the flux density distribution (on the image plane) of a point Sun as formed by

Table 3.3. Hermite functions and coefficients to order 5.

Order n,m	$H_n(\frac{x}{\sigma_x}) H_m(\frac{y}{\sigma_y})$	C_{nm}	Comments
0,0	1	1	normalized
1,0	$x/\sigma_x \cdot 1$	0	centralized
0,1	$1 \cdot y/\sigma_y$	0	centralized
$n+m=2$	$\neq 0$	0	standardized
$n+m=3$	$\neq 0$	$\neq 0$	but odd moments of S and G are zero
4,0	$(x^4/\sigma_x^4 - 6x^2/\sigma_x^2 + 3) \cdot 1$	$\mu_{40}^D/\sigma_x^4 - 3$	
3,1	$\neq 0$	$\neq 0$	see above
2,2	$(x^2/\sigma_x^2 - 1)(y^2/\sigma_y^2 - 1)$	$\mu_{22}^D/\sigma_x^2\sigma_y^2 - 1$	
1,3	$\neq 0$	$\neq 0$	see above
0,4	$1 \cdot (y^4/\sigma_y^4 - 6y^2/\sigma_y^2 + 3)$	$\mu_{04}^D/\sigma_y^4 - 3$	
$n+m=5$	$\neq 0$	$\neq 0$	see above

From the law of convolution for moments (3.13):
$\mu_{00}^D = \mu_{00}^S \mu_{00}^G = 1 \cdot 1$ by normalization of S and G
$\mu_{11}^D = 0$ as do all odd i, j due to symmetric Sun
$\mu_{20}^D = \mu_{20}^S + \mu_{20}^G = \sigma_u^2 + \rho_u^2 = \sigma_x^2$
$\mu_{02}^D = \mu_{02}^S + \mu_{02}^G = \sigma_v^2 + \rho_v^2 = \sigma_y^2$
$\mu_{40}^D = \mu_{40}^S + \mu_{40}^G + 6\mu_{20}^S\mu_{02}^G \approx \mu_{04}^D$
$\mu_{22}^D = \mu_{22}^S + \mu_{22}^G + \mu_{20}^S\mu_{02}^G + \mu_{02}^S\mu_{20}^G$

an error-free concentrator, provides the intensity distribution of the image of the finite solar disc formed (on an image plane) by the imperfect concentrator.

3.3.4 The Error Function for the Concentrator

The appropriate description of $G(x,y)$ depends on the application and on a number of other factors. A proper description can become quite complex, although at about the point the complexity becomes overwhelming one can often refer to the Central Limit Theorem and approximate the entire degradation function by a Gaussian, with little loss in accuracy. Still some care is required. Fortunately, the nature of large tracking collectors leads to a reasonably standard attribution of errors, from the microscopic level to the system level (Fig. 3.2).

Specularity. Surface reflector imperfections at the microscopic level are discussed in terms of loss of specularity. When a colinear incident beam (as from a laser) reflects from a surface, a fraction $R_{2\pi}(\lambda)$ will be reflected (hemispherical reflectance). Relative to the nominal reflected beam angle θ, each photon will be scattered by an amount $\Delta\theta$ due to local inhomogeneities, surface imperfections, etc. Experimentally [31] it has been found that the resulting beam profile may usually be reasonably approximated by a simple Gaussian:

$$R(\Delta\theta) \propto \exp(-\Delta\theta^2/2\sigma^2). \tag{3.15}$$

However, for rolled or polished substrates, a broader Gaussian distribution due to roll marks or polishing scratches is often superimposed. The resulting distribution is simply described by the sum of two Gaussians:

$$R(\Delta\theta) \propto R_1 \exp(-\Delta\theta^2/2\sigma_1^2) + R_2 \exp(-\Delta\theta^2/2\sigma_2^2) \tag{3.16}$$

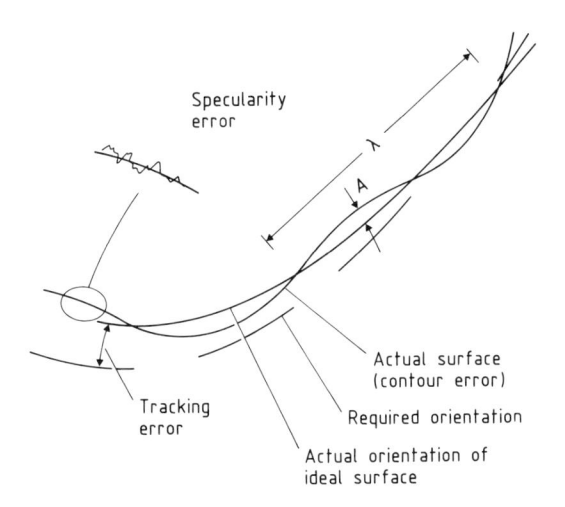

Fig. 3.2. A graphical representation of the several classes of mirror errors (discussed in the text).

with the constraint that, for each wavelength $R_1 + R_2 = R_{2\pi}(\lambda)$. While rolled stock gives somewhat different values for R and σ along axes parallel and perpendicular to the rolling marks, these differences are small, and probably not worth the added complexity. Representative values obtained by Pettit [31] for several typical solar materials are given in Tab. 3.4.

Table 3.4. Specular reflectance of mirrors (representative nominal values). $R_{2\pi}(S)$ is the solar averaged hemispherical reflectance, while R_1 and R_2 are tabulated for $\lambda \approx 500$ nm.

Material/Substrate	$R_{2\pi}(S)$	R_1	σ_1 (mrad)	R_2	σ_2 (mrad)
Polished Bulk Al Reflector:					
Alcoa Alzak, $\perp$ to roll	0.85	0.56	0.42	0.33	10.1
$\parallel$ to roll	0.85	0.62	0.29	0.27	7.1
Kingston Ind. Kinglux	0.85	0.66	0.40	0.22	17.0
Metal Fab. Bright Alum.	0.84	0.44	1.4	0.43	10.3
Silver Reflector:	all second surface				Support thickness
Float Glass	0.83	0.83	0.15		2.7 mm
Drawn Glass, $\perp$ to draw	0.90	0.92	0.40$^{(a)}$		3.35 mm
$\parallel$ to draw	0.90	0.92	< 0.05		
3M Scotchcall 5400	0.85	0.86	2.0		0.1–0.2 mm
3M FEK – 163	0.85	0.86	0.85		0.1–0.2 mm
3M ECP 300					
Stretched film	0.95	0.95	0.37		0.1–0.2 mm
Backing material$^{(b)}$					
• float glass		0.95	0.28 ± 0.02		
• painted Al		0.95	1.09 ± 0.22		
• bare Al		0.95	1.94 ± 0.60		

$^{(a)}$ away from striations, which produced a double image
$^{(b)}$ note 'print through' of surface roughness

Slope, or contour errors. Contour errors can arise from manufacturing errors, thermal or gravity effects on the support structure, design approximations to the correct optical surface, etc. Combined with ripples and distortions in the reflector substrate, these can cause sizable displacement of the surface from the desired location. However, as these displacements only introduce a small fractional change in the distance to the focal region, they can generally be ignored under the guidelines of concentrator optics. Small distortions causing slope errors, on the other hand, deflect the beam through a large lever arm (the focal distance) and must be considered carefully.

Slope or contour errors are measured relative to the ideal figure of the mirror. In this frame the ideal reflector would give a radius of curvature R of infinity. Typically, however the substrate introduces ripples having a 'wavelength' of 0.01 to 1 m. Frequently these can be approximated locally as being sinusoidal, with an amplitude A and wavelength λ. Thus, if $z = -A\cos(2\pi x/\lambda) = -A\cos kx$, the local slope is just $dz/dx = kA\sin kx$. The normal projected from a point x on the surface ($x \ll \lambda$) intersects the z axis at $(x, z) = (0, R)$, giving an effective radius of curvature equal to R. Thus, R can be written

$$R = \frac{x}{dz/dx} = \frac{1}{Ak}\frac{x}{\sin kx} = \frac{1}{Ak^2}\left(\frac{kx}{\sin kx}\right) \tag{3.17}$$

where $kx/\sin kx$ is just the reciprocal of the well known diffraction function which is 1 when $kx = 0$, decreases quadratically for small kx, and first passes through zero at $kx = \pm\pi$. Thus, the trough of each ripple approximates a parabolic cylinder with focal length $f = R/2 = 1/(2Ak^2)$, and aberrations for off-axis rays are defined by the diffraction function. This property explains the striations frequently observed in the near field of a mirror. Thus, a ripple of amplitude ± 1 mm and wavelength 1 m (or 0.3 m) will provide a focal *line* at 15 m (or 1.5 m).

Alternatively the ripples can be treated as a slope error. The amplitude is just $dz/dx = Ak\sin kx$ giving maximum slope errors of $\pm Ak$, or ± 6 mrad (± 18 mrad) in the above example. The associated probability distribution is $P = (k^2 A)^{-1}\sec kx$, which has a r.m.s. value $c = Ak/\sqrt{2}$. If one assumes no comparable effects in the y direction it is straightforward to calculate moments in the form of Tab. 3.2. Normalizing to $\mu_{00} = 1$ gives $\mu_{20} = c^2 = k^2 A^2/2$, $\mu_{40} = 1.5c^4$, and $\mu_{60} = 2.25c^6$; all other moments are zero to 6th order. Of course, there will be variations in the wavelength and amplitude from region to region on the mirror, so probable values of A and k should be used.

Due to reflection at the mirror surface the local beam deviation is twice the slope error. Thus, if one describes the surface contour error using a function with a second moment μ_{20} of σ_c^2 and consequently a standard deviation in the direction of the normal of σ_c, the standard deviation in the beam is $2\sigma_c$.

The local surface normal vector may be resolved into components perpendicular to ($\epsilon_\perp$) and lying in ($\epsilon_\parallel$) the optical plane (the plane containing both incident and reflected beams). Local slope errors are measured relative to the local surface normal and thus deflect the reflected beam away from the ideal impact point. To first order in $\epsilon_\perp$ and $\epsilon_\parallel$ (which are small), the resulting beam deflections relative to the error free reflected beam are [34]

$$\delta_\parallel = 2\epsilon_\parallel |\hat{i}\cos\theta + \hat{k}\sin\theta| = 2\epsilon_\parallel \tag{3.18}$$

and

$$\delta_\perp = 2\epsilon_\perp |\hat{j}\cos\theta + 0| = 2\epsilon_\perp \cos\theta. \tag{3.19}$$

In addition to the factor of 2 from reflection, these equations contain a rather non-intuitive result. In a plane perpendicular to the reflected ray the error pattern becomes an ellipse with the axis which projects out of the optical plane ($\delta_\perp$) foreshortened by cosine θ. In contrast, the image of a (flat) mirror itself is foreshortened along the other axis (the ordinary incidence angle cosine effect). In a first approximation, the combined surface errors, $\epsilon_\perp$ and $\epsilon_\parallel$, can be represented by their standard deviations, thus we have

$$\sigma_\parallel = 2\sigma_{\text{surface}\parallel} \tag{3.20}$$

and

$$\sigma_\perp = 2(\cos\theta)\sigma_{\text{surface}\perp}. \tag{3.21}$$

Tracking errors. Tracking errors are the final contribution to the error equation. For a single collector the instantaneous tracking error produces a well defined degradation in instantaneous performance. Note that if instantaneous performance of a single collector is of interest, as in a performance test, it is appropriate to determine the loss in performance associated with the measured tracking offset at each instant and not to include tracking as a random error. However, on a time average or system average basis, tracking errors produce a distribution function which will give an equivalent average performance. Although individual tracking errors (arising from on-off tracking, drift, error in aim point, etc.) are less likely to be intrinsically Gaussian, the average over the many collectors in a system, and over time, assures that a Gaussian approximation represented by σ_T will be a good r.m.s. representation of the error function.

The correct way to incorporate σ_T into the overall error expression depends on the details of the specific system involved. *Heliostat systems or fixed-receiver point focus*: the reflected beam is deviated by $2\sigma_T$. *Monolithic point focus*: as above, but the receiver is supported by the collector and so moves in the same direction as the collector by an amount σ_T, so the effective beam deviation is only σ_T. *Line focus*: one must carefully apply the above arguments; in addition one must distinguish between deviations parallel to the tracking axis ($\sigma_{T\|}$) and perpendicular to it ($\sigma_{T\perp}$).

Typically, line focus systems track the Sun only about their cylindrical axis. As the Sun moves, foreshortening of the aperture by $\cos\theta_\|$ is accepted. However, parabolic trough optics is such that the focal position remains at the receiver, even as the slant range from the reflector to the receiver is elongated by $1/\cos\theta_\|$. The focal spot is enlarged by this projection, but the focus is not destroyed. The effect of the tracking error is also enhanced. Thus, in the $\perp$ direction, the deviation is 2 (or 1 for a monolithic system) $\sigma_{T\perp}$ while in the parallel direction it is 2 (or 1) $\sigma_{T\|}/\cos\theta_\|$. This phenomenon suggests that the effects of contour errors will be similarly affected, and they are. The peculiar fact is that this $1/\cos\theta_\|$ effect exactly cancels the $\cos\theta_\|$ reduction in the beam deviation due to contour errors described earlier. Thus, for line focus collectors the algorithm for contour errors becomes 2 (or 1) σ_T.

Combining beam errors. We have described beam errors arising from several sources, and each type of error itself is usually due to many individual contributions. Properly, one should generate the convolution of all the individual errors to finally convolve with the solar distribution to obtain the final distribution of the degraded Sun. In fact, under the conditions above, the Central Limit Theorem allows us to attribute a Gaussian shape to the sum of the errors added in quadrature. Thus, we have:

$$\sigma_B^2 = \sigma_{\text{beam}}^2 = \sigma_{\text{specular}}^2 + \sigma_{\text{contour}}^2 + \sigma_{\text{mechanical}}^2 + \sigma_{\text{tracking}}^2 + \sigma_{\text{receiver}}^2 \qquad (3.22)$$

by paying careful attention to factors of 2 and $\cos\theta$ in the definition of the individual beam error contributions as described above, σ_B is appropriate for monolithic or fixed-receiver systems, for $\|$ or for $\perp$ components, and for line or for point focus systems.

Then $D = S * B$ describes the degraded Sun, so to the lowest order $(2, 0$ or $0, 2)$,

$$\sigma_D^2 = \sigma_S^2 + \sigma_B^2. \qquad (3.23)$$

3.3.5 Flux Density and Concentration

The concepts of image formation and phase coherence characteristic of imaging optics are abandoned in concentrator optics. They are replaced with concern about flux density levels and concentration ratios. Each element of the concentrator contributes to the flux density on an appropriate area of the receiver. To determine the flux density at any point it is thus necessary to sum over all elements of the concentrator, taking full account of the degraded

Sun, of the local angle of incidence on the receiver and of the displacement of the aim point of each element relative to the computation point. Further complications arise, such as vignetting (blocking the field of view by an aperture in front of the absorbing surface) as in a cavity receiver, or additional reflections if a terminal concentrator is present. In the simplest cases, reasonable approximations to the receiver flux density distribution can be achieved in analytical form, but the functions are usually evaluated on a computer even then. In general, methods of computer ray tracing are combined with a convolution of the degraded Sun with moments of the ray-traced mirror images in order to generate a flux density distribution. Such methods allow determination of the peak flux density (and its amelioration by special aiming or distortion of reflector elements). While peak flux densities of several hundred Suns may be achieved in line focus systems and several tens of thousands of Suns in point focus systems, more typical values are 50 Suns and 2500 Suns. One Sun corresponds to the local value of direct-beam insolation, thus 'Suns' is really a measure of concentration rather than flux density. Thus, a peak concentration ratio of 50 means that the peak flux density is 50 Suns, so if the direct beam insolation is 800 W/m², the peak flux density is 4000 W/m². An optimistic but simple conversion is to substitute kW/m² for *Suns*.

Several terms are commonly used:

- Geometric concentration ratio:
$$C_g = \frac{\text{collector aperture}}{\text{receiver aperture}}. \qquad (3.24)$$

There are several caveats: 1) The receiver aperture must be just large enough to intercept most of the flux. Otherwise arbitrarily large (or if the receiver is oversized, small) fictitious geometric concentration ratios can be quoted. 2) The collector aperture is really the effective aperture, after accounting for open or opaque spaces between mirrors, cosine effects on the aperture, etc. 3) The geometric concentration ratio as defined is actually an average concentration over the receiver. 4) Beam absorption (in reflector, atmosphere, etc.) between incidence and receiver is not accounted for.

- Flux density concentration ratio:
$$C_f = \frac{\text{flux density at receiver}}{\text{direct beam insolation}}. \qquad (3.25)$$

Note that in contrast to C_g, C_f incorporates the effects of losses such as mirror absorption, atmospheric attenuation of reflected beams etc., and varies from point to point on the receiver.

3.3.6 Cassegranian Optics

In many cases it is advantageous to locate the receiver nearer to the collector than one focal length. For parabolic primaries this reduces the assembly moment of inertia and may even allow a heavy or complex receiver to be ground mounted while the collector tracks around it. The tall tower of central receivers is an immediate target for shortening, with significant savings in pumping power and piping as additional incentives. The Cassegranian design allows for this configuration, but there are penalties and difficulties which have so far prevented practical use of this concept in solar applications.

The obvious penalty is the added reflection. The associated 5–10% loss must be recovered by a savings of 5–10% of the system cost above the cost of the reflector itself. An obvious difficulty is that for a practical design the area of the secondary reflector must be much (5–100 times) smaller than that of the primary or the design makes no sense at all[5]. The difficulty

---

[5] The exception is a solar furnace such as those at Odeillo, France, where cost is secondary to performance. Then a flat primary is used in the true sense of *heliostat* and the beam is directed horizontally or vertically into a stationary parabolic secondary.

3.3 Concentration Optics

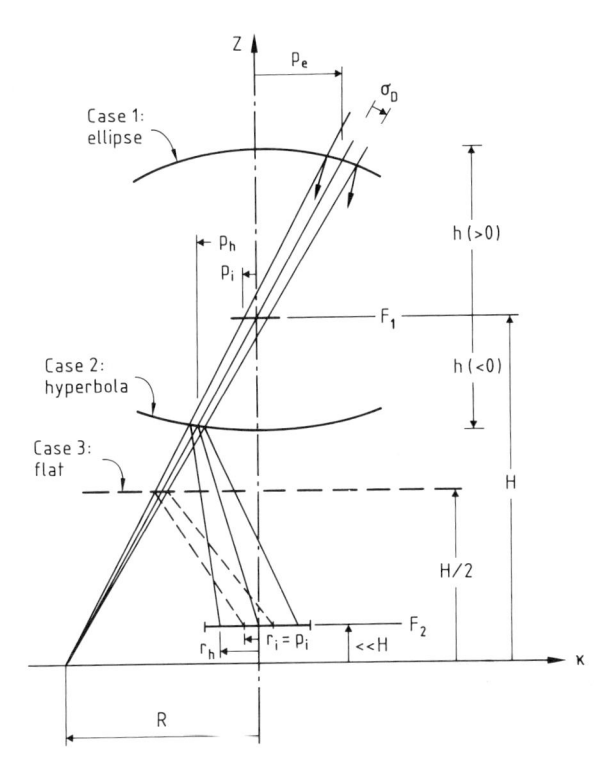

Fig. 3.3. Secondary mirror geometries. The aim point at H serves as one focus (F_1) of the ellipse or hyperbola, the other focus being at the receiver located near the ground (at F_2).

is that the secondary must now tolerate a flux density of 5–100 Suns with almost normal incidence. Aside from photodegradation effects, a 10% reflection loss requires the reflector to dissipate 0.5 to 10 kW/m², resulting in heating, thermal stress, thermal deterioration, and perhaps a need for active cooling. Thus the cost goes up, especially as the surface must retain nominal optical imaging quality or the image, and thus receiver aperture (and thermal loss), will increase in size.

Now consider the geometrical optics involved. We will restrict ourselves to point focus concentrators, but the extension to line focus is trivial. There are three cases (Fig. 3.3): 1) An elliptic reflector beyond the focal point. 2) A hyperbolic reflector below the focal point, i.e. between the primary and its focus. 3) A flat reflector with it and the receiver placed so that the optical path length is unchanged.

Case 1. This is of no interest as it requires a support even longer than the focal length and increases the optical path, and hence image size, without any compensating advantage.

Case 2. This is the conventional Cassegranian geometry. The primary continues to aim all central solar rays at the focus F_1, a distance H above the primary. The reflector, a hyperbola of revolution a distance h below F_1, re-images the central rays at F_2, which is characteristically in or near the plane of the primary. Thus, the (virtual) image distance is h, the object distance is $H - h$, and the linear magnification is $M = (H - h)/h = (A_2/A_1)^{0.5}$. The secondary must intercept the entire cone of rays from the primary, so its radius is at least $p_h = Rh/H$. Thus, if the secondary mirror is much smaller than the primary, the secondary image is proportionally larger than the primary image. When $h = H/10$, the magnification of the final image due to the secondary is 81, and the area of the secondary is 1/100 that of the primary. Other values are $H/4$, 9, 1/16; and $H/2$, unity and 1/4, corresponding to a flat secondary which simply folds the upper half of the converging cone of rays back to the plane of the primary. The large magnification of most *reasonable* designs implies very low concentration ratios and low receiver flux densities at the receiver located at F_2 for most of these designs. A terminal concentrator can regain some of the lost concentration but at a further cost in material and reflector losses.

Case 3. As in the Case 2, a flat secondary does not affect the flux density distribution on the receiver, other than to reduce it 5–10% due to absorption, and increase the dispersion somewhat. Thus, a flat secondary can be inserted at any height ($l < H = F_1$) above the primary so long as the receiver is elevated to $H - l$, keeping the optical path length the same. The secondary radius is given by the larger of $2\sigma_B l/\cos^2\psi$ (to intercept the entire beam) or $Rh/H = R(H - l)/H$. Thus, a moderately large flat mirror must be supported l m above the primary but the receiver can be relatively near the ground.

Although often recommended, when subjected to a comprehensive design and costing analysis, all such concepts have failed to date.

3.4 Ideal Concentrators

3.4.1 Conceptual Framework

Each point on the surface of the Sun radiates uniformly into 2π steradians (sterad). As discussed in Chap. 2, the aperture of an ideal concentrator will intercept and redirect a portion of the (nearly parallel) light reaching the earth in such a way as to fill its exit aperture with light diverging into 2π sterad. Neglecting atmospheric attenuation, reflection losses, etc. the resulting aperture brightness will equal that of the Sun, in keeping with the Second Law of Thermodynamics. As shown in Chap. 2, the concentration ratio required to accomplish this is just:

$$C_{\text{ideal}}\left(1, \frac{\pi}{2}, \alpha\right) = \frac{1}{f_s} = \left(\frac{D_{\text{ES}}}{R_S}\right)^2 = \left(\frac{1}{\sin\alpha}\right)^2 = 46,200. \quad (3.26)$$

It is interesting to note that if the terminal concentrator and its aperture is filled with a material with index of refraction n (relative to 1 for the vacuum surrounding the Sun) the entering light will be refracted into a narrower cone angle [42]. Thus, according to Snell's law $\sin\alpha = n\sin\alpha'$ and $\sin\alpha'$ is the effective divergence angle so

$$C_{\text{ideal}}(n, \frac{\pi}{2}, \alpha) = \left(\frac{n}{\sin\alpha}\right)^2 = n^2 C_{\text{ideal}}\left(1, \frac{\pi}{2}, \alpha\right) \leq 46,200\, n^2. \quad (3.27)$$

More generally, if the concentrator accepts radiation incident on its aperture within a cone half-angle of θ_a and funnels the radiation through to its exit, where it fills a cone of half-angle θ_e in a material of relative index n, the concentration ratio is:

$$C_{\text{point}} < C_{\text{ideal}}(n, \theta_e, \theta_a) = \left(\frac{n\sin\theta_e}{\sin\theta_a}\right)^2. \quad (3.28)$$

The same logic applied to line focus concentrators results in:

$$C_{\text{line}} < \sqrt{C_{\text{ideal}}} = \left(\frac{n\sin\theta_e}{\sin\theta_a}\right) \leq 215\,n \quad \text{for sunlight.} \quad (3.29)$$

These relations have many ramifications, presuming the requisite concentrator geometry can be identified. In Tab. 3.5 we show C_{ideal} for a number of cases. It is important to note that to achieve the indicated concentration ratio, the entire aperture must be illuminated uniformly over the entire range of angles $\leq \theta_a$, and the concentration measured within the dielectric (if $n > 1$).

The first 6 cases represent single stage ideal concentrators showing the high concentrations theoretically possible. They also demonstrate that limiting the exit aperture to 60°, to reduce incidence angle losses on an absorber, does not reduce the concentration markedly. However increasing the acceptance angle to accommodate a degraded Sun does. The last 5 cases reveal

Table 3.5. Examples of ideal non-imaging concentrators, θ_e = exit aperture, θ_a = entrance aperture half angle.

Case	1	2	3	4	5	6	7	8	9	10	11	12
n	1	1.5	1.5	1	1	1	1	1.5	1	1	1	1.5
θ_e	90	90	90	90	60	60	90	90	90	90	90	90
θ_a	0.266	0.266	0.53	0.53	0.53	1	14.5	14.5	30	45	60	30
C_{point}	46,200	104,000	26,000	11,550	8,660	2,430	16	36	4	2	1.33	9
C_{line}	215	322	161	107	93	49	4	6	2	1.4	1.15	3

the performance of second stage terminal concentrators. Here the entrance aperture has been enlarged to accommodate a converging beam, e.g. from a parabolic dish of rim angle θ_a. The concentration of the ideal concentrator is no longer very large, but it multiplies the concentration ratio of the first stage. Recall that parabolas with small rim angles (such as 15°–30°) have very small coma or cosine effects at the focal zone. Thus, they can produce a nearly ideal image of the Sun, degraded primarily by their contour errors etc. but not by aberrations. Recently Winston has coupled such a dish to a nearly ideal terminal concentrator, like case 8, and achieved an overall concentration ratio of 60,000 within the dielectric at the exit aperture, including the effects of losses.

Two special geometries have been developed which can achieve the ideal concentration for line focus and approximate it in three dimensions.

3.4.2 Compound Parabolic Concentrator

By forsaking the idea of image formation entirely it is possible to approach the thermodynamic limit of concentration discussed in Sect. 2.3. The resulting concentrators are often called *ideal*, although this term is better reserved for the mathematical idealization of (3.26) and (3.27). The compound parabolic concentrator, usually referred to as a CPC, is ideal in the 2-D case, and nearly so in 3-D.

The two-dimensional (line focus) CPC geometry is derived [42] by requiring that all rays incident on the aperture at the limiting acceptance angle are reflected directly to the edge of the exit aperture A (or B) (Fig. 3.4). This defines a section of a parabola BC (or AD) with focal point at A (or B) and axis inclined at the angle θ_D (or $\theta_C = -\theta_D$). Any rays incident on the aperture with a smaller angle either pass directly through the exit aperture or are reflected to reach the exit aperture at a smaller displacement than A, and so will pass through the aperture. In contrast, all rays incident at a larger angle bounce back and forth between the reflectors until they are finally rejected or absorbed.

Thus, collimated light incident along θ_D (θ_C) is focused to a line at A (B), producing a relatively high flux density along that line, even for low concentration ratio collectors. Light with intermediate angles of incidence produces aberrated foci between A and B. Only when the entrance aperture is uniformly illuminated, both in space and in incident angle, is maximum concentration ratio obtained, under which condition the exit aperture is uniformly filled with Lambertian radiation, i.e. light diverging into the full 2π steradian solid angle. Clearly the absorber must be designed with these properties in mind.

For small acceptance half angles the height of the parabolas becomes many times the aperture, resulting in very poor use of the expensive mirror area, as shown in Fig. 3.5 [34]. As suggested by the figure, the concentration ratio (for a uniformly illuminated aperture) is little affected by eliminating the top portion of the parabolas. In fact reducing the reflector

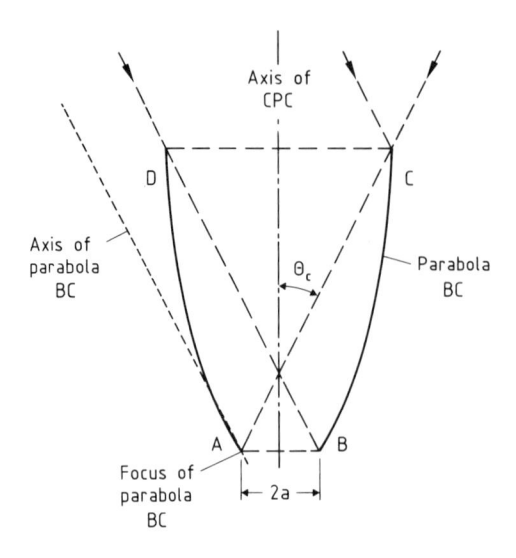

Fig. 3.4. Compound parabolic concentrator. Exit aperture is $\overline{AB}$, entrance aperture is $\overline{DC}$, and the acceptance angle $\theta_a = |\theta_c|$.

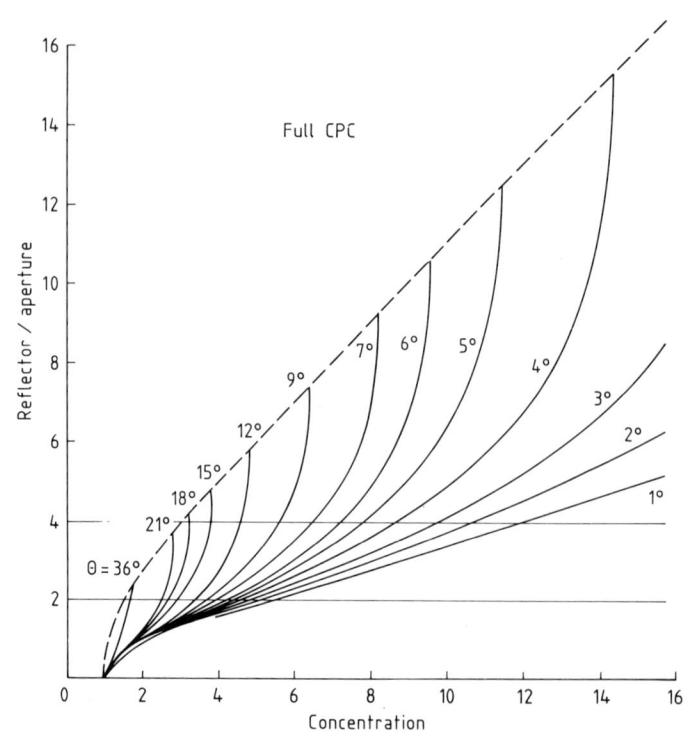

Fig. 3.5. Reflector/aperture ratio as function of concentration for full and for truncated CPCs θ = acceptance half angle.

area by half reduces the concentration ratio by only about 15%. It can be shown that skew-rays, i.e. those with a component of velocity along the cylinder axis, are effectively focused

3.4 Ideal Concentrators

by a parabolic cylinder. Thus, the above discussion shows that CPCs are indeed maximally concentrating, or ideal, 2-D concentrators. Because the skew-ray argument does not apply for 3-D CPC's (formed by rotating the off-axis parabola about the vertical axis) they do not quite achieve maximal concentration ratio, but for most applications the deficit is ignored.

Other CPC-like secondaries have been studied for optimal coupling to specific absorber shapes, such as cylinders or fins [34,42]. Such devices with relatively large acceptance half angles can be used very effectively to enhance the performance of parabolic cylinder concentrators employing tubular collectors. Parabolic dish concentrators, in contrast, typically employ cavity receivers and the standard CPC of revolution can be quite effective, especially for dishes where a typical rim angle is 45° (where $C_{\text{cpc}} = 1/\sin^2 45° = 2$). The cavity must, of course, be designed to handle the Lambertian radiation passed by the aperture, and the CPC must be designed to accommodate the peak flux density from the parabola during *walk-off* events such as start up or loss of tracking power.

While we have continued to use the term CPC in the last paragraph, the appropriate ideal concentrator shape for use with a finite source, such as a parabolic primary concentrator, consists of a pair of ellipses (the parabola is simply an ellipse, one of whose foci is 'infinitely' far away). Each ellipse is defined by applying the edge ray principle [42]. Thus, (Fig. 3.6) light leaving V (V') at one end of the virtual source (i.e. one edge of a parabolic primary) is reflected by an elliptical surface BC (AD) to the edge of the absorber A (B) provided both points are foci of the ellipse BC (AD). The useful section of the ellipse extends from the edge of the absorber A to the point where it is intersected by the direct ray from V to B. The exit aperture of the compound elliptical concentrator (CEC) will be filled with Lambertian radiation provided every point on VV' illuminates all of CD.

All terminal concentrators introduce an added system loss due to absorption by the reflective surface. In [34], the average number of reflections $\langle n \rangle$ for various CPCs has been computed, and it was suggested to compute the loss approximately by $\rho \cdot \langle n \rangle$, with $\langle n \rangle$ estimated as $1 - 1/C$ and $\rho = \langle \rho(\theta) \rangle$. With reasonable truncation and full illumination of the acceptance angle by a primary concentrator with a rim angle exceeding 10°, this approximation is quite reasonable as many rays pass directly through the aperture without reflection, especially for small C.

CPCs and CECs are designed to be mounted with their inlet aperture replacing the receiver. This means the actual receiver is moved back from the focal plane so its entrance aperture coincides with the exit aperture of the CPC. In contrast, the hyperbolic concentrator (next section) is mounted in front of the focal point so the receiver location is not changed, only its aperture is reduced by the concentration factor.

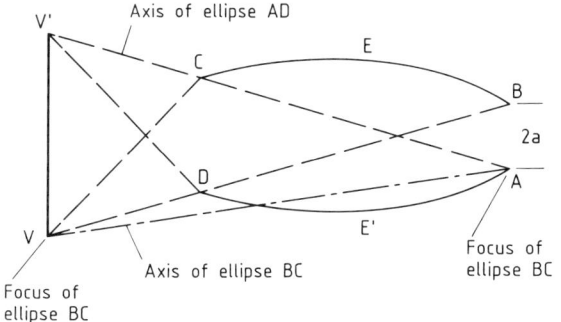

Fig. 3.6. A compound elliptical concentrator for the object VV' at a finite distance.

3.4.3 Flow-line or Trumpet Concentrators

Recently a new class of ideal nonimaging concentrators was discovered [41,43]. They are based on a new concept, the geometrical vector flux density J, which follows from the conservation of Etendue, or throughput, in a loss free optical system. The components of J are comparable to the configuration factors of radiation transfer theory. Consider the two principal cases of interest, a line absorber (or Lambertian radiator) of width $2c$, or a disc absorber (or Lambertian radiator) of diameter $2c$ lying in the X, Y plane (Fig. 3.7). For the strip absorber, lines of constant vector flux density are found to lie on confocal hyperbolae given by

$$y/a = \pm [1 - z^2/(c^2 - a^2)]^{\frac{1}{2}}. \tag{3.30}$$

The strip, $y \pm a$, at $z = 0$ represents the exit aperture of the *flow-line concentrator* formed by replacing the flow line defined by c and a with a mirrored surface. The equivalent case for the disc is found by rotating the same hyperbolae about the z axis.

The asymptotes of the flow-line concentrator are defined by

$$\tan \theta_a = (dy/dz)_\infty = a(c^2 - a^2)^{-1/2}. \tag{3.31}$$

From this it follows immediately that $\sin \theta_a = a/c$ and the flow line concentrator is ideal, as

$$C_{\text{ideal}} = \quad 1/\sin \theta_a = c/a \quad \text{line focus} \tag{3.32}$$
$$C_{\text{ideal}} = \quad 1/\sin^2 \theta_a = (c/a)^2 \quad \text{point focus} \tag{3.33}$$

which is exactly the geometric concentration of the specified geometry, ignoring reflective losses or truncation.

The properties of this concentrator are that all light, initially directed at the strip (or disc) $\pm c$ and within the cone of the reflector, will finally pass through the exit aperture $\pm a$. Thus, the strip (or disc) cc' forms a virtual source. Points c and c' are the foci of the hyperbolae having vertices at a and a'. The complete hyperbolae extend to infinity, but if truncated at

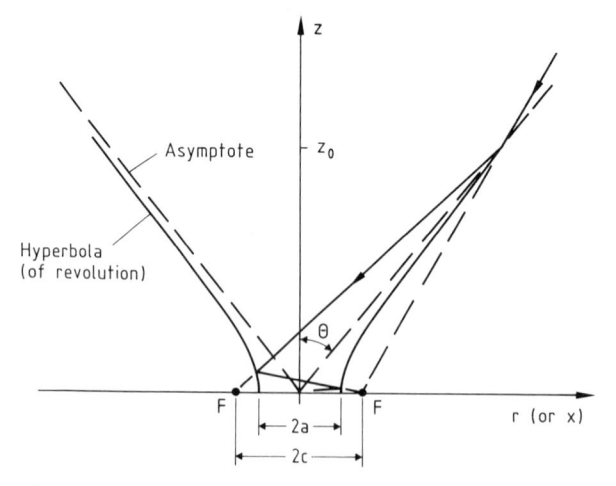

Fig. 3.7. Flow line (trumpet) concentrator truncated at $z = z_0$. The asymptotes to the hyperbolic surface have a half angle θ relative to the z-axis. For the hyperbola, $1/\sin\theta = C/a$ so the concentration is C/a (or C^2/a^2 for a hyperboloid of revolution). Flux from a trough (dish) of rim angle $\psi_r \leq \theta$ and aimed between $\pm C$ will be concentrated on to $\pm a$.

$z = z_0$ an aperture of width (or diameter) $2S$ is formed. Most rays entering this aperture well within $\pm S$ and well within the acceptance angle pass directly through the exit aperture without any reflections. On the other hand, the edge rays originally directed at c or c', i.e. at the foci, are reflected toward the other focus repeatedly until after infinitely many reflections they reach the exit aperture at grazing incidence. Less extreme rays suffer fewer reflections and fill the 2π sterad required at the exit aperture of an ideal concentrator. More extreme rays suffer multiple reflections until their z component of momentum is reversed and they are returned to the aperture.

To avoid the poor performance of the edge rays, the asymptotes are usually set a few degrees larger than required by the actual source. Thus, the concentration ratio is reduced somewhat, but losses due to multiple reflections are also. The value of z_0 at which to truncate is also of interest. Clearly, when used as a secondary, one should truncate when the flux density on the mirrored surface of any element is exceeded by that intercepted on its back. In practice material and structural costs are even more restrictive and will require a substantial gain in collected flux to justify each element of the concentrator. In higher concentration applications the exit aperture, i.e. the throat of the trumpet-shaped hyperbolic concentrator, may require cooling, because of absorption of solar and also of IR radiation from the hot absorbing surface of the receiver.

Light concentrated by a well-designed flow-line concentrator will typically suffer fewer reflections than for a CPC because more rays pass directly through the aperture. An added advantage is that the flow-line concentrator is simply inserted in front of the receiver. Thus, it does not lengthen the optical path nor cause an increase in the support lengths or moment of inertia of the receiver. While the maximum diameter of the hyperbolic secondary is larger than that of a CPC, it can usually be constrained to that of the cavity receiver or engine already at the focal point so there is in fact no additional shadowing effect. If not, the above truncation considerations apply.

3.4.4 Conical Flux Density Redirector

A simple alternative to the ideal nonimaging flow-line concentrator was proposed by Brumleve in 1972 [36]. Although not an ideal concentrator in the Winston sense, the function of the conical flux density redirector (CFR) is much the same as that of a hyperbolic (trumpet) concentrator.

The CFR is designed to overcome one of the primary problems encountered when using a flat plate or cavity receiver. For larger rim angles (exceeding $\approx 45°$) the edge rays are greatly elongated when projected onto the aperture. The beam diameter to the 90% point is about $4\sigma_D L$ where $\sigma_D = (\sigma_S^2 + \sigma_B^2)^{1/2}$, and the slant length L is $2f/(1 + \cos\psi)$ for a parabola and $f/\cos\psi$ for a Fresnel reflector. The projected solar image thus produces an ellipse having a well behaved minor axis of $4\sigma_D L$, but the major axis is elongated to $4\sigma_D L/\cos\psi$, which is twice the minor axis for a 60° acceptance angle and 4 times for 75°. To accommodate this image, the optical figure of the primary (or aim point of the Fresnel facet) may be adjusted so the lower edge of the beam just enters the far side of the aperture of diameter $2\sigma_D f$. A conical reflector with exit aperture diameter $2\sigma_D f$ is then placed to fold the upper portion of the beam into the aperture. So long as the aim point remains within the aperture, the concentration ratio gain due to such a device is $C_t/C_0 = 4/(1 + \cos^2\psi)^2$.

Clearly the CFR enhances the concentration ratio at higher rim angles and thus moves the rim angle for maximum concentration ratio to higher primary rim angles. For rim angles exceeding 60° the cone becomes excessively long and must be truncated at 10–15% of the collector radius [3]. Because only the Gaussian tails of most rays reach the cone, reflection losses and the associated heating are not serious problems.

The same design logic applied to a cylindrical receiver results in a parasol-like configuration with the CFR extending downward from the top of the receiver. This configuration has been developed in some detail [35]; it was shown that the parasol CFR performs better than the simple CFR for primaries with acceptance half angles exceeding 55°, and is competitive between 50° and 55°. Essentially the parasol allows a reduction of cylindrical receiver length by folding the upper edge of the beam horizontally.

3.5 Parabolic Geometries

3.5.1 General Considerations

The simplest and most common solar concentrator consists of a parabolic reflector. As the definition of a parabola provides that all paraxial rays striking a parabolic reflector are reflected toward the focus, a parabola is an imaging device and consequently provides a lower concentration ratio than is possible with the ideal concentrators of Sect. 3.4. However, the parabolic shape is straightforward to analyze, easy to build, conservative of material, provides adequate concentration ratio for many applications, and can be supplemented with an 'ideal' terminal concentrator when required. As a consequence, it is the most commonly used solar concentrator.

First, let us define the parabolic geometry. Let z point toward the Sun. Then a parabola in the x-z plane is defined by (Fig. 3.8)

$$z = (x^2/4f) + cf \qquad (3.34)$$

where f is the focal length of the parabola and c is determined by the choice of coordinate system origin:

$$c = +1 \quad \text{origin at directrix}$$
$$c = 0 \quad \text{origin at vertex}$$
$$c = -1 \quad \text{origin at focal point.}$$

The preferred coordinate system puts the focal point at the origin. Thus at $z = 0$ we have $x^2/4f = f$, so ($x = \pm 2f$, $z = 0$) defines the points on the rim of a 90° rim angle collector.

A parabolic cylinder (trough, 2-D geometry) is generated by translating this parabola along the y axis. A paraboloid of revolution results if the parabola is rotated about the z axis. Paraboloids are commonly referred to as dishes or three-dimensional (3-D) collectors. In our discussion the following relations based on the properties of a parabola (Fig. 3.8) will be useful. We direct the z axis toward the Sun and place the origin at the focus ($c = -1$). Then, taking advantage of the polar coordinate system, we have the relations in Tab. 3.6.

As we will see in Chaps. 5 and 8, concentration ratio plays a dominant role in performance analysis for solar collectors. For a parabolic collector there are two basic approaches. The simpler one leads to an average, or geometrical concentration ratio, while a more detailed analysis provides the actual flux density distribution over the receiver.

3.5 Parabolic Geometries

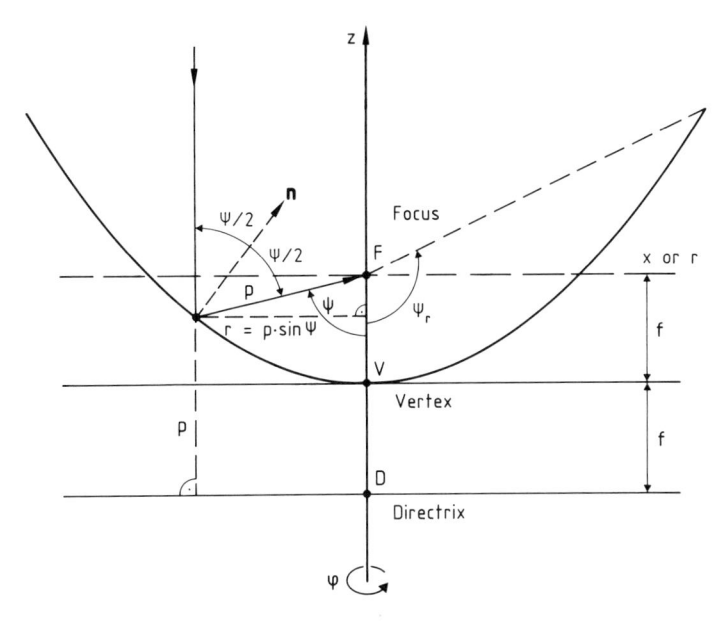

Fig. 3.8. Parabolic geometry. ψ is the cone half angle, ψ_r is the rim angle or aperture angle, and n is the local normal to the surface.

Table 3.6. Geometric relationship for 2-D and 3-D parabolas. In this table, $\theta_\parallel$ is the out-of-plane or skew angle of the incoming ray, p is the projection distance from a point on the mirror surface to the focal line (or point), ψ_r is the aperture angle, and f/D corresponds to the standard numerical aperture, or f-number.

2-D cylindrical geometry y-axis is focal line		3-D spherical geometry z-axis is axis of paraboloid	
p	$= p_\perp / \cos\theta_\parallel$	r^2	$= x^2 + y^2$
x	$= p_\perp \sin\psi$	r	$= p\sin\psi$
z	$= -p_\perp \cos\psi$	z	$= -p\cos\psi$
$z + f$	$= x^2/4f$	$z + f$	$= r^2/4f$
p	$= (x^2 + y^2 + z^2)^{1/2}$	p	$= (r^2 + z^2)^{1/2}$
	$= (4f(z+f) + y^2 + z^2)^{1/2}$		$= (4f(z+f) + z^2)^{1/2}$
	$= ((2f+z)^2 + y^2)^{1/2}$		$= 2f + z$
$p_\perp$	$= p\cos\theta = z + 2f$	p	$= 2f - p\cos\psi$
	$= 2f/(1 + \cos\psi)$		$= 2f/(1 + \cos\psi)$

ψ_r	0°	30°	45°	60°	90°	135°
$p_\perp/f$	1.00	1.07	1.17	1.33	2.00	6.83
f/D	1/0	1/1.07	1/1.67	1/2.3	1/4	1/9.67

3.5.2 Geometric Concentration Ratio

The geometric concentration ratio, defined in (3.24)

$$C_g = \frac{\text{collector aperture}}{\text{receiver area}} \qquad (3.35)$$

implies that the receiver intercepts nearly all the flux density (say 95%) redirected by the collecting aperture. Further, thermal loss analysis assumes that *receiver area* encompasses all exposed heated surfaces, such as the back of a flat absorber unless it is adequately insulated (which is usually the case) or the entire circumference of a tubular absorber. As there are several types of collectors and several common receiver shapes, we have tabulated the relevant relationships in Tab. 3.7.

Thus, for a 3-D parabolic dish with circular facets, a flat receiver, and a degraded Gaussian Sun with $\sigma_{\text{optical}} = 5$ mrad, we get from Tab. 3.7 and Fig. 3.8,

$$C_g = \frac{A_C}{A_R} = \frac{\pi p_r^2 \sin^2 \psi_r g^2}{\pi(p_r \sin \sigma_D)^2 / \cos^2 \psi_r} \qquad (3.36)$$

$$= \left(\frac{g \sin \psi_r \cos \psi_r}{\sin \sigma_D} \right)^2 \qquad (3.37)$$

where σ_D is to be evaluated using (3.23), noting that $\sigma_{\text{beam}} = 2\sigma_{\text{optical}}$.

The rim angle giving maximum concentration ratio occurs when $dC_g/d\psi = 0$, thus

$$\frac{1}{C_g} \frac{dC_g}{d\psi} = \frac{2 \cos \psi}{\sin \psi} + \frac{2(-\sin \psi)}{\cos \psi} = 0 \quad \text{at} \quad \psi = 45° \qquad (3.38)$$

and inserting 45° and the appropriate values for the example above,

$$C_g(45°) = \frac{0.9}{10^{-6}} \frac{(0.71)^2 (0.71)^2}{(4.65)^2 + (10.0)^2} = 1860. \qquad (3.39)$$

It is most important to note that this is not a system (or even a subsystem) optimum, but only the maximum value. Considerations such as are described in Sect. 3.8 must be used to define the optimum rim angle in all cases.

If Fresnel optics were used, the primary difference would be that p_r is about 20% larger, resulting in collector facets distributed over a larger area and requiring a larger receiver (1.44 times). The significant factor affecting C_g would be that the mirror fill factor would be decreased by a factor $g^2 \approx 0.64$, resulting in $C \approx 1200$. The larger receiver will have higher losses and a greater cost, but perhaps the Fresnel mirror will provide sufficient cost reduction or allow sufficient scale up in size to overcome these disadvantages.

Other configurations are analyzed similarly with reference to (Tab. 3.7). Note that f, p_r and l all cancel out in the final expression for C_g. Thus, combining terms we have for a flat receiver

$$C_{3D} = \left(\frac{C_{2D}}{g_l} \right)^2 = \left(\frac{g \sin \psi_r \cos \psi_r}{\sin \sigma_D} \right)^2 ; \quad \psi_{\text{max}} = 45° \qquad (3.40)$$

$$C_{3D}(45°) = \left(\frac{0.5g}{\sin \sigma_D} \right)^2. \qquad (3.41)$$

Similarly, for a cylindrical or spherical receiver

3.5 Parabolic Geometries

Table 3.7. Relationships for geometric concentration ratios.

$$C_g = \frac{\text{effective collector aperture}}{\text{exposed receiver area}} = \frac{A_C}{A_R}$$

Collector aperture A_C at normal incidence:

$$\begin{aligned} A_C &= l(2p_r \sin \psi_r)g_l g & \text{for 2-D geometry} \\ &= \pi(p_r \sin \psi_r)^2 g^2 & \text{for 3-D geometry} \end{aligned}$$

where l is the length of cylindrical collector, ψ_r is the half angle of rim, and p_r is the projection distance from rim of mirror to the receiver. The mirror-to-receiver projection distance p is

$$\begin{aligned} p &= 2f/(1 + \cos \psi) & \text{for parabolic geometries} \\ &= f/\cos \psi & \text{for flat Fresnel geometries} \end{aligned}$$

g_l, g, g^2 are mirror fill factors (useful area/aperture area), which become

$$\begin{aligned} g^2 &= 1 & \text{for continuous parabolas} \\ g^2 &= 0.90 & \text{for circular facets} \\ g^2 &= 0.78 & \text{for circular facets, chess board packing} \\ g &\approx 0.8 & \text{for 45° rim angle Fresnel} \\ g &\approx 0.5 & \text{for 75° rim angle Fresnel} \\ g_l &= 1 - f \tan \theta / l & \text{to account for end effects in 2-D geometry} \end{aligned}$$

Receiver area A_R to intercept the reflected light:

$$\begin{aligned} A_R &= 4lr/\cos \psi_r & \text{for 2-D uninsulated flat receiver} \\ & 2lr/\cos \psi_r & \text{for 2-D back-insulated flat receiver} \\ & 2\pi rl & \text{for 2-D cylindrical receiver} \\ & \pi r^2 / \cos^2 \psi_r & \text{for 3-D flat or cavity receiver} \\ & 4\pi r^2 & \text{for 3-D spherical receiver} \end{aligned}$$

where r is the effective radius of the solar image projected to the receiver. The value of r becomes:

$$\begin{aligned} r_0 &= f \sin \xi & \text{for paraxial rays} \\ r_r &= p_r \sin \xi = 2f \sin \xi / (1 + \cos \psi_r) & \text{for edge rays.} \end{aligned}$$

ξ is the divergence angle assumed for the reflected solar beam. Edge ray interception, η^*, for some characteristic values of ξ is:

ξ	ξ-Value	η^*_{2D}	η^*_{3D}	Comment	
				(Sun)	(optics)
α_s	4.65 mrad	1	1	pillbox	perfect
σ_s	$\approx \alpha_s/2$	0.608	0.25	pillbox	perfect
σ_s	$\approx \alpha_s/\sqrt{2}$	0.683	0.608	Gaussian	perfect
$2\sigma_s$	$\approx \alpha_s$	0.954	0.87	Gaussian	perfect
σ_D	$(\sigma_s^2 + \sigma_B^2)^{0.5} \approx \alpha_s$	0.683	0.608	Gaussian	real ($\sigma_B = 0.71\alpha_s$)

The integral over the collector will be much better as $r < r_r$ for all but edge rays. Note that while the 4th and 5th examples have the same numerical value for ξ, case 4 corresponds to the 2σ point on the Gaussian for an undegraded Sun, while case 5 is the 1σ point after accommodating imperfect optics.

$$C_{3D} = \left(\frac{\pi}{2g_l}C_{2D}\right)^2 = \left(\frac{g\sin\psi_r}{2\sin\sigma_D}\right)^2 ; \quad \psi_{\max} = 90° \tag{3.42}$$

$$C_{3D}(90°) = \left(\frac{0.5g}{\sin\sigma_D}\right)^2. \tag{3.43}$$

For flat (Fresnel) collectors, inserting a reasonable angular variation of g results in $\psi_{\max} \approx 38°$ and $\approx 52°$ in the above cases. Again, $\psi_{\max}$ is simply the rim angle giving maximum concentration.

Note that if the back of the flat receiver is not insulated C_{2D} is reduced by a factor of 2, while if it is insulated a quantity k equal to (receiver width/aperture width) should be subtracted from C_{2D} in the above expressions, due to shadowing.

A most interesting fact [33] is that $C_{2D}\sin\sigma_D$ and $C_{3D}\sin^2\sigma_D$ effectively parameterize the concentration ratio calculations for parabolic collectors.

Frequently trough collectors only track on one axis so the Sun is incident on the aperture plane at an angle $\theta_\parallel$. Although we have noted that the image remains focused at the receiver, the path length is elongated by $1/\cos\theta_\parallel$. Consequently the receiver should be enlarged by $1/\langle\cos\theta_\parallel\rangle$, reducing the concentration ratio by $\pi/2$ for a trough with an east-west axis, assuming the diurnal insolation can be approximated by $\cos\theta_\parallel$. With a north-south axis the instantaneous value of $\theta_\parallel$ is approximately: latitude − north elevation angle of axis + solar declination. Near the equator, or for a polar-axis-tracking trough, $\theta_\parallel$ never exceeds 23° so the value of $\langle 1/\cos\theta_\parallel\rangle$ averaged over the year is only 1.04 or so. In the more usual case of a horizontal trough at 45° latitude, $\theta_\parallel$ varies from 22° to 68° giving $\langle 1/\cos\theta_\parallel\rangle \approx 1.62$. In addition, end effects for each collector module cause a fractional loss of $\approx f\langle\tan\theta_\parallel\rangle/l$ where $\langle\tan\theta_\parallel\rangle$ would be ≈ 1.12. Part of this loss can be recaptured by *end mirrors* on each segment, so long as the reflected energy does not cause local overheating of the receiver.

All of these complications disappear if a fully tracked trough is used, and considerably more energy is collected over the year. However, the M.A.N. experience [17] with such a system at IEA-SSPS revealed even more complications and losses associated with the required thermal collection system. Consequently fully tracked trough collectors, or even polar mounts, are seldom used.

Once a system level decision to adopt a fully tracking collector is made, the natural conclusion is to also adopt a point focus geometry. This is easily justified on the basis of the higher concentration ratio and consequent lower losses or higher temperatures available with dish collectors.

As we discussed in Sect. 3.4, a terminal concentrator can be fitted to the flat or cavity receiver in either the trough or dish configuration by using the rim angle (plus a few degrees to capture divergent rays) as the acceptance angle for the terminal concentrator. At $\psi_r = 45°$ the increased concentration ratio of the terminal concentrator is, at most, 1.41 (or 2 for a dish), but more impressive results are obtained at smaller rim angles (Tab. 3.4). Care must be taken in these cases that the terminal concentrator does not significantly shadow the reflector, or that material requirements do not become excessive, especially for smaller rim angles. Energy overload at the throat of the TC is also a concern for high concentration ratio systems as both incident solar and radiated infra-red energy must be dissipated without over-heating the reflective surface.

3.5.3 Local Concentration Ratio: Flux Density Distribution

Writing an analytic expression for the flux density at a point on the receiver of a parabolic collector is conceptually simple. However, one immediately becomes involved in three-dimensional geometry, complicated trigonometric or vector expressions for distances, angles of incidence, etc. Finally, very inconvenient expressions for the limits of integration are required to assure integration over the solar disc. If the degraded Sun concept is used the integrals can extend to the limit of the receiver, but now one has also to deal with integrating a Gaussian distribution combined with all the trigonometry. Several very careful analyses have been reported [6,7,11,12,15,19,21,30] and in general a first integral can be obtained. The resulting expression must be integrated numerically over the collector rim angle for all but the most trivial cases. In fact, in all such cases a very simple approximation which accounts for the principal optics can also be carried out. The resulting first order approximations provide useful insight into the nature of the more complex solutions, but are not adequate for detailed design analysis involving flux density distributions, interception factors, detailed receiver sizing, required collector accuracy, etc. Such questions require numerical evaluation of the first integrals, or a detailed ray-trace analysis.

Flat receivers. An approximation to the flux density at the center of a flat receiver for a uniformly bright Sun can be easily obtained using the properties of parabolae, and the inverse square law of optics. The solid angle within which the solar beam is intercepted by an element of the collector is $d\Omega = (dl \cos\theta_i) \cdot d(p\sin\psi) = p\cos\theta_i dl d\psi$ where l represents a collector strip at constant ψ. In this approximation, the integral over dl is simple. For a trough it runs parallel to the axis over the length of the strip illuminating the image point, i.e. between $\pm p\sin\alpha$ returning, $2p_\perp \sin\alpha / \cos\theta_i$, while for a dish it runs over the ring at radius $p\sin\psi$ returning $2\pi p \sin\psi$.

For the uniformly bright Sun, the irradiance at any field point is

$$E = \frac{d\Phi}{dA} = \int_{\text{Sun}} L \cdot d\Omega = L\pi \sin^2 \alpha. \tag{3.44}$$

However, the diverging rays from a point on the reflector are projected over a distance $p \ (= p_\perp / \cos\theta_i$ for a trough) and intercept the flat receiver at the angle θ_0, so

$$L_R = L_0 \cos\theta_0 / p^2 \tag{3.45}$$

where $\cos\theta_0$ equals $\cos\psi$ for a dish and $\cos\theta_i \cos\psi$ for a trough.

Combining all this information for any point x on the receiver, measured relative to the focal line (or point), and noting that the local concentration ratio is just the local flux density divided by the insolation (neglecting multiplicative factors such as the reflectivity) we have, for a trough collector with a flat receiver

$$\begin{aligned} C_{2D} &= \frac{E_{\text{Receiver}}}{E_{\text{Collector}}} = \int_{\text{Collector}} L_R \cdot d\Omega \bigg/ \int_{\text{Sun}} L \cdot d\Omega \\ &= \int_{-\psi_m}^{\psi_m} \left(\frac{L_0 \cos\theta_i \cos\psi}{p_\perp^2 / \cos^2\theta_i} \right) \left(\frac{2p_\perp \sin\alpha}{\cos\theta_i} \right) p_\perp d\psi \bigg/ L_0 \pi \sin^2\alpha \, . \end{aligned} \tag{3.46}$$

Simplifying the integrand we obtain a simple integral, which can be evaluated immediatly for the special case $x = 0$, thus

$$C_{2D} = 2 \int_0^{\psi_m} 2\cos^2\theta_i \sin\alpha \cos\psi d\psi \bigg/ \pi \sin^2\alpha \tag{3.47}$$

$$C_{2D}(x=0) = \frac{4\cos^2\theta_i}{\pi \sin\alpha} \sin\psi_m.$$

The behavior for small x, i.e. near the focal line, can be investigated by noting that the $\sin\alpha$ term in the integrand of (3.47), extends to the solar limb. For non-paraxial rays this becomes a chord of the circular Sun so we should replace $\sin\alpha$ by

$$\left[\sin^2\alpha - \left(\frac{x}{p_\perp/\cos\theta_i}\right)^2\right]^{1/2} = \sin\alpha\left[1 - \left(\frac{\cos\theta_i}{\sin\alpha}\frac{1+\cos\psi}{2f}x\right)^2\right]^{1/2}. \tag{3.48}$$

Applying the binomial expansion and integrating we achieve an approximation valid for $|x| < x_0 = f\sin\alpha/\cos\theta_i$;

$$\frac{C_{2D}(|x|<x_0)}{C_{2D}(x=0)} \simeq 1 - x^2\left(\frac{\cos\theta_i}{2f\sin\alpha}\right)^2\left(1 + \frac{\cos\psi_m}{2} - \frac{\sin^2\psi_m}{6}\right) \tag{3.49}$$

$$- x^4\left(\frac{\cos\theta_i}{2f\sin\alpha}\right)^4\left(1 + \frac{7}{16}\cos\psi_m - \frac{1}{3}\sin^2\psi_m + \frac{1}{8}\cos^3\psi_m + \frac{1}{40}\sin^4\psi_m\right)$$

where the first expression in ψ_m ranges from 1.5 to 0.84 and is 1.27 at $\psi_m = 45°$; the second ranges from 1.56 to 0.69 and is 1.19 at $\psi_m = 45°$. The local concentration ratio drops off quadratically for small x. For $|x| \approx 2f\sin\alpha/\cos\theta_i(1+\cos\psi_m)$ the local concentration ratio must approach zero abruptly, a result of the sharp solar limb used in the model. For a degraded Sun the approach to zero would be exponential and extend well beyond this value of x.

Proceeding in a similar fashion for a dish collector with a flat receiver

$$C_{3D} = \int_0^{\psi_m}\int_0^{2\pi}\left(\frac{L_0\cos\psi}{p^2}\right)(p\sin\psi d\phi)pd\psi \bigg/ L_0\pi\sin^2\alpha$$

$$= \int_0^{\psi_m}\frac{2\cos\psi\sin\psi d\psi}{\sin^2\alpha}. \tag{3.50}$$

In this case each receiver point within the radius $x = f\sin\alpha$ 'sees' the Sun in the entire collector, so for a uniformly bright Sun the local concentration ratio is

$$C_{3D}(|x|<f\sin\alpha) = \left(\frac{\sin\psi_m}{\sin\alpha}\right)^2 \tag{3.51}$$

while for sufficiently large x the Sun is not visible in the collector at all, so

$$C_{3D}\left(|x| > \frac{2f\sin\alpha}{(1+\cos\psi_r)\cos\psi_r}\right) \approx 0. \tag{3.52}$$

The disc of uniform illumination becomes peaked if a more realistic sunshape or a degraded Sun is used. Also, one should then replace $\sin\alpha$ in (3.52) by σ_D.

Cylindrical or spherical receivers. A very simple approximation can be obtained for a round (cylindrical or spherical) receiver, if we assume a point Sun. While this is a brutal assumption, considerable insight can be gained as long as we remember that any features of the resulting flux density distribution would be smeared out if a more representative degraded Sun were used. For a point Sun the local concentration ratio is just the ratio of collector area to receiver area for each element of solid angle projected from the focus. Clearly, the receiver radius, R_R, must be specified relative to that required to effectively intercept the degraded non-point Sun projected from the rim of the concentrator as discussed previously. Then for a trough, with $d(p\sin\psi) = pd\psi$,

$$C_{2D}(\text{point Sun}) = \frac{dA_C}{dA_R} = \frac{\cos\theta_i dl d(p\sin\psi)}{dl R_R d\psi}$$

$$= \frac{p\cos\theta_i}{R_R} = \left(\frac{f\cos\theta_i}{R_R}\right)\left(\frac{2}{1+\cos\psi}\right) \tag{3.53}$$

where R_R should be sized to accommodate the nominal value of θ_i, i.e. $R_R = p\sin\alpha/\langle\cos\theta_i\rangle$, recovering the $\approx \cos^2\theta_i/\sin\alpha$ dependence observed for the flat receiver. For a dish concentrator we have

$$C_{3D}(\text{point Sun}) = \frac{2\pi p\sin\psi d(p\sin\psi)}{2\pi R\sin\psi d(R\psi)}$$

$$= \frac{p}{R}\frac{pd\psi}{Rd\psi} = \left(\frac{p}{R}\right)^2 = \left(\frac{f}{R}\right)^2\left(\frac{2}{1+\cos\psi}\right)^2 \tag{3.54}$$

which is just the square of that of the 2-D concentrator; after noting that $\cos\theta_i = 1$ for the 3-D case.

3.5 Parabolic Geometries

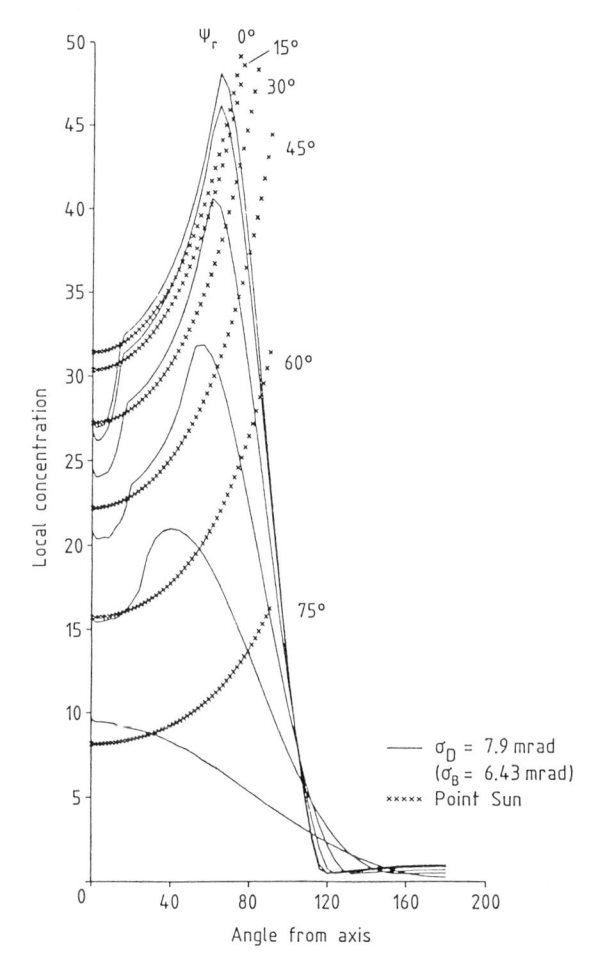

Fig. 3.9. Concentration ratio calculated for a hypothetical point source and a round 2-D concentrator. Dips near axis are due to receiver shadow.

Considerably more realistic cases are handled in the literature, but the resulting expressions invariably require computerized numerical analysis. Some comparative results given in [22] are shown in Figs. 3.9 and 3.10. For the round receiver it is clear that the exact results show the primary features of the point Sun results, but the features are smeared out by the finite Sun effects. Essentially, replacement of the point Sun by the real degraded Sun represents a convolution process, but as each ray is projected over a different distance and the divergence projects it to different points on the cylindrical or spherical receiver, no simple approximations are meaningful. For the flat receiver the initial y^2 dependence is retained but the ψ_r dependent coefficient is increased by use of the more realistic sunshape.

The other alternative is to simply establish the geometry on a computer and do detailed ray trace or fast fourier analysis. Any one of these approaches can meet the needs of a design analysis, but the simple methods above provide the insight required to properly utilize the more accurate computerized methods.

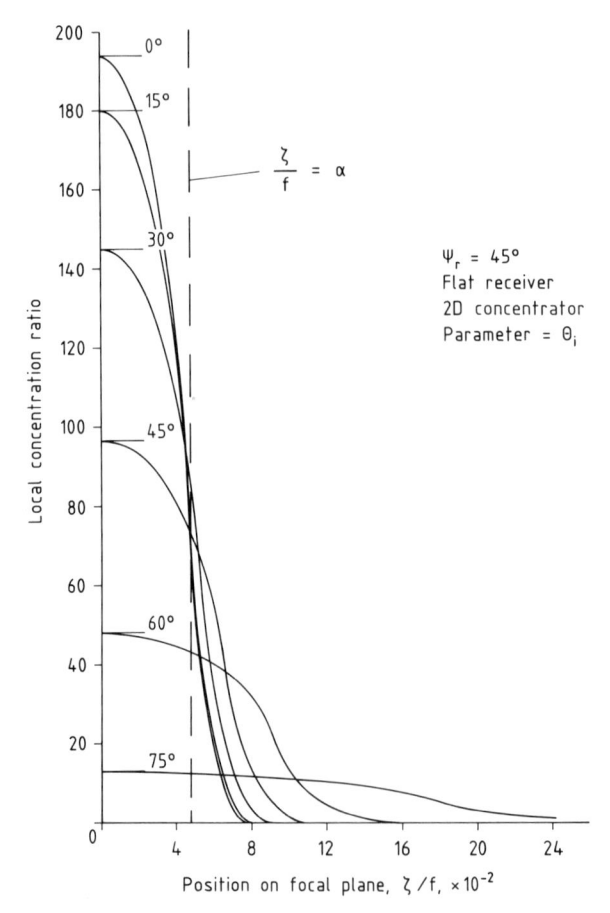

Fig. 3.10. Distribution of the local concentration ratio at the focal plane of a parabolic trough concentrator for various incident angles and a solar radius of 16 minutes.

3.5.4 The Iso-Intensity Problem

In some cases, such as a photovoltaic collector or possibly a receiver serving as a chemical reaction vessel, it is desirable to produce a uniform or otherwise defined flux density across the receiver. From the previous section it is clear that the local flux density depends upon the shape of the receiver. The local flux density distribution can also be modified by changing the optical figure of the collector: essentially *aiming* each region of the collector at a specific region on the receiver. Clearly this is not a well posed problem as many configurations might give the same result, while a seemingly simple flux density distribution might not be achievable.

The equivalent problem in the Central Receiver technology is dealt with by adjusting the aim points and focal length of each heliostat. A similar approach with *parabolic mirrors* has the additional constraint that the mirror must remain continuous, i.e. it may be distorted from the parabolic shape, but the distortion of one element moves other elements. Several rather elegant responses to this problem have appeared in the literature, and the interested reader is referred to them [27,45].

3.6 Other Concentrating Geometries

3.6.1 Introduction

A variety of optical alternatives to the parabolic geometry exist, and several have been developed as alternate solar concentrators. In general, these alternatives do not produce as good an optical image as does the parabolic geometry, but often the differences are insignificant, particularly since the objective is energy collection, not image quality. To be successful the alternative must produce the required concentration ratio and provide an additional advantage over the parabolic geometries. Thus, there are concentrator designs which utilize stationary reflectors with moving receivers, stationary receivers and support structures with tracking facets, Fresnel mirror and lens systems, etc.

The effective collector cost depends upon size, number of components, complexity, installation and operating costs, reliability, and performance. In addition, required modification to the balance of the system; receiver, support, and heat transport, must be considered. While most concepts proposed perform adequately, many of them prove to be noncompetitive as a result of one of the other properties mentioned.

3.6.2 The Hemispherical Bowl Concentrator

Because of spherical aberrations a reflective concave hemisphere forms a line focus along the radius, parallel to the incident light. A receiver, moved to remain in this focal line, can absorb the collected solar energy. From symmetry considerations it is clear that the nature of this focus is independent of the angle of incidence, consequently there is no need to track a hemispherical mirror, so a stationary mirror as large as is desired can be constructed using identical spherically curved facets. Of course, the aperture of such a fixed collector suffers an incident-angle cosine loss, so little energy is collected at low Sun elevations [9]. This can be partially overcome by tilting the aperture toward the equator, or by mounting *outrigger* mirrors that can be moved from the west rim in the morning and to the east rim in the afternoon, but this forfeits much of the *nontracking* advantage. Mounting, aligning and maintaining mirrors in a spherical configuration requires a monolithic structure (not a simple earthen, or even concrete, hole) complete with mirror supports, alignment fixtures, and a means for access. Reasonably large-scale examples are the 10 m diameter French *Pericles* [4] and the 20 m diameter U.S. *Crosbyton Gridiron* [28,44].

For a sphere, paraxial rays focus at $f_p = R/2$, but as the aperture increases the focal region for each successive ring moves further away from the center of the sphere according to $R/2\cos\psi$, where ψ is measured relative to a line through the Sun and the center of the sphere. Then $f(\psi) = 0.5R(2 - 1/\cos\psi)$, and a linear receiver, pivoted at the center of the sphere and moved to always point exactly away from the Sun, will always stay in the focal line. As $f(60°) = 0$, the ring of mirrors with $\psi \approx 60°$ all focus at the surface of the hemisphere at $\psi = 0$ (the anti-Sun position), resulting in a moving local *hot spot* of 200–300 Suns which can cause mirror breakage. For $\psi > 60°$ multiple reflections occur, resulting in a series of intense bands of light on the receiver defined by $\psi_C = 90° - 90°/2n + 2$, and located at the distance $R - R\cos(90°/2n + 2)$ away from the mirror along the focal line.

The principal solar images are projected over a distance ranging from $R/2$ (paraxial rays) to R (at $\psi = 60°$). It follows that a cylindrical receiver of radius $\approx 2\alpha R$, or a truncated cone tapering from $\approx \alpha R$ at f_p to $\approx 2\alpha R$, is appropriate (If beam errors are significant one should use σ_D in place of α). This receiver must tolerate the focal bands near the reflector surface, as well as substantial time-dependent circumferential variations as it tracks the focal line from the west to the east edge of the dish. Of course, average concentration ratio is much less than that provided by a tracking dish because of the line focus.

3.6.3 The Line Focus Fixed Mirror Collector

This concept is based on the observation that a series of strip mirrors, mounted as elements of a cylinder, can focus parallel incident light onto the surface of the cylinder [13]. This requires only that the mirrors are rotated by an angle $\psi/2$ relative to the central plane of the cylinder. As the angle of incidence, θ, changes, the resulting focal line sweeps over the cylinder. Thus a linear receiver pivoted about the center of the base circle, on an arm with length R equal to the radius, can track the focal line. Again a monolithic structure is required to support the mirrored strips at the appropriate angle. The collector suffers a cosine-of-incidence-angle loss as well as shading losses between the facets for lower Sun angles. Typically the aperture of the system is tilted toward the equator by approximately the latitude angle to minimize these effects. Then each element of the collector projects the solar image a distance of $2R\cos(\psi - \theta)/2$, where ψ is measured from the axis of the cylinder. The image formed on a flat receiver is more uniform than in the case of a parabola of equal aperture where the projection distance may be written $f/(\cos\psi/2)^2$. In terms of the aperture, the projection distance is 2 to 3 times the value occurring in the parabolic case, so the concentration ratio is lower. Also, unless the slats are curved to focus at $\approx 2R$, or are very narrow, they also contribute to image enlargement. Of course, if the slats are very narrow they increase both the parts count and fabrication and maintenance costs. Heat transfer from the moving receiver is also a problem, requiring a long section ($\approx 2R$) of flexible piping between each module to prevent breakage in case one stops tracking.

3.6.4 Tracking Facet Distributed Receiver Systems

An alternative to the previous two concepts is to hold the receiver fixed and track the collector relative to the fixed receiver. This can easily be done for the standard parabolic geometry with cylindrical or spherical receiver by pivoting the monolithic mirror assembly about the focal point. Fixing the receiver doubles the effect of tracking errors (see Sect. 3.3.4) and also introduces the possibility of relative motion between focal point and receiver, but the optical analysis of the parabolic geometry remains unchanged. The obvious advantages are simplified mounting of the receiver and thermal transport (no flex connectors or swivel joints).

Obviously it would be still more desirable to fix both the collector and the receiver, but the variation in apparent Sun position precludes a concentration ratio greater then the 2–3 available with a CPC. The remaining alternative is to assemble steerable mirror elements on a monolithic structure. If each facet is initially adjusted to redirect incoming sunlight (direction vector $-\vec{n}_S$) toward the receiver, it will continue to do so if it is rotated by half the change in the Sun's direction vector. Thus, if every mirror is turned by $(\vec{n}'_S - \vec{n}_S)/2$ the focus will be preserved. A number of techniques have been developed to accomplish or approximate this task.

Line focus tracking facets. Russian, Italian, French, U.S., Japanese, Swiss and undoubtedly other versions of this concept have been demonstrated.

Mirrored strips are arrayed on a flat surface, and axles parallel to their long axes are connected to a common drive. For any axis orientation the required tracking rate is just half that required by a parabolic trough in the same configuration. The strip spacing is set to simultaneously minimize optical interference and spillage losses. To first order the spacing can be estimated by segmenting a parabola and projecting the resulting strips along rays from the receiver to the plane. Relative to the parabola the projection distance for the reflected light is 1.2 p (at 45°) and 2.0 p (at 70°), consequently the concentration ratio is reduced. Incidence angle losses due to $\theta_\perp$ may be reduced by tilting the collector assembly toward the latitude angle, but the usual end effect and $\cos\theta_\parallel$ losses persist.

If flat strips are used it is clear that the concentration ratio is less than the number of strips, so 30 or more strips are required. Also the receiver must accommodate the strip width plus the projected degraded solar disc. Higher concentration ratio and reduced sensitivity to strip width can be achieved if each strip is curved, either to focus at $p(\psi)$ or at the average value of $p_\perp$. In any case, as $p = p_\perp/\cos\theta_\parallel$ the nominal image size increases significantly for $\theta_\parallel > 45°$.

A large-scale (10–100 MW$_e$) version of this concept has been studied [10], using large linear heliostats and a series of receivers 61 meters long (≈ 100 m/MW) mounted on towers. It seems to offer no advantages over the central receiver concept to compensate for the losses mentioned above and for the lower concentration ratio of the line focus.

Point focus tracking facets. Several design teams have modified the line focus strip concentrator to produce higher concentration ratio. In the U.S., Power Kinetics [8] scaled up the concept to produce several hundred thermal kilowatts. They curved each strip along its axis to produce a short line-focus. The strips were then aligned to superimpose the short focal lines, producing a nominal point focus. The strips were ganged and driven about their long horizontal axis to track the Sun in elevation. As off-axis aberrations destroy the imaging properties of this design, the entire assembly (including receiver) tracks the Sun in azimuth, and the collector array is tilted up at approximately the latitude angle. While reasonably economical to construct on a hand built basis, the design contains a multitude of structural and moving parts and does not seem to lend itself to economical mass production (including assembly and installation on site). Performance of the several prototypes has been marginal, producing a much lower concentration ratio than a parabolic dish of even modest quality.

For the Sunshine Project [37] in Nio, Japan, a row of twenty flat mirrors, each 1.5 m wide by 3 m tall and provided with a common drive, project sunlight into the 3.8 m aperture of a 3.6 m long parabolic trough. Five tiers of twenty ganged mirrors illuminate five adjacent fixed parabolic segments operating as secondary concentrators. All five parabolic segments focus onto a single long tubular receiver 54 mm in diameter. Thus the nominal concentration ratio is

$$C \approx 20\,(1.5/3.6)\,(3/0.054\,\pi)\,(\langle\cos\theta_i\rangle) \approx 150\langle\cos\theta_i\rangle. \tag{3.55}$$

Twenty five arrays were combined to provide 370°C steam to operate a 1 MW$_e$ turbine, providing very reasonable performance. The support structure for the tiered array appears to be overwhelming for a commercial installation, especially if component sizes were scaled up to reduce the parts count per MW. Each heliostat mount supports only 4.5 m^2 of reflector, but each drive unit controls 90 m^2. In fact, all five tiers on a stand could be controlled by a single drive forming a 450 m^2 multifaceted unit.

The Nio concentrator is an interesting modification of the multiple heliostat solar furnace, such as in Odeillo, France or Tokyo, Japan. In these furnaces the heliostat mirrors all remain parallel while tracking the Sun, simply filling a fixed parabolic dish with a beam of parallel sunlight.

A less singular, but perhaps more sophisticated point focus tracking facet concentrator concept is due to Francia [16]. A north-south shaft is driven at 15°/hour. This shaft is directly coupled to 10-30 East-West shafts, each of which is directly coupled to drive 10-30 polar-axis drives at 15°/hour. Francia devised a special kinematic actuator which automatically bisects the angle between the fixed receiver and the moving Sun (represented by the 15°/hour drive). Declination variations are accommodated by daily or weekly adjustments to each kinematic drive. Thus, at the Georgia Tech 400 kW Advanced Components Test Facility, a single constant speed drive has been used to control up to 550 mirrors, each 1 m^2 in area and all mounted on a rigid framework. With flat mirrors and a 20 m focal length for the array, a concentration ratio of $\approx$ 500 Suns is obtained over a 0.8 m diameter circle. However a concentration ratio of $\approx$ 1500 Suns was easily obtained by stress curving each mirror by pulling back on its center against a support ring. As the rim of the mirror tends to remain flat, special edge tensioning jigs were added to produce accurately spherical facets resulting in a peak concentration ratio of $\approx$ 2200 Suns.

On a much larger scale an individual drive for each mirror becomes appropriate and they may be supported from individual foundations. However, this represents the central receiver concept, which is discussed in detail in Sect. 3.8. Detailed techniques for calculating shadowing and blocking by adjacent facets as a function of their separation is also discussed in Sect. 3.8. Techniques and criteria for determining the optimum spacings are also described. While the conditions are somewhat different for these smaller systems due to the common drive and mount, the techniques are readily adaptable.

3.6.5 Fresnel Reflectors

A Fresnel reflector is generated by cutting or forming mirrored facets in an essentially flat surface. Incident light will be reflected to a defined focal point (or focal line) in accordance with Snell's law, so long as the normal to each facet is oriented to bisect the angle between the incident light and the focal point. Thus, Fresnel reflectors emulating spherical, elliptical, parabolic or off-axis parabolic mirrors can easily be defined.

Unlike the tracking facet collectors of the previous section, the Fresnel reflector turns as an entity to track the source. The most common and simplest Fresnel reflector emulates a one-axis parabolic mirror so parallel incident light will be reflected to a focal line (or point if a 3-D parabolic dish is being emulated). In this case each element is tilted toward the axis of the base parabola by the angle $\psi/2$, just as is each element of the base parabola itself. However, as the Fresnel elements are on a plane surface the projection distance is $f/\cos\psi$, compared to $2f/(1+\cos\psi)$ for the parabola. By symmetry the Fresnel elements are strips parallel to the focal axis in the 2-D case, and rings concentric with the axis of revolution in the 3-D case.

In fact, the figure of the reflecting surfaces cannot be generated by simply cutting strips or rings out of a parabolic reflector and remounting them on a flat surface. The correct optical figure requires that each facet be cut from a successively larger parabola. Conceptually, construct a nested set of parabolae such as shown in Fig. 3.11, all positioned to have a common focal point F, but with focal lengths given by

$$f_i = f\,(1+\cos\psi_i)/2\cos\psi_i \tag{3.56}$$

Then extract a thin slice perpendicular to the optical axis of the nest and mount it on a rigid substrate. The resulting array of oriented strips (or rings) will produce a focal plane image at F similar to that of the base parabola. However, the image will be degraded because of the larger projection distance (and consequent enlarged solar images) occurring for the facets with larger ψ, due to their increased focal length $f_i\;(= 1.5f$ at $\psi_i = 60°$). With the orientation and location of a facet defined by its value of ψ, the remaining parameters are its width, height, and curvature.

Width. If the figure of each facet is that of its respective parabola the width is not limited by optical considerations. For extremely narrow facets diffraction could be important, but a diffraction width of ±2 mrad

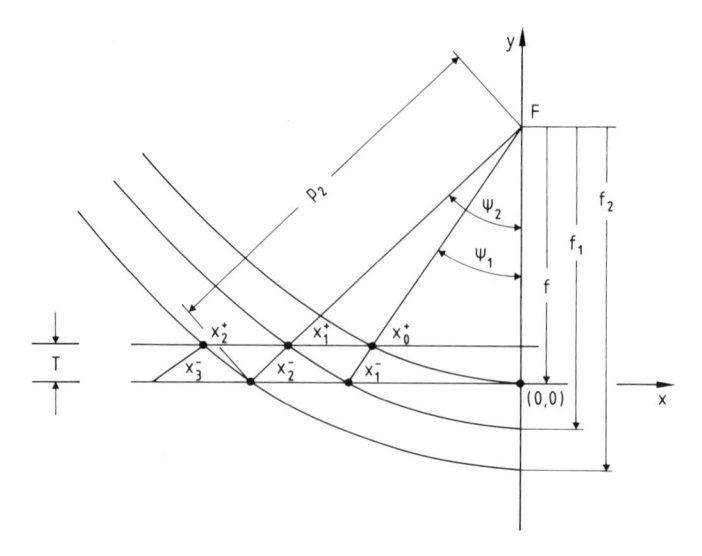

Fig. 3.11. Nested parabolas for the Fresnel mirror geometry. Relative to F, $y = x^2/4f_i - f_i$; relative to vertex at f_0 of base parabola, $y = x^2/4f_i - f_i + f_0$, as $\rho_2 = 2f_2/(1+\cos\psi_2) = f_0/\cos\psi_2$; then in general, $f_i = f_0(1+\cos\psi_i)/2\cos\psi_i$.

corresponds to a facet width of 1/4 mm for $\lambda = 0.5$ mm. Such narrow facets are only likely to be encountered in the outer portions of small reflectors such as are used to concentrate sunlight on individual solar cells.

Wide facets pose no problem unless each facet is composed of a flat surface. In this case the width must be limited to somewhat less than the apparent width of the receiver or spillage will result. Thus a restriction of strip width to less than $2\alpha f$ is reasonable. For small ψ this may require that strips narrower than otherwise allowed may be required. This poses no problem in machined or cast reflectors, and if the reflector is assembled from facets one need only use a few more closely spaced facets of the allowable width.

Thickness. The thickness T of the reflector assembly essentially dictates the depth of groves or height of the top edge of facets above a base plane. Typical values might be $f/30$ to $f/3000$, but let us use $T = f/100$ in the following discussion as a concrete example.

The lens is composed of slices of a series of parabolas, all focused at F but referred to a common plane, which is the vertex of the base parabola. Thus, referring to Fig. 3.11 it is convenient to write the height of the mirror as a function of x in the form

$$y + f_i - f = x^2/4f_i. \tag{3.57}$$

Now inserting the earlier expression for f_i in terms of f and ψ_i, we can solve for x

$$\frac{x}{f} = 2\left(\frac{f_i}{f}\frac{y}{f} + \frac{f_i^2 - ff_i}{f^2}\right)^{1/2} \tag{3.58}$$

$$= 2\left[\frac{1+\cos\psi_i}{2\cos\psi_i}\frac{y}{f} + \left(\frac{\tan\psi_i}{2}\right)^2\right]^{1/2}. \tag{3.59}$$

In this equation y varies between 0 and T in each slice. The angle ψ_i is defined as the angle at which $y = T$ for the previous slice, thus from Fig. 3.11

$$\psi_i = \tan^{-1} x_{i-1}(\max)/(f-T). \tag{3.60}$$

For the first facet $\psi_0 = 0$ so $x_0^-/f = 0$ and $x_0^+/f = 2\sqrt{T/f} = 0.2$ (for $T = f/100$) then

$$\psi_1 = \tan^{-1}[0.2\,(f/f - 0.01\,f)] = \tan^{-1} 0.202 = 11.42°$$

and

$$x/f = 2[1.01\,y/f + (0.202/2)^2]^{1/2}.$$

Thus at $y/f = 0$, $x_1^-/f = 2(0.101) = 0.202$ and at $y/f = T/f = 0.01$, $x_1^+/f = 2(0.0101 + (0.10101)^2)^{1/2} = 0.285$. The facet width is thus $x_1^+ - x_1^- = 0.083f$ and $\psi_2 = \tan^{-1}(0.285f/f - 0.01f) = \tan^{-1} 0.2878 = 16.06°$ etc.

One can also address the nominal width of a slice at any value of ψ, say 60°. Thus

$$\frac{x}{f} = 2\left(1.5\frac{y}{f} + \frac{3}{4}\right)^{1/2} = 1.7320 \quad \text{at} \quad y = 0$$
$$= 1.7493 \quad \text{at} \quad y = 0.01f$$

so the facet width is the difference, or $0.0173f$, and the length of the slanted reflecting surface is $0.0197f$. The final value of x/f for the previous facet is also easily obtained from the formula for ψ_i and is 1.7147, resulting in a *shoulder* width of $0.0173f$. Thus the effective area at $= 60°$ is about 50%. From this we can conclude that, although the mirror surface is still used effectively at $\psi = 60°$, the area of support structure per unit of mirror is increasing twice as fast as it would for the base parabola and is larger than it would be for an alternative Fresnel reflector design with a longer focal length and smaller aperture angle. So again we come to system trades to achieve the optimum design based on image size, flux density distribution, support structure, costs and other requirements.

Curvature. From the example it is clear that curvature is important for the first few facets which have a width (for $T = 0.01f$) of $0.20f$ and $0.083f$. Such facets may either be cut to approximate the shape of the appropriate section of their particular parabola, or if only flat facets are allowed, subdivided to facets of width $\approx 2\alpha f \approx 0.01f$. Note that for $T = 0.01f$ such subdivision would be required even at $\psi = 60°$. This is so because $T = 0.001f$ is a more realistic value (1 mm height for a 1 m focal length reflector).

In a more detailed design, facet slope and/or curvature can be modified to control the flux density distribution at the receiver. Thus the center facets (which produce the smallest images) could be aimed near the edge of the receiver to reduce the peak flux density and produce a more uniform receiver flux density distribution without significantly increasing spillage. As the reflectors are typically stamped or extruded, such *figuring* or *blazing* of the die facets is not more difficult than machining a simple die, although the design computations may be substantial.

3.6.6 Fresnel Lenses

The Fresnel lens is a refractive optical device of negligible thickness. It is, at least conceptually, produced by removing non-essential flat slabs of dielectric material from a conventional lens. The dielectric surfaces of the resulting thin facetted lens intercept the rays of light with the same incidence angle the original lens would have, and so perform an identical refractive function. A familiar example is the stage lens in an overhead projector.

The design procedure for a Fresnel lens is much the same as that described in Sect. 3.6.4 for a Fresnel mirror, although adapted to refractive optics. The essential difference is that the index of refraction of most practical materials is wavelength dependent, resulting in chromatic aberration. Thus, incident photons following identical paths will be refracted by different amounts according to Snells' law of sines.

Large arrays of Fresnel lenses can be stamped (or, if linear for 2-D optics, extruded) directly on a substrate. Alternatively, thin films of embossed dielectric can be bonded onto a flat substrate. Such arrays are frequently used with individual solar cells, and have the obvious advantage over the Fresnel reflector that the cells and support structure do not shadow the mirror. In fact, the location behind the lens enables the receiver, of any type, to be completely enclosed, which will protect it from the environment and suppress forced convection.

Some concepts relevant to the design of Fresnel lenses follow:

1. In Sect. 3.2 we mentioned that spherical aberration and coma are both reduced for a thin lens if both R_1 and R_2 are negative, that is, if the lens is domed over the focal point.
2. Each individual facet of a Fresnel lens is, in fact, a prism. If a prism is operated under the condition of minimum deviation, the error in the angle by which it bends incident light rays is quadratic in variations from the design angle. Consequently, it is very tolerant to orientation errors.
3. For incidence angles exceeding 30° the total reflection $(r_\parallel^2 + r_\perp^2)$ increases significantly with angle. The sum of the back and front surface reflection losses is minimum when the incident and exit angles are equal, as in a minimum deviation prism.

Recently these ideas were combined and the constraint was added [29] that the outer surface of the lens be smooth to facilitate cleaning, etc. The result is a unique Fresnel lens configuration. The use of minimum deviation prisms implies equal incident and exit angles. This specifies the slope of each surface in terms of the cone angle measured from the receiver. The projection distance is defined relative to the adjacent facets by the requirement of a smooth outer surface. The width of each facet is limited by the allowed lens thickness or, if flat facets are used, by the image size at the receiver.

A particularly nice feature of the final design is that the individual prism shoulders need not actually cause any losses, because the adjacent prism refracts light away from the shoulder (see Fig. 3.12). The interior angles between adjacent prisms must, however, be as sharp as possible to minimize scattering of light. Collectors using these design principles and having a 45° rim angle have been built and tested with good results. The constraints require that the lens be rather strongly curved, having a radius of curvature about equal to the focal length of an equivalent flat Fresnel lens. However, because of the minimum deviation condition the refracted solar image is not enlarged, nor do prism angle or lens contour errors have significant effect on the projected image. Even dispersion (chromatic aberration) is less than it would be in a comparable flat Fresnel lens. While the initial design was for a line focus collector, a point focus lens using similar design principles has been defined.

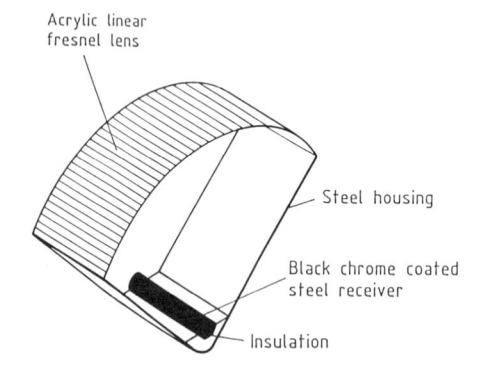

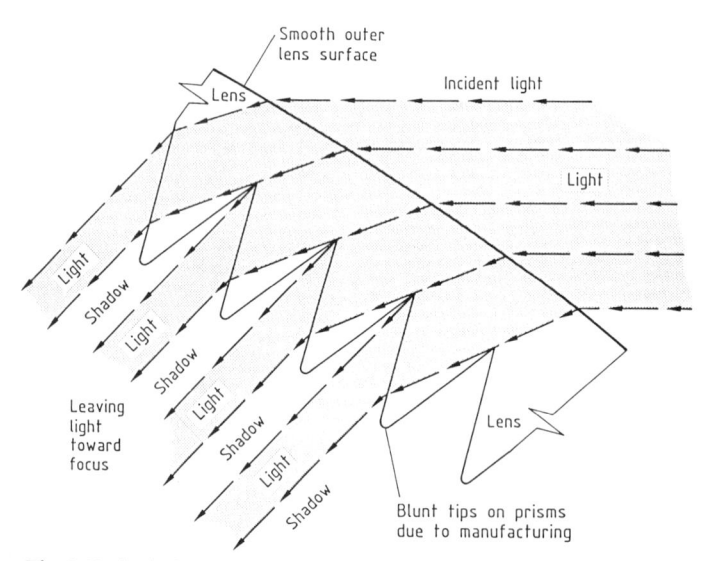

Fig. 3.12. Optimized minimum deviation of Fresnel lens [29].

3.6.7 Other Optical Configurations

The number of unique, novel, or *improved* designs for concentrating solar collectors is endless. Each can be analyzed according to the methods of this chapter, and most will perform approximately as predicted. However, more is required. To have any significant impact a new design must show compelling improvements in some combination of mass productibility, capital cost, operating cost, maintenance costs, reliability, simplicity, survivability or annual performance under realistic solar and climatic conditions. Unless some subset of these conditions is met such that a real economic advantage occurs to the new design relative to a viable competitor, substantial work is not justified. Proponents are entitled to point out the strong points of a new concept, but they must do it in a system context so that the disadvantages can also be easily evaluated.

3.7 Central Receivers

3.7.1 Introduction

In a central receiver system [20] hundreds or thousands of individually aimed mirrors (heliostats) reflect the direct-beam solar energy striking them toward a common focal zone. To avoid excessive interference with the redirected beam by adjacent heliostats, the focal zone must be elevated well above the plane of the heliostats. Thus, the system resembles a large facetted Fresnel mirror, which tracks the apparent motion of the Sun by steering the normals to its individual facets to bisect the angle between the Sun and the fixed focal zone. An appropriately sized receiver supported in the focal zone by a tower will intercept 10 to 1000 MW of sunlight, which has been transmitted optically to this central region. In the design of such a system there are important problems to be solved, many of which require optical analysis:

1. How much power is delivered to the receiver over the course of a typical year?
2. What is the instantaneous distribution of flux density on the receiver from the field of heliostats?
3. What is the flux density distribution from a single heliostat, and how does it depend on the *figure* or contour of the mirror.
4. How do the interactions between neighboring heliostats (shading and blocking) depend on their spacing?
5. How do the shape and size of the heliostats affect items 3 and 4 ?
6. How should the heliostats be spaced in the field to achieve the most cost effective design?
7. For a given design point power, how big and how high should the receiver be?

Basically these problems all involve a combination of ray tracing, to determine angles or intersection points, and analysis to determine interactions, flux density, etc. Although the geometry involved is straightforward, the interactions between thousand of multi-sided heliostats must be tracked through hundreds of points in time throughout a year, and the integrated flux density distribution defined on the receiver at each time. Consequently, computer analysis is mandatory. In fact, even with a large computer, computational time and storage must be carefully husbanded if system optimization is an objective, or even if a reasonable number of designs is to be studied. Therefore, let us first investigate the question of scaling relationships.

3.7.2 Scaling Relationships

It is conventional and convenient to select the system focal length (vertical distance from the plane of mirror centers to the centerline of the receiver aperture) as a standard of distance, let us call it HT. Then all significant descriptors of the system can be expressed either as ratios to HT, or as angles, or as constants (examples are receiver diameter, heliostat beam error, and receiver absorptivity respectively). For a system defined in this way, the optical analysis is independent of HT, i.e. of system size. Thus, an accurate hand sized model can light cigarettes with performance factors identical to those of a 1000 MW system.

One can also identify a second important scale factor, namely a characteristic heliostat dimension. Let us choose the reflector diagonal or diameter and call it D_m. To just avoid shading and blocking losses one must position neighboring heliostats so the shadows cast by both the Sun and the receiver (shading and blocking) just touch, but do not fall on the representative heliostat. Clearly these shadows are proportional in length to the characteristic heliostat dimension. They also depend upon the angular location of both the Sun and the receiver, but not on their distance. Thus, independent of the system scale, shading and blocking calculations at each position in the field are also independent of heliostat size, so long as spacings are specified in units of D_m. Note that *position in the field* is defined in terms of HT and not in meters.

3.7.3 Shading and Blocking Calculations

As one consequence of these scaling relationships it is clear that a shading, blocking, and $\cos \theta_i$ data base can be generated which is valid for any size heliostat and for any size system (so long as it involves several hundred heliostats). Annual average values of shading and blocking performance vary smoothly as the representative heliostat is moved around the field. It is therefore possible to generate a data base for several hundred representative heliostat locations and to interpolate between these to define the performance of heliostats of any size in a system of any size. Of course, for a different latitude, heliostat shape, or insolation/cloud model a new data base is required.

Shading and blocking are most easily evaluated in a reference plane containing the reflective surface of the representative heliostat [25], i.e. the heliostat whose efficiency is being evaluated. The boundaries of neighboring heliostats in front of that plane are projected backward, along both the line from the Sun (shading) and the line from the receiver (blocking) into the reference plane. If any of these images overlap that of the representative, the fractional loss is calculated. For reasonably well designed heliostat fields the loss events seldom overlap one another. Thus double counting is usually not a significant problem when the individual fractional losses are summed to determine the total shading and blocking loss at any instant. In fact, major double counting is most likely to occur when a primary shading and a primary blocking event from the two nearest neighbor heliostats overlap. Only for very low Sun elevation angles when a *careful* analysis will consider events from more distant neighbors does double counting become a problem, and then the available solar energy is so low it is not very important. In any case, it is a good strategy to eliminate from consideration all heliostats aligned beyond the nearest neighbors, as any events they may cause are almost sure to overlap those of the nearest neighbors.

3.7.4 Flux Density Distribution at the Receiver

The receiver flux density at each point is just the properly projected sum of the flux density produced at that point by each heliostat in the field. This is frequently approximated by properly augmenting the flux density of a smaller number of carefully selected representative heliostats. Three or four hundred representatives are enough unless a complex aim strategy is invoked, which may require that five to ten times that number of images be evaluated. If ten aim points are required, time can be saved by generating the image of one representative

3.7 Central Receivers

and evaluating it at each aim point, counting it as representative of ten nearby heliostats, or twenty if each flux density is doubled, etc.

There are several approaches to evaluating the flux density distribution due to a single heliostat. To save effort and computer time, it is usual to first generate a *degraded Sun* using the methodology of Sect. 3.3.2. A uniform brightness Sun (subtense = 4.65 mrad) is modified by limb darkening in the solar photosphere and scattering in the Earth's atmosphere. This gives an approximate Gaussian distribution having a second moment of ≈ 2.2 mrad. In addition there is an exponential tail out to 5° or so representing the solar aureole. The resulting *sunshape* as it is called, is fitted by a sum of Gaussian functions, or the moments of the sunshape may be evaluated up to perhaps 6th order. The *degraded Sun* is then generated by adding in quadrature the moments of the beam error function for the heliostat as detailed in Sect. 3.3.3. Typically the moments of a simple Gaussian error function having σ = 2–3 mrad are appropriate. If the heliostat is perfectly focused at the receiver the above model is complete and the flux density may be calculated at an image plane centered at the receiver and perpendicular to the optical axis of the heliostat. Projections for short distances (1/2 a beam diameter) parallel to the optical axis, from the image plane to the receiver surface, are valid.

If the heliostat focus is not perfect (due to off axis aberrations or a focal distance unequal to the slant range from the heliostat to the image plane) a further contribution to the image size occurs. This contribution is most simply evaluated by tracing rays. Remember the simple relationships for transverse and sagittal rays, expressed in terms of the slant range SR for spherical mirrors at an incidence angle θ:

$$\frac{D_t}{D_m} = t \cos\theta \left| 1 - \frac{SR}{f \cos\theta} \right| \tag{3.61}$$

$$\frac{D_s}{D_m} = s \left| 1 - \frac{SR \cos\theta}{f} \right| \tag{3.62}$$

where t and s represent the chord of the mirror in the transverse and sagittal directions divided by the characteristic dimension D_m. The absolute values are required to prevent negative values of mirror image area from occurring between the location of the transverse focal line (at $SR = f \cos\theta$) and the sagittal focal line (at $SR = f/\cos\theta$). Assume the mirror has a perfectly defined parabolic shape and that any local variations from this figure were counted as slope errors when the beam errors were evaluated. By tracing rays from the center of the Sun through the corners of the heliostat, an image of the heliostat can be formed in the image plane. For a flat heliostat, this image is an exact cosine foreshortened replica of the heliostat, for normal incidence on a perfectly focused heliostat it is a point, etc. For a reasonably well focused heliostat the image size may be 1/5 to 1/10 the heliostat dimension in height and width, depending on the angle of incidence. The various moments of this image may be calculated and added (in quadrature) to those of the degraded Sun in a straightforward extension of the methods of Sect. 3.3.3 to get the final convolution of Sun shape, beam error, and mirror size [40].

Alternative approaches combine these features using an analytic integration over the mirror [26], fast Fourier transform methods (Helios) [39], or a rigorous Monte Carlo calculation (Mirval) [23]. In addition, many 'simple' algorithms have been developed which do a reasonable job of approximating the image formed by a heliostat. While there is little interest in the exact flux density distribution (can Sunspots be resolved? Who cares?), it is necessary to know that the representation is valid under all likely circumstances. Both allowable flux density constraints and the requirement to evaluate spillage with good accuracy put rather stringent requirements on simple empirical or model equations. By the time they are *improved* enough to be reliable they are probably beyond hand calculation anyway so it is just as easy to program up one of the more detailed models described above.

If a heliostat has several facets, they may be canted at one focal length and focused to another. The image from each facet may be evaluated separately after determining the impact point for its central ray. Alternatively, if the central ray impact points are close enough together that the final image is compact (as is usually the case), a single set of moments for all the facets may be evaluated simultaneously from the combined impact points of all the corners. Of course, if the central rays nearly superimpose in the focal plane of the facets, the entire array can be treated as a single mirror of equal focal length. This is often a good approximation.

Optical transmission of the reflected energy from the heliostat to the receiver introduces one additional loss element – absorption and scattering in the intervening air. The equivalent depth of the atmosphere is only 8431 m at *STP* (Chap. 2). As typical added transmission paths range from 5–30% of this, a corresponding loss of 2–7% might be expected on a moderately clear day. A very simple approximation to the loss is given by

$$L(\%) = [1 + (7 + \rho/4.5 \text{ g H}_2\text{O}/\text{m}^3)(SR/1000 \text{ m})](50 \text{ km}/VR)^{1/2} \quad (3.63)$$

where ρ is the local water vapor density, SR is the slant range of propagation in meters, and VR is the local visual range in km. A definitive analysis in [32] of detailed radiation transfer calculations carried out in [38] provides a much more reliable and versatile, but more cumbersome, expression for $L(\%)$, along with the relations required to estimate the required parameters.

3.8 Design Issues and Constraints

There is a wide range of issues which, while not strictly optics, directly constrains the optical designer. While such issues arise in any optical system design, we will address them here using central receivers as a specific example.

3.8.1 Preliminary System Level Considerations

Design objective – minimum levelized energy costs. Solar power systems must be designed to compete economically with alternative energy sources, and eventually to replace fossil fuels. To win this competition they must be designed to have low lifecycle costs per unit of energy produced. Evaluation measures which satisfy this requirement are levelized energy cost (LEC), or cost/benefit ratio. While LEC calculations are relatively complex, the cost/benefit ratio may be simply defined as (construction cost plus present value of operations and maintenance over the design life)/(energy produced for sale in a typical or average year). Present value of operations and maintenance (PVO&M) is the capital which must be invested at today's interest rate to provide money to pay for the (inflated) O&M cost in the future. The series can be summed to produce the present value multiplier:

$$PVM = \frac{(1+e)}{(d-e)} \left(1 - \frac{(1+e)^n}{(1+d)^n}\right) \quad (3.64)$$

where: e is the inflation rate, d is the discount rate (cost of money), n is the number of years (design life).

In the simple case of interest rate equal to inflation, $PVM=$ years of design life, while with a true interest rate, $d-e$, of 3%, $PVM=$ 19.6 for a 30 year design life. In this case, if annual O&M equals only 5% of invested costs, PVO&M equals the original investment. O&M costs are typically a higher percentage of investment costs for small systems because of scheduling problems, travel costs, or inefficient use of manpower. Thus, systems of 300 to 1000 MW_t are usually more cost effective than smaller systems unless the smaller solar systems are integrated into the operations of an existing plant.

To produce the most cost effective design the annualized cost/benefit ratio (or equivalently the LEC) must be used as the objective function in a system optimization. The alternatives of designing for maximum efficiency at design point, or even on an annual basis, will produce high performing plants with excellent solar efficiencies, etc. but the cost per unit of energy produced over a year will be excessively high. Although the solar energy is free, the collectors are expensive and must be used effectively. On the other hand, the balance of plant, and operating it, is also expensive and that investment must not be wasted. Consequently the annualized cost/benefit ratio (or LEC) must be the function to minimize.

Design priorities. For distributed collectors the usual procedure is to design the most cost effective collector/receiver combination for the required output temperature. Rim angle, concentration ratio, receiver enclosure, etc. are selected. Then the thermal collection system is designed and the insulation, pipe sizes, and collector spacings are co-optimized with respect to piping costs, thermal losses, operating hours and collector shading interactions.

In contrast, for central receiver systems the most cost effective heliostat (with acceptable beam quality) and the most cost effective receiver are designed. Receiver dimensions and tower height are estimated in light of design point power requirements and receiver flux density limitations. Then heliostat spacings, number, and field boundaries are defined considering shading, blocking, $\cos \theta_i$ effects, spillage, receiver thermal losses, and the overall system cost. After a number of iterations of tower height, receiver dimensions, and collector field configuration the most cost effective design is determined. Both receiver redesign and heliostat size, accuracy, and cost may be reconsidered in these iterations.

Systems designed in accordance with the above considerations have tended to have relatively large optical components (50–150 m² aperture) in order to reduce mounting and tracking costs. Such elements are, almost universally, mirrors. Large Fresnel refractor systems have been designed for certain applications, but conventional lenses are not used. While the collector must be low in cost, the receiver operates under relatively harsh conditions of temperature, flux density, cycling, and corrosion. Thus, it often imposes constraints on the collector to produce a uniform and high but limited flux density at the receiver. While there are no image quality requirements (nor even a requirement to form an image), the collector is required to provide a well behaved and predictable flux density distribution at the receiver and to control the amount of reflected energy which misses the receiver altogether. To satisfy these requirements one needs to perform detailed optical analysis and design of the collector.

End use of energy. The end use defines the required plant size, the delivery temperature, and requirements for capacity factor (and hence solar multiple and thermal storage). The above requirements tend to limit the heat transfer fluid to only a few options. Thus, above $\approx 800°$ C, air or helium are the only alternatives other than a direct chemical reaction cycle. Between 400° C and 600° C the choice between water/steam, nitrate salts, sodium, or Na-K eutectic salts, depends strongly on the required capacity factor, frequency of cloud transients, and technology available to the designer. For our purposes we can assume the end use defines a plant size, the required delivery temperature, and the heat transfer fluid.

Receiver. The receiver must intercept sunlight reflected by the heliostat field, absorb the incident sunlight, convert it into heat, transport the heat to the heat transfer fluid, and continually replenish the working fluid. It must do this without excessive radiative or convective loss of the absorbed energy, without using excessive parasitic energy or without loss or degradation of the working fluid. It must survive daily deep thermal cycles over a 20 to 30-year design life, and three to five times this number of shallow transients (due to cloud passage etc.) without suffering failure due to thermally induced distortions, thermal fatigue or low cycle creep fatigue. The solar absorbing paint and the primary heat exchanger surface must tolerate high solar flux densities without failure or rapid photodegradation.

The receiver designer must take these constraints into consideration and design a receiver which is not excessively costly, heavy, or large and which does not impose impossible constraints on the heliostat field.

End use and other design constraints generally define the design point power and design output temperature required from the receiver. As described in Chap. 5, these requirements are combined with a desire to minimize radiation and convection losses, provide a reliable least cost receiver, and minimize constraints on the heliostat field. The two primary configurations which have resulted are cavity and external receivers.

The poor heat transfer associated with high-temperature gas-cooled receivers requires special measures to reduce radiation losses from the relatively large heat exchanger. The result is a protective enclosure (cavity receiver) or a volumetric receiver. The aperture of the cavity receiver is faced toward the most effective area of the heliostat field, i.e. to the north in the northern hemisphere and 30–60° downward. The aperture area is selected during system optimization to minimize the sum of radiative and spillage losses. Following a preliminary system optimization, the heliostat aim points and the boundary of the heliostat field, along with the interior configuration of the cavity are subject to modification to satisfy peak flux density constraints and heat distribution problems on the volumetric receivers or within the cavity, with minimum impact on the cost-benefit ratio of the system.

If the sum of radiative and convective loss per m^2 is less than about 20% of the allowable peak flux density, an external receiver is probably more cost effective. It is surely smaller and lighter than the protective enclosure for the cavity receiver, and it imposes less constraint on the heliostat field. Thus, a water/steam, molten salt or liquid metal receiver in the 300° C to 600° C temperature range will use an external receiver. Typically this will be in the form of a cylinder with height 1–1.5 times diameter and sized to produce an average flux density 0.5–0.6 times the allowable peak flux density. The associated heliostat field will surround the receiver, but will have 60–70% of its heliostats north of the tower because of the superior performance there compared to the south field. Again, the receiver will be sized to optimize system performance in conjunction with the heliostat field, but final modifications of both the field and the receiver may be required to satisfy all system constraints, particularly the allowable maximum flux density level.

Heliostats. The primary system requirement on the heliostats is that they redirect sunlight onto the receiver without excessively degrading the beam quality, and do so at the lowest cost. The heliostats designed, built, and tested in response to these requirements have ranged in area from 40 m^2 to 200 m^2 and in beam quality from $\approx$ 1 mrad to $\approx$ 5 mrad. Projected cost ranges from 500 \$/m^2 for a noncompetitive small purchase to $\approx$ 100 \$/m^2 for guaranteed buys of 20,000 to 50,000 heliostats over 5 years. Round, square, and rectangular heliostats have been considered, and all offer a means to preset or adjust the focal range to any reasonable design value. Developmental heliostats have tended to be of a size larger than the 40–50 m^2 of the various pilot plants because this reduces the cost per m^2 of the pedestal, drive, controls etc. Quality of the focused beam can be maintained in these larger heliostats. However, off-axis aberrations increase as $Dm(1 - \cos\theta)$ and can add 5 m to the diameter of the focused spot for a 200 m² round reflector at 45° incidence angle, and up to 6.5 m for a rectangular heliostat with an aspect ratio of 2–1. Consequently, larger heliostats are unlikely to be cost effective.

The heliostat parameters in order of importance to system optimization are thus beam quality, cost, average lifetime, reflectivity, shape, size, reliability, and ratio of shading area to reflecting area. Details of facetting, tracking axis, cleaning cycles, etc., may also be employed in the analysis but have no significant effect on the optimization other than their impact on heliostat cost.

3.8.2 System Optimization

Preliminary considerations. According to the discussion above, the objective function to be optimized is the system capital cost, including present value of operation and maintenance, divided by the total energy produced in a typical year. That is

$$F = \frac{C}{E}. \qquad (3.65)$$

The condition for the optimum is zero variation in F for all consistent variations in the parameters determining C and E [24]

$$\delta F = 0 \quad \text{implies} \quad \frac{\delta C}{C} = \frac{\delta E}{E}, \quad \text{or} \quad \frac{\delta C}{\delta E} = F. \qquad (3.66)$$

Thus, for the optimized system a small change in any parameter resulting in a decrease δE in energy will be accompanied by a decrease in life cycle cost of $\delta C = F \delta E$. Two problems are immediately obvious:

Costs. The cost of the entire system must be expressed in terms of the pertinant parameters. Although not easy, this can be accomplished by expressing the cost of each component or subsystem as an equation involving all parameters appropriate to the system under optimization. Typical cost elements are permits and access roads, feed pumps, vertical piping, receiver, tower, heliostats, land, wiring, and operation and maintenance costs of each of the above items, including parasitic power, expressed as a present value. For example, the feed pump cost for a given working fluid depends on the tower height and on the design point flow which is related to the design point power level. Its annual O&M is perhaps 5% of its capital cost while parasitic power cost depends on the value of electricity, tower height, fluid density, and the annual energy produced. Alternatively the parasitic power can be divided by the plant electric conversion efficiency and subtracted from the delivered thermal power. Substantial care must be used to assure that all appropriate cost dependencies are included in each cost function.

Heliostat locations. There are several thousand heliostats in a typical system, and land and wire costs and the performance associated with each heliostat depend upon both its location and the location of all nearby neighbors. Also the variation in neighbors' positions associated with optimizing one heliostat location will perturb the performance of that neighbor and others. To convert this to a well posed problem requires that the field be decomposed into an array of independent cells. A representative heliostat is placed at the center of each cell and surrounded by a defined array of neighbors.

While the array of neighbors could be randomly spaced, arrayed on N-S and E-W lines, the radial-staggered array is well defined and gives the best performance nearly everywhere in the field. Close to the tower circles of closely spaced heliostats seem to be preferable, say for a receiver elevation $> 40°$. The heliostats are thus placed around the representative in receiver-centered circles. Alternate circles are staggered, i.e. shifted azimuthally by half a heliostat spacing. This configuration doubles the distance to the major blocking heliostat on the radial line to the tower and allows the representative to *peek through* between heliostats in the intervening staggered circle. Each cell is considered to be filled with heliostats with average performance equal to that of the representative. For a defined radial and azimuthal spacing in each cell, an insolation weighted annual average of shading loss, blocking loss, and cosine foreshortening (SBC) can be computed. An adequate sample is 19 times on one average day of each month for each of 200–300 cells. To facilitate the optimization process a SBC data base may be generated providing this data for 16 heliostat configurations in each cell: 4 variations each about the nominal azimuthal separation and radial separation, i.e. $-3d/2$, $-d/2$, $+d/2$ and $+3d/2$. Recall from the scaling arguments that this data base can be used with any size heliostat of the same shape and for any sized system so long as the dimensions of each cell scale with the tower height. Recent increases in computer speed suggest that a 5×5 or a 6×6 array should be considered for future development.

For the same cell array an image from each representative heliostat is projected to the receiver and the fraction of the energy which is intercepted by each of several hundred nodes on the receiver is computed. The resulting *node file* can be used to calculate the interception fraction for each representative heliostat. Following optimization the number of heliostats located in each cell is known, so the same node file can be used to generate flux density maps on the receiver.

The optimization process. The variational analysis applied to the objective function defined in terms of this cellwise decomposition of the heliostat field returns the requirement that a cell matching parameter, $\tilde{\mu}$, be the same for every occupied cell in the field [24]. Numerically

$$\tilde{\mu} = \frac{C_H}{Fa\overline{S}} \approx 0.6\text{--}0.8 \tag{3.67}$$

where C_H is the cost/m² of heliostats including present value of O&M,
F is an estimate of the final cost/benefit ratio,
a includes all multiplicative losses such as mirror reflectivity, receiver absorptivity, etc.,
$\overline{S}$ is the annual direct beam insolation during operational hours (e.g. Sun elevation greater than 10°).

Clearly the process is iterative, as F is not determined until the trial run is complete. Fortunately, convergence is quite fast. Actually, in addition to heliostat spacings in each cell and the field boundary, the process returns design point power, cost, annual energy, and cost/benefit ratio for the specified tower height, receiver dimensions and F.

If a design point power or annual energy constraint exists, one can choose trial values of F converging to that desired result at some increase in the cost/benefit ratio. Further trials with different choices of receiver size or tower height will produce the minimum value of the cost/benefit ratio.

The variational analysis yields three additional equations for the constant $\tilde{\mu}$, which must all be satisfied in every cell containing heliostats

$$\tilde{\mu} = \eta_c(\lambda_c + f_c\partial\lambda_c/\partial f_c)/(1+\beta_f)_c \tag{3.68}$$

$$\tilde{\mu} = \eta_c(\partial\lambda_c/\partial t_c)/(\beta_t)_c \tag{3.69}$$

$$\tilde{\mu} \leq \eta_c\lambda_c f_c/\psi_c \tag{3.70}$$

3.8 Design Issues and Constraints

where c is an index which runs over all cells; thus the 3 relations are satisfied independently for each cell;
η_c is the interception fraction for cell c;
λ_c is the SBC performance for the values of R and Z in question;
R is the radial separation in mirror diameters;
Z is the azimuthal separation in mirror diameters;
f_c $= 1/RZ$ is the ground coverage fraction;
t_c $= 0.5/(R^2 - Z^2)$ is a variable orthogonal to f_c;
β_f is the variation in land and wiring cost associated with a change in heliostat density (small);
β_t is the variation in wiring cost associated with a change in the orthogonal variable t (very small);
$\psi_c - 1$ is the ratio of land and wiring cost to the cost of heliostats in a given cell. (0.03 to 0.2).

The solution procedure consists of evaluating equations (3.68) and (3.69) for values of R and Z in the range covered by the data base, and searching for a point of common solution. Then, if (3.70) is satisfied for the resulting R and Z, the cell is within the boundary of the field and its performance is evaluated and summed to the receiver.

The partial derivative in (3.68) is always negative, and corresponds to the added shading and blocking losses suffered by the heliostats already in a cell when their spacings are decreased to add additional heliostats. In (3.69), β_t is a very small term, so the derivative is nearly zero at the solution, that is, a constant density change in heliostat configuration must (almost) not result in a change in the performance at the optimum point. As f/ψ is the ratio of heliostat to total field related costs, (3.70) places the field boundary where land and wiring costs become excessive or where the performance ($\eta\lambda$) is inadequate.

Result – an optimized design. The optimization process provides the boundary of the field, the heliostat spacings in each cell, and the receiver elevation and dimensions. The existing SBC data base and node file allow immediate evaluation of the design point power and receiver flux density distribution, as well as the annual performance E, the system cost C, and the cost/benefit ratio.

Response to flux density limitations. The receiver is always limited by a maximum allowable flux density constraint. Frequently the optimum heliostat field produces a peak flux density which exceeds the allowable at some time during the day. The best alternative is to modify the focus conditions and aim points of the heliostats to ameliorate this condition without inducing significantly larger spillage. The receiver design should also be checked for the possibility of altering flow paths or heat exchanger geometry to accommodate the excess flux density at little or no cost.

If significant *hot spots* remain there are several more costly possible solutions. One can accept a reduction in receiver life. One can *defocus* a portion of the heliostat field any time a hot spot occurs, at a significant cost in collected energy. One can oversize the receiver to the extent an adequate aim strategy can be developed. One can *de-optimize* the heliostat field by arbitrarily moving heliostats illuminating characteristically *hot* areas to locations where they illuminate characteristically *cool* areas. One can modify the cell matching parameter, $\tilde{\mu}$, to disfavor or favor the areas identified above and repeat the optimization process. As this last step will increase the cost of the heliostat field, it is appropriate to simultaneously consider larger receivers. The least-cost solution to the excess flux density problem will result from an optimum application of all these techniques.

3.8.3 System Performance

To obtain clear-day diurnal power curves or hourly system efficiencies, it is necessary to recompute the shading and blocking for the specified geometry at each time point, and to project the resulting energy onto the receiver. An adequate array of efficiencies requires evaluations for about 7 times on an afternoon of each of 7 months (taking maximum advantage of symmetry). A reasonable representation of system output can be obtained by defining an average clear day for each month, computing the system output, and multiplying the resulting values by the average monthly cloud cover.

To achieve a more realistic estimate of output, one must use a 365 day weather tape with 15–30 minute sampling intervals along with interpolated clear day field efficiencies from the above array. True performance calculations require a sophisticated logic system incorporating component or subsystem time constants, operation and dispatch strategies, cognizance of storage levels, and implementation of partial field defocusing under intense Sun or other conditions when power delivery exceeds either the receiver or the storage capability. Several such detailed evaluation codes have been written. The latest upgrade to the Sandia developed code, SOLERGY, does well at emulating Solar One performance [2] and has been modified to represent the salt or sodium cooled external receiver system of the Utility Study recently performed in the U.S.

3.8.4 Layout

The cellwise model discussed in Sect. 3.8.2 defines the spacings of heliostats at each cell center, but provides no means of realizing the heliostat field this represents. Fortunately, for the radial-stagger array a plot of heliostat spacing vs receiver elevation angle, $\Theta = 90° - \psi$, $\Theta < 60°$, results in a well behaved curve. A typical result for a site at 35° latitude using low cost (80 \$/m^2) round heliostats, and with Θ expressed in radians, is

$$R/Dm = 1.007/\Theta - 0.420 + 1.016\Theta = 2.035 \text{ at } 30°$$
$$Z/Dm = 1.834 - 0.7448\Theta + 0.788\Theta^2 = 1.660 \text{ at } 30° \quad (3.71)$$

The associated field boundary extends to 7.5° in the north, 9° on the sides and 16° on the south. Moderate cost (120 \$/m^2) square heliostats result in

$$R/Dm = 1.009/\Theta - 0.063 + 0.4803\Theta = 2.115 \text{ at } 30°$$
$$Z/Dm = 2.170 - .6589\,\Theta + 1.247\,\Theta^2 = 2.167 \text{ at } 30°, \quad (3.72)$$

(The $\Theta = 30° = 0.524$ radians values are given only as a convenient reference.) A more detailed analysis shows there is also a modest azimuthal variation in both R and Z in each case, but we will ignore this for the present discussion and use the above azimuthally averaged values.

The above functions define heliostat spacings at any point in the field, so we are now freed of the cellwise constraints. However, we have a new constraint. Both Z and R are defined everywhere, but the radial-stagger condition also specifies the azimuthal spacing required for heliostats on a second circle at a radius defined by the above condition. To exactly retain the *peek through* feature the azimuthal separation of subsequent circles must be exactly proportional to their radius, in contradiction to the spacing defined by Z.

One can resolve this impasse by recalling that the optimization process really defines the required performance in each cell, and then determines R and Z to satisfy this constraint. Thus, for each radial position there is a curve of required radial separations as a function of the defined azimuthal separation (i.e. defined by the requirement of the radial stagger condition). One can use these relationships to define successively smaller circles, each with the same number of heliostats on it, until the deviation from the optimum ground coverage becomes excessive. At that point it is necessary to define a *slip plane* where the number of heliostats in a circle is markedly reduced – say to 3/4 or 6/7 of its previous value. As this will destroy the exact *peek through* feature of the radial stagger, one may either delete an entire circle of heliostats, or just those which can not achieve a reasonable approximation to the nominal alignment. A range of 5-10 zones containing 5-10 circles of 40–400 heliostats each is adequate for a wide range of systems.

Close to the tower ($\Theta > 30$-$45°$) shading tends to dominate blocking, and the *peek through* feature of the radial-stagger array becomes less important. In this region, circles of closely spaced heliostats obeying essentially the same R/Dm function may well be more cost effective.

The above discussion applies as well, with very minor modification, to the north field heliostat configuration associated with a cavity receiver. For the same power level the tower will typically be 50% higher and only a 90-150° north segment of the field will be used. In both cases, the boundary of the field is defined via the optimization process and a smoothed boundary is used to define those portions of each circle to be used.

While the last few sections have not really been *optics*, they are the direct result of optical analysis, define the configuration of the optical surfaces, and are not subject to discussion under any other heading. Essentially they comprise the unique field of central receiver optics and analysis.

3.9 System Sizing

Preliminary sizing of a solar concentrating system to deliver a given design point power level is reasonably straightforward. The end use will define the delivery temperature, working fluid, and required power level. These properties drive the receiver design and typically result in an allowable flux density specification and thermal loss/m² for the receiver, as outlined in the examples of Chap. 5. Selecting a rather arbitrary series of parameters we here demonstrate how various systems may be sized. Thus, let design point power $P=300$ MW thermal at equinox noon, insolation $I=900$ W/m², reflectivity $\rho = 0.92$, interception $\eta = 0.93$, absorptivity $\alpha = 0.94$, and thermal loss $L_R=10$kW/m² for a low temperature trough system and 30kW/m² for 500°C dish or central receiver systems. The distributed systems will also have a piping efficiency $\eta_p = 95\%$, although dishes with receiver mounted electric generators will avoid this loss. For central receivers $\eta_p = 99.9\%$ and so may safely be neglected.

The sizing relationship for the distributed system is then

$$A_c = \frac{(P + A_R L_R)/\eta_p}{I\rho\eta\alpha \cos\theta_\|} \qquad (3.73)$$
$$= P/(I\rho\eta\alpha \cos\theta_\| \eta_p - L_R/C)$$

where $\cos\theta_\| = 1$ for dishes and E-W axis troughs and $\theta_\|$=latitude (at equinox noon) for N-S axis troughs.

The first task is to estimate the receiver area and thermal loss which are determined by the systems concentration ratio, estimated using Tab. 3.6 or concentrator specifications. Thus, for an N-S axis trough system at 41.4° latitude ($\cos\theta_\| = 0.75$) and a geometric concentration ratio of 50

$$A_c = 3 \cdot 10^8/(900 \cdot 0.92 \cdot 0.93 \cdot 0.94 \cdot 0.75 \cdot 0.95 - 10^4/50) \qquad (3.74)$$
$$= 3 \cdot 10^8/(515.7 - 200.0) \qquad (3.75)$$
$$= 950,000 \text{m}^2 \qquad (3.76)$$

and the land area required to prevent shading losses for solar elevation angles greater than 20° is 2.75 times this or 2.6 km². Larger spacings would collect little more energy but have higher piping losses and costs, so the optimum spacing will result from a trade study. A comparable dish system would have $\cos\theta_\| = 1$, and

$C_g \approx 2500$. Thus, $A_c = 3 \times 10^8/(687.6-12) = 444,000$ m^2 and the land area required to minimize energy losses both at midday in the winter (Sun elevation $\approx 25°$) and on summer evenings (20°) is again ≈ 2.6 km^2.

For an external central receiver system, the peak allowable absorbed flux density is an important constraint, which, for a salt-cooled receiver may be 0.85 MW/m^2. With a safety margin of 10% and a ratio of average/peak flux density of 0.6, the required receiver area is

$$300/0.85 \times 0.9 \times 0.6 = 654 \text{ m}^2. \tag{3.77}$$

Selecting an aspect ratio of 1.25, the cylindrical receiver is 12.9 m diameter and 16.1 m tall. This will provide excellent interception for heliostats ≈ 1 km distant if their $\sigma_D \approx 3.5$ mrad. For central receiver systems, shading and blocking are typically very small at design point, but the average cosine factor on the heliostats is 0.82 for a surround field and 0.95 for a north field (annual average values are 0.77 and 0.87, respectively).

Thus, for our surround field

$$\begin{aligned} A_c &= \frac{3 \cdot 10^8 + 654 \cdot 3 \cdot 10^4}{900 \cdot 0.92 \cdot 0.93 \cdot 0.94 \cdot 0.82 \cdot 1.0} \\ &= 538,000 \text{ m}^2 \end{aligned} \tag{3.78}$$

and with a nominal land coverage of 0.25, the land area required is 2.154 km^2, or adding 4% for road and power plant, 2.24 km^2 resulting in a nominal field radius of 0.844 km. In a typical optimized field the tower is offset 20% to the south, and the extreme rim angle is about 8°. Thus, the tower height is approximately $1.2 \cdot 0.844$ km$\cdot \tan 8° = 142$ m. Using a reheat turbine at 42% efficiency, $300 \cdot 0.42 = 126$ MW$_e$ can be generated and, allowing 10% for parasites, 113 MW$_e$ of electricity delivered. Alternatively, with a solar multiple of 1.8 and 6 hours of storage, 70 MW$_e$ can be generated, and 60 to 65 MW$_e$ of electricity delivered for more hours per day.

Notice that in no case did the size of the individual concentrators (troughs, dishes, or heliostats) enter into these calculations.

3.10 Appendix: Frequently Used Symbols

2-D	line focus system, single axis tracking
3-D	point focus system, two axes tracking
c	vacuum speed of light
C_{nm}	coefficient in two dimensional expansion in Hermite polynomials
C	concentration ratio
C_f, C_g	flux density, geometric concentration ratio
CPC	Compound Parabolic Concentrator
D	degraded solar distribution function $S * G$
D, Dm	diameter of optical element, characteristic dimension of heliostat
f	focal length
g	mirror fill factor
G	error function of concentrator
HT	focal height in a tower system
H_n	Hermite function of order n
$\hat{i}$	unit vector of incident beam
l	length of cylindrical collector
L	solar radiance function
LEC	Levelized Energy Cost
M	Mirror distribution function of image plane
n	index of refraction
$\hat{n}$	local surface normal
$\vec{n}$	unit vector of local surface normal
p, p_r	projection distance to focal point from minor element, from rim
PVM	Present Value Multiplier
v	phase velocity of light wave
r	reflection coefficient

$\hat{r}$	unit vector of reflected beam
R	radius of curvature
s	chord of mirror in sagittal optical plane
S	solar brightness distribution function
SR	slant range, central receiver equivalent of p
t	transmission coefficient, optical depth
t	chord of mirror in transverse optical plane
$\hat{t}$	unit vector of transmitted beam
$*$	the convolution operator
$\|$	optical plane, contains incident and reflected or transmitted beams
$\|$	(as subscript) component parallel to optical plane
$\perp$	(as subscript) component perpendicular to optical plane
$\langle\ \rangle$	weighted average value
α	half angle of solar disc
η,η^*	interception factor for beams, for edge ray
$\mu_{i,j}$	i,j th moment of the distribution function
ρ	reflectance, reflected intensity
τ	transmittance
$\tilde{\mu}$	cell matching parameter
θ, θ_i	angle of incidence (from normal)
Θ	arcsine HT/SR, receiver elevation angle
θ_a, θ_e	acceptance, exit half angle of a nonimaging concentrator
ξ	angle measured relative to center of Sun
ξ	divergence angle of solar beam
ψ,ψ_r	cone half angle to minor element, to rim
σ	standard deviation μ_{20} from a Gaussian distibution
σ_c, σ_t	standard deviation due to contour errors, to tracking errors
σ_B, σ_D	standard deviation due to beam error, to degraded Sun
ρ	μ_{20} from an arbitrary distribution
λ	wavelength

Bibliography

[1] Allen, C. W.: Solar Limb Darkening. In *Astrophysical Quantities*, University of London: The Athlone Press, 1964
[2] Alpert, D. J.; Kolb, G. J.: *Performance of the Solar One Pilot Plant as Simulated by the SOLERGY Computer Code*. Technical Report SAND88-0321, Sandia National Laboratories, Albuquerque/NM, 1988
[3] Athavaley, K.; Lipps, F.; Vant-Hull, L. L.: An Analysis of the Terminal Concentrator Concept for Solar Central Receiver Systems. *Solar Energy*, 22 (1979) 493–504
[4] Authier, B. F.: P.E.R.I.C.L.E.S. Design of a Stationary Spherical Collector. In *Proc. ISES Solar World Congress, New Delhi/India*, de Winter, F.; Cox, M. (Ed.), pp. 1235–1243, Oxford (UK): Pergamon Press, 1978
[5] Bendt, P.; Rabl, A.: *Effect of Circumsolar Radiation on Performance of Focussing Collectors*. Technical Report SERI/TR-34-093, Golden/CO, Solar Energy Research Institute, 1980
[6] Bendt, P.; Rabl, A.: *Optical Analysis of Point Focus Parabolic Radiation Concentrators*. Report SERI/TR 631-336, Golden/CO, Solar Energy Research Institute, 1980
[7] Bendt, P.; Rabl, A.; Gaul, H. W.; Reed, K. A.: *Optical Analysis and Optimization of Line Focus Solar Collectors*. Report SERI/TR 34-092, Golden/CO, Solar Energy Research Institute, 1979
[8] Bilodeau, E. A.: Power Kinetics SCSE#2 Molokai, Hawaii. In *Proc. Solar Thermal Technology Conference*, Tyner, C. E. (Ed.), pp. 130–141, Sandia National Laboratories, Albuquerque/NM, 1987
[9] Clausing, A. M.: The Performance of a Stationary Reflector/ Tracking Absorber Solar Concentrator. In *Sharing the Sun, Solar Technology in the Seventies, Proc. Joint Conf. American Sect. ISES and SES of Canada, Winnipeg/Manitoba*, Boer, K. W. (Ed.), pp. 304–326, Newark/DE: American Section of ISES, 1976

[10] Di Canio, D. G.; Treytl, W. J.; Jur, F. A.; Watson, C. D.: *Line-Focus Solar Thermal Central Receiver Research Study*. Final Report 1977–79 DOE/ET/20426-1, FMC Corporation, Santa Clara/CA, 1979
[11] Duff, W. S.; Lameiro, G. F.: A Performance Comparison Method for Solar Concentrators. In *Proc. ASME Winter Meeting, New York*, American Society of Mechanical Engineers, New York: 1974. Paper 74-WA/SOL-4
[12] Duran, J. C.; Nicolás, R. O.: Development and Applications of a Two-Dimensional Optical Analysis of Non-Perfect Cylindrical Concentrators. *Solar Energy*, 34 (1985) 257–269
[13] Eggers, G. H.; Housman, J. J.; Openshaw, F. L.; Russell, J. L.: *Solar Collector Field Subsystem Program on the Fixed Mirror Solar Concentrator*. Final Report GA-A14209 (Rev), San Diego/CA, General Atomic, 1976
[14] Etievant, C.; Amri, A.; Izygon, M.; Tedjiza, B.: *Central Receiver Plant Evaluation, Vol. 1–5*. Technical Report SAND86-8185, SAND87-8182, SAND88-8101, SAND88-8100, SAND88-102, Sandia National Laboratories, Albuquerque/NM, 1988
[15] Evans, D. L.: On the Performance of Cylindrical Parabolic Solar Concentrators with Flat Absorbers. *Solar Energy*, 19 (1977) 379–385
[16] Francia, G.: Pilot Plants of Solar Steam Generation Stations. *Solar Energy*, 12 (1968) 51–64
[17] Grasse, W.: *SSPS Results of Test and Operation 1981–1984*. Technical Report IEA-SSPS SR7, Köln (D), DLR, 1985
[18] Grether, D. F.; Evans, D.; Hunt, A.: Application of Circumsolar Measurements to Concentrating Collectors. In *Proc. ISES Solar World Congress, Atlanta*, Boer, K. W.; Glenn, B. H. (Ed.), Newark/DE: American Section of ISES, 1979
[19] Harris, J. A.; Duff, W. S.: Focal Plane Flux Distributions Produced by Solar Concentrating Reflectors. *Solar Energy*, 27 (1981) 403–411
[20] Hildebrandt, A. F.; Vant-Hull, L. L.: Power with Heliostats. *Science*, 198 (1977) 1139–1146
[21] Jeter, S. M.: Analytical Determination of the Optical Performance of Practical Parabolic Trough Collectors from Design Data. *Solar Energy*, 39 (1987) 11–21
[22] Jeter, S. M.: Calculation of the Concentrated Flux Density Distribution in Parabolic Trough Collectors by a Semifinite Formulation. *Solar Energy*, 37 (1986) 335–345
[23] Leary, P. L.; Hankins, J. D.: *User's Guide for MIRVAL*. Technical Report SAND77-8280, Sandia National Laboratories, Albuquerque/NM, 1979
[24] Lipps, F. W.; Vant-Hull, L. L.: A Cellwise Method for the Optimization of Large Central Receiver Systems. *Solar Energy*, 20 (1978) 505–516
[25] Lipps, F. W.; Vant-Hull, L. L.: Shading and Blocking Geometry for a Solar Tower Concentrator with Rectangular Mirror. In *Proc. ASME Winter Annual Meeting, New York*, New York: 1974. Paper 74-WA/SOL-11
[26] Lipps, F. W.; Walzel, M. D.: An Analytic Evaluation of the Flux Density Due to Sunlight Reflected from a Flat Mirror Having a Polygonal Boundary. *Solar Energy*, 21 (1978) 113–121
[27] Mijatovic, M.; Veselinovic, V.; Dimitrovski, D.: Inverse Iso-Intensity Absorber Problem. *Solar Energy*, 37 (1986) 25–29
[28] O'Hair, E. A.; Simpson, T. L.; Green, R.: Results from Operation of the Crosbyton Solar Bowl. In *Proc. 8th ASME Solar Energy Div. Conference, Anaheim/CA*, Ferber, R. R. (Ed.), pp. 205–209, American Society of Mechanical Engineers, New York: 1986
[29] O'Neill, M. J.: Analytical/Experimental Study of the Optical Performance of a Transmittance-Optimized Linear Fresnel Lens Solar Concentrator. In *Proc. 1980 Annual Meeting, Phoenix/AZ*, Franta, G. E.; Glenn, B. H. (Ed.), pp. 510–514, American Section of ISES, Newark/DE: 1980
[30] O'Neill, M. J.: Measured Performance for the First Three Years of Operation of the DFW Airport 27 $kW_{electric}$/120 $kW_{thermal}$ Photovoltaic & Thermal (PVT) Concentrator System. In *Conf. Rec. 19th IEEE PV Specialist Conference, New Orleans*, p. 1249, New York: IEEE, 1987
[31] Pettit, R. B.: Characterisation of the Reflected Beam Profile of Solar Mirror Materials. *Solar Energy*, 19 (1977) 733–741
[32] Pitman, C. L.; Vant-Hull, L. L.: Atmospheric Transmittance Model for a Solar Beam Propagating Between a Heliostat and a Receiver. In *Proc. 1982 Annual Meeting, Houston/TX: The Renewable Challenge*, Glenn, B. H.; Kolar, W. A. (Ed.), pp. 1247–1251, Boulder/CO: American Solar Energy Society, 1982
[33] Rabl, A.: *Active Solar Collectors and Their Applications*. New York: Oxford University Press, 1985
[34] Rabl, A.: Optical and Thermal Properties of Compound Parabolic Concentrators. *Solar Energy*, 18 (1976) 497–511
[35] Schoeffel, U.; Sizmann, R.: Optimization of High Flux Density Terminal Concentrators. In *Proc. ISES Solar World Congress, Kobe 1989: Advances in Solar Technology*, Oxford (UK): Pergamon Press, 1990

[36] Skinrood, A. C.; Brumleve, T. D.; Schaefer, C. T.; Yokomigo, C. T.; Leonard, C. M.: *Status Report on a High Temperature Solar Energy System*. Technical Report SAND74-8017, Sandia National Laboratories, Livermore/CA, 1974
[37] Tanaka, T.: Solar Thermal Electric Power Systems in Japan (Review Paper). *Solar Energy*, 25 (1980) 97–104
[38] Vittitoe, C. N.; Biggs, F.: Terrestrial Propagation. In *Proc. 1978 Annual Meeting, Denver/CO: Solar Diversification*, Boeer, K. W.; Franta, G. E. (Ed.), pp. 664–668, Newark/DE: American Section of ISES, 1978
[39] Vittitoe, C. N.; Biggs, F.: *A User's Guide to HELIOS – A Computer Program for Modeling the Optical Behavior of Reflecting Solar Concentrators, Appendices Concerning HELIOS-Code Details*. SAND81-1562 (Part III) and SAND81-1180 (Part I), Sandia National Laboratories, Albuquerque/NM, 1981
[40] Walzel, M. D.; Lipps, F. W.; Vant-Hull, L. L.: A Solar Flux Density Calculation for a Solar Tower Concentrator Using a Two-Dimensional Hermite Function Expansion. *Solar Energy*, 19 (1977) 239–253
[41] Welford, W. T.; Winston, R.: Geometrical Vector Flux and Some New Nonimaging Concentrators. *Opt. Soc. Am.*, 69 (1979) 532–535
[42] Welford, W. T.; Winston, R.: *High Collection Non-Imaging Optics*. New York: Academic Press, 1989
[43] Welford, W. T.; Winston, R.: Ideal Flux Concentrators as Shapes That do not Disturb the Geometrical Vector Flux Field: A New Derivation of the Compound Parabolic Concentrator. *Opt. Soc. Am.*, 69 (1979) 536–539
[44] Williams, T. A.; Dirks, J. A.; Brown, D. R.; Antoniak, Z. I.; Alleman, R. T.; Coomes, E. P.; Craig, S. N.; Drost, M. K.; Humphreys, K. K.; Nomura, K. K.: *Solar Thermal Bowl Concepts and Economic Comparisons for Electricity Generation*. Technical Report PNL-6129, Richland/WA, Battelle National Laboratories, 1988
[45] Zakhidov, R. A.: Mirror System Synthesis for Radiant Energy Concentration – an Inverse Problem. *Solar Energy*, 42 (1989) 509–513

4 Aspects of Solar Power Plant Engineering

W. Grasse, H. P. Hertlein and C.-J. Winter [1]

4.1 Introduction

Having presented physical and optical properties of the solar resource in Chaps. 2 and 3, principles, concepts and terminology relevant for *solar power plant* (SPP) systems will now be addressed.

First we shall discuss front-end and energy end-use aspects differentiating SPP from conventional fossil- or nuclear-fueled power plants. Next, collection principles and subsystems of SPPs will be reviewed. Finally, SPP system design aspects and operating characteristics will be discussed. This sets the stage for presenting the technical considerations applicable to SPPs in the subsequent chapters.

4.2 Solar and Conventional Power Plants: Similarities and Differences

At first glance, SPP seem functionally like conventional power plants. Depending on end-use objectives, their output may be thermal energy for direct use in heat processes (e.g. at low temperature as in the food or textile industry, or at high temperature as in the chemical industry), or it may be electricity to be used in an autonomous network or to be fed into a utility grid. Electricity may be generated either directly from solar radiation using *photovoltaic modules* in *photovoltaic power plants* (photovoltaic SPPs), or via intermediate thermomechanical conversion in *thermal solar power plants* (thermal SPPs).

On second look, however, it is apparent that the nature, quality and availability of input energy to a SPP differ significantly from those of conventional power plants. Input to a fossil or a nuclear power plant is combustible matter (coal, oil, gas, even biomass) or nuclear fuel which, after extraction from geological deposits, becomes an article of commerce. They are transported to wherever power plants are located, where they constitute a stock of concentrated (but depletable) *primary energy* raw material in chemical or nuclear form, and can be utilized if, whenever, and to the extent end-use requirements dictate. In comparison, solar radiation as primary energy input into SPPs has no raw material form, is dilute and is terrestrially accessible only during daylight hours. Its availability hinges on latitude, season,

---

[1] Wilfried Grasse and Carl-Jochen Winter, Deutsche Forschungsanstalt für Luft- und Raumfahrt (DLR), Pfaffenwaldring 38-40, D-7000 Stuttgart 80
Hansmartin P. Hertlein, Deutsche Forschungsanstalt für Luft- und Raumfahrt (DLR), Linder Höhe, D-5000 Köln 90

4.2 Solar and Conventional Power Plants: Similarities and Differences

time of day, morphology of the location and momentary meteorological conditions. The most distinguishing and limiting feature of solar radiation is that it cannot be stored directly for later use, and that it is not abundant in most places where bulk end-energy demand prevails. On the other hand, solar radiation is free, indigenous, renewable and, being free of toxins or radioactivity, of inherently very low risk.

The repercussions on SPP plant technology, operation and utility are several. First, 'mining' and supply functions must be carried out at the site of a SPP, i.e. solar energy input must be 'harvested' by *collectors*. The power rating of the plant then becomes directly proportional to the effective collector surface area. Hence, SPPs require more on-site land area than conventional power plants; on the other hand, no off-site land area is needed for energy raw material mining, processing, handling and transport, or for the disposal of (sometimes hazardous) residues.

Including land requirements for plant access roads, fuel storage and cooling towers/ponds – but excluding those for fuel mining, processing and transport – typical area requirements per kW_e installed capacity for conventional power plants range from

- 1–2 m^2/kW_e (0.2–0.4 $m^2/MWh_e \cdot a$) for coal-fired plants;
- 0.2 m^2/kW_e (0.04 $m^2/MWh_e \cdot a$) for nuclear plants;
- 0.1 m^2/kW_e for oil-fired plants;

Including open-pit or strip mining operations, the land area needed for a mine-mouth lignite power plant is in the order of 5–8 m^2/kW_e.

Land area requirements for solar/renewable energy facilities depend on local resource conditions, conversion efficiency and tracking/non-tracking configurations, but range at present typically from

- 100–1,000 m^2/kW_e for wind parks, depending on site morphology and converter size;
- 20–35 m^2/kW_e for concentrating thermal SPP (central receiver);
- 36–80 m^2/kW_e for photovoltaic SPP (tracking and fixed configurations);
- 100–2,000 m^2/kW_e for hydropower plants.

Secondly, energy must be collected by converting solar radiation into heat by *absorbers*, or into electricity by *photovoltaic cells* or *modules*. If radiation concentration is not employed, absorbers or modules immediately constitute the *collector field*. If radiation is to be concentrated before conversion, then reflective surfaces become the collectors in the thermal case, and radiation is converted to heat in separate elements (*absorber tubes; receivers*). If the concentrating collectors are refractive devices, conversion of radiation to electricity takes place in *concentrator solar cells* located within a few centimeters of the refractive elements.

Thirdly, local meteorological forecast probabilities, seasons and time of day determine the degree of certainty to which solar energy can be counted on to be available as input for a SPP. Solar energy comes 'naturally' and no authority is exercised over availability in time, quality and quantity. In comparison to fossil/nuclear power plants, SPP operating times and strategies are dictated by solar radiation availability, and special measures in SPP design become necessary if a flexible response to output demand is desired.

This limitation poses little conflict as long as consumer needs correlate in time to when solar energy is available, and as long as the demand for energy does not exceed that available from the SPP. The same holds true if output from a SPP supplements fossil/nuclear plant generation in a larger grid, so that any slack in solar generation can be taken up by conventional generation. SPPs operate essentially as *fuel savers* in this case, substituting for fossil/nuclear fuels which would have had to be spent otherwise.

If, however, capability of a SPP to generate in anticipation of or response to demand is essential, then

- either enough solar energy must be accumulated during times of sunshine, kept in intermediate storage in appropriate secondary form, and tapped at times when output is needed, whether solar radiation is present or not;
- or a second backup energy source 'auxiliary' to solar must be made available, and its utilization incorporated in the design of the SPP (*hybrid solar plant*). Auxiliary energy sources for SPPs may be gas or oil, to be combusted for generating output, either for augmenting insufficient levels of solar energy input, or for temporarily substituting solar energy[2].

Backup energy sources may have significant technical or economic advantages and benefits for operating SPPs, depending on application and energy end-use (see Chaps. 6 and 10).

4.3 Engineering Aspects

4.3.1 Collection

Several methods for the collection and conversion of solar radiation into heat are feasible (Fig. 4.1). Concentrating configurations in combination with liquids (oil; molten salts/metals) or water/steam as primary heat transfer fluids (HTF) have been favored concepts in most of the thermal SPPs which exist today. The high temperature heat achievable with radiation concentration is useful immediately for industrial process heat purposes, and is attractive for driving chemical reactions. Furthermore, efficient thermodynamic conversion becomes feasible in high temperature regimes, and prime movers can be used which are established technology in conventional power plants.

Nonetheless, the potential of lower investment costs may make non- or low-concentrating configurations attractive thermal SPP concepts, provided operating viability and economy of such systems can be verified under field conditions. Likely candidate concepts are the *solar pond* or the *vacuum tube collector*. Both concepts are suited in principle to supply low- to medium-temperature process heat, or to generate electricity in combination with a low-temperature *organic Rankine cycle* (ORC) engine. The solar pond is capable of producing even at off-sunshine hours, due to considerable thermal inventory in the pond. Experimental solar pond power systems have been built in Israel (Fig. 4.2), Australia and in the U.S. [11,17,18].

The assembly of collectors, consisting of either *point-focussing* parabolic dishes, *line-focussing* parabolic troughs, vacuum tube or non-concentrating flat plate collectors, constitute a *collector field*. Thermal SPPs using such fields of collectors are referred to as *distributed collector*, distributed *receiver*, or *farm* systems.

Point-focussing thermal SPPs concentrate direct irradiation on an absorber (i.e. on a *receiver*) of relatively compact volume. Whereas thermal energy is collected by a multitude of such receivers in parabolic dish farm systems, a field of mirrors (*heliostats*) with large focal length is used to 'transport' radiation optically to one centrally located receiver in *central receiver* or *tower* SPPs.

Another solar-thermal concept, utilizing a completely different non-concentrating principle, is the *solar chimney* consisting of translucent, large-area above-ground panes with a

---

[2] While solar-derived energy can be stored in the form of heat, the ultimate approach for matching need to supply is to shift energy demand to time periods when solar radiation is available. Such extreme adaptations in energy end-use appear to be neither urgent nor viable at present in grid-connected operation. Until such actions become necessary, priority focus must lie in fully exploiting the yet marginally tapped potential for rational and efficient energy end-use.

4.3 Engineering Aspects

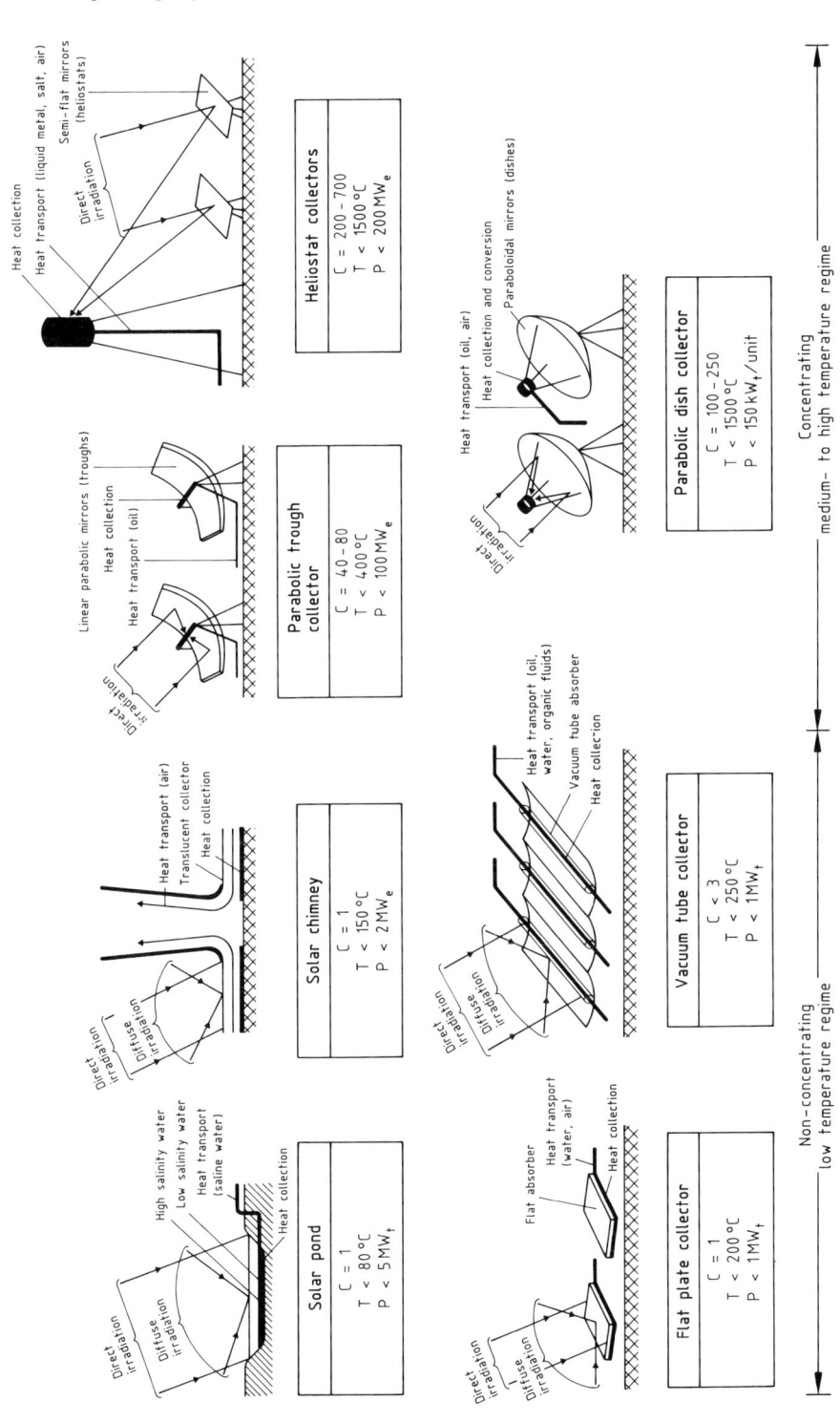

Fig. 4.1. Principal methods of solar radiation collection, absorption and conversion. With $C < 3$, total irradiation is converted into thermal energy or electricity (flat-plate PV collectors not shown). With $C > 500$, high temperatures well above 1,000 °C are achievable. Concentration in PV systems (by Fresnel lenses; not shown) is utilized only to achieve cost advantages relative to non-concentrating PV collectors. Typical state-of-the-art values are indicated. The optimal combination of parameters is a function of envisaged performance and energy production costs.

Fig. 4.2. Aerial view of the solar thermal pond power plant in Beit Ha'Arava/Israel. 5 MW$_e$ nominal rating; 250,000 m^2 pond surface; power conversion using an organic Rankine cycle (ORC); inaugurated in 1984 (Courtesy of Ormet Turbines Ltd., Yavne, Israel).

centrally attached vertical pipe. Energy is extracted by a wind turbine at the neck of the tunnel from the upward motion of air heated underneath the translucent panes. An experimental plant of this type has been built and investigated (Fig. 4.3).

Photovoltaic SPPs are coarsely analogous to distributed-receiver thermal SPPs, i.e. they harvest energy in a multitude of collecting units arranged in a collector field. The configuration of photovoltaic SPPs is modular: two-terminal *modules* (i.e. flat laminates of strings of solar cells) are attached to supporting structures and are electrically interconnected into *arrays*. Non- or low-concentrating modules and arrays are usually mounted in a fixed tilted position corresponding to latitude. If the additional gain of energy and energy value justifies the added expense for tracking, modules can be mounted to follow the Sun in one- or even two-axes.

Fig. 4.3. Aerial view of the first experimental solar thermal chimney power plant in Manzanares, Spain [14]. 50 kW$_e$ nominal rating; 194.6 m chimney height; 10.16 m chimney ID; 10 m turbine rotor diameters; 100 rpm nominal speed; 44,000 m^2 collector area; constructed in 1982, chimney destroyed in 1989 by hurricane due to guy wire break (Courtesy of Schlaich, Bergermann and Partner, Stuttgart, Germany).

Radiation collection and conversion takes place in the same plant element, e.g. in the solar cell within the module.

If Fresnel lenses of short focal length are employed in photovoltaic SPPs, the Fresnel PV modules must track the Sun in one or two axes (depending on Fresnel type). The collector field configuration remains similar to the tracking flat-module case, but the functions of collection and conversion are spatially separated (see Chap. 8 for details).

4.3.2 Energy Conversion

Electricity generation is – at present – the primary output objective of SPPs; solar process heat facilities are regarded as application opportunities for the future.

Thermal SPPs produce electricity from thermal energy quite conventionally by thermodynamic and electromechanical conversion. Thermal SPPs also inherently provide the capability of storing energy in thermal form if thermal energy is not converted directly, or used immediately for heat purposes (or in endothermic chemical reactions). Photovoltaic SPPs, in contrast, convert solar radiation into electricity without intermediate steps but lack the capability for cost-effective bulk electric energy storage.

As they are able to utilize only direct radiation, concentrating SPPs are restricted to 'sunbelt' regions where direct irradiation prevails. As heat can be stored in concentrating thermal SPPs, continuous operation through cloud transient and into non-sunshine hours is possible. On the other hand, non-concentrating photovoltaic SPPs are capable of also converting diffuse irradiation which is less restricted regionally and seasonally.

In a thermal SPP, electricity is usually generated centrally by one *power conversion unit* (PCU), i.e. the functional combination of a generator with a turbine as *prime mover*. Alternatively, parabolic dish concentrators may be equipped individually with Stirling engine (or gas turbine) generator units; several such dish/Stirling modules may be combined to form one thermal SPP of larger capacity.

Thermodynamic conversion has been introduced and was discussed in general physical terms in Chap. 2. Photovoltaic conversion will be discussed in Chap. 8. However, as thermodynamic cycle processes have profound impact on operating domains and generating characteristics of thermal SPPs, some considerations associated with radiation-to-electricity conversion via thermal energy are of basic interest.

Several thermodynamic conversion processes exist, each with distinct operating regimes and characteristics [19]; of these, two processes have attained importance in thermal power plant engineering for large installations: the Rankine cycle with water/steam as phase-change medium in a closed loop, and the Brayton cycle with air/gas as working medium in an open loop. For the operating conditions in conventional power plants, thermodynamic conversion technology is mature and the experience broad based. This does not necessarily hold true, however, if such prime movers have to be operated under the non-steady-state, frequently variable input conditions characteristic of SPPs.

Other thermodynamic cycles, developed only in small unit sizes (< 30 MW), are also of practical significance for thermal SPPs. These are the closed-loop Rankine cycle with inorganic or organic phase-change liquids common in refrigeration technology, such as chlorofluorocarbon (CFC) and hydrochlorofluorocarbon (HCFC) compounds, and the closed-loop Stirling cycle with gas (e.g. hydrogen, helium) as a working medium. Rankine cycles with working fluids other than water/steam usually are referred to as organic Rankine cycles (ORC). Small ORC turbine units are not well developed, and their future development is questionable due to the hazards of working fluids to the ozone layer. On the other hand, Stirling and Brayton cycle engines using air or hydrogen as working fluid are well suited for external heat input by solar radiation and are candidates for application in parabolic dish concentrators. However, the technology for small, efficient, closed-cycle Rankine, Stirling or Brayton converters is considerably less developed than for small steam Rankine turbines. The state diagrams of thermodynamic cycles relevant for thermal SPP applications in Fig. 4.4 illustrate the principal physical differences of these cycles.

The Carnot efficiency (2.59) defines the physical limits for conversion of heat, Q, into mechanical work, W, merely as a function of highest to lowest cycle temperature. The conversion quality of thermodynamic cycles achieved in practical settings is expressed in relation to this limit, usually at nominal steady-state operating conditions. Off-nominal operating conditions reduce this relation, sometimes significantly, depending on cycle and converter type, capacity and operation parameters. The thermodynamic conversion quality (*thermodynamic cycle efficiency*) at steady-state nominal operating conditions is broadly categorized in Fig. 4.5 as a function of upper cycle temperature [19], for specific ranges of cycle capacities.

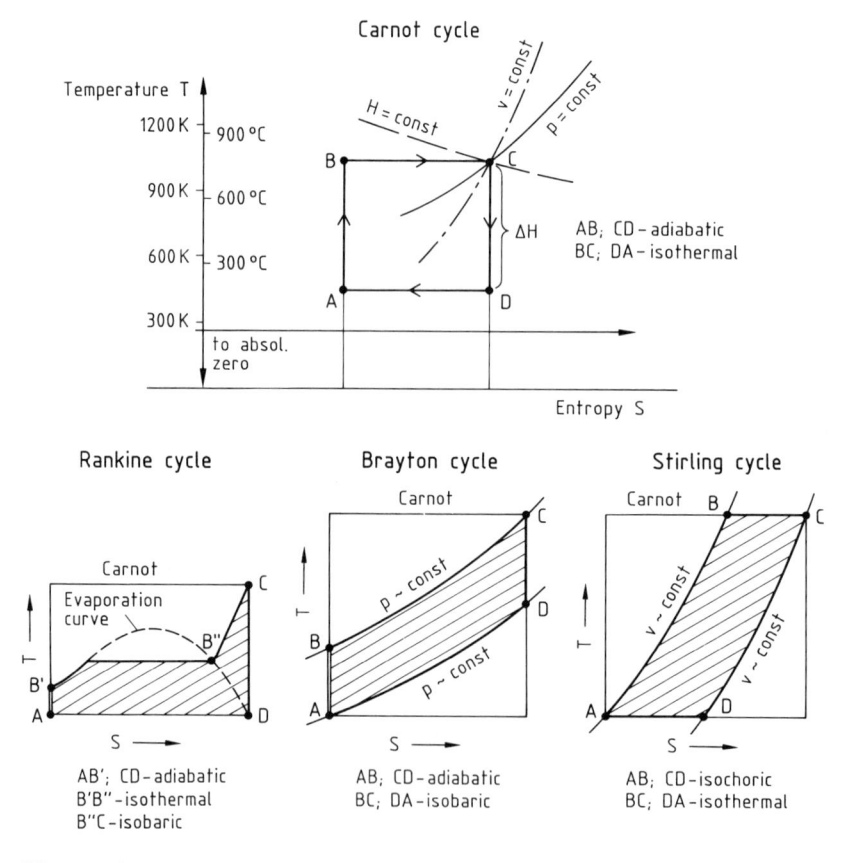

Fig. 4.4. Ideal state diagrams of thermodynamic cycles of practical significance for thermal SPPs. Carnot is a hypothetical loss-free process of four infinitesimally slow changes of state: isentropic (or adiabatic) compression, isothermal heat addition, isentropic expansion, and isothermal heat rejection. The enclosed area is indicative of conversion efficiency, depending only on cycle temperature differences (in K). For given temperatures, a Stirling cycles is the most efficient as state changes occur isothermally and at constant volume (isochoric), the latter approaching adiabatic conditions.

4.3 Engineering Aspects

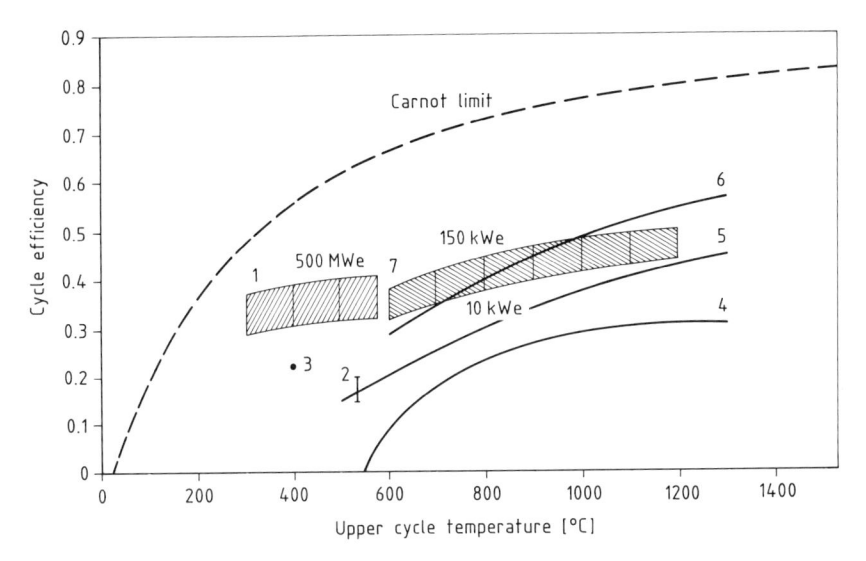

Fig. 4.5. Steady-state conversion efficiencies of real thermodynamic cycles (at rated operating conditions and 300 K ambient temperature [19]). Cases and conditions: (1) steam Rankine cycle in the range of 300 MW$_t$ (100 MW$_e$); (2) small steam Rankine cycle cycle about 3 MW$_t$ (0.5 MW$_e$); (3) ORC at about 100 kW$_t$ (25 kW$_e$); (4) open Brayton cycle without heat recuperation; (5) closed Brayton cycle with intake-exhaust pressure ratio of 4 and heat recuperation; (6) steam Rankine combined with open Brayton topping cycle; (7) Stirling cycle in the range of 0.03 to 0.4 MW$_t$ (0.01 to 0.15 MW$_e$).

Although practical cycle efficiencies mirror the increase in Carnot efficiency with upper cycle temperature only to a degree, they may be expressed, at specific operating conditions, as a fraction f of Carnot

$$f = \frac{\eta_{\text{real}}}{\eta_{\text{Carnot}}}. \tag{4.1}$$

Mature large capacity Rankine cycles (> 200 MW$_t$) may reach f-values of 0.5 to 0.7 at nominal steady-state operating conditions; small capacity Rankine cycles (< 1 MW$_t$), on the other hand, may have difficulty exceeding f-values of 0.3 at the same upper cycle temperature (Fig. 4.5).

Other observations can be drawn from Fig. 4.5. While large steam Rankine cycles may obtain high steady-state conversion efficiencies (for instance 42% at 570°C and 200 bar [19]), Brayton cycles require an operating temperature about 300°C higher and optimal configuration (i.e. closed loop, heat recuperation and intercooling) to reach the same efficiency level, and a temperature about 500°C higher to achieve the same f-value. Also, steam Rankine cycles are not operated above about 560°C, because added expense for higher temperature metal alloys does not compensate for the incremental gain in efficiency (the onset of thermal dissociation of water is another limiting factor above 800°C). Thermal stability of HCFCs limits the upper operating temperature of organic Rankine cycles to below 300°C. Stipulating equal lower process temperature for all cycles, the longer span between upper and lower operating temperatures (600 to 900°C) is a major contributor to the superior Stirling conversion efficiency. Closed-loop operation and advantageous thermodynamic changes of state are further factors (a 35.7% peak gross electrical output at about 650°C for a 25 KW$_e$ Stirling engine/generator unit has been demonstrated [3]). Unfortunately, Stirling converters of > 300 kW$_t$/140 kW$_e$ are not commercially available, Stirling prototypes in the MW range do not exist, and the present development base for Stirling technology is very small (as compared to that for steam Rankine or Brayton converters). Still, the available Stirling converters of modest size are already very attractive for application in combination with single dish concentrators.

Application of thermodynamic conversion technology to thermal SPPs may not be trivial. Steam Rankine cycle equipment must technically and operationally be adapted to the cyclic, frequently part-load and often variable thermal input conditions in thermal SPPs which lack compensating provisions[3]. Providing operating conditions which remain optimal for steam Rankine cycle operation thus becomes a design objective for thermal SPPs, particularly for those with large-capacity and high-efficiency cycles operating at high nominal temperatures (> 500°C) and pressures (> 100 bar). Good part-load efficiency is critical irrespective of Rankine cycle capacity (a 60% increase in turbine gross heat rate at 20% of nominal load is indicative of the efficiency decay at part-load [1]). Small-size steam Rankine power conversion units, i.e. with less than 20 MW_t, are advantageous in terms of operating flexibility and response capability, but are neither honed for high efficiency nor are very common for power generation.

Brayton cycles are conventionally operated in open-loop configuration, mostly for short-duration peak generation. Applied to a thermal SPP, only closed Brayton configurations operating at high temperatures appear viable for efficiency reasons, again particularly at part-load. Nonetheless, lower complexity and cost may still make open cycle solutions attractive in solar applications [2].

Radiation-to-thermal and thermodynamic cycle efficiencies limit the attainable and practical thermal SPP system efficiency. Such efficiency regions are calculated in Fig. 4.6 and Fig. 4.7, under the assumptions specified, for the three major combinations of thermal SPP concepts (parabolic trough, central receiver, parabolic dish) and thermodynamic cycles (Rankine, combined Brayton/Rankine, Stirling). The upper edge of these regions is indicative for peak thermal SPP performance, steady-state operation and a technically viable combination of collection concept and conversion cycle. Temperature limits of the heat transfer media (water/steam, synthetic oils) or materials determine the upper process temperatures, and hence the attainable cycle efficiencies.

In photovoltaic SPPs, conversion to electrical energy takes place in semiconductor material immediately at the collecting front-end, and only DC electricity may need to be processed before output. Such conversion is performed in electrical/electronic devices and does not involve a change in the physical form of energy. For example, a *power conditioning unit* may adapt DC voltage and current levels, e.g. for battery charging, and an electronic *power conversion unit* (PCU) may invert DC to AC electricity. As in thermal SPPs, part-load performance of power conditioning/conversion devices constitutes a key engineering issue also in photovoltaic SPPs (an efficiency reduction by 10–15% when operating at 10% of nominal rating is a typical figure for demonstrated PCU performance [16]).

4.3.3 Characterization and Physical Properties of Solar Power Plants

System diagram. Any power plant can be reduced to and functionally represented as a *block diagram*. The functional blocks correspond individually or in combination, depending on plant concept, with the major subsystems. The block diagram is characteristic, although quite similar for thermal and photovoltaic SPPs (see Fig. 4.8).

---

[3] An intermediate thermal storage unit can mitigate the cycling effect of cloud transcients on the turbine performance; a larger storage capacity can furthermore increase daily operating hours by 50–60%, but the thermal SPP remains subject to daily start-up procedures, to be repeated 300 to 350 times a year.

4.3 Engineering Aspects

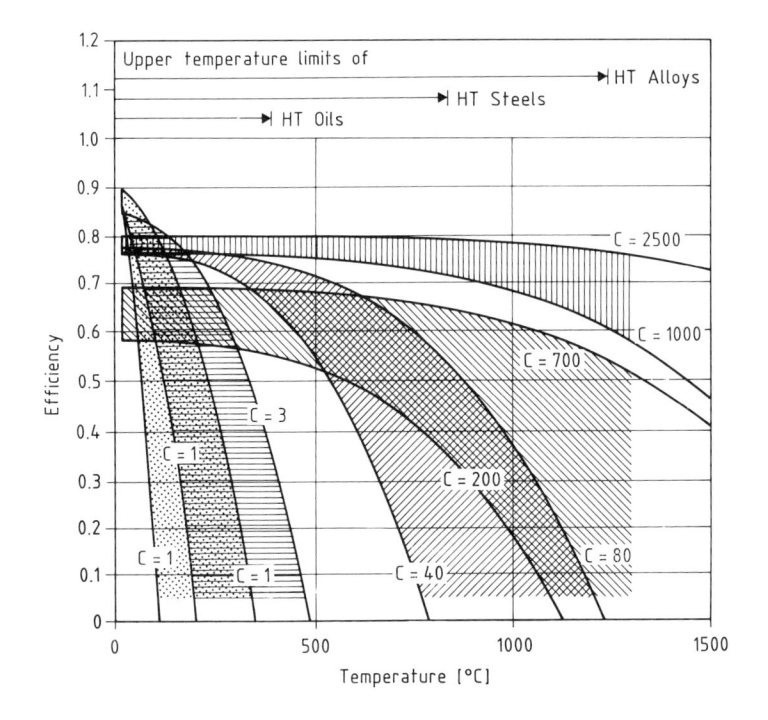

Fig. 4.6. Practical limits of solar thermal collector efficiency and operating temperature. Using (2.113) and neglecting convective and conductive losses, regions for radiation-to-heat conversion for principal concepts of radiation collection and/or concentration can be defined (depending on concentration, surface absorption/emission, and reflection/transmission). Performance close to the upper end of the bands is attainable with high-quality technology.

Assumptions:

Collectors	C	α	α/ε	η_{TR}
Flat-Plate	1	0.95	1.0	0.95
	1	0.95	10.0	0.95
Vacuum	1	0.90	3.3	0.95
	3	0.90	10.0	0.95
Trough	40	0.90	2.5	0.86
	80	0.90	5.0	0.86
Heliostat	200	0.90	2.0	0.65
	700	0.90	3.3	0.77
Dish	1000	0.90	2.0	0.85
	2500	0.90	3.3	0.89

Radiation is the primary energy input to SPPs (the *solar fuel*). Optionally, an auxiliary fossil-fueled source may provide backup in case of inadequate or lacking solar energy input. Usually a gas boiler (*auxiliary boiler*) provides the alternate source for thermal input in a thermal SPP. In large photovoltaic SPPs, a diesel generator may be the backup for electrical plant output. If a Diesel/generator is used in a thermal SPP, exhaust heat of the diesel engine can be useful in addition for covering part of parasitic thermal stand-by requirements.

The collecting, concentrating and radiation-to-heat conversion elements constitute the *collector/receiver subsystem* of a thermal SPP. In photovoltaic SPPs, the assembly of several photovoltaic modules constitutes an *array*, one or several arrays being the *array field*.

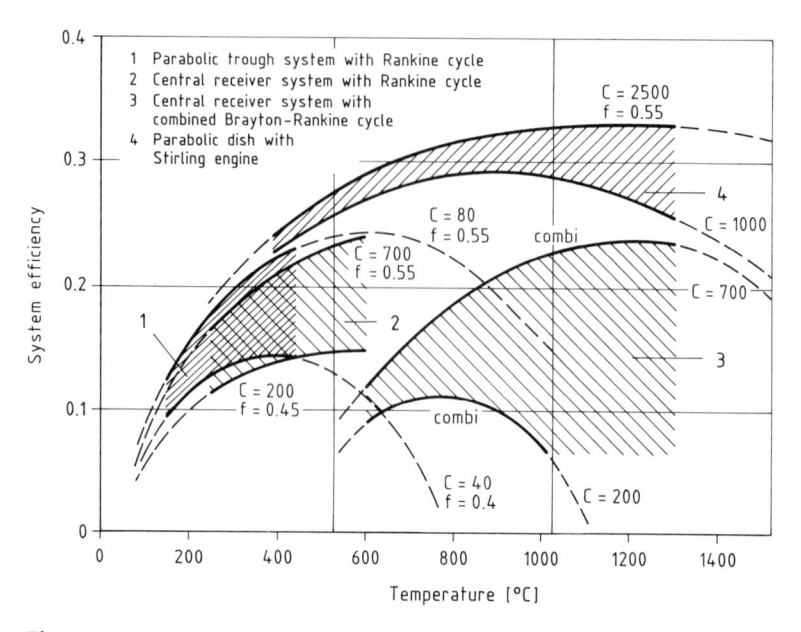

Fig. 4.7. Regions of practical peak conversion efficiency and temperature for thermal SPPs. Operating regimes are bound by radiation-to-heat conversion efficiency, attainable concentration, and temperature limits. Radiation-to-heat conversion modeled as in Fig. 4.6, and cycle efficiency as in Fig. 4.5. (1) $\eta_{cycle}=40\%$ at C=40, and 55% at C=80, temperature limited by thermal oils; (2) $\eta_{cycle}=45\%$ at C=200, and 55% at C=700, temperature bound by material cost at $\approx 560°C$; (3) $\eta_{cycle}=45\%$ at C=200, and 55% at C=700, temperature bound by material cycle fatigue; (4) $\eta_{cycle}=40\%$ at C=1,000, and 55% at C=2,500, temperature bound by material cycle fatigue.

Once converted into sensible and/or latent heat of a suitable heat transfer fluid (HTF), energy in thermal SPPs is processed downstream in the *heat transfer subsystem* (similar to practice in standard thermal power plants). No equivalent subsystem exists in the photovoltaic SPP; after conversion to electricity, plant-internal energy transport takes place by electric conduction.

Intermediate storage of solar-derived energy (*storage subsystem*) is quite feasible in thermal SPPs, but only in principle in photovoltaic SPPs. Its obvious main function is to shift utilization of solar energy into off-sunshine time periods. It provides an essential second function in stand-alone operations in steadying plant output under fluctuating radiation input conditions. In thermal SPPs, the beneficial further function of storage is the shielding of thermodynamic cycle operation from fluctuations in solar energy input, thus stabilizing operating conditions and improving conversion efficiency. Additionally, the thermal inventory provided by storage can serve as a ready reservoir for covering parasitic thermal needs, for instance for thermal SPP start-up.

The key functional distinction between thermal and photovoltaic SPPs is their capability of bulk energy storage. In thermal SPPs, thermal energy storage in large quantities is feasible by storage of hot HTFs (although not yet demonstrated on a commercial scale in thermal SPPs). Electrochemical battery energy storage in photovoltaic SPPs, however, is not cost-effective unless demand in stand-alone applications so dictates.

Usually, more detailed *system diagrams* are used for technically describing SPP systems. Such diagrams trace the main energy flow through the piping network and subsystems in a

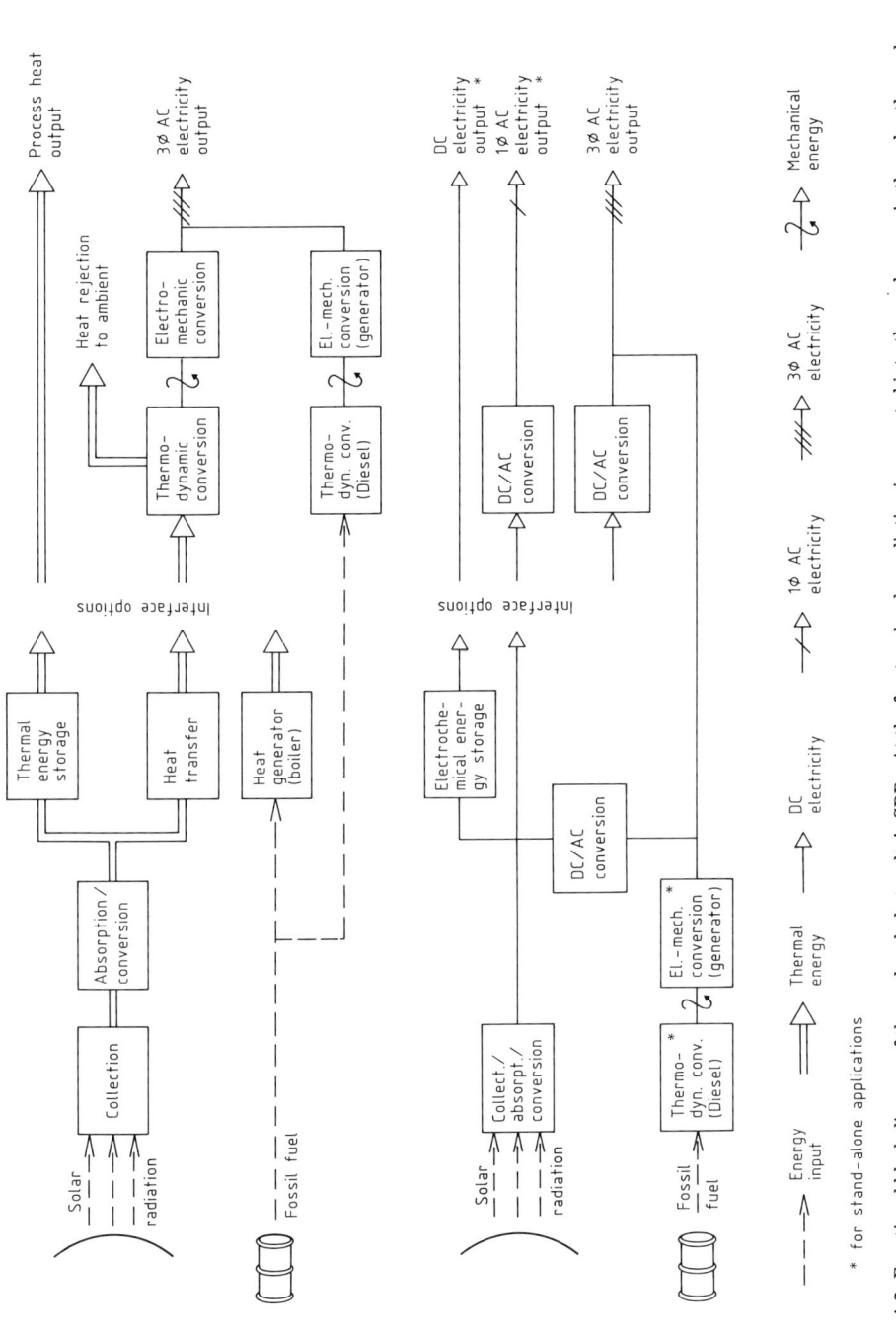

Fig. 4.8. Functional block diagram of thermal and photovoltaic SPPs. At the front end, solar radiation is converted into thermal energy in the solar thermal case, and into DC electricity in the photovoltaic case. Hence, energy in solar thermal systems is processed via mass transport of heat transfer media, while energy is handled by electrical/electronic means in photovoltaic systems. Fossil energy input auxiliary to solar is possible both in solar thermal (by heat boiler; Diesel/generator set) and photovoltaic systems (Diesel/generator set). Thermal energy is amenable to being stored in significant quantity in solar thermal systems, while bulk energy storage by electrochemical batteries in a photovoltaic system is not viable for cost reasons.

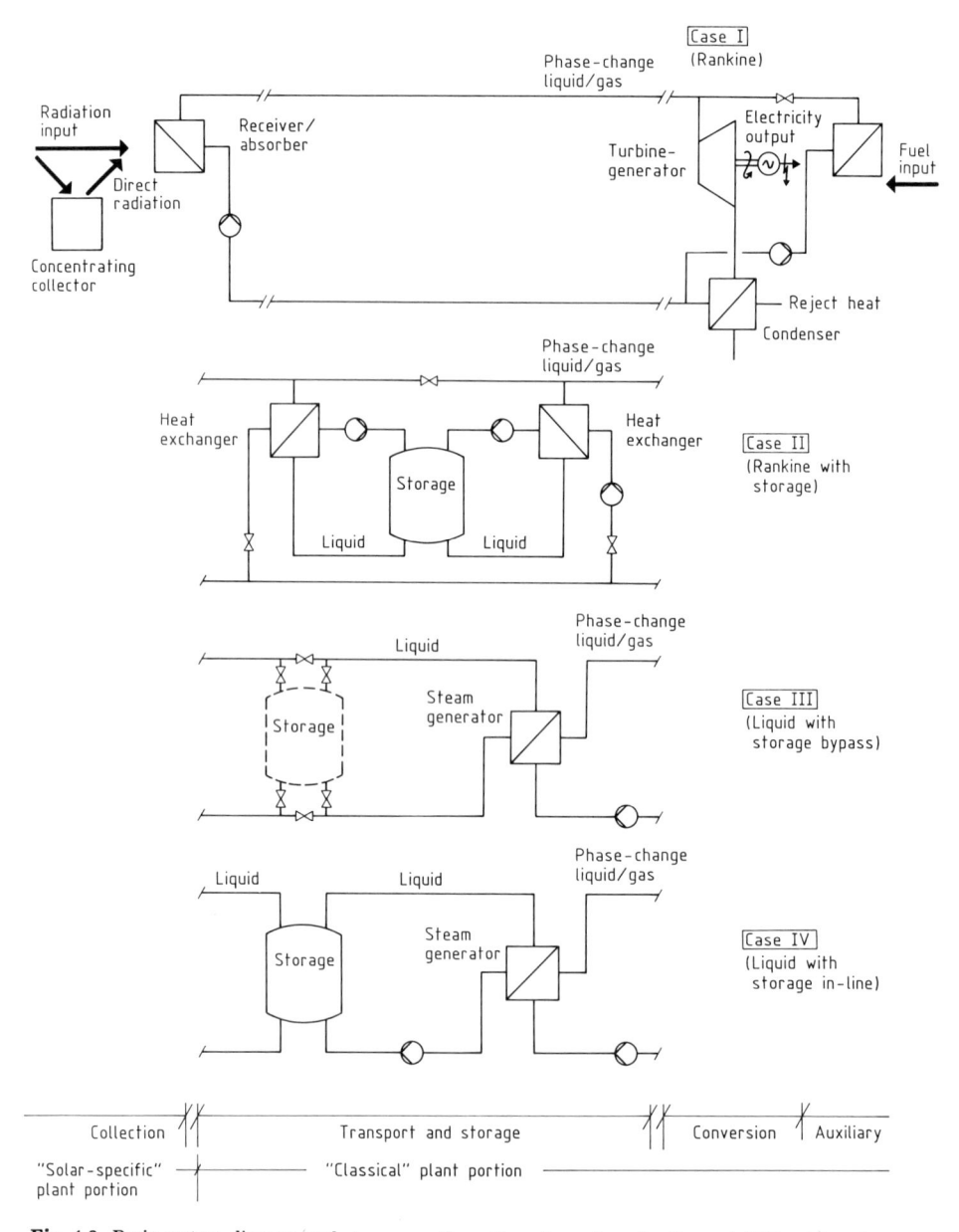

Fig. 4.9. Basic system diagram and storage configuration alternatives for thermal SPPs. After absorption, thermal energy is transported by phase-change media (water; organic fluids), liquids (mineral oils; molten salts) or gases (air; hydrogen). By sensible heat, thermal energy can be stored and retrieved later (except in dish/Stirling systems with moving dish and engine). In-line storage (Case IV) shields power conversion from the immediate impact of radiation input fluctuations. For storage charging/discharging a heat exchange is needed if media for transport, storage and power conversion differ (Case II and III). In Stirling systems, absorption, heat transport and conversion are integral within one engine block.

thermal SPP. System diagrams in photovoltaic SPPs become cabling or electrical interconnection diagrams. Such system diagrams are commonly used on the detailed engineering level. In Fig. 4.9, the system diagram of a thermal SPP is shown in its simplest form, together with configuration alternatives for the thermal transport and storage subsystems. Depending on the primary and secondary heat transfer fluid (HTF) and storage medium, storage can be arranged in in-line or bypass configuration. Frequently, intermediate heat exchangers become necessary. In-line storage can be instrumental for shielding operation of the thermodynamic conversion cycle from fluctuations in solar energy input. In dish/Stirling combinations, however, thermal energy storage usually is not viable for physical and technical reasons.

Energy flow diagrams. Absorption, transport and conversion of energy is unavoidably associated with losses. It is instructive therefore to examine the flow of energy through the system in the form of a Sankey (*loss tree*) diagram, or in a cascade (*waterfall*) diagram. A generalized loss tree for thermal SPPs is shown in Fig. 4.10 and, for comparison, for photovoltaic SPPs in Fig. 4.11. Such representations illustrate the performance of the system at each level in terms of energy, i.e. of energy output in relation to energy input over a specified time period. Such representations are valid for steady-state operation under specified conditions, for instance for operation under peak, nominal or off-nominal conditions. Frequently, loss trees are used also, although imprecisely, to illustrate time-averaged energy values over time periods which include transient operating episodes; hence, as inertia of thermal systems is usually not negligible, and since inertia effects usually do not average out over the time period considered, energy loss tree representations need careful interpretation.

Several factors influence the performance of a SPP and must be accounted for in a design process. Such factors are, for example,

- inoperability for reasons of planned or unexpected outages and repairs, because irradiance stays below a certain threshold needed for operation, or because the energy balance is negative, i.e. if irradiance stays below the level needed for generating enough output to cover system losses and auxiliary energy requirements (over the time period considered).
- Restrictions or losses incurred for physical and/or technical reasons, such as
  - the ability to utilize only direct irradiation in concentrating thermal and photovoltaic SPPs,
  - optical losses (i.e. imperfect reflection caused by dirty reflectors, or reduced absorption at large incidence angles), atmospheric losses (e.g. beam attenuation by haze or dust under large focal lengths), or spillage losses (e.g. imperfect tracking and beam aiming under windy conditions);
  - thermal reradiation, conductive and convective losses at receivers/absorbers in thermal SPPs,
  - thermal conductivity (heat dissipation) and non-recoverable heat inventory in the heat transport and storage subsystems in thermal SPPs, or ohmic losses in the field cabling in photovoltaic SPPs,
  - cosine losses in fixed-collector or 1-axis-tracking SPPs,
  - thermodynamic and electromechanical conversion losses (in thermal SPPs, and radiation-to-electricity conversion losses in the semiconductor material together with power conditioning losses in photovoltaic SPPs).
- The thermal and electrical parasitic needs for operating a SPP, or for keeping it operational over periods when solar input is not, or only inadequately, available.

The existence, amount and relative significance of these losses may vary in individual cases, depending on the collection principle employed, on full or part-load operating regimes, on the time period considered, and on plant type, design and size. For instance, immediate thermal

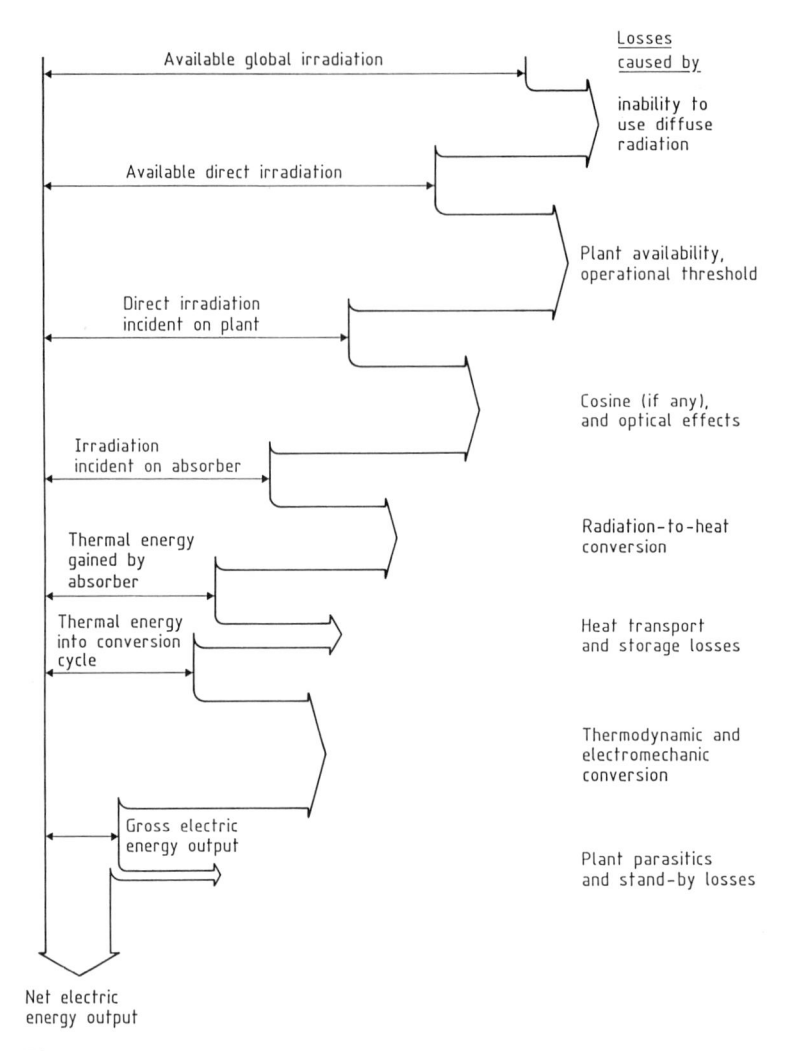

Fig. 4.10. Simplified Sankey-diagram (or 'loss tree') of a thermal SPP (not to scale). Three kinds of energy 'losses' can be differentiated. First, loss of solar radiation if the plant is inoperable during sunshine hours, or if only direct irradiation can be utilized. Secondly, losses which are unavoidable for physical/technical reasons during collection, conversion and processing operations. Thirdly, operational stand-by losses induced by keeping the plant operational overnight and over (possibly extended) inclement weather periods.

end use obviates thermodynamic conversion losses, or low plant availability may overshadow the significance of the other loss elements.

Efficiency. Efficiency, as defined in Chap. 2 in general terms, implies – and efficiency measurements require – steady non-changing boundary conditions ($dQ/dt = dW/dt = 0$). Although energies are used in the definition, time is not an explicit factor. Under steady-state conditions, ratios of energy and ratios of power become the same.

4.3 Engineering Aspects

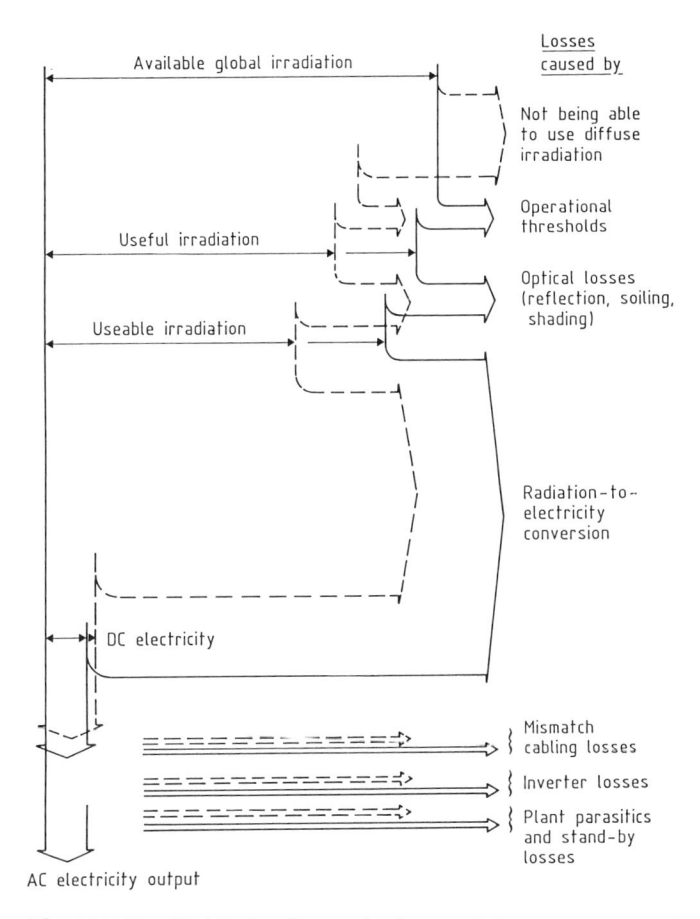

Fig. 4.11. Simplified Sankey-diagram (or 'loss tree') for concentrating and non-concentrating photovoltaic SPPs (not to scale). The same categories of energy 'losses' apply as for the thermal SPP case (see Fig. 4.10).

It is established practice in engineering, therefore, to define and to interpret efficiency as the ratio of output power (kW_{out}) to input power (kW_{in}). The tacit assumption remains, however, that steady-state operating conditions prevail, independent of any length of time which might be considered. The attractiveness of these assumptions is that system-level efficiency becomes readily calculable, i.e. by multiplying the efficiencies of successive subsystems which make up the total system. The efficiencies thus used and defined would more appropriately be termed *power ratios*. Power ratios can be zero but can never become negative.

In any practical setting, however, time and time-dependent events are indeed factors which influence the energy throughput of a system. Under transient regimes, efficiency in terms of energy (*energy efficiency*) may differ significantly from power ratio; specifically, system-level efficiency no longer is the simple product of subsystem power ratios. Energy-based efficiency under transient conditions becomes ambiguous because technical factors such as energy inventories and storage, thermal inertia, operating parameters, etc. and the time period over which energy flow is accumulated, all influence the numerical value of the energy input-output ratio.

Any energy balance of a thermal SPP has to include its thermal inventory. For example, energy flows into the system but no output is generated during morning warm-up of a thermal SPP. In terms of input and output, power ratio and efficiency would be zero over this time period. At the other extreme, output can be produced solely by depleting a thermal inventory, i.e. the thermal storage (solar energy input being absent). Such an operation mode cannot be described in simple terms of power or energy input-output ratios. A third example might be the case where the amount of energy output generated over some time period is less than the parasitic energy needed for keeping the system operational; efficiency in terms of energy output to solar energy input would be negative, and hence would be without meaning.

Using the term efficiency as an indicator for the capability of a SPP to convert and to transfer energy hence requires qualifying annotations[4].

A further distinction has to be made when assessing SPP performance: the time periods on which energy efficiency are based may be restricted to periods when output actually has been generated; to periods of sunshine hours (whether or not the plant is operating); or to the entire 24-hour day period. This distinction is important because net generation is affected noticeably by auxiliary and parasitic energy needs for plant operation and stand-by[5]. The significance of different time bases is best illustrated by way of an actual example (Fig. 4.12). Four bases for calculating input/output relationships are in use [3] and are to be differentiated:

1. *Gross* or *net power efficiency*. Ratio of output power, divided by the active collector area (kW_t/m^2; kW_e/m^2), at a given instant in time, to the total or direct irradiation (kWh_t/m^2) incident on the collector surface at the same instant in time (depending on SPP type).

2. *Operating hours energy efficiency*. Ratio of energy output, divided by the active collector area (kWh_t/m^2; kWh_e/m^2), over a given time span during which the SPP is in operation, to the total or direct irradiation incident on the collector surface (kWh_t/m^2) over the same time span.

3. *24-hours energy efficiency*. Ratio of 24-hour average energy output, divided by the active collector area (kWh_t/m^2; kWh_e/m^2), to the total or direct irradiation (kWh_t/m^2) incident on the collector surface area from sunrise to sunset. This ratio may also be averaged over several 24-hour periods.

4. *Annual energy efficiency*. Ratio of the total of output generated over a year (kWh_t/m^2a; kWh_e/m^2a), to the aggregate yearly total or direct irradiation (kW_t/m^2a) incident on the collector surface.

For given site-specific meteorological conditions, the annual net energy efficiency (or annual energy yield) is the most significant criterion for evaluating SPP technical/operational performance at the design stage, and to assess operational performance/reliability in relation to predicted performance. It is also the key performance baseline figure for judging value and cost-effectiveness of a SPP scheme. As different meteorological conditions/statistics will result in different net annual energy efficiencies, the same SPP configuration may perform and produce differently at various sites.

Input-output characteristics. In detailed design, annual output performance is predicted by computational simulation, using a model of the SPP dynamic characteristics and irradiation

---

[4] Using *energy yield factor* to denote the ratio of SPP energy output to solar energy input, and to differentiate it from the customary use of the term efficiency, could add to clarity of argument in technical discussions.

[5] With the exception of peaking plants, most conventional power plants generate rather continuously over extended time periods; parasitic power needs then range in the order of < 5% to 10% of rated power. Operating parasitics of thermal SPPs are of similar magnitude, but SPPs are typically in stand-by mode roughly 2/3 of the time. Unless stand-by parasitics are kept small, they may consume a considerable fraction of gross SPP output generated during insolation intervals.

4.3 Engineering Aspects

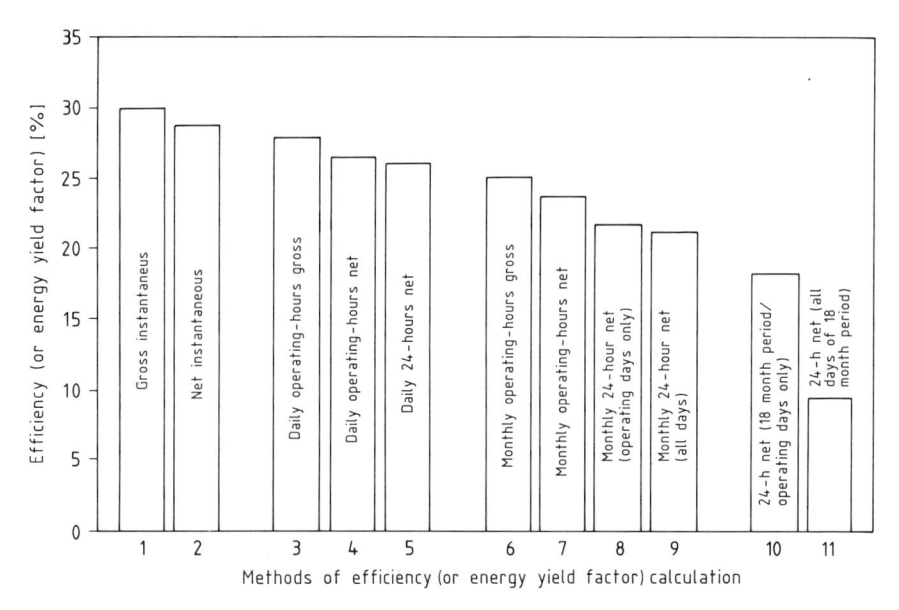

Fig. 4.12. Solar power plant efficiencies calculated on different bases [3]. This example of a dish/Stirling concentrator system experiment demonstrated record-setting peak power efficiency, but net energy yield over an 18 month operation period was significantly lower. Depending on the period of time over which performance is averaged, significant differences may result.

data representative of the plant location. Measured long-term irradiation data, averaged over appropriate small time increments (minutes, hours), are the best basis for such a numerical analysis. The performance predicted on this basis is well suited for analytically comparing the calculated performance of alternate SPP configurations for the same irradiation conditions at the same site, even for extended time periods.

As many of the early SPPs were built and operated as experimental/prototype facilities (excepting the Solar One facility in Barstow/CA which was operated for 2–3 years in a mode resembling utility-like operation), only limited long-time operational performance data at system level have been accumulated and reported in the open literature. This fact renders it difficult to correlate actual with predicted output performance, and to compare different plant concepts. Methods simpler than extensive computer runs would be useful for assessing SPP performance.

In the past, approaches for presenting plant performance of different SPP configurations varied. In 1985, it was suggested [4] to characterize and to approximately evaluate the performance of a thermal SPP using input-output functions. Such transfer functions relate daily plant output, Q, to solar energy, I, which has been available for input over the same day (both input and output normalized with respect to the effective collector area). Solar energy input may be defined in terms of direct normal irradiation (DNI) for concentrating systems, and in terms of total irradiation for other SPP configurations.

Input-output curves are compiled by linear regression of representative data sets consisting of daily plant output (selecting days with 24 hours trouble-free operation) and correlated aggregate daily irradiation. In a first approximation, the data sets from which input-output functions are derived need neither to comprise 365 days nor to represent consecutive operational days[6].

A set of generalized linear input-output functions (Fig. 4.13) comprises, in the thermal SPP case, functions for heat produced, Q_{therm}, and for gross and net electric output, ($Q_{el,gross}$, $Q_{el,net}$), all functions exhibiting irradiation thresholds before positive output can be attained.

Input-output functions in the case of photovoltaic SPPs reduce to gross and net electrical transfer functions, as useful thermal energy is not produced. The slope of the input-output curve is less steep in comparison, but insolation threshold for positive output is lower and – in the case of non-concentrating photovoltaic SPPs – energy input is total rather than only direct irradiation; in addition, the slope of the input-output curve is temperature sensitive (for effects of solar cell temperature at high insolation levels, see Chap. 8).

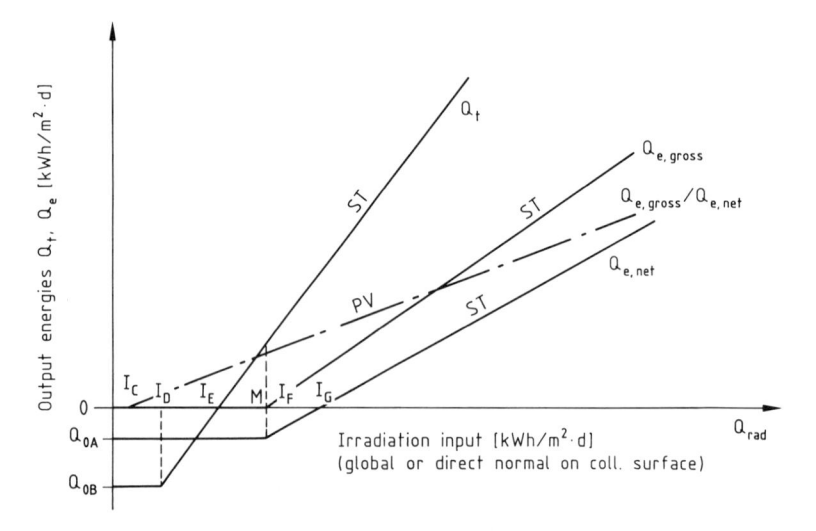

Fig. 4.13. Daily energy input-output relationships for thermal (ST) and photovoltaic (PV) SPPs. Below direct irradiation of I_E, thermal losses of magnitude Q_{OB} occur during warm stand-by of thermal SPPs (partially covered above I_D). Gross/net electric output occurs above I_F/I_G. Electric parasitics correspond to Q_{OA} (covered partially above I_F). Output slope is indicative of conversion efficiency. In photovoltaic SPPs, net DC electric output is generated already above (direct or total) irradiation I_C but slope of the output curve is lower.

Averaging irradiation daily (or monthly) and establishing a distribution of the number of days falling within certain categories of daily average irradiation (Fig. 4.14), a first-order approximation of yearly plant output performance can be estimated using the input-output curve. The approach produces upper-bound performance estimates, as the influence of plant availability on performance is neglected; the impact of weather and meteorological statistics on performance can only be estimated in the case of thermal SPPs, for instance the effects due to weather-induced irradiation transients, or due to prolonged periods of energy input below the threshold for net output.

Nonetheless, input-output representations of SPP performance in terms of location-specific irradiation distribution offer a simple method for visualizing operational behavior, and for

---

[6] Recent data [10] suggest that a non linear regression of the set of daily input-output data could more truthfully represent the performance of thermal SPPs. Validation of this hypothesis, however, requires further analysis and a larger, statistically more significant performance data base.

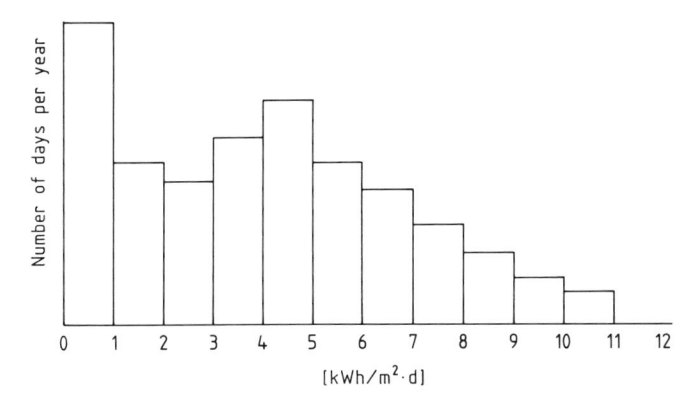

Fig. 4.14. Yearly distribution of daily (total or direct normal) irradiation per m² of collector area (kWh/m²·d). No net output from solar energy input can be generated during days with average irradiation less than the threshold for output. Low thresholds for net output generation, as well as a maximum of days in the year with daily irradiation levels above this theshold, are essential for optimal SPP operation and performance.

comparatively assessing, in a first-order approximation, performance of different SPP types and configurations under different irradiation conditions[7].

4.4 Design Aspects

Design of conventional power plants, in general, starts with a decision about supply needs, plant size and operational strategy (e.g. peak, intermediate, base load or load-following supply). The next steps focus on the selection of the primary energy source and on means and options for heat rejection (in case of thermodynamic cycles). Definition of the heat transfer and/or working medium, as well as the selection of a prime mover unit which best satisfies supply needs, are other key decisions. Hence the procedures for correlating size and capacity of subsystems with requirements, and the evaluation of system performance, are well established; in comparison, design procedures for SPPs are still being refined and are not yet standardized.

There is no question as to the primary energy source for a SPP. A key decision will be whether intermediate energy storage and/or auxiliary (fossil) energy sources must be incorporated to satisfy output requirements. However, because of the nature and quality of solar radiation, SPP design procedures differ in some essential aspects from the conventional design approach. Concentrating thermal SPPs are dependent on good direct normal irradiation (DNI) conditions, typically found in desert locations. In a desert environment, however, availability of water for cycle cooling can become critical. If wet cooling is not feasible for lack of water, heat must be rejected at ambient air temperatures, diminishing the advantage of a desert location with good DNI availability.

---

[7] Such assessment may become questionable if radiation distribution and meteorological statistics are distinctly different from those on which the derivation of the transfer function has been based.

4.4.1 Terminology

Design point. A conventional thermal plant facility is usually sized on the basis of nominal (nameplate) output conditions. Operation at these conditions may be maintained for extended time periods (base load supply).

As solar energy input is periodic and fluctuating, however, design of a SPP must start by specifying energy input conditions, i.e. assuming a specific irradiance to be present at a specific time during a particular day of the year. This so-called *design point* along with a specification of the output power capacity under rated conditions provides the basis for technically sizing a SPP, its subsystems and elements. The design point irradiation (800 W/m² for example) is frequently chosen at 10:00 a.m. or at noon at equinox, but sometimes also at noon of summer solstice. Rated (nameplate) power capacity of SPPs is usually stated in terms of design point operating conditions. Comparison of power capacity between SPP alternatives on the basis of nominal performance, therefore, is only of limited significance as design points usually are not identical, and because irradiation and ambient conditions may be different at the locations where SPPs are sited.

Performance. SPP power capacity under design point conditions conveys only an idea of the potential of a SPP to provide energy output over extended time periods. Actual SPP *energy output performance* (over a year, month, week, or a day) must be calculated by adding up average energy output over incremental time intervals (e.g. minutes). Key factors influencing SPP energy output performance are the irradiation as determined by the meteorological situation, the SPP operating strategies aiming at economizing plant parasitics and losses including those for storage (if any), as well as the readiness of a SPP to operate whenever solar input is available. Plant output performance may be improved significantly if an auxiliary energy source can be exploited for providing energy output in the absence of (or in combination with) solar energy input, or, in the thermal SPP case, if thermal storage can be used as a thermal buffer or as a tool for time-shifting solar energy input.

Solar multiple. To obtain better SPP performance on 'average irradiation days' and over the year, a larger-than-nominal collector subsystem may be desirable, even if energy storage capacities are small, or if PCU output possibilities are exceeded under (short-term) favorable input conditions. Furthermore, if plant output and charging of thermal storage are to be accomplished simultaneously in a thermal SPP, the collector subsystem must be oversized relative to that needed for rated (design point) output.

By definition, *solar multiple* (SM) is the ratio of collector subsystem output power at design point conditions, to that needed by the PCU for generating nominal output. With SM > 1, the excess of energy collected by the collector subsystem can be used to charge the thermal storage. With appropriately large storage and SM > 2.4, 24-hour operation at rated output would theoretically become possible (see Chap. 6).

Capacity factor. For reasons of maintenance, service or cost, power plants are never operated 8,760 h/a at rated capacity. Plants meeting a fraction of total demand within an interconnected grid for significant portions of the year (*base load plants*) usually operate near nameplate rating until they undergo scheduled service. In Europe, for example, nuclear plants may reach full-load operating hours of approx. 6,500 h/a, while coal-fired plants in intermediate load mode are operated only about 4,500 h/a equivalent at nominal power. At the other end, plants with high response capability are used to satisfy extremes in demand (*peaking plants*) and may be operated only some hundred hours per year at widely fluctuating output levels. To a utility, the economic value of peaking capacity may be up to two or three times that base load capacity.

With finite storage capacity and only solar radiation as input, the capability is finite to keep output of a thermal SPP continuous for extended periods[8]. The average of output power of SPPs over multi-day time periods is affected by irradiation conditions and plant availability, and therefore is lower than the rated capacity at design point. The ratio of yearly average output to the hypothetical yearly output if operating continuously at nameplate rating is defined as the *capacity factor* (CF). The higher the CF, the better the utilization of the plant. A high CF improves the revenue situation for the financial investment which a SPP represents. Three factors principally affect the CF, the location-specific irradiation conditions, the SPP type and system configuration, and the SPP operating reliability during times of irradiance availability. Measured CFs, if different from calculated ones, are performance indicators for the maturity of the SPP in question[9].

Field-receiver ratio. To increase the number of hours at which the receiver can operate at its design point rating, the converter may be oversized at the design point by 10–15%. While some energy must be spilled for a few hours on the best solar days, the receiver will be used more effectively on an annual basis.

4.4.2 Factors Influencing Power and Energy Performance

The objective in SPP system design is to maximize – at minimum cost – the collection of solar energy and the benefit which is derived from energy output (in the form of electricity, heat, or both). Some physical, technical, location-specific or economic factors are of influence for such an objective. Key factors are irradiation, ability to utilize it, losses incurred by operating the SPP, and operational strategies.

Irradiation conditions. Reliable information about local insolation conditions (in quantity and quality) is crucial for assuring adequate solar resource availability. Insolation is given usually in terms of yearly distribution of hourly averages of gobal, direct and/or direct normal (DNI) irradiance values (kW/m^2), or in terms of yearly distribution of daily averages of global or direct irradiation ($kWh/m^2 d$). Both representations are useful for rationally setting the SPP design point, considering that power and energy thresholds are specific for SPP configurations[10].

In the course of ongoing meteorological observations, global irradiance measurements are routinely carried out, and a considerable long-duration data base has been accumulated for

---

[8] SPP operation may mistakenly be viewed as similar to that of a peaking plant. By the very nature of the input source, a SPP can satisfy supply needs only to the degree irradiation (and/or energy in the storage system) is available during peak load periods.

[9] Under irradiation conditions of the California Mohave Desert, the following CFs are reported: up to 27.2% p.a.(equivalent to 2382 h/a operation at rated output) and up to 34.5% per month for two-axes tracking photovoltaic SPP without storage [16]; up to 35% p.a. (equivalent to 3060 h/a operation at rated output) for one-axis tracking hybrid thermal SPP without storage, but 25% of annual output generated from natural gas [9,10]; up to 38% p.a. (SM=1.8; 6 hours salt storage) calculated for a tower SPP of 100 MW_e rating (SOLERGY calculation results).

[10] An attempt has been made to identify criteria for minimal local irradiation conditions. For normal operation of concentrating thermal SPPs with SM=1.0 and without auxiliary input source, a minimum need of about 300 $W(DNI)/m^2$ daily average irradiation, and > 3 h above 600 $W(DNI)/m^2$ for net output was reported [7]. Photovoltaic SPPs need a minimum of irradiance of about 50 $W(total)/m^2$, and 150 $W(DNI)/m^2$ for concentrating PV [16]. Data from experimental SPPs [6,8] suggest energy input thresholds for operation with SM=1.0 range in the order of 0.02–0.5 $kWh(global)/m^2 d$ for non-concentrating photovoltaic SPP; 0.05–1.2 $kWh(DNI)/m^2 d$ for concentrating photovoltaic SPP; 1.0–1.5 $kWh(DNI)/m^2 d$ for parabolic dish/Stirling thermal SPP; 2.0–3.0 $kWh(DNI)/m^2 d$ for parabolic trough thermal SPP; 3.0–4.0 $kWh(DNI)/m^2 d$ for central receiver thermal SPP. Presumably, advanced thermal SPPs will have smaller thresholds. The low energy thresholds for photovoltaic SPP are a result of low inertia and fast response capability on system level.

most land mass regions. Direct normal irradiance measurements of high reliability, not being of immediate importance for meteorological surveys, are carried out in far fewer locations; hence the amount of DNI data available is correspondingly smaller. The discrepancy between the need for and availability of DNI information is aggravated by the fact that data measurement, acquisition and reliability is not uniform, and that DNI tends to be very location-specific due to local cloud interference and air quality (e.g. air-borne dust, aerosols, haze, etc.).

In the past, the amount of technically usable solar radiation tended to be overestimated and, although seemingly obvious, the significance of reliable and statistically relevant multi-year local irradiation information tended to be underestimated. This caused disappointments about quantity and quality of irradiation actually available at the locations chosen for some early solar experiments (of particular significance where direct normal irradiation was the resource needed).

In the absence of reliable long-term local irradiation information, advance multi-year measurement and analysis of irradiation conditions therefore is a prerequisite, particularly for concentrating SPP configurations, for proper site selection, and SPP design and performance prediction. Advance local insolation observations remain unavoidable until methods for determining local conditions from satellite information (see Chap. 2) become further refined, widely verified and accepted as a tool for siting and design decisions.

Energy storage size. Storage capacity which is large enough to ensure, with high probability, 24 hours of SPP output generation is feasible in principle, but not practical. For cost reasons, the size of storage and collector subsystem must be matched and optimized. Consequently, yearly operating hours and CF of SPP without auxiliary fossil energy input will probably not exceed, for the time being, those of intermediate-load conventional power plants, i.e. 30–50%.

Energy inventory, operating flexibility and thresholds. In thermal SPPs – but not in photovoltaic SPPs – energy inventory and thermal capacity are important issues. Besides energy stored in a dedicated storage device, *thermal inventory* is associated with any thermal SPP element heated above ambient temperature during operation. This thermal inventory (sometimes inprecisely called *thermal mass*) can be helpful to operationally bridge short interruptions in solar energy input caused by passing clouds. But it is also detrimental to the capability of swiftly adjusting to changing input conditions. The smaller the thermal inertia, the higher the probability of net output generation under non-steady-state operating conditions. On the other hand, the rate of change of temperature or pressure, or both, for critically stressed thermal SPP plant elements may be subject to material-specific limitations (thermal, or low cycle fatigue). In any case, thermal losses must be continuously replenished during operation and stand-by; thermal energy at temperatures lower than usable in thermodynamic cycle operation can perhaps be used to warm the cold side of the thermal storage unit, otherwise it is unretrievable.

Energy is needed continuously for operating a thermal SPP (e.g. to operate feed water pumps or fans), or for keeping the system in operational stand-by (e.g. to trace-heat critical components or to keep electric controls active). These autoconsumed energies are termed *parasitics*. An amount of energy equal to that consumed during operational stand-by must be obtained during solar operation and must be stored intermediately, or it must be covered by some other source. The parasitic power and energy needs are characteristic for each SPP. As SPPs cycle naturally between some maximum of power output during daylight and operational stand-by (or shut-down) between days, low parasitic needs and low operating thresholds are key design objectives for every SPP; they are of particular relevance for thermal SPPs. For photovoltaic SPPs, parasitic energy needs during operation are considerably smaller, and may approach zero during stand-by.

Maturity and reliability. Whenever irradiation above operating threshold becomes available, it can be utilized for energy generation to achieve the highest possible output yield. This requirement calls for high operational availability of SPPs during every such irradiation period, and for optimized operational strategies/skills in the case of thermal SPPs. Availability, in turn, hinges on technical reliability of subsystems, with reliability again closely connected with technical maturity.

Many solar-specific elements (collectors, solar cell modules, heliostats, receivers) have proven more reliable than expected in experimental and pilot SPP systems, on occasion even in a power plant operating environment. It is also probable that operational *availability* of SPPs of modular configuration (e.g. dish/Stirling farms or photovoltaic SPPs) may be higher since failure of one or more modules out of many only degrades performance but does not disable the entire plant. On the other hand, experience on multi-year SPP system availability is only beginning to be accumulated, for instance for thermal SPPs with parabolic trough collectors, or photovoltaic SPPs with flat-plate modules. This is an information deficit of high importance; hence, accumulating long-term SPP operating experience at the system level with different technologies and configurations should be a priority objective in any program of support of SPP technology development.

4.4.3 Design Objectives

Any power plant constitutes a considerable capital investment risk which is to be recouped from revenues accumulated over the life of the facility. Hence, the overriding objective in a SPP design is to generate output at the least possible unit cost. The goal of SPP development efforts is to reduce the cost of production so as to render SPPs more competitive with other alternatives.

There are several widely accepted models and techniques for calculating and judging the attractiveness of investment propositions. One basic approach is to relate costs and expenditures with benefits, suitably aggregated over lifetime and converted to a yearly basis. This cost/benefit approach compares, in the case of power plants, the cost for capital investment plus present value of recurrent costs over plant lifetime (for operations, maintenance and consumables) with the present value of lifetime revenues from the sale of plant output (*net present value*).

For the relative ranking between power plant alternatives, frequently *specific energy cost* or *levelized energy cost* (LEC) is the preferred indicator, defined as the average annual cost divided by the average expected yearly energy output. Ranking by LEC presupposes that economic assumptions used in the calculations are consistent, and that service provided by the SPP alternatives is equivalent. LEC may also be viewed as the constant amount of revenue per unit of energy output which is necessary to recover the full cost of the system over its lifetime, provided of course that the stipulated amount of annual energy is actually produced, and that plant lifetime predictions are actually achieved. LEC and the associated present value multiplier have been introduced in Chap. 3, and are used in Chap. 10 to rank SPP alternatives.

Definition, detailed design and comparative assessment of SPP configurations require a techno-economic optimization with the objective functions either of

- least LEC (LLEC) if SPP energy output can be utilized whenever it is produced, tacitly assuming that revenue produced per unit of energy output is fixed, or of
- least cost/benefit ratio for specific applications of interest, in the case when SPP energy output can be controlled (by intermediate storage of solar energy, or by auxiliary fossil

energy input) and shifted to the time of day when highest revenue per unit of energy output can be achieved.

The two objectives – least energy cost or higest energy value – may not necessarily result in the same design solution because of different time-of-day and/or time-of-year energy value variations.

Assumptions about technical reliability, operational availability and lifetime, as well as amount of O&M over lifetime, influence the calculated energy unit cost. Precision in quantifying these parameters increases with accumulating SPP construction and operating experience (see Chaps. 7 and 8). On the other hand, energy value strongly depends on the application environment and on the economic outlook, i.e. the value of energy depends on whether it is available during periods of need, and on future energy price expectations. If energy value is the driving function, this objective, possibly in combination with time-related output delivery and assurance requirements, may lead to quite distinct SPP designs. For example, expectation of a low escalation of fossil fuel cost may favor hybrid SPPs with minimal SM and minimal storage capacity, while high future fuel cost expectations may lead to quite opposite SPP solutions.

4.4.4 Design Process and Parameters

Since many parameters and factors must be considered which influence the SPP design and the cost of annual energy collected, analytical design tools were developed and refined for the conceptual design of most types of SPP and SPP technologies [5,12,15] (some public-domain and proprietary design codes are outlined in Chaps. 7 and 8).

Design process. Optimum cost/performance is the principal optimization criterion in a conceptual design process. For comparing design alternatives, quantifiable factors such as system efficiency at design point, annual energy yield, or cost and value of the energy produced may be used. Qualitative assessments may enter the process, for instance the assessment of technical risk and reliability.

The conceptual design process usually involves distinct stages. These stages may be illustrated using the central receiver systems as the most complex example for a thermal SPP design [5]. The design stages are in succession:

- Plant definition – defining and specifying the basic plant configuration and the nominal performance at design point;
- Energy collection optimization – for given insolation statistics, sizing of the tower and receiver, layout and configuration of the heliostat field, with particular regard to output temperature and receiver and heat transfer fluid characteristics, with LEC as objective functions for the thermal energy produced in the system (in this case at the base of the tower); and
- Energy utilization optimization – selecting the optimal turbine-generator power block, thermal storage capacity and/or auxiliary energy input. Depending on dispatch strategy, the objective function may be optimal service by assuring, to a high degree, that output can be made available when needed, or it may be maximum revenue for one year and/or over lifetime.

Design parameters. Major design parameters in each of the three design phases are listed in Tab. 4.1 for the three major types of thermal SPP. The variables in the first column must be specified to initiate the design process. Of these, CF is not a true independent parameter; it is indicative of the type of service the thermal SPP can provide. If any of these parameters

4.4 Design Aspects

Table 4.1. Major technical parameters impacting thermal SPP conceptual design, based on [5]

	Design stages		
	Plant definition	Energy collection optimization	Energy utilization optimization
General parameters	Insolation site Design point Design point power Heat transfer fluid Storage type/medium Output temperature Backup source	Solar multiple Peak thermal power Collector aperture area Land area	Storage size Auxiliary source size Annual energy yield Plant cost Design point variation
Additional parameters for dish or trough plant	Field configuration Conversion centrally or decentrally	Collector rim angle α/ε ratio Convection loss suppression	
Additional parameters for central receiver plant	Field configuration Receiver configuration	Tower height Receiver peak flux Receiver dimensions	

are changed, for instance by selecting a different site with different irradiation conditions, a new conceptual design is required. For comparative studies of design alternatives, a multitude of designs must be prepared and the results compared.

Energy collection optimization requires, in addition to the plant parameters discussed, specification of the performance and cost factors associated with transforming site-specific irradiation into thermal energy. The first column of Tab. 4.2 lists significant physical factors which influence the energy optimization. In many cases, analytical models are needed to represent the influence of these factors properly (see second column of Tab. 4.2) on the optimization process, for instance the site-specific, long-term irradiation. The third column of Tab. 4.2 lists significant cost-related sizing issues which are functionally related to the physical plant parameter which must be sized. In the fourth column, additional cost-related issues are shown which are not dependent on plant size but reflect more closely assumptions about hardware/equipment costs and construction-related factors.

The influence of one factor on energy collection optimization may be compensated by the influence of another; the effect of these factors, grouped in terms of those favoring larger and smaller elements of the thermal SPP collector subsystem, are listed in Tab. 4.3.

The last stage of the design process is the optimization analysis to select the best combination of turbine/generator size, thermal storage capacity and (possibly) fossil fuel input to achieve greatest revenue at least cost. Important variables in this process are energy dispatch strategy, storage medium and capacity, and type/size of auxiliary boiler [15].

By this conceptual design approach, performance and cost figures for each technical alternative can be compared. However, important factors influencing the analysis exist which generally are not considered explicitly. These are, amongst others,

- detailed operations and maintenance (O&M) cost,
- operator/item plant availability and reliability,
- plant lifetime,
- technical risk.

Table 4.2. Physical, cost and cost-related factors affecting energy collection optimization of a central receiver thermal SPP [5].

Factor	Physical issues	Cost issues	Cost-related considerations
Sun	Solar disc representation		
	Site-specific, long-term insolation		
Collector field	Heliostat layout pattern		
Heliostat	Size and shape	Mirror surface area	Mass production assumption
	Number and configuration of reflective facets		Factory location and transportation costs
	Facet cant and curvature		
	Mirror surface waviness		
	Tracking accuracy		
	Gravity and wind-induced deflection		
Heliostat image	Analytical procedure for flux calculator		
Shading and blocking	Analytical representation of process		
Atmospheric attenuation	Form, magnitude		
Energy losses	Heliostat reflectivity		
	Receiver absorptivity		
	Receiver reradiation		
	Receiver convection		
Land	Slope	Land area	Raw land cost
			Site preparation costs
Wiring		Heliostat spacing	Cable costs
		Number of heliostats	Trenching method and costs
			Cable routing
Receiver	Maximum allowable flux	Absorber area	Configuration (cavity or external)
	Thermal losses		Heat transfer fluid
Tower		Height	Site specific factors (soil bearing strength; foundation design; wind speeds; seismic activity; type of tower)
Pump	Compatibility with heat transfer fluid	Tower height	Reference pump configuration and costs
		Thermal power	
Piping	Freeze control	Tower height	Basic pipe costs
		Receiver thermal rating	Pipe support equipment
Balance of plant		Plant size	
Operation and maintenance		Plant size	Number of shifts

Table 4.3. Compensating trends of influencing factors observed in the conceptual design of central receiver thermal SPPs [5].

Favors larger fields	Favors smaller fields
expensive receiver	expensive heliostats
low cost heliostats	low cost receiver
inexpensive land and/or field wiring	expensive land and/or field wiring
low atmospheric attenuation	high atmospheric attenuation
	restricted area

Favors larger receivers	Favors smaller receivers
low receiver cost/m^2	high receiver cost/m^2
low receiver losses/m^2	high receiver losses/m^2
large flat heliostat	high performance heliostat
severe heliostat aberrations	smaller heliostat
large beam spread	high peak flux limit
low peak flux limit	

Favors taller towers	Favors shorter towers
large fixed cost	low fixed cost
low tower cost	high tower cost
restricted or expensive land	inexpensive land
expensive heliostats	low cost heliostats
	large beam spread

Bibliography

[1] Alpert, D. J.; Kolb, G. J.: *Performance of the Solar One Pilot Plant as Simulated by the SOLERGY Computer Code*. Technical Report SAND88-0321, Sandia National Laboratories, Albuquerque/NM, 1988
[2] Becker, M.; Boehmer, M. (Ed.): *GAST – The Gas-Cooled Solar Tower Technology Program*. Berlin, Heidelberg, New York: Springer, 1989
[3] Droher, J. J.: *Performance of the Vanguard Solar Dish-Stirling Engine Module*. Technical Report EPRI AP-4608, Electric Power Research Institute, Palo Alto/CA, 1986
[4] Etievant, C.; Amri, A.; Izygon, M.; Tedjiza, B.: *Central Receiver Plant Evaluation, Vol. 1-5*. Technical Report SAND86-8185, SAND87-8182, SAND88-8101, SAND88-8100, SAND88-102, Sandia National Laboratories, Albuquerque/NM, 1988
[5] Falcone, P. K.: *A Handbook for Solar Central Receiver Design*. Technical Report SAND86-8009, Sandia National Laboratories, Livermore/CA, 1986
[6] Grasse, W.: *Design Basics for Solar-Thermal Power Plants – Results and Experiences from Operating Experimental Facilities* (in German). Volume 704 of *VDI-Berichte*, Düsseldorf (D): VDI-Verlag, 1988
[7] Grasse, W.: *SSPS Results of Test and Operation 1981-1984*. Technical Report IEA-SSPS SR7, Köln (D), DLR, 1985
[8] Hertlein, H. P.: Bulk Electricity Generation – Prospects for PV and Solar-Thermal Systems. In *Proc. 8th E.C. PV Solar Energy Conference, Florence*, Solomon, I.; Equer, B.; Helm, P. (Ed.), p. 451, Dordrecht (NL): Kluwer, 1988
[9] Kearney, D.; Gilon, Y.: Design and Operation of the Luz Parabolic Trough Solar Electicity Plants. p. 53, Düsseldorf (D): VDI-Verlag, 1988
[10] Klaiss, H.; Geyer, M.: Economic Comparison of Solar Power Electricity Generating Systems. In *GAST – The Gas-Cooled Solar Tower Technology Program*, Becker, M.; Boehmer, M. (Ed.), Berlin, Heidelberg, New York: Springer, 1989

[11] Lin, E. I. H.: *A Review of the Salt-Gradient Solar Pond Technology.* Technical Report DOE/SF-11552-1, Pasadena/CA, Jet Propulsion Laboratory, 1982
[12] Menicucci, D. F.; Fernandez, J. P.: *User's Manual for PVFORM: A Photovoltaic System Simulation Program for Stand-Alone and Grid-Interactive Applications.* Technical Report SAND85-0376, Sandia National Laboratories, Albuquerque/NM, 1988
[13] Pitman, C. L.; Vant-Hull, L. L.: *The University of Houston Solar Central Receiver Code System: Concepts, Updates and Start-Up Kits.* SAND88-7029, Houston/TX, University of Houston, 1989
[14] Schlaich, J. M.; Schiel, W.; Friedrich, K.: Solar Chimney Power Plants – Technical Design, Experience and Development Potential (in German). p. 145, Düsseldorf (D): VDI-Verlag, 1988
[15] Stoddard, M. C.; Faas, S. E.; Chiang, C. J.; Dirks, J. A.: *SOLERGY – A Computer Code for Calculating the Annual Energy from Central Receiver Power Plants.* SAND86-8060, Sandia National Laboratories, Livermore/CA, 1987
[16] Stokes, K. W.; Risser, V. V.: *Photovoltaic Field Test Performance Assessment: 1987.* Technical Report EPRI-GS-6251, Electric Power Research Institute, Palo Alto/CA, 1989
[17] Swift, A.; Reid, R. L.; Boegli, W. J.; Sewell, M. P.: Operational Result for a 3355 Square Meter Solar Pond in El Paso, Texas. *Solar Engineering*, (1987) 287–293
[18] Tabor, H. Z.: Israel Pioneers Solar Ponds. *Modern Power Systems*, (1987) 66–69
[19] Winter, C.-J.; Nitsch, J. (Ed.): *Hydrogen as an Energy Carrier – Technologies, Systems, Economy.* Berlin, Heidelberg, New York: Springer, 1988

5 Thermal Receivers

M. Becker and L. L. Vant-Hull [1]

The optical system collects and concentrates direct (beam) solar radiation and delivers it to the receiver. The receiver must be designed to intercept the energy effectively, absorb it efficiently, and convert it to thermal energy at the temperature required by the conversion process. The design must account for thermal losses and should be such as to mitigate them to the extent possible.

The conflicting role of temperature becomes a major design issue. The efficiency of mechanical conversion cycles is essentially proportional to the driving temperature, while chemical reaction rates increase exponentially with temperature. However, receiver radiation losses increase with the fourth power of temperature and materials problems become much more severe. Even for lower temperature systems where thermal losses are less critical, the receiver must be cost effective. This usually implies minimal heat exchanger area, primarily limited by the allowable flux density/temperature level. A small allowable flux density leads to a large receiver area and, consequently, to considerable radiative and convective losses unless design features can protect the absorbing surfaces and thus decrease the losses (cavity or vacuum-jacketed receiver concept). A high allowable flux density decreases the receiver area and consequentially leads to low losses (external receiver concept). In any case, the heat exchanger material must be able to withstand the heat load, and the properties of the heat transferring fluid must be such as to allow the required thermal energy transport.

In this chapter, these effects and the approaches dealing with conflicting requirements in the design of receivers are considered.

5.1 Introduction

In a thermal solar power plant (SPP), the receiver plays the key role of intercepting concentrated sunlight and converting it into a form appropriate for the specific application[2]. This process generally involves high temperatures and high radiant flux levels. Therefore, high technology is usually involved in the design of receivers which, in fact, are the focus of technological development and of most trade studies of thermal SPP.

The trade studies arise from conflicting requirements, and the receiver is subject to many of the following requirements:

---

[1] Manfred Becker, Deutsche Forschungsanstalt für Luft- und Raumfahrt (DLR), Linder Höhe, D-5000 Köln 90;
Lorin L. Vant-Hull, Energy Laboratory, University of Houston, 4800 Calhoun Road, Houston, TX, 77004, USA.

[2] Photovoltaic conversion of concentrated sunlight will be discussed in Chap. 8.

- Nominal operating temperature is usually set by application needs, but total system efficiency (and cost-effectiveness) are dependent on the extent nominal temperature can be attained.
- Material degradation rates (and costs) increase with temperature for absorber, receiver, and heat transfer fluid (HTF).
- The choice of HTF (steam, Na, air, He, $KNO_3/NaNO_3$, oil, 'sand', process chemicals, etc.) affects the general design, the corrosion of receiver containment, heat transfer rates (and thus the allowable flux density), and the range of practical receiver concepts.
- Receiver thermal losses increase with effective receiver area and receiver temperature.
- For a given size receiver, spillage is determined by the convolution of the solar image with beam errors resulting from reflector aberrations, mirror surface errors, mechanical errors, and tracking errors. Thus, one can trade the reduction of these errors (more costly heliostats) against increased receiver size (increased cost and thermal loss).
- Allowable maximum flux density ranges from 2.5 MW/m^2 for liquid sodium and some direct absorption types of receivers to 0.2 MW/m^2 for an air cooled tube receiver. For a high temperature receiver thermal losses per m^2 can be very high, so the receiver must be carefully configured to maintain a small effective area and to enhance the effective allowable flux density.
- Cavity receivers may be more efficient in some cases, especially in systems employing high temperatures combined with low allowable flux density. Compared to external receivers, their field of view is limited. For a given energy requirement this limitation leads to longer focal length systems. For central receivers this means that cavity receivers generally require tall towers, which may be very expensive if the cavity receiver is heavy.
- Medium temperature parabolic trough receivers achieve lower concentrations, and consequently require protective devices against convection losses. This requirement is usually met by applying a cylindrical glass enclosure with the special property of transmitting the incoming radiation and absorbing most of the infrared back radiation.

These conflicting aims must be resolved in system design studies. While there are many options and alternatives, usually the application (defining the operating temperature) and selection of the HTF (defining allowable flux levels and corrosion problems) are the main independent choices. Other selections can largely be made on economic grounds following the result of trade studies.

5.2 Principles and Concepts for Energy Transfer

The receiver is that element which receives the concentrated solar radiation and transforms it to thermal energy to be used in subsequent processes. The radiative-to-thermal heat exchange process in the receiver may involve one of three possibilities, i.e.

1. The use of *tubes* which receive the radiation on the outside, conduct the energy through the wall and exchange it to the HTF carried on the inside. In a high-temperature application, the tubes may be protected from thermal radiation loss by placing them in a cavity. In a medium temperature environment, equivalent loss protection is achieved by placing the tubes inside of infrared reflecting glass envelopes. The tube receiver operates as a *recuperative heat exchanger.*
2. The use of wires, foam, or appropriately shaped walls within a volume to enhance the absorptive surface area which receives the radiation, converts it to heat and transfers

5.2 Principles and Concepts for Energy Transfer

thermal energy by convection to air passing by. The volumetric receiver operates as a *convective heat exchanger*.

3. The use of fluids or particle streams which receive the direct radiation and immediately absorb it in their volume or on their surfaces. The fluid/particle receiver operates as a *direct heat exchanger*.

The most essential and characteristic features[3] of these possibilities are sketched in Fig. 5.1.

In conventional power plants, the heat receiving element is the boiler which is heated by the combustion of coal, oil or gas. The boiler is heated by radiation and convection. The distribution of heat is controllable if not uniform. In a solar environment, radiation concentration results more or less in a Gaussian distribution of energy on the receiver, if all Sun rays are concentrated on one aim point. This is attractive as the available solar image is the smallest in this case. Small geometry and high concentration is equivalent to high efficiency and, as a consequence, high technical usefulness.

If there is a constraint of maximum allowable flux density, this can be met by modifying the distribution of incoming radiation. The flux distribution can be controlled by defining several aim points for the heliostat field. Such attempts have to face the detrimental effects of larger geometries required to intercept the flux, and of the associated higher radiative and convective losses. Thus, receiver optimization is a true engineering challenge.

The reception of radiation poses few problems if concentrated peak flux densities, say in the order of 1 MW/m^2, can be accommodated by the material of the tubes or of the wall, and by the HTF. In this case, there may be no need for protection against losses because the heated area can be held very small, and an *external receiver* can be used.

If, however, materials and HTF do not permit rapid heat transfer, usually a protective device like an orifice (defining the so-called *cavity receiver*) is needed. In the enclosed region behind the orifice, a diffusive ray portion and a definitely larger absorptive area can be tolerated. In this case, the average heat flux density at the absorber may range as low as 100 kW/m^2.

The upper half of Fig. 5.1 illustrates the structures of several tube type receivers: the cavity receiver with a large-area, low-flux-density absorptive surface, irradiated by and protected only by a small aperture (orifice); and the external receiver with a small-area, high-flux-density absorptive surface. The actual values of performances and losses can be comparable for both receiver types at equal throughput rating. The tube materials may be metallic or ceramic, depending on the temperature range in which the receiver operates.

The third receiver in Fig. 5.1 represents the *volumetric receiver* type. It consists of a volume, filled with a multitude of porous interlocking shapes, wire packs, foam, or foil arrangements, built from metallic, ceramic or other suitable materials. The advantage of such a concept is the absence of bending stress limitations which determine low cycle fatigue and lifetime for tubes. In volumetric receivers, the concentrated radiation heats the material of the volume. At the same time, HTF passes through the volume, and is heated up convectively. Higher than atmospheric pressure could be used in a closed loop, if the receiver were protected by a window. Transparency, stability and durability of such a window, however, present problems yet to be solved. R&D investigation of this concept is presently in progress. For small pressure differences, air curtains might be considered to function as protective devices.

A more elegant method of transferring radiative energy is to expose falling fluid films or solid particles to a flux of photons. This can be achieved in a *direct absorption* process with or without chemical reactions taking place within the fluid, or the particles and is shown as

---

[3] For general consideration, the most variable case, the central receiver power plant, was chosen. Receiver versions for dish units look alike, except for smaller scales and emphasis on circular concepts. Receiver versions for distributed collectors are always glass-protected tubes.

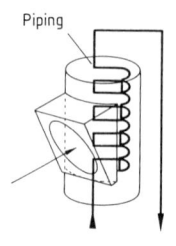

Cavity

Conduction through tube wall

HTF	$\left(\dfrac{\dot{Q}}{A}\right)_{avg}\left[\dfrac{kW}{m^2}\right]$	$\dot{q}_{max}\left[\dfrac{kW}{m^2}\right]$
Gas	50 to 100	200
Liquid	300	900

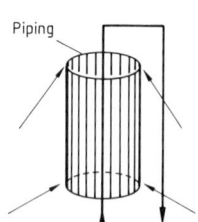

External

Conduction through tube wall

HTF	$\left(\dfrac{\dot{Q}}{A}\right)_{avg}\left[\dfrac{kW}{m^2}\right]$	$\dot{q}_{max}\left[\dfrac{kW}{m^2}\right]$
Liquid	800	2500

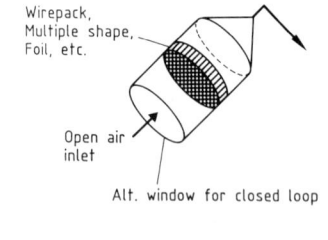

Volumetric

Convection back from wall

HTF	$\left(\dfrac{\dot{Q}}{V}\right)_{avg}\left[\dfrac{kW}{m^2}\right]$	$\dot{q}_{max}\left[\dfrac{kW}{m^2}\right]$
Gas	unknown	1000
Open air	unknown	(2500)

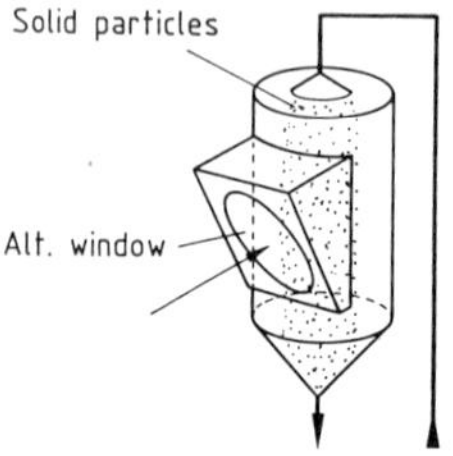

Particles and Liquids

Direct absorption — Heat transport without reaction
— Heat transport with reaction

HTF	$\left(\dfrac{\dot{Q}}{V}\right)_{avg}\left[\dfrac{kW}{m^3}\right]$	$\dot{q}_{max}\left[\dfrac{kW}{m^2}\right]$
Solid particles	unknown	(2500)
Liquids	unknown	(2500)
Reacting mixture	unknown	(2500)

Fig. 5.1. Basic receiver concepts [20] (bracketed values denote design goals).

the last example in Fig. 5.1. If a chemical process is to be incorporated, the receiver has to be equipped with windows. While small scale experiments have been completed, further R&D of and experiments with this concept are needed.

Both cases displayed in the lower half of Fig. 5.1 can use heat flux densities in the order of 1 MW/m². Peak loads up to 2.5 MW/m² are conceivable. The average ratio of transferrable power $\dot{Q}$ [4] to absorbing surface A or volume V could be a cost criterion. For volumetric and direct absorbing receivers, these ratios are expected to be very favorable in comparison to tube receivers. In the case of direct absorption the transfer mechanism is definitely very complex. The parameters in this case are optical densities, refractive indices, and reaction

---

[4] Following engineering tradition within this chapter, the heat flow is designated as $\dot{Q}$ and the power produced as P (both units in watts). Chapter 2, following the standards of Physics, displays differing definitions.

temperatures. While receivers of the first two classes have been extensively tested at the pilot plant level, the volumetric and direct absorption receivers are still at the R&D level.

5.3 Thermal and Thermodynamic Basis for Receiver Design

In optimizing a concentrating solar collector, losses at the receiver play a key role, along with receiver costs. End use requirements generally dictate the receiver inlet and outlet temperature while the choice of HTF defines the maximum allowable flux. Within these constraints the receiver is configured to provide thermal energy at the lowest cost per unit.

Basic loss mechanisms are illustrated in Fig. 5.2 and 5.3. The basic relationships among the various energy transports and loss mechanisms at the receiver are:

concentrated energy	$\longrightarrow$	atmospheric attenuation + spilled + intercepted energy
intercepted energy	$\longrightarrow$	reflected + absorbed energy
absorbed energy	$\longrightarrow$	reradiated + convected + conducted + delivered energy
delivered energy	$\longrightarrow$	sensible energy (local temperature increase)
local temperature increase	$\longrightarrow$	absorber rise + wall + fluid film + local HTF temperature increase
HTF temperature	$\longrightarrow$	inlet temperature + sensible temperature $(\Sigma \dot{q} A / \dot{m} C_p)$.

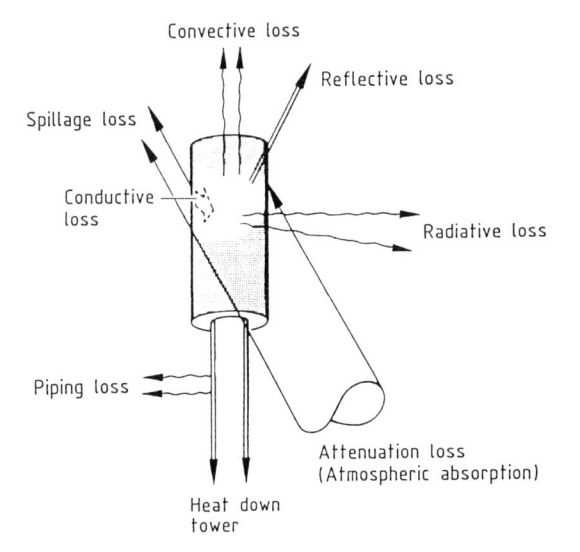

Fig. 5.2. Optical and thermal losses [4,15].

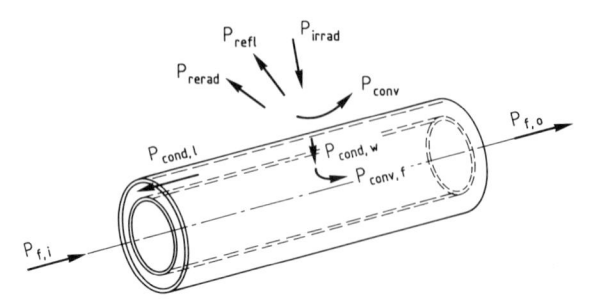

Fig. 5.3. Overall loss balance.

P	power
cond	conduction
conv	convection
irrad	incident radiation
refl	reflected radiation
rerad	reradiation
l	longitudinal
w	wall
f	fluid
i	inlet
o	outlet

Of these, atmospheric attenuation and spillage were discussed in Chap. 3. Spillage and allowable flux density constraints call for larger-area receivers, thus increasing costs and thermal losses. For a specific design, losses may be estimated and the performances of various designs compared.

The other individual loss mechanisms and phenomena involved in these relationships will be discussed in the following paragraphs.

Reflection. For a receiver employing a glass or quartz envelope or a window, the known Fresnel equations apply to each surface. For normal incidence, the reflection coefficient for visible light at each surface as given in Chap. 3 is

$$\rho = (n-1)^2/(n+1)^2 \tag{5.1}$$

where n is the index of refraction, giving e.g. $\rho = 0.04$ for $n = 1.5$, and $\rho = 0.06$ for $n = 1.65$. Special antireflective treatment can reduce this loss somewhat. Multiple reflections and envelope absorptivity have only a minor effect on the incident sunlight; so at normal incidence the window reflectivity (two surfaces) is 8 to 12%. At incidence angles more than 40° from normal, the Fresnel equations predict increasing reflectivity, reaching about twice the above amount at 60° and then increasing rapidly to unity at grazing incidence.

A window may be used to isolate the interior of a cavity from the environment. Both the irradiation at the window and the angle of incidence can be reduced by using domes rather than flat windows. Outward domes are effective for evacuated or ambient pressure receivers while inward domes can withstand several atmospheres. The larger windows required for a central receiver would need a mosaic of meter-sized domes. Protecting and cooling the support structure then becomes a problem.

For a tubular transparent enclosure (as used with parabolic trough receivers), the 'effective aperture' is equal to the absorber width in all cases, even including multiple reflections [29]. The incidence angle on the enclosing tube only exceeds 60° if the receiver exceeds 87% of the tube diameter for on-axis rays, and even at 45° off-axis this only decreases to 70%.

For an external tube wall receiver the local reflectivity is simply $\rho = 1 - \alpha$. For rough sandblasted iron alloys the absorptivity α for oxidized surfaces is about 0.88. A thin (50–100 μm) layer of absorbing paint such as Pyromark© on the same surface increases the absorptivity to 0.96 at maximum. A long term average value of 0.92 to 0.94 seems likely for such a surface with occasional (5–10 years) resurfacing[5].

Sunlight entering the aperture of a cavity receiver will suffer multiple reflection/ absorption scatterings within the cavity. The effective reflectivity can be estimated from the respective areas and absorptivities. If the direct sunlight strikes a highly absorbing (heat exchanger)

---

[5] Pyromark© consists of finely divided particles of transition metal oxides in a silicate binder which is cured at 600°C after application.

surface, the fraction $\rho = 1 - \alpha$ will be reflected, of which a fraction due to $A_{\text{aperture}}/A_{\text{wall}}$ is likely to reach the aperture again and be scattered outward. As these numbers are typically 1/10 and 1/6 respectively, only a few percent of reflective loss occur. The balance of the reflected light (6–8%) will be absorbed on adiabatic surfaces along with any direct sunlight striking them. Such surfaces may become even hotter than the heat exchange surface (so they can radiate energy to it). Hence they also contribute to convective receiver loss. Essentially the cavity traps reflected sunlight, but will lose some of the advantage of this through enhanced convective losses or, if windowed, by window reflection.

In some cases the cavity is designed to intercept most of the incident radiation on a diffusely reflecting adiabatic wall (or the heat exchanger is made diffusely reflective) in order to provide for an even distribution of flux on the heat exchanger. Thus, if the absorptivity in the cavity is 1/3, the average photon will strike three walls before being absorbed. If the wall area is six times the aperture area, this will result in a reflective loss of approximately $(1 - \rho)^{-1} \cdot A_{\text{aperture}}/A_{\text{wall}} \approx 1/2$. Clearly, care must be taken with such designs. In general a detailed view factor calculation is required for all but the simplest cases.

An empirical correction to the absorptivity for off-normal incidence is:

$$\text{for } \theta > 60°: \quad \alpha(\theta) = \alpha_\perp (0.395 + 1.21 \cos\theta) \quad (5.2)$$
$$\text{for } \theta \leq 60°: \quad \alpha(\theta) = \alpha_\perp$$

where θ is the angle of incidence on the receiver. θ is greater than 60° only for the outer 13 % of a cylinder (on the sides) or for the outer heliostats of a 120° sector field. In fact, this is a local relationship and could be integrated over the circumference of tubes on a tube wall. Enhanced absorption near the cusps will increase the normal absorptivity, while at a 60° angle of incidence adjacent tubes obscure half of the 'good' section of the tube (where $\theta < 60°$) resulting in an average absorptivity about 97% of $\alpha_\perp$: this will be enhanced to about 98 % of $\alpha_\perp$ by absorption of scattered light in the cusp. Even at 85° incidence where the value of $\alpha(85°)$ is 0.5 $\alpha_\perp$, the integral over a tube wall is 0.84 $\alpha_\perp$. Of course, for a heliostat beam incident on a cylindrical receiver, only a small fraction of the energy exceeds an incidence angle of 60°.

The reflection from direct absorption receivers can not be discussed in general. The process may involve a blackened fluid film flowing over a metal wall, a falling sheet of 'sand', a fluidized bed of catalytic particles, a gas containing submicron carbon particles, gas flowing through a wire pack or a ceramic matrix, etc. Each case will require a determination of reflective losses, as well as the requirement for a cavity, a window or other special containment provisions.

Surface temperature. The rest of the thermal receiver losses depends on temperature and is somewhat more difficult to estimate accurately. Fortunately, in a well-designed system thermal losses run about 5 to 15%; so a reasonable estimate using the mean HTF temperature is adequate to determine system performance within a few percent. To deal more accurately with thermal losses in high flux density receivers, the local effective surface temperature should be calculated at all points of the receiver. For radiative losses the local temperature is used, while for convective losses the mean film temperature, the wind velocity, and an estimate of the buoyancy are required in determination of the heat transfer coefficient h. An iterative procedure is usually required because the surface temperature depends on the local heat flux and on the local HTF temperature. This temperature is obtained by integrating along the entire flow path to determine the required mass flow to achieve the defined outlet temperature, and then to the point of interest to determine the local HTF temperature.

For the first iteration any reasonable value of loss can be used, 0, $\dot{q}/10$, 30 kW/m² etc. For a fluid-in-tube receiver the heat transfer coefficient h depends on the fluid, the flow velocity and the temperature. Then the local surface temperature is

$$T_s = T_{\text{fluid}} + \dot{q} \bigg/ \left(\frac{1}{h} + \frac{r_0 \ln(r_0/r_i)}{k_{\text{wall}}} + \frac{t}{k_{\text{coat}}} \right) \quad (5.3)$$

where, for the thin absorber film $t = r\ln(r+t)/r$, as t is $\ll r$. Here h is the heat transfer coefficient, r_o and r_i are the outer and inner tube diameters, k_{wall} its thermal conductivity and k_{coat} is the thermal conductivity of the thin absorber coating. This gives the crown temperature for the tube which, since it is highest, dominates the losses. For reasonably thin walled tubes, conduction around the tubes is not very important. Although the absorbing film thickness is quite small (typically $t \simeq 50\ mum$) it is generally an insulator (oxides or ceramic) so k_{coat} is about 100 times smaller than k_{wall} for a metallic tube.

Volumetric receivers, which draw air or another working fluid along the irradiation surface, are designed to present a low effective surface temperature to the outside. The HTF and the absorbing/convecting medium thus only reach operating temperature deep inside the heat exchanger matrix where thermal or radiative transfer to the exterior is ineffective.

Surface temperatures for a direct absorption receiver depend on mixing within the moving HTF, the optical depth of the sand, blackened salt, submicron dispersion, or fluidized bed, perhaps of catalytic particles, and on design-specific details.

If windows are used to contain the receiver materials or HTF, their temperature defines the losses because most practical window materials are opaque in the infrared. Thus, essentially the window stands at equilibrium between thermal losses to the outside and heat gains from the inside. The window may also be cooled and cleaned by impinging HTF. Estimation of the effective loss temperature is a major part of the design of these special receiver types.

Cavity receivers utilize adiabatic surfaces to complete the enclosure. These surfaces receive infrared radiation from their surroundings and may or may not be subject to direct and reflected solar irradiation; in either case their local temperature is defined by thermal equilibrium.

Frequently the last two terms in (5.4) are ignored in carrying out the cavity radiation balance calculation because they are difficult to calculate. Furthermore, at the higher temperature regime of cavity receivers, these effects often do not contribute more than about 10% of total losses. It is now understood that convective losses should be calculated for the entire interior area of the receiver below the top of the aperture. If a substantial area is above that point it may also contribute via a convective roll with gaseous interchange between the two regions, approximating a high temperature 'roof' in the interchange zone.

$$\phi_S + \sum \rho_i \phi_i \mathcal{F}_{i \to J} A_i + \alpha^{IR} \sum \epsilon_i \sigma T_i^4 \mathcal{F}_{i \to J} A_i \qquad (5.4)$$
$$= (1-\alpha^s)\phi_J + (1-\alpha^s)\sum \rho_i \phi_i \mathcal{F}_{i \to J} A_i + \epsilon \sigma T_J^4 + \bar{h}(T_J - T_\infty - \Delta T) + (T_J - T_\infty)\frac{k}{l}$$

where
- ϕ = concentrated solar intensity, W/m²
- α^s = solar band absorptance
- ρ_i = solar reflectivity of ith element = $1 - \alpha^S$
- α^{IR} = infrared absorptance
- $\mathcal{F}_{i \to J}$ = real body view factor from element i (of area A) to element J
- ϵ_i = infrared emissivity of ith element at temperature
- T_J = surface temperature of the convecting region
- $\bar{h}$ = average convection coefficient for the cavity
- T_∞ = temperature of the environment
- ΔT = estimated temperature rise of gas convecting in cavity at the point in question, $\Delta T = 0$ at lower lip, $\Delta T = \Delta T_{\max}$ at upper lip
- k/l = thermal resistance to 'ground'.

5.3 Thermal and Thermodynamic Basis for Receiver Design

The view factors $\mathcal{F}_{i \to J}$ are generally difficult to evaluate and, with hundreds of nodes, equilibrium temperatures are determined by computerized matrix inversion or iterative approximation. Only in simple cases with a high degree of symmetry can analysis lead to useful results, e.g. cylindrical cavity receiver for a symmetric parabolic dish. Even then, tracking leads to a tilted receiver and that, or winds, causes convective losses which are not symmetric. Also, poor tracking causes the insolation to be asymmetric; so computer analysis is still needed. Heat pipe receivers have a particular simplicity because the entire working surface is at the boiling temperature of the fluid, e.g. Na, K, H_2O. Thus, the outer surface temperature varies only slightly with the local solar concentration.

Radiation. For each node of an external receiver (flat or cylindrical)

$$d\dot{Q} = \epsilon \sigma (T^4 - T_\infty^4) dA. \tag{5.5}$$

Here ϵ ranges from 0.88 to 0.96, T is the surface temperature discussed earlier, and T_∞ is typically 273° to 310°C depending on the climate.

For a deep cavity receiver, ϵ is unity because of the black body nature of the aperture. For a shallow cavity a value intermediate between ϵ_{wall} and unity is appropriate, such as

$$\epsilon_{avg} = \epsilon / [\epsilon + (1 - \epsilon) A_{\text{aperture}} / A_{\text{wall}}]. \tag{5.6}$$

(5.6) is exact for a three-walled tetrahedron. The radiation temperature T for a cavity is based on the interior surface temperature, where the view factor for each node is weighted by T_i^4. Thus, nodes directly visible through the aperture tend to define the effective temperature.

Volumetric receivers tend to have very small radiative losses if the exposed surface temperature is kept relatively close to ambient. However, the outer layer of a wire pack or ceramic matrix, for instance, is exposed directly to the total solar flux. Because air and most gases are relatively poor heat transfer agents the outer looking surface temperature may be very high and radiate effectively to the surroundings. Effective designs depend on this hot surface area being very small – e.g. 1% for each layer in a wire pack or perhaps 15% in a porous ceramic absorber.

Radiant loss in a direct absorption receiver originates directly from the HTF itself. The appropriate temperature is not that at the surface, but should be averaged over an infrared optical depth. It therefore may be higher than the 'surface' temperature which may be reduced by convective losses, particularly in the case of the external receiver or other exposed designs.

Conduction. The general law of conductive heat transfer in the Fourier form is

$$\dot{Q} = -kA\nabla T \tag{5.7}$$

with $\nabla = \hat{i}\frac{\partial}{\partial x} + \hat{j}\frac{\partial}{\partial y} + \hat{k}\frac{\partial}{\partial z}$ as the gradient operator, and the thermal conductivity k as proportionality factor having the dimension of W/mK. In one-dimensional treatment this is simplified to

$$\dot{Q} = -k\frac{dT}{dy}A.$$

For tubes, the heat per length in the radial direction can be calculated as

$$\frac{\dot{Q}}{L} = -2\pi k \frac{T_i - T_0}{\ln r_0/r_i}$$

with the local temperature within the tube wall given by

$$T = T_i - \frac{(T_i - T_0)\ln r/r_i}{\ln r_0/r_i}.$$

Conduction within the tubes of a receiver does not necessarily contribute to losses but rather is a means of equalizing a non-uniform temperature distribution. However, piping to the outside of the receiver or contacts with supporting elements contribute to conductive losses, so conductive losses are highly dependent on geometric details, and – for a carefully performed design – are very small in relation to other losses.

Convection. An analysis of convective heat transfer simultaneously takes into account the heat conduction between positions of different energy (temperature) and the transport of energy by HTF movement. The degree of fluid motion dictates whether heat transfer takes place mainly by temperature gradients (at low velocities), or predominantly by fluid mixing under laminar or turbulent flow conditions. Mass flow towards or along a body induces *forced convection*; temperature difference between the heated body and its surroundings induces *natural convection*.

Heat transfer by natural convection is responsible for a considerable portion of external receiver losses. More essentially, heat transfer from heated tubes to the HTF flowing within tubes is governed by the laws of forced convection. Such convectional heat transfer can simply be calculated as

$$\dot{Q} = A\bar{h}(T_w - T_\infty) \tag{5.8}$$

where $T_w - T_\infty$ is the temperature difference between wall and undisturbed conditions, and $\bar{h}$ in $W/m^2 K$ is the heat transfer coefficient. The bar denotes the average value over the total surface of the body under consideration; coefficients without a bar indicate local values. An integration over all local values results in the average heat transfer coefficient.

$$\bar{h} = \frac{1}{L} \int_0^L h(x)\,dx. \tag{5.9}$$

Generally, the heat transfer coefficient is obtained from a dimensionless relationship, giving the Nusselt number as

$$Nu = \frac{hx}{k}. \tag{5.10}$$

A rich literature (e.g. [12,21,22,25]) exists on how to evaluate

$$Nu = f(x, D/L; v; \rho, C_p; k, \nu, \mu)$$

representing geometry, motion, thermal and transport properties with exponents. In terms of appropriate selections of dimensionless groups this gives

$$Nu = f(Re, Pr, Gr, D/L). \tag{5.11}$$

Forced convection is determined by combinations of Reynolds (Re) and Prandtl (Pr) numbers while natural convection is characterized by variations of Prandtl and Grashoff (Gr) numbers.

The above used terms stand for:

$Re = \frac{\rho v x}{\mu} = \frac{v x}{\nu}$ the ratio of inertia ($\rho v^2 A$) to viscous forces ($\mu v A/x$)

$Pr = \frac{\mu C_p}{k} = \frac{\nu}{a}$ the ratio of kinematic viscosity (ν) to thermal diffusivity (a)

$Gr = \frac{g\beta(T_w - T_\infty)x^3}{\nu^2}$ the ratio of buoyant to viscous forces.

The bulk expansion coefficient β is close to $1/T$ for most gases. All the above mentioned numbers are evaluated at a mean temperature determined by

$$T_m = \frac{T_w - T_\infty}{2}.$$

5.3 Thermal and Thermodynamic Basis for Receiver Design

The determination of Nusselt numbers for plates, cylinders and tubes is described in standard text books, especially in [22] and [25]. The following lines give a short survey on Nusselt number determination *for forced convection* at

Plates:

$$Nu = 0.664 Re^{0.5} Pr^{0.333} \text{ for laminar flows} \tag{5.12}$$
$$Nu = 0.037 Re^{0.8} Pr^{0.333} \text{ for turbulent flows} \tag{5.13}$$

Cylinders (for $T_w < 1,000°C$):

$$Nu = C Re^m (0.785 \, T_w/T_\infty)^{m/4} \, 1.167 Pr^n \tag{5.14}$$
$$\text{with} \quad n = 0.45 \text{ for gases at} \quad 0.5 < Pr < 10$$
$$\text{and} \quad n = 0.33 \text{ for fluids at} \quad 1.0 < Pr < 1,000$$

Re			C	m
1	to	4	0.891	0.330
4	to	40	0.821	0.385
40	to	4,000	0.615	0.466
4,000	to	40,000	0.714	0.618
40,000	to	400,000	0.0239	0.805

Tubes:

$$Nu = \left[3.66 + \frac{0.0668 Re Pr D/L}{1 + 0.045 (Re Pr D/L)^{0.667}}\right] \left(\frac{\mu_m}{\mu_w}\right)^{0.14} \tag{5.15}$$
for laminar flows

$$Nu = 0.0235(Re^{0.8} - 230)(1.8 Pr^{0.3} - 0.8)\left[1 + (D/L)^{0.667}\right]\left(\frac{\mu_m}{\mu_w}\right)^{0.14} \tag{5.16}$$
for turbulent flows.

The Nusselt numbers for *natural convection* are usually not distinguished by geometric forms but rather by orientation of the heated body as for

Vertical bodies:

$$Nu = 0.59(Gr Pr)^{0.25} \text{ for laminar flows} \tag{5.17}$$
$$\text{with} \quad 10^4 < Gr Pr < 10^9$$
$$Nu = 0.13(Gr Pr)^{0.333} \text{ for turbulent flows} \tag{5.18}$$
$$\text{with} \quad 10^9 < Gr Pr < 10^{12}$$

Horizontal bodies:

$$Nu = 0.50(Gr Pr)^{0.25} \text{ for laminar flows} \tag{5.19}$$
$$\text{with} \quad 10^4 < Gr Pr < 10^9$$
$$Nu = 0.14(Gr Pr)^{0.333} \text{ for turbulent flows} \tag{5.20}$$
$$\text{with} \quad 10^9 < Gr Pr < 10^{12}$$

In Sect. 5.10 the evaluation of natural convection of horizontal cylindrical absorbers of a parabolic trough collector, and of vertical cylinders such as the external receivers in a tower SPP are outlined. A more detailed relationship for an isothermal horizontal cylinder is given in [11] in the form of:

$$Nu = 0.36 + \frac{0.518(GrPr)^{0.25}}{[1+(0.559/Pr)^{0.5625}]^{0.444}} \quad (5.21)$$

$$\text{for } 10^4 < GrPr < 10^9$$

$$Nu = \left\{0.60 + 0.387\left[\frac{GrPr}{[1+(0.559/Pr)^{0.5625}]^{1.778}}\right]^{0.1667}\right\}^2 \quad (5.22)$$

$$\text{for } 10^6 < GrPr < 10^{12}$$

Investigations of natural heat transfer mechanisms at vertical cylindrical receivers for tower SPP required the analysis of novel boundary conditions, i.e.:

- 'cylinder walls' composed of 12–50 mm tubes forming a cylinder of several meters diameter (surfaces macroscopically and microscopically rough);
- size and operating temperature of the receivers were such that Grashoff numbers were 10 to 100 times larger than available in the literature ($Gr \approx 10^{14}$).
- although data on forced or natural convection exists, no data is available for mixed forced and natural convection (for instance when wind is superimposed on natural convection).

A series of experiments were funded in the US. Data were obtained, starting from the shaded areas of Fig. 5.4. From these data the relationship for the combined convection coefficient was determined to be

$$h = (h_{\text{nat}}^a + h_{\text{forced}}^a)^{1/a} \quad (5.23)$$

with $a = 3.2$ for external receivers, and $a = 1.0$ for cavity receivers. The Nusselt numbers for smooth glass cylinders were determined in [33] and [35] as

$$\frac{h_{\text{nat}}H}{k} = Nu_{\text{nat}} = 0.098\, Gr^{0.333}\left(\frac{T_\infty}{T_w}\right)^{0.14} \quad (5.24)$$

$$\frac{h_{\text{forced}}D}{k} = Nu_{\text{forced}} = 0.00239\, Re^{0.98} + 0.000945\, Re^{0.89}. \quad (5.25)$$

with $h_{\text{rough}} = \frac{\pi}{2}h_{\text{smooth}}$. This correlation is valid for diameter-related Reynolds numbers between $3.7 \cdot 10^4$ and 10^7.

Natural convection of cavities turned out to be more complex than expected. This was proven in [33] by data from the IEA–SSPS experience, as well as by data from the GAST-20 project [7] (obtained by wind tunnel tests and smoke simulation methods). Apparently, various parts of the cavity participate in a changing way in the heat loss mechanisms. The prevailing rest volume is probably filled with intricate recirculation zones. Influencing factors are the details of receiver design, such as aperture area relative to that of the absorber surfaces, cavity volumes, and aperture positions (tilting angle of the receiver opening) as well as wind velocities and their directions relative to the aperture area (stagnation, injection, suction).

5.4 Physical Interactions

The receiver is the transfer unit between incoming concentrated solar radiation and convectively transported thermal energy within heat exchanging tubes or at suitable surfaces within

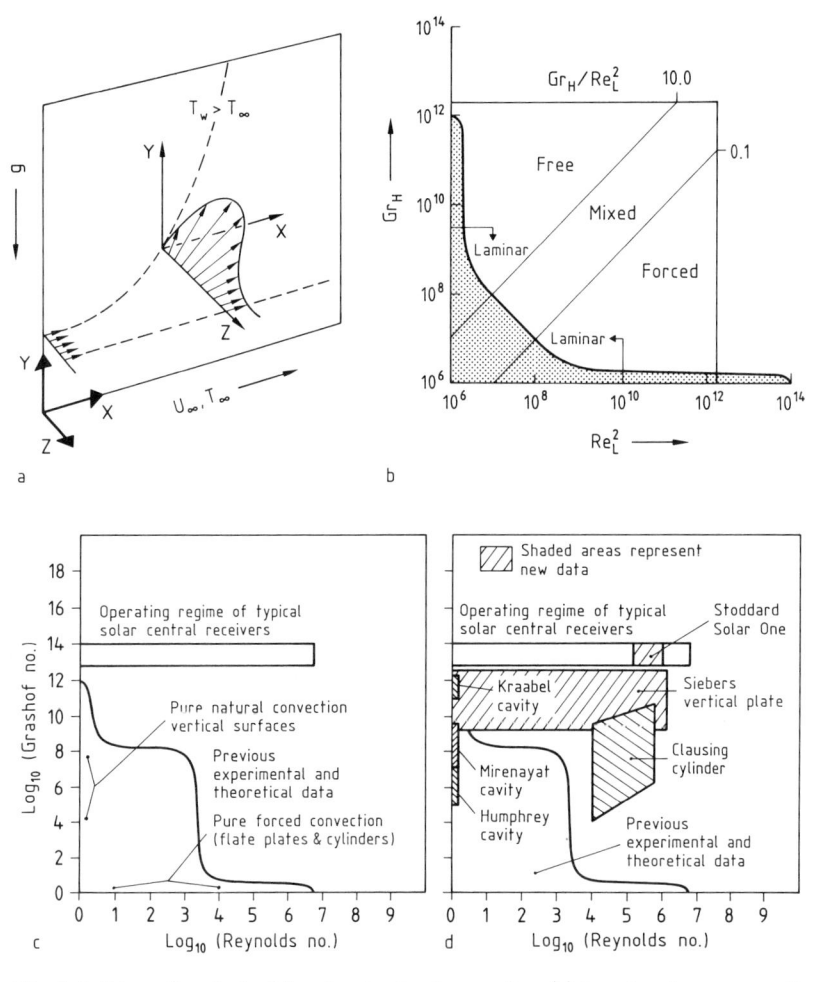

Fig. 5.4. Schematics of mixed forced and natural convection. (a) boundary layer on a surface [33]; (b) flow regions of interest [33]; (c) operating domains status 1979 [35]; (d) status 1985 [35].

volumes. As solar energy is originally matterless radiation, conventional heat exchanger designs are not necessarily the best solution for solar receivers. The volumetric receiver concept offers an alternative solution. The direct absorption principle also carries a reasonable chance of being realized as a receiver unit, allowing much higher temperatures than possible with the conventional tube technology.

To investigate the potential and the limitations of receivers operating with concentrated sunlight is the objective of this section[6]. It is derived to some extent from the study of Boese and colleagues [9]. The interaction between incoming radiation and heat transferring mechanism is the topic; losses from reflection, natural convection or conduction in pipes or containers are not considered.

---

[6] A more theoretically detailed approach is given in Chap. 2

For idealized conditions, the flux entering a system is defined by

$$\dot{Q}_{in} = A\alpha C\phi - A\epsilon(T)\sigma T^4 \tag{5.26}$$

with A being the heat transfer area, α the absorptance, C the ratio of concentration, ϕ the solar radiation, $\epsilon(T)$ the temperature dependent emissivity and T any relevant temperature. Clearly, the energy (or power) transferred into a system *decreases* with *increasing* temperature.

The process of utilizing thermal energy in a subsequent ideal system is described by the Carnot efficiency

$$\eta_{\text{Carnot}} = \frac{T - T_a}{T} \tag{5.27}$$

with T_a being the ambient or lowest possible reference temperature. This efficiency *increases* with *increasing* temperature.

Combining both effects then gives the maximum thermal power as

$$\dot{Q}_t = \eta_{\text{Carnot}} \dot{Q}_{in} \tag{5.28}$$

resulting in the definition of an optimal temperature regime. Inserting (5.26) and (5.27) into (5.28) and assuming $C\phi = \sigma T_s^4$ with T_s as the temperature on the Sun's surface, (5.28) delivers as maximum value

$$\dot{Q}_{t,\max} = A\sigma(\alpha T_s^4 - \epsilon(T)T^4)\frac{T - T_a}{T}.$$

With the simplification of $\alpha = \epsilon$, the optimum temperature follows from

$$\frac{\partial \dot{Q}_{t,\max}}{\partial T} = 0 \Rightarrow 4T^5 - 3T^4 T_a - T_s^4 T_a = 0 \tag{5.29}$$

resulting in $T_{\text{opt}} = 2,462$ K when assuming $T_s = 5,760$ K and $T_a = 300$ K.

In reality, however, heat losses and irreversibility mechanisms lead to lower optimum temperatures, ranging down to half of the above mentioned optimum value. This fact can be explained by taking the following considerations into account:

- Instead of Carnot efficiency, a technical efficiency has to be used,

$$\eta_{\text{technical}} = \frac{\bar{T} - T_a}{\bar{T}},$$

with

$$\bar{T} = \frac{T_2 - T_1}{\ln T_2/T_1}.$$

T_1 and T_2 are inlet and outlet temperatures of the HTF (for order-of-magnitude calculations, $T_2 = 1.5 T_1$ is often used resulting in $\bar{T} = 1.233 T_1$). This relationship accounts for the non-uniformity of temperatures in the heat receiving unit. Within the heat transfer medium, non-uniform mass flow might also exist, additionally complicating an optimization.
- The ambient temperature T_a in most cases is larger than 300 K.
- The maximum concentration factor $C_{\max} = 46,000$, corresponding to the maximum solar flux density of 62 MW/m^2, is only an upper bound value. Also, solar radiation in terrestrial applications changes annually, daily and with atmospheric attenuation. Consequently, realistic values of $C\phi$ should be used instead of σT_s^4.
- Finally, the possibility should be mentioned that the optimum temperature range can even be higher with selectivity ($\alpha/\epsilon > 1$), although selectivity is difficult to achieve at higher temperatures.

Taking these facts into consideration a new optimum temperature can be determined, by changing (5.29) to

$$4\bar{T}^5 - 3\bar{T}^4 T_a - \frac{\alpha C\phi}{\epsilon\sigma}T_a = 0.$$

The essence of this approach is that the optimum temperature depends critically on the ratio $\alpha C\phi/\epsilon$. The following examples using $T_a = 300$ K illustrate this:

	C	α/ϵ	ϕ_S [W/m²]	$\alpha C\phi_S/\epsilon$ [MW/m²]	$\bar{T}_{out}$ [K]
Ideal	46,000	1	1,353	62.64	2,462
Space	5,000	2	1,353	13.53	1,828
Terrestrial	1,000	1	700	0.7	1,033
Improved	1,000	2	700	1.4	1,180
	3,000	2	700	4.2	1,457

In conclusion, it is worth pointing out that a concentration of 1,000, as achievable today for tower SPP, is a very small fraction ($\approx 2\%$) of the limiting theoretical value. That represents future potential for R&D (for experimental purposes, a single-point aiming strategy is able, even today, to produce very high concentration ($C > 3,000$) in the center region of a Gaussian-like distribution.).

Note also that the optimum temperature is computed at peak concentration. Because Carnot efficiency increase becomes quite flat near the optimum temperature, deviations of 100°C or so from the optimum can be expected to have relatively little effect on SPP performance or economics.

Rather than driving a Carnot cycle, the solar energy may be used in an endothermic reaction that drives a chemical process. Typically then a specific input temperature is required so the foregoing analysis does not apply. Instead, (5.26) can be used to estimate the concentration needed to provide heat at the required temperature with good first-law efficiency.

5.5 Engineering Methods of Computation

The receiver of line- or point-focus systems (parabolic trough, parabolic dish, heliostat/tower) is but one element of a total SPP. Yearly performance of a tower SPP is simulated and evaluated, using the SOLERGY computation code [2], which is based on the most recent experience in the US with tower SPP. SOLERGY models both water/steam- and single phase fluid-receivers, with start-up and shut-down algorithms and a generalized representation of thermal loss.

For reasons of simplicity, receiver calculations are based on the law of energy conservation. More detailed calculations of thermodynamics and low cycle creep fatigue have to be performed independently, and must take into account the receiver design features.

The most severe operations environment occurs at high temperature air receivers. Therefore, the designers of those heat exchangers have to master the toughest problems. The receiver required for the German/Spanish GAST-20 program [7] described in [1] and shown in Fig. 5.12, is not only exposed to high temperatures (e.g. up to 840°C for metallic materials or up to 1,100°C for ceramic materials) but also to frequent temperature changes as a result of the non-steady state radiation input. In addition stress interrelationships (creep, fatigue) complicate the calculations. Nevertheless, finite element analysis is an efficient method which enables the designer to predict and evaluate resulting loads.

A counter flow geometry for the HTF would provide the most effective energy transfer for a receiver. Unfortunately, considerations of minimal material stress and the physics of conversion of concentrated radiation modify this approach. To minimize material stress, a rather uniform heat flux distribution would be beneficial. For concentrated sunlight, a Gaussian-like heat flux distribution has to be accepted. Not considering complicated spiral-designs or multipass solutions, the once-through tube with specific flux profiles and peak wall temperatures (Fig. 5.5) has been computed by [26].

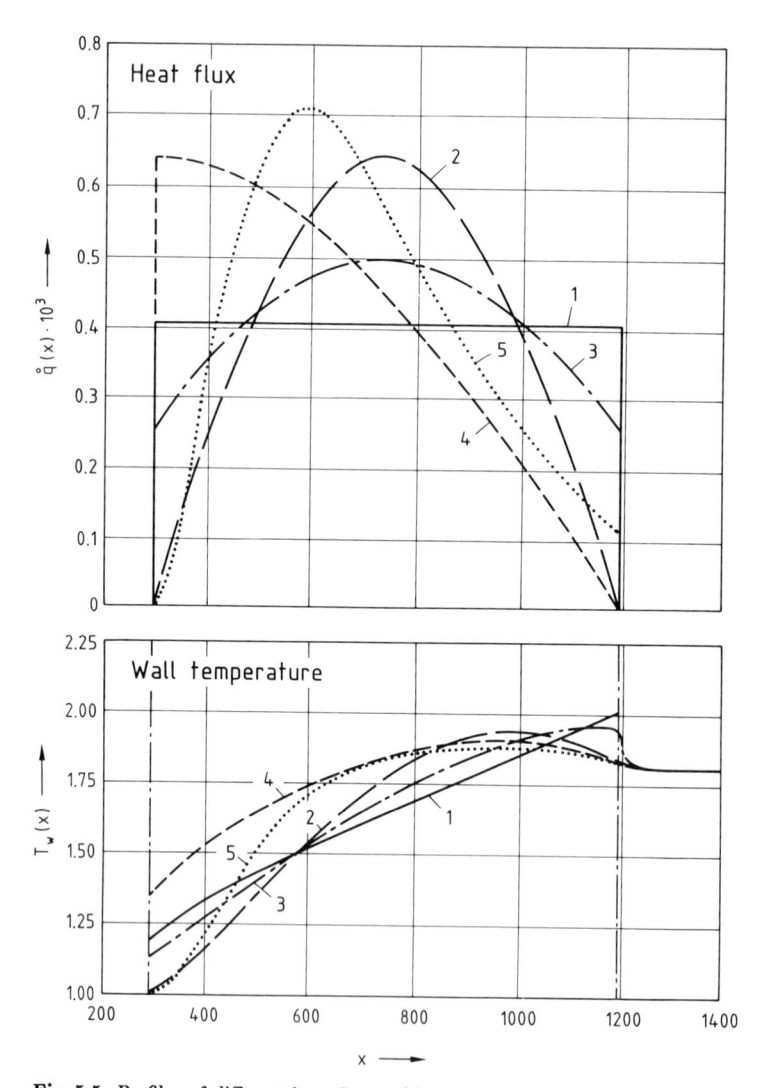

Fig. 5.5. Profiles of different heat fluxes. (1) uniform; (2) cosine; (3) tailored cosine; (4) half cosine; (5) non-symmetric Gaussian; and resulting profiles of wall temperatures along the tube length (all values normalized) [26].

The calculations were performed for constant tube inlet and outlet temperatures, $T_i = 625\text{ K} = 352°\text{C}$ and $T_o = 1,143\text{ K} = 870°\text{C}$. For all cases the energy added by solar radiative flux is the same. Therefore, the power gain in the tube is

$$\dot{Q} = \dot{m}(C_{p,o}T_o - C_{p,i}T_i). \tag{5.30}$$

Case	Flux distribution	Peak Temp. (°C)
1	uniform	983
2	full cosine	942
3	tailored cosine	955
4	half-cosine	917
5	non-symmetric (Gaussian-like)	906

Uniform flux density results in the highest wall temperature, while half cosine flux distribution and non-symmetric Gaussian deliver optimum values of wall temperature. The half-cosine condition might be very attractive from a thermodynamical point of view (counter flow with highest flux density value near the 'cold' entrance); however, the material stress resulting from sudden heat flux changes can probably not be sustained for long periods. Consequently the less extreme non-symmetric Gaussian flux distribution seems to be preferable.

5.6 Receiver Designs

Receivers for thermal SPP do not benefit significantly from mass fabrication. Their design depends strongly on the application, the kind of heat transfer medium, the range of working temperatures, the material used and the method of energy transfer. From every point of view, the receiver is a most critical component of thermal SPP, and its design and operation must be investigated very thoroughly (see [5,6] for a good survey on receivers for tower SPP).

5.6.1 Tube Receiver Concept (Central Receiver)

Highly concentrated radiation is directed onto tubes configured as an external vertical cylinder, a flat vertical plate or arranged within a protected cavity. In all cases, the HTF is heated through metal or ceramic tubes, either within one state of phase (liquid or gaseous) or, if liquid, eventually phase changed (water into steam).

Tube receivers entail pressure losses associated with HTF flow. The flow rates depend on the length of the loops; trade-off considerations are the cost for many low-flow/short loops and the parasitic power needs for few high-flow/long loops. Usually, the tube sizes are selected to achieve high flow rates with resultant high heat flux transfer.

Three tube receivers for tower SPP will be described in some detail, i.e. the

- 560°C steam receiver of the Solar One facility at Barstow.
- 530°C sodium receiver of the SSPS–CRS plant in Almería.
- 800°C/1,000°C GAST-20 air receiver tested at Almería.

Present 'conventional' tube receivers (with steam, sodium or salt fluids) are suitable for temperatures up to 500–600°C. Using gases as HTF, metallic materials can tolerate maximum temperatures in the range of 800–900°C. They are limited because of mechanical erosion/corrosion resistance. Hence, the maximum gas temperature in solar-thermal heated tubes cannot exceed this limit. The only way to overcome this limit is to use ceramic materials, preferably based on silicon carbide[7]. It can be stated that all receivers built to date could be operated successfully.

Solar One steam receiver. Power production operated during the years 1984 to September of 1988; test and start-up operation 1983–1984.

The external receiver has twenty-four panels, each approximately 1 m wide and 14 m long (Fig. 5.6). Six panels on the south side of the receiver are feedwater preheat panels while the remaining eighteen are once-through-to-superheat boiler panels. Each panel consists of seventy tubes (12.7 mm OD; 6.9 mm ID) of Alloy 800. The tubes are welded to each other over the full length within each panel (Fig. 5.7). All tubes are coated with Pyromark© black paint. The receiver, along with its associated valves, piping and controls, is designed to produce steam of 516°C at 100 bars. The achievable power is 42 MW, which was converted into 10 MW electrical energy. This is the largest existing solar-thermal central receiver.

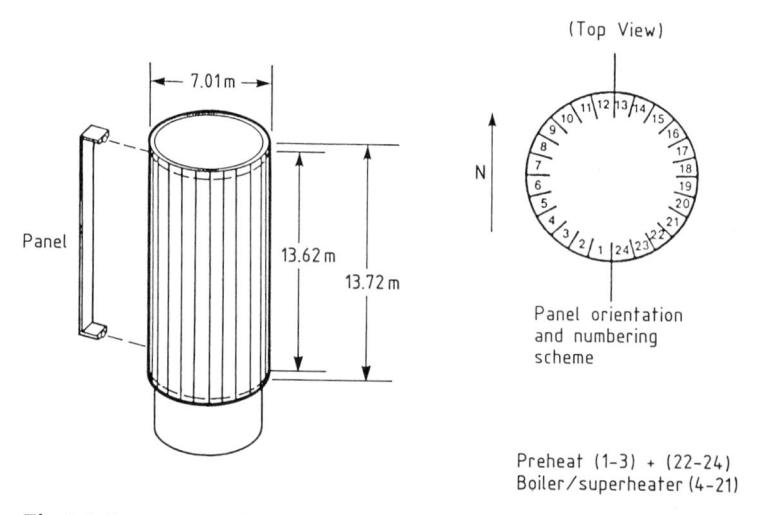

Fig. 5.6. Barstow external steam receiver [3]. The 24 panels from a 7.01 m ⌀ by 13.62 m tall absorbing area.

IEA-SSPS sodium receivers. These were investigated in the years 1982 to 1986.

Two receivers (Figs. 5.8 and 5.9) were analyzed. The first was a north-facing cavity type, having a vertical octogonally shaped aperture. The absorber panel is a 120° segment of an upright circular cylinder of 4.5 m diameter. Sodium flows in six horizontal parallel tubes (38 mm OD; 35 mm ID) which wind in a serpentine from the inlet header on the bottom of the cavity to the outlet header at the top. The tubes are not welded along their length but are individually supported. Sodium enters the inlet header at 270°C and exits through the outlet header at 530°C. The location of the absorber panel inside the cavity is such that the peak heat flux density is about 0.63 MW/m^2, with 0.16 MW/m^2 as average value.

The second receiver, the Advanced Sodium Receiver (ASR), is an external type and consists of five panels arranged to form a rectangular absorber. Each panel consists of a tube bundle with thirty-nine tubes (14 mm OD; 12 mm ID). The top header can move vertically to

---

[7] The panel experiments demonstrated that a mean gas temperature of 800°C can be achieved when using Incoloy© material, or 1,000°C when using silicon infiltrated silicon carbide.

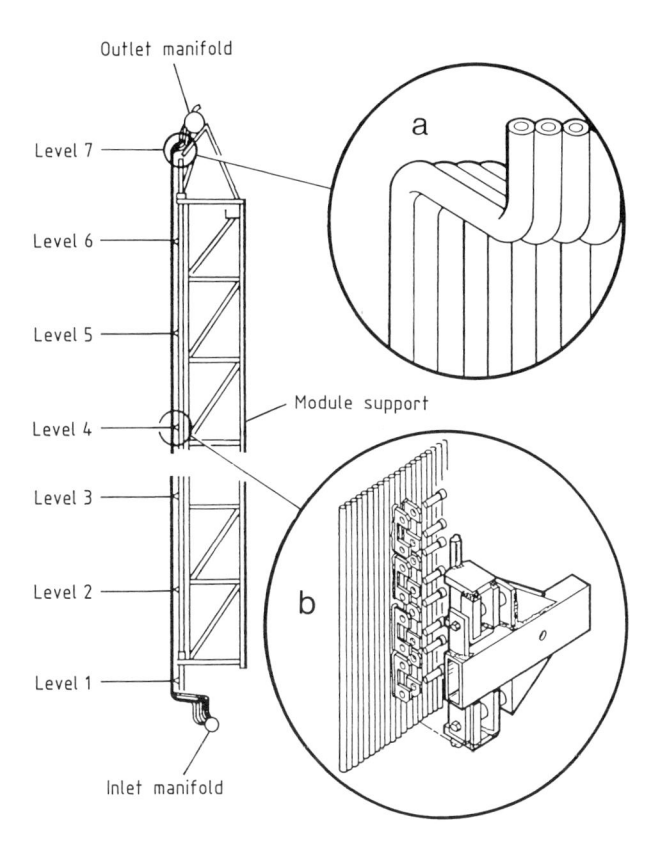

Fig. 5.7. Barstow steam receiver details [3]. (a) stress relief bends on approach to manifold; (b) mounting attachment to allow for thermal growth of panel.

accommodate the thermal growth of the panel. The irradiated tubes are assembled in groups of three. These triplets are connected to the panel framework by means of pins so that the tubes can grow axially with respect to the frame and can also rotate.

Unavoidable gaps between the tubes allow some concentrated solar flux to pass along the tubes and impinge on the back-wall structure. Therefore, a double shielding of high refractory alumina-based material is located behind the tube bundle system to protect the back structure from the incident radiation and to reflect and radiate this energy back to the tubes. The panel was designed for a peak heat flux density of 1.38 MW/m^2, with 0.35 MW/m^2 as average value.

Both receivers were able to produce 2.4–2.5 MW$_t$ of power which was converted to 0.5 MW$_e$ electrical power [19,24]. In addition, the second advanced receiver was tested up to its very limits by exerting a peak flux density of 2.5 MW/m^2 on its absorbing surface; the results are displayed in Fig. 5.10. This test demonstrated the highest flux values sustained by a solar receiver. No detrimental effects, degradations nor changes were observed; details of computations including consequences of stress analysis and final results were documented in [32], and summarized in [6]. It is apparent from these results (Fig. 5.11) that receiver efficiencies between 80 and 90% are achievable both for cavity and external receiver concepts.

GAST-20 air receiver. This was tested with metallic (in 1985/86) and ceramic (in 1987) materials.

The 20 MW$_e$ reference concept features about 2,000 heliostats of about 52 m^2 each and two side-looking cavity receivers on top of a 170 m high tower, each receiver providing 28 MW$_t$

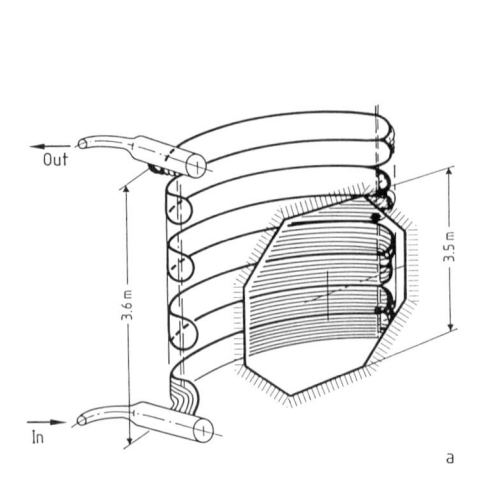

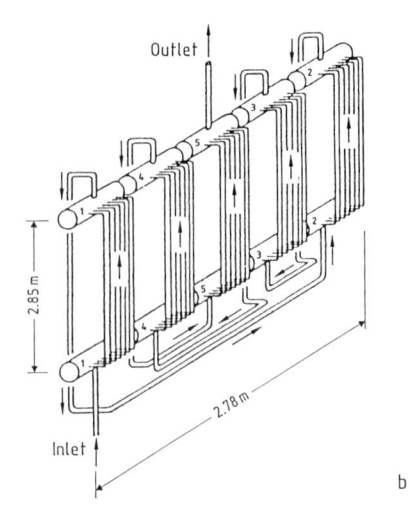

Fig. 5.8. The IEA-SSPS sodium test receivers [19].

Specifications	Cavity Sodium Receiver (a)	External Sodium Receiver (b)
No. of tubes	6	5×39
Tube diameter	38 mm	14 mm
Tube wall	1.5 mm	1 mm
Tube material	AISI 304 H	AISI 316 L
Aperture area	9.7 m^2	7.9 m^2
Single flow path length	87 m	23.5 m
Active surface	17 m^2	
Total surface	62 m^2	
Total tube weight	710 kg	300 kg
Coating: Pyromark©	2,500	2,500
Peak heat flux	0.63 MW/m^2	1.38 MW/m^2
Average heat flux	0.16 MW/m^2	0.35 MW/m^2
Inlet/outlet temp.	270/530°C	270/530°C
Mass flow (design)	7.3 kg/s	7.3 kg/s
Pressure	2.6 bar	6 bar
Pressure drop		1.5 bar
Input/output power	2.8/2.4 MW	2.8/2.5 MW

power for a subsequent Brayton cycle. Air was selected as HTF operating at mean exit temperatures of 800°C with metallic materials (Incoloy© 800 H), and up to 1,000°C with ceramic materials (silicon infiltrated SiC). A gas turbine cycle was foreseen to be bottomed by a steam cycle.

Figure 5.12 sketches the main idea of such a receiver design; Fig. 5.13 details the flux interaction and transfer to the tubes [7].

For receiver development, two panels were built that characterized the most stringent and loaded receiver elements, the exit portion of the metallic version and a feasibility unit of the ceramic version. The main metallic and ceramic test data of the panels, both tested in Almería, are shown in Tab. 5.1 (see [7]):

Figure 5.14 shows a schematic of the first test panel system mounted at the top of the CESA-1 tower. From the successful performance of these experiments it is concluded that

5.6 Receiver Designs

Fig. 5.9. Views of the IEA-SSPS receivers [5,8]. (a) Cavity receiver (Interatom/Sulzer); (b) Inside view of cavity receiver (Interatom/Sulzer; courtesy of Interatom); (c) External receiver (Franco Tosi/Agip Nucleare).

high-temperature gas-cooled receivers can be designed and indeed be built up to the projected performance levels to 20 MW_e or more (Fig. 5.15)[8]. However, development for optimization and qualification procedures are still required to finalize the joining techniques of long ceramic tubes.

The status as described for the central receiver also holds for the receiver units of parabolic dish systems, though these are of smaller scale but with even higher temperature potential.

---

[8] The details of the GAST-20 goals and achievements are given in [1] and [7]. Optimization of the test and evaluation of its results have been supported by finite element computing, stresses were exerted on the components by the solar-specific operating conditions. Unsteady thermal losses at high temperatures were evaluated

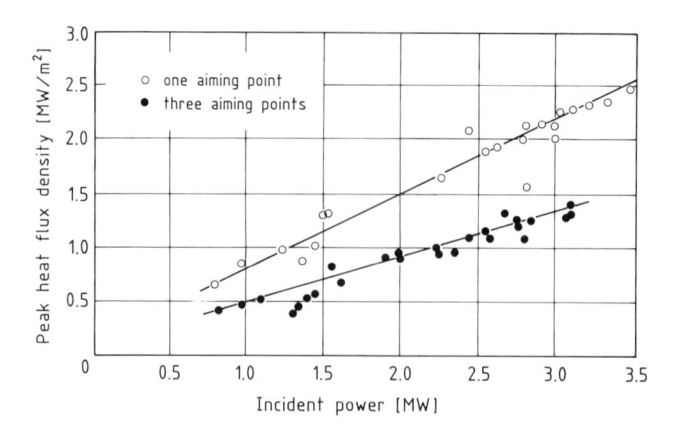

Fig. 5.10. Peak fluxes of an external sodium receiver (IEA–SSPS) [32].

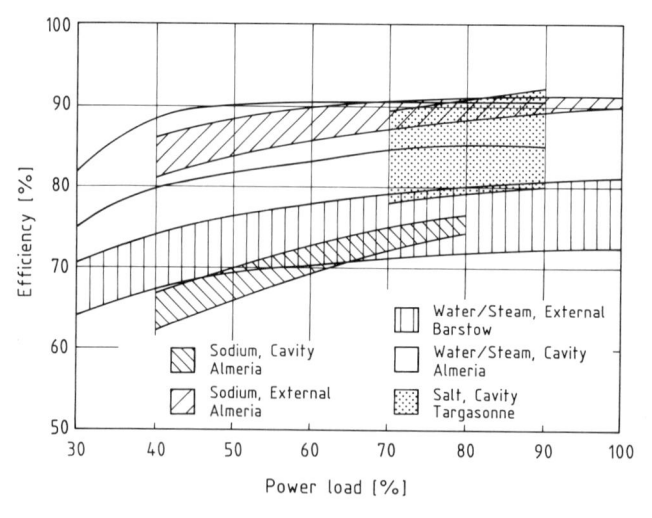

Fig. 5.11. Receiver efficiencies tested for water/steam, salt, and sodium receivers [6].

5.6.2 Tube Receiver Concept (Parabolic Trough)

In comparison to tower SPP receivers, the receiver technology for the parabolic trough/line focus collectors is well advanced. Early receivers – as tested in the IEA-SSPS-project [19], (Fig. 5.16) – already represented a fairly mature solution. The energy-receiving black chrome coated tubes (38 mm diameter) were protected by a surrounding glass annulus (64 mm diameter). To eliminate convection losses between these cylinders, the pressure had to be below about 10^{-4} bar. The tube length ranged in the order of 6 m, with a support in between.

In recent years, the parabolic trough development was fostered by building large-scale plants in the US (Barstow and Kramer Junction, California). These receivers are considered to be fully matured now. Figure 5.17 shows the characteristics and technical data of the LS-2 collector as well as a picture of the 5 m wide reflector and the 100 mm diameter windowed receiver tube, belonging to a 30 MW unit.

5.6 Receiver Designs

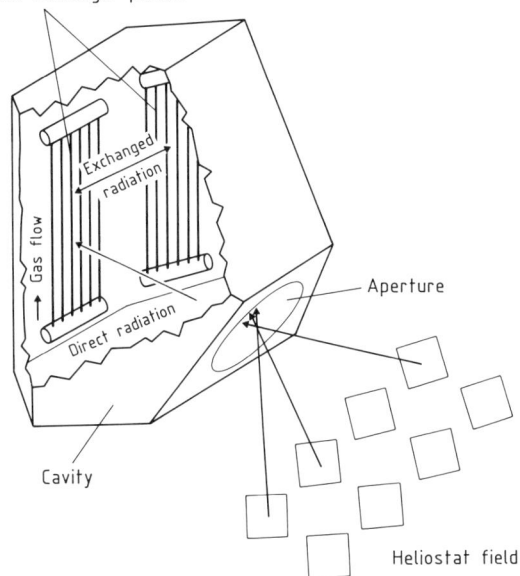

Fig. 5.12. Concept of a $2 \cdot 28$ MW$_t$ gas-cooled receiver (GAST-20) for a 20 MW$_e$ tower SPP [7]. Losses are: Emissions = 5.7%, reflection = 1.5%, convection = 6.1%, and conduction = 1.7%.

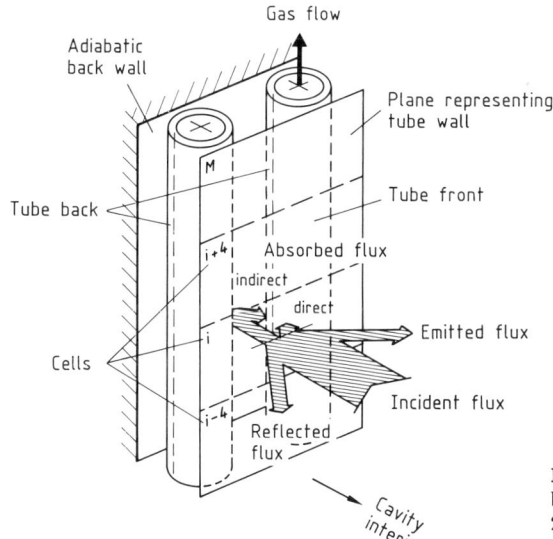

Fig. 5.13. Interaction model of tubes and back wall for flux computations (GAST-20) [7].

5.6.3 Volumetric Receiver Concept

The volumetric receiver is a system of structures arranged to fill a volume. The structures might be a wire grid, foil, foam or another alternative system. Two tests were performed, one testing a metal wire mesh up to 700°C [17], the other a ceramic wire grid, which sustained temperatures higher than 1,100°C [28].

The schematic of Fig. 5.18 exemplifies the volumetric receiver concept. Concentrated solar energy hits several planes of ceramic (SiC) fibres which absorb the incident radiation and

Table 5.1. Metallic and Ceramic Test Data.

	Metallic	Ceramic
Inlet temperature, °C	625	350
Inlet pressure, bar	9.3	9.3
Outlet temperature, °C	800	1,000
Mass flow, kg/s	2.45	0.36
Outer tube diameter, mm	42	41
Tube wall thickness, mm	2.1	5
Number of tubes	18	10
Material of tubes	X 10 Ni Cr Al Ti 3220H	RBSC (SiC)
Irradiated length, m	8	4.1
Power output, kW	489	270

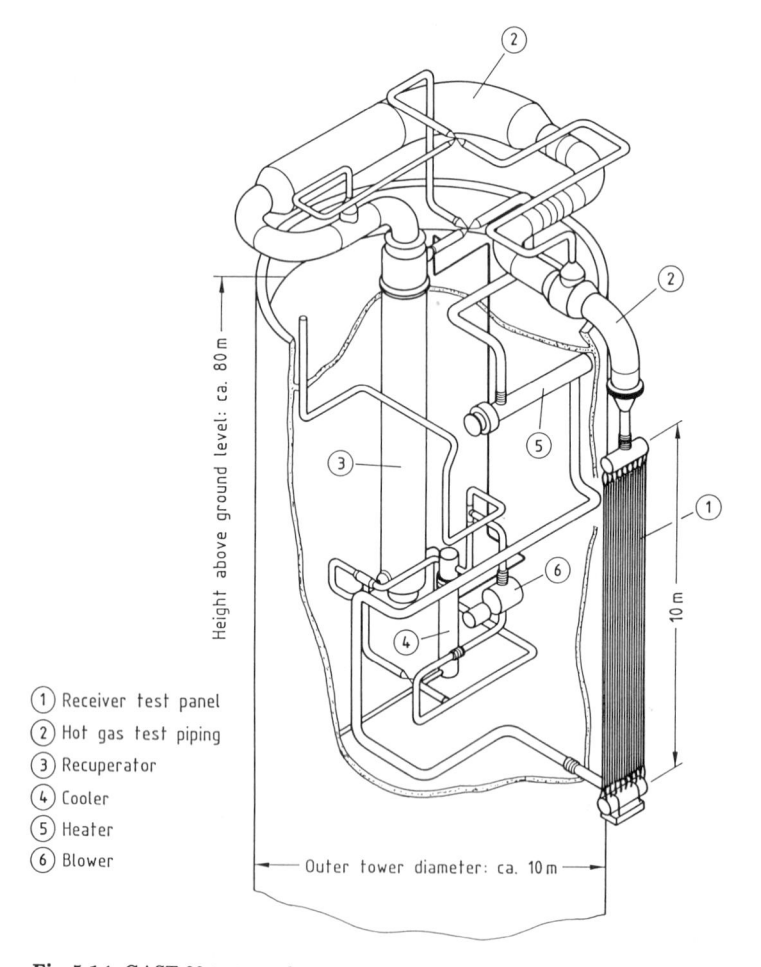

① Receiver test panel
② Hot gas test piping
③ Recuperator
④ Cooler
⑤ Heater
⑥ Blower

Fig. 5.14. GAST-20 test panel mounted at the top of CESA-1 in Almería [7].

5.6 Receiver Designs

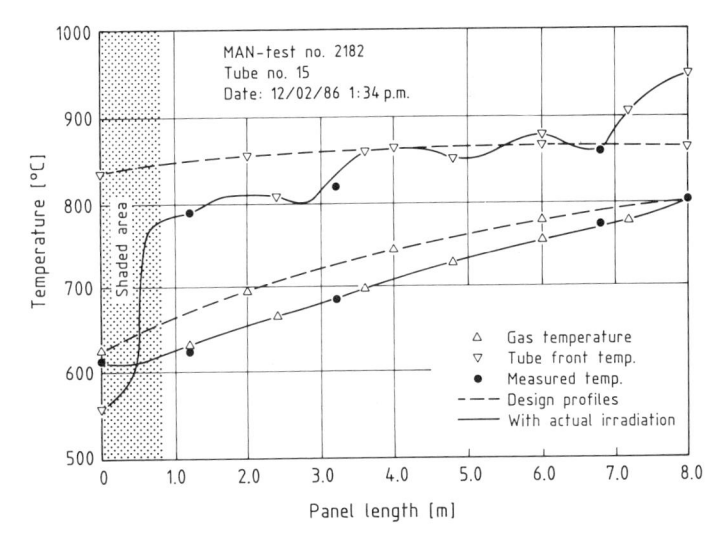

Fig. 5.15. Comparison of calculated design temperatures with temperatures measured under actual irradiation for the metallic GAST-20 panel [7].

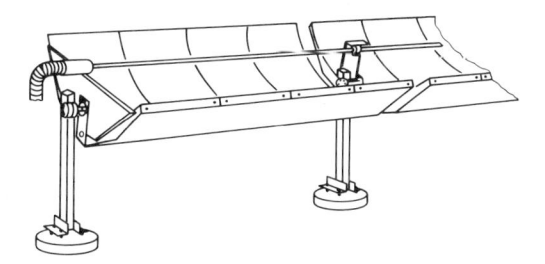

Fig. 5.16. Acurex type 3001 collector [19].

Technical data (for one Module)

Aperture area	5.58 m²	Collecting/non-collecting tube length ratio (a)	1:0.3
Reflectivity (design)	0.92		
Reflector	0.6 mm glass/steel laminate	Min/max temperature	225/295°C
Receiver α/ε (design)	0.96/0.27	Weight (b)	32.1 kg/m²
Absorber diameter	32.8 mm	Wind constraints	
Glass tube diameter	50 mm	– Full operation	30 km/h
Tube material	steel	– Reduced operation	50 km/h
Coating	black chrome	– Survival	161 km/h
Tracking accuracy	0.25°	(a) including field piping	
Collecting efficiency		(b) including reflector (129 kg), support posts (26 kg each) and motor drive equipment (136 kg)	
– Design	58%		
– Max. measured	59%		

Fig. 5.17. View and technical data of the Luz LS-2 collector at Kramer Junction [18] (Courtesy of Flachglas Solartechnik GmbH, Cologne, Germany).

Technical Data (for one LS-2 collector)

Aperture width	5 m	Coating	cer./met. ex.
Aperture length	47.1 m	Absorptivity at 350°	0.97
Aperture area	235 m^2	Emissivity at 350°	0.15
Rim angle	85°	Overall optical efficiency	0.75
Support distance	8 m	Concentration factor	71
Reflectivity	0.94	Geometric concentration	71/π
Reflector (silvered)	4 mm white glass	Min/max temperature	293/391°C
Window diameter	95 mm	Thermal efficiency	0.66
Window transmittance	0.98 (AR coated)	Wind constraints	
Absorber diameter	70 mm	– Operation up to	56 km/h
Tube material	steel	– Survival up to	128 km/h

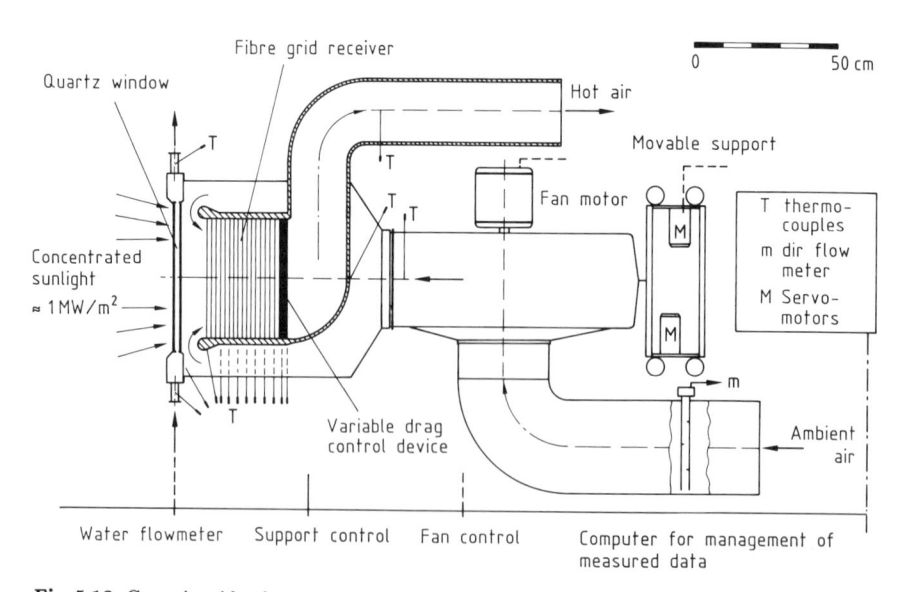

Fig. 5.18. Ceramic grid volumetric receiver (tested at DLR from 1986 to 1988) [20,28].

convectively heat the air passing by. At the end of the ceramic grids, an adjustable drag device tailors the local mass flow relative to the flux distribution. The development of a quartz window at the entrance is essential to the further development of volumetric receivers, as high pressure is required for cost-effective configurations in power applications, as well as for containing chemical process reactions. The piping on the right side of the schematic diagram accommodates the inlet of ambient air including pressurization and the hot air duct to a possible user. The movable support structure is necessary to accommodate thermal growth of the ductwork.

5.6.4 Direct Absorption Receiver Concept

Direct absorption of highly concentrated radiation allows immediate transfer of radiative energy into a medium which is either a liquid suspension or a cloud of particles. The thermal energy produced is used for heating a process or driving a chemical reaction.

The concept of particle heat transfer, originally studied in France [30], is currently under investigation in the US (SANDIA, SERI, LBL) and in Germany (DLR); it shows some very promising features [13,23]:

- high temperatures up to $2,000°C$, or more
- high fluxes up to 2 MW/m^2, or more
- selective behavior of particles
- no walls, merely containment requirement.

The temperature potential is shown in Fig. 5.19 for large particles ($R_p > 1$ μm) to be well above 1,500 K, and for small particles with selective properties ($R_p < 0.1$ μm) to be even more than 2,500 K [13]. In this figure, the equilibrium stagnation temperature is plotted versus solar flux intensity and concentration factor.

The alternative means of direct absorption, a blackened liquid flowing as a sheet over a nearly vertical backing plate, continues to receive attention in the US. Preliminary research was performed with a large-scale water flow test apparatus to evaluate turbulent flow pro-

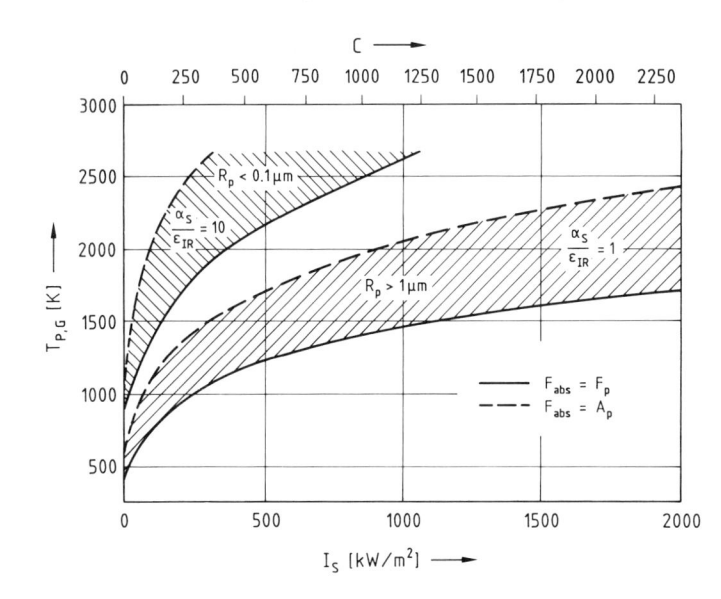

Fig. 5.19. Characteristics of solid particle heat transfer receiver [13,20].

cesses, wave formations and detachments [10]. Cold water simulates the flow of molten salt at 350°C–800°C, as their viscosity and surface tension are comparable[9].

5.7 Relationships Between Design and Type of Application

A wide range of SPPs have been studied and numerous experimental facilities were built and operated. For low temperature process heat (to about 400°C), oil-cooled parabolic troughs have been successfully operated in power ranges from 200 kW_e to 80 MW_e. With natural gas superheat as supplement, 250 MW (eight systems in California) are operating in 1990. Many smaller systems have been built and operated successfully (see Chap. 7). While a market niche exists for simple small trough SPPs, their operation must be very reliable and their operating and repair costs must be competitive.

Several dish thermal SPPs (thermal energy carried to a central point) have been operating up to 800°C. Flexible inlet and outlet piping required for each dish reduces the practicability of such SPPs, unless a fixed receiver is used. Still, a high temperature collection manifold is required. This problem is avoided by operating an engine-generator at *each* receiver. 25–50 kW_e organic Rankine cycle, Brayton cycle and Stirling cycle engines have been studied for this purpose. Receivers employing either a pumped heat transfer fluid or a directly coupled heat pipe show promise. Over 29% overall efficiency has been demonstrated for a dish/Stirling system.

Solar-driven chemical reactions using individual reactors and central receiver reactors have been conceived for chemical production, for fuel production and for an efficient means of transporting thermal energy to a central site [27]. Methane reforming[10] experiments are under investigation for a small scale reactor [16] and for a central receiver system [27]. The large-scale proposal of a cavity receiver/reactor [31] is shown in Fig. 5.20 in a three-dimensional view.

Parabolic dish receivers offer a range of design options, depending on the process needs. For moderate temperatures up to 500 or 600°C, a simple external receiver can be used. This can be a helical close-wound tube, or a heat sink enclosure containing a heat exchanger tube. An average flux of 0.2–0.3 MW/m^2 is achievable with acceptable thermal losses of approximately 10%. For higher temperatures (or a receiver with a lower allowable flux, say a receiver-reactor) it is necessary to reduce the losses by using a cavity design. The cavity reduces reflective and radiative losses and, because the aperture predominantly faces downward (away from the Sun), convective losses are suppressed as well. If the allowed flux is too small and the interior surface becomes excessive, conductive losses grow, and forced convection may become dominant because of the large heat exchange surface. A window will suppress this effect and seals the interior. Such configuration would make direct absorption possible, e.g. absorption in small particles or in a gaseous photo-catalyzed chemical reaction.

Adding a terminal concentrator to a long focal length dish allows achievement of very high concentrations, e.g. more than 20,000 Suns. While no requirements for such a configuration exist, this is a research item with future potential.

---

[9] Salt flow tests have begun on a 1 m wide by 7 m tall panel at Sandia National Laboratories; the detailed performance behaviour will be reported soon.

[10] Methane reforming: $CH_4 + H_2O \rightleftharpoons CO + 3H_2$ and $CH_4 + CO_2 \rightleftharpoons 2CO + 2H_2$.

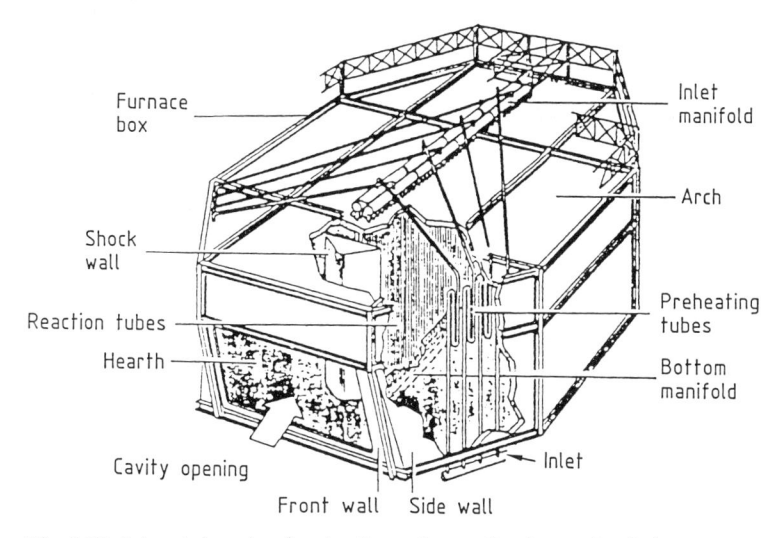

Fig. 5.20. Integrated receiver/reactor for methane reforming as chemical process reaction with substantial thermal energy input [31]. The reaction tubes hang from the inlet manifold on the centerline of the receiver.

If the allowable flux at the receiver surface exceeds about 0.5 to 1.0 MW/m² at temperatures around 600°C, an external receiver is the most cost effective solution. Thermal losses and costs are lower than for a cavity receiver, and the absence of optical constraints can further reduce system costs. Thus, systems delivering heat to a conventional turbine might best use external receivers. Conversely, the higher temperatures and lower allowed flux levels of gas cooled receivers driving Brayton or Stirling cycle engines call for a cavity receiver and a north heliostat field. Similarly a tube receiver/reactor can typically accept only 0.1 to 0.3 MW/m² at 800 to 1100°C. Alternative receivers capable of producing similar or higher temperatures would be superior (volumetric or direct absorption) if they were developed to their full maturity.

5.8 Status and Prospects

The state-of-the-art receiver technology for central receiver systems can be described as follows:

- Tube receivers can be readily designed, constructed and operated with metallic materials up to HTF mean exit temperatures of 800°C. Their lifetime is not yet verified as long time operating data (for low cycle creep fatigue) are available only for a few receivers, i.e. for steam (Solar One) sodium (IEA-SSPS) and salt (Themis [14] and Sandia [34] – Tests[11]) as HTF. Significant test results for air receivers exist from CRTF (Boeing) and Almería. Ceramic element tests for air temperatures up to 1,000°C have also been successfully carried out [7].

---

[11] Molten salt cavity receivers have been tested recently at CNRS, Targasonne, France and Sandia National Laboratories, Albuquerque/NM and performed very well at a power range of 9 MW with output temperatures up to 440°C and 4.5 MW with temperatures up to 560°C, respectively.

- Volumetric receivers are in an early state of development. Functional tests with wire meshes and ceramic grids have taken place [17,28][12].
- Experiences with particle direct absorption receivers do not yet exist. The direct absorption process is still in a research phase, with first feasibility experiments performed [13,23].

Concerning future prospects, the conventional heat exchange process utilizing tubes has been proven. Further development tends towards silicon infiltrated silicon carbides when higher fluid temperatures are required (for chemical processes), or if optimization considerations require temperatures as high as $1,000°C$. Since the efficiency optimum is relatively broad, slightly higher or lower temperatures will be acceptable from an efficiency point of view.

However, for other reasons such as design and configuration (loss mechanism), heat transfer fluid, materials available, reliability and safety, investment, maintenance, variable and fixed operating cost, it might be more favorable for SPP electricity production to stay within the presently well proven regimes of HTF temperatures around 400 to 500°C, associated with material temperatures up to 600°C.

The prospects for volumetric receivers are positive because of the opportunity for geometric flexibility, with heat flux densities considerably higher than in the high temperature case of tube receivers (where restrictions are governed by pressure and leakage problems). The prospects are theoretical at this point, as further developments are required to verify the possibilities (this applies particularly to the aperture window).

Fluid film direct absorption receivers are presently in the advanced R&D phase. First tests with a panel (1 m by 7 m) for investigating the heating efficiency of running nitrate salt films are carried out at Sandia National Laboratories/NM. Future prospects are closely tied with the outcome from these tests.

The outlook for particle direct absorption receivers depends very much on the outcome of substantial research. Research is required in the area of radiative transfer into particle clouds, of the containment problem, and for the verification of the high pressure tolerance of windowed configurations. Also, suitable chemical processes have to be identified. Then, the temperature regime is clearly dictated by the application.

Concerning parabolic trough collector SPP, progress by using very large units (aperture width more than 10 m) also indicates receiver enlargement. Substantial research items will be the direct use of steam within the receiver tubes (direct steam generation) and improvement of the selective absorber surface to archieve both stability and higher values of α/ϵ at higher temperatures.

5.9 Measurement Techniques

Of particular interest in this context are details on optical methods for the measurement of concentrated radiation. The Heliostat/Receiver Measurement System HERMES (see [32]) is an example of a system capable of measuring the flux characteristics of a single heliostat or of a total heliostat field, either at the aperture of a cavity receiver or near the absorptive area of an external receiver.

The non-intrusive measurement system consists of three measuring elements:

1. A video camera for measurement of the incident flux distribution in the visible range at the diffuse reflecting surface of a quickly moved traversing bar.

---

[12] Several foil and foam receiver concepts are presently under investigation in Almería.

5.9 Measurement Techniques

2. An infrared camera for measurement of the absorber surface temperature distribution in the infrared atmospheric window at 8–12 μm (IR).
3. A Kendall radiometer for calibration measurements (to which the meteorological conditions as global and beam radiation have to be related) together with a data acquisition system and a process computer with a specially developed code.

Figure 5.21 shows the schematic set-up for the visible and the infrared range; the traversing bar is indicated at the receiver entrance.

A typical picture generated by the infrared imaging system is shown in Fig. 5.22. The set of surface temperatures – reaching well above 650°C – is shown in the figure. On the right side certain false colors are artificially adjoined to the measured infrared intensities which are proportional to the temperature of the surface of origin.

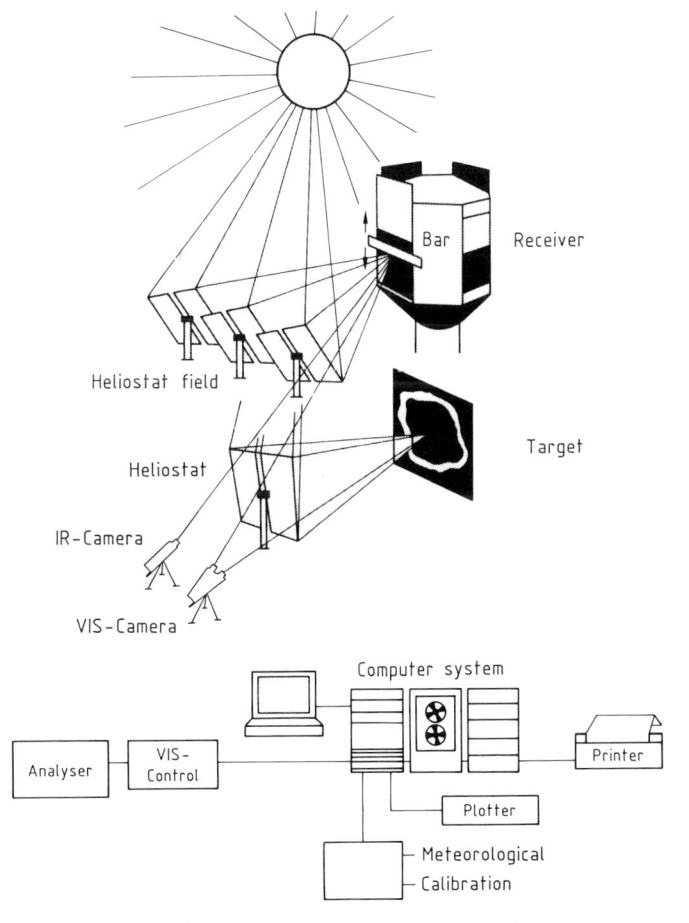

Fig. 5.21. Heliostat/receiver measuring system (HERMES) for visible radiation and infrared temperatures [32].

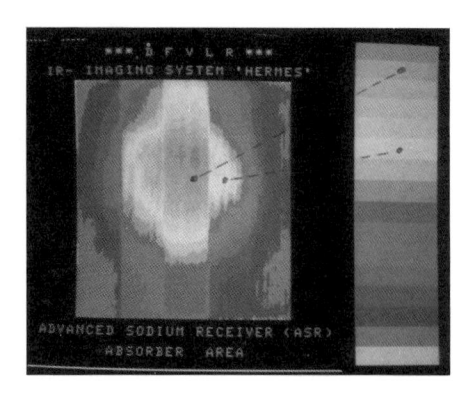

Fig. 5.22. Infrared picture of HERMES system. Correspondence is indicated by the connection lines [32].

5.10 Receiver Loss Calculation Examples

Example 1: Receiver for tower SPP.

At the tower SPP in Barstow/California, heliostats reflect and concentrate the Sun's radiation onto a heat exchanging receiver. The mean flux density at the cylindrical receiver can be assumed to be $\dot{q}_m = 160 \text{ kW/m}^2$. The absorptivity $\alpha \, (= \epsilon)$ is estimated with 0.94. The mean wall temperature is $T_{w,m} = 447°\text{C}$ while the mean ambient air temperature is $T_{a,m} = 17°\text{C}$, and the mean wind velocity is $v_{a,m} = 5 \text{ m/s}$ (see also Fig. 5.7).

Determine[13] (a) the incident flux $\dot{Q}_{\text{in}}$; (b) the reflected flux $\dot{Q}_{\text{refl}}$; (c) the reradiated flux $\dot{Q}_{\text{rad}}$; (d) the convected flux $\dot{Q}_{\text{conv}}$ (conduction is assumed to be negligible); (e) the thermal flux $\dot{Q}_t$ to the heat transfer fluid; and (f) the receiver efficiency η.

(a)
$$\dot{Q}_{\text{in}} = \dot{q}_m A = \dot{q}_m \pi D H = 160 \, \frac{\text{kW}}{\text{m}^2} \cdot \underbrace{\pi \cdot 7.01\text{m} \cdot 13.46 \text{ m}}_{A = 296.42 \text{ m}^2} = 47.43 \text{ MW}$$

(b) $\rho = 1 - \alpha = 1 - 0.94 = 0.06$

$$\dot{Q}_{\text{refl}} = \rho \dot{Q}_{\text{in}} = 0.06 \cdot 47.43 \text{ MW} = 2.85 \text{ MW}$$

(c) with the assumption that $\int T_\omega^4 dA = T_{\omega,m}^4 A$,

$$\dot{Q}_{\text{rad}} = A\epsilon\sigma(T_{w,m}^4 - T_{a,m}^4) = 296.42 \text{ m}^2 \cdot 0.94 \cdot 5.67 \cdot 10^{-8} \, \frac{\text{W}}{\text{m}^2\text{K}^4} \cdot (7.2^4 - 2.9^4) \cdot 10^8 \text{ K}^4 = 4.13 \text{ MW}$$

(d) for $T_m = (T_{w,m} + T_{a,m})/2 = 232°\text{C} = 505 \text{ K}$,

$$k_m = 41.16 \cdot 10^{-3} \, \frac{\text{W}}{\text{mK}}; \quad \nu_m = 38.93 \cdot 10^{-6} \, \frac{\text{m}^2}{\text{s}}; \quad \beta = 1.99 \cdot 10^{-3} \, \frac{1}{\text{K}}$$

$$Re = \frac{v_{a,m} D}{\nu_m} = \frac{5 \text{ m/s} \cdot 7.01 \text{ m}}{38.93 \cdot 10^{-6} \text{ m}^2/\text{s}} = 0.9 \cdot 10^6$$

$$Gr = \frac{g\beta(T_{w,m} - T_{a,m})H^3}{\nu_m^2} = \frac{9.81 \, \frac{\text{m}}{\text{s}^2} \cdot 1.99 \cdot 10^{-3} \frac{1}{\text{K}} \cdot 430 \text{ K} \cdot 13.46^3 \text{ m}^3}{(38.93 \cdot 10^{-6})^2 \, \frac{\text{m}^4}{\text{s}^2}} = 1.35 \cdot 10^{13}$$

Following (5.25) gives

$$\begin{aligned} Nu_{\text{forced}} &= 0.00239 \, Re^{0.98} + 0.000945 \, Re^{0.89} \\ &= 0.00239 \, (0.9 \cdot 10^6)^{0.98} + 0.000945 \, (0.9 \cdot 10^6)^{0.89} \\ &= 0.00239 \cdot 0.68416 \cdot 10^6 + 0.000945 \cdot 0.19919 \cdot 10^6 \\ &= 1635.1 + 188.2 = 1823 \end{aligned}$$

---

[13] Thermal and transport properties for air can be obtained from standard reference books, e.g. [21,22,25]

5.10 Receiver Loss Calculation Examples

for the rough tube wall $h = \pi/2 h_{\text{smooth}}$, thus

$$\bar{h}_{\text{forced}} = \frac{\pi}{2} \frac{Nu_{\text{forced}} k_m}{D} = \frac{\pi \cdot 1{,}823 \cdot 41.16 \cdot 10^{-3} \frac{W}{mK}}{2 \cdot 7.01 \text{ m}} = 16.81 \frac{W}{m^2 K}$$

$$\frac{T_\infty}{T_{w,m}} = \frac{290 K}{720 K} = 0.4028 \implies \left(\frac{T_\infty}{T_{w,m}}\right)^{0.14} = 0.88$$

Following (5.24) gives

$$Nu_{\text{nat}} = 0.098\, Gr^{0.333} \left(\frac{T_\infty}{T_{w,m}}\right)^{0.14} = 0.098 \cdot (1.35 \cdot 10^{13})^{0.333} \cdot 0.88 = 2033$$

$$\bar{h}_{\text{nat}} = \frac{\pi}{2} \frac{Nu_{\text{nat}} \cdot k_m}{H} = \frac{\pi \cdot 2{,}033 \cdot 41.16 \cdot 10^{-3} \frac{W}{mK}}{2 \cdot 13.46 \text{ m}} = 9.77 \frac{W}{m^2 K}$$

Following (5.23) gives

$$\bar{h} = \left(\bar{h}_{\text{forced}}^{3.2} + \bar{h}_{\text{nat}}^{3.2}\right)^{1/3.2} = (8{,}353 + 1471)^{0.3125} = 17.68 \frac{W}{m^2 K}$$

$$\dot{Q}_{\text{conv}} = A\bar{h}(T_{w,m} - T_{a,m}) = 296.42 \text{ m}^2 \cdot 17.68 \frac{W}{m^2 K} \cdot 430 \text{ K} = 2.25 \text{ MW}$$

(e)

$$\dot{Q}_t = \dot{Q}_{\text{in}} - \dot{Q}_{\text{refl}} - \dot{Q}_{\text{rad}} - \dot{Q}_{\text{conv}} - \dot{Q}_{\text{cond}}$$
$$= 47.43 \text{ MW} - 2.85 \text{ MW} - 4.13 \text{ MW} - 2.25 - 0 \text{ MW} = 38.2 \text{ MW}$$

(f)
$$\eta = \dot{Q}_t/\dot{Q}_{\text{in}} = 38.20 \text{ MW}/47.43 \text{ MW} = 0.81$$

Example 2: Line focus receiver for parabolic trough.

The farm SPP at Kramer Junction, California concentrates the solar radiation such that the mean flux density at the surface of the receiver tube is $\dot{q} = 15.82 \text{kW/m}^2$. The absorptivity α is 0.97 and the effective emissivity of the black chrome coating ϵ reaches 0.15 by applying a protecting glass tube. The outer diameters are $D_R = 70$ mm for the receiver-tube and $D_G = 95$ mm for the glass-tube. Near the outlet of a collector string, a condition with receiver temperature $T_{R,W} = 350°$C, glass tube temperature $T_{G,W} = 60°$C, ambient temperature $T_a = 17°$C and negligible wind should be considered.

Determine for a tube unit length of 1 m: (a) the incident flux $\dot{Q}_{\text{in}}$; (b) the reflected flux $\dot{Q}_{\text{ref}}$; (c) the reradiated flux $\dot{Q}_{\text{rad}}$; (d) the convected flux $\dot{Q}_{\text{conv}}$ (conduction is assumed to be negligible); (e) the thermal flux $\dot{Q}_t$ to the heat transfer fluid; (f) the receiver efficiency η.

(a)
$$\dot{Q}_{\text{in}} = \dot{q}A = \dot{q}\pi DL = 15.82 \text{ kW/m}^2 \cdot \pi \cdot 0.070 \text{ m} \cdot 1.0 \text{ m} = 3.479 \text{ kW}$$

(b)
$$\dot{Q}_{\text{refl}} = (1 - 0.97)3.479 \text{ kW} = 0.104 \text{ kW}$$

(c)
$$\dot{Q}_{\text{rad}} = A\epsilon\sigma(T_{R,W}^4 - T_a^4) = \pi D_R L \epsilon \sigma (T_{R,W}^4 - T_a^4)$$
$$= \pi \cdot 0.070 \text{ m} \cdot 1.0 \text{ m} \cdot 0.15 \cdot 5.67 \cdot 10^{-8} \frac{W}{m^2 K^4} \cdot (6.23^4 - 2.90^4) \cdot 10^8 \text{ K}^4$$
$$= 0.269 \text{ kW at exit of collector string}$$

(d) for $T_{G,W} = 60°$ C,

$$k = 28.94 \cdot 10^{-3} \frac{W}{mK}; \quad \nu = 18.88 \cdot 10^{-6} \frac{m^2}{s}; \quad \beta = 3.007 \cdot 10^{-3} \frac{1}{K}; \quad a = 27.4 \cdot 10^6 \frac{m^2}{s};$$

$$Gr = \frac{g\beta(T_{G,W} - T_a)D_G^3}{\nu^2} = \frac{9.81 \frac{m}{s^2} \cdot 3.007 \cdot 10^{-3} K^{-1} \cdot (60°C - 17°C)(0.095)^3 m^3}{(18.88 \cdot 10^{-6})^2} = 3.051 \cdot 10^6$$

$$Pr = \frac{\nu}{a} = \frac{18.88 \cdot 10^{-6}\frac{m^2}{s}}{27.4 \cdot 10^{-6}\frac{m^2}{s}} = 0.69;$$

$$Nu = 0.50(GrPr)^{0.25} = 0.50(3.051 \cdot 10^6 \cdot 0.69)^{0.25} = 19.05;$$

$$\bar{h} = \frac{Nuk}{D_G} = \frac{19.05 \cdot 28.94 \cdot 10^{-3}\frac{W}{mK}}{0.095 \text{ m}} = 5.803 \frac{W}{m^2 K} \quad \text{(for smooth glass tube)}$$

$$\dot{Q}_{\text{conv}} = A\bar{h}(T_{G,W} - T_a) = \pi D_G L \bar{h}(T_{G,W} - T_a)$$

$$= \pi \cdot 0.095 \text{ m} \cdot 1 \text{ m} \cdot 5.803 \frac{W}{m^2 K}(60°C - 17°C) = 0.074 \text{ kW}$$

(e)

$$\dot{Q}_t = \dot{Q}_{\text{in}} - \dot{Q}_{\text{refl}} - \dot{Q}_{\text{rad}} - \dot{Q}_{\text{conv}} - \dot{Q}_{\text{cond}}$$

$$= 3.479 \text{ kW} - 0.104 \text{ kW} - 0.269 \text{ kW} - 0.074 \text{ kW} - 0 = 3.032 \text{ kW}$$

(f)

$$\eta = 3.032 \text{ kW}/3.479 \text{ kW} = 0.872$$

5.11 Appendix: Frequently Used Terms

A	area	[m^2]
a	thermal diffusivity	[m^2/s]
ASR	Advanced Sodium Receiver	
C	concentration ratio	
C_p	specific heat	[J/KgK]
CRTF	Central Receiver Test Facility	
CRS	Central Receiver System	
D	diameter	[m]
F	real body view factor	–
GAST	Gascooled Solar Tower	
Gr	Grashoff number	–
h	heat transfer coefficient	[W/m^{2}K]
HTF	Heat Transfer Fluid	
IEA	International Energy Agency	
k	thermal conductivity	[W/mK]
L	length	[m]
$\dot{m}$	mass flow	[kg/s]
n	index of refraction	–
Nu	Nusselt number	–
P	power	[W]
Pr	Prandtl number	–
$\dot{Q}$	heat flux	[W]
$\dot{q}$	heat flux density	[W/m^2]
r	radius	[m]
Re	Reynolds number	–
SPP	Solar Power Plant	
T	temperature	[K,°C]
t	thickness	[m]
V	volume	[m^3]
v	velocity	[m/s]
x,y,z	coordinates	[m]
α	absorptivity, absorptance	–
β	expansion coefficient	–
ϵ	emissivity	–
η	efficiency	–
θ	angle	[degree]

μ	dynamic viscosity	[kg/ms]
ν	kinematic viscosity	[m²/s]
ϕ	solar intensity	[W/m²]
ρ	density	[kg/m³]
	reflectivity	–
	reflection coefficient	–
σ	Stefan-Boltzmann constant	$5.67 \cdot 10^{-8}$ [W/m²K⁴]

Bibliography

[1] Agatonovic, P.; Dogigli, M.: Thermomechanical Stress on a High-Temperature Heat-Exchanger Unit under Solar-Specific Operating Conditions. *Engineering Computations*, 1 (1984) 161–172

[2] Alpert, D. J.; Kolb, G. J.: *Performance of the Solar One Pilot Plant as Simulated by the SOLERGY Computer Code*. Technical Report SAND88-0321, Sandia National Laboratories, Albuquerque/NM, 1988

[3] Baker, A. F.: *10 MWe Solar Thermal Central Receiver Pilot Plant Receiver Performance Final Report*. Technical Report SAND88-8000, Sandia National Laboratories, Livermore/CA, 1988

[4] Battleson, K. W.: *Solar Tower Design Guide*. Technical Report SAND81-8005, Sandia National Laboratories, Livermore/CA, 1981

[5] Becker, M. (Ed.): *Proc. 3rd Intl. Workshop on Solar Thermal Central Receiver Systems, Konstanz*, Berlin, Heidelberg, New York: Springer, 1986

[6] Becker, M.; Boehmer, M.: Achievements of High and Low Flux Receiver Development. In *Proc. ISES Solar World Congress, Hamburg 1987*, Bloss, W. H.; Pfisterer, F. (Ed.), pp. 1744–1783, Oxford (UK): Pergamon Press, 1988

[7] Becker, M.; Boehmer, M. (Ed.): *GAST – The Gas-Cooled Solar Tower Technology Program*. Berlin, Heidelberg, New York: Springer, 1989

[8] Becker, M.; Ellgering, H.; Stahl, D.: *CRS Construction Report*. Technical Report IEA-SSPS SR2, Köln (D), DLR, 1983

[9] Boese, F.; Huber, P. F.; Kappler, H. W.; Lammers, J.: Considerations and Proposals for the Future Research and Development of High Temperature Solar Processes. In *Solar Thermal Energy Utilization*, Becker, M. (Ed.), pp. 169–299, Berlin, Heidelberg, New York: Springer, 1987

[10] Chavez, J. M.; Tyner, C. E.; Couch, W. A.: Direct Absorption Receiver Flow Testing and Evaluation. In *Solar Thermal Technology – Proc. 4th Intl. Symposium, Santa Fe/NM, 1988*, Gupta, B. P.; Traugott, W. H. (Ed.), pp. 645–654, New York: Hemisphere Publ. Co., 1990

[11] Churchill, S. W.; Chu, H. H. S.: Correlating Equations for Laminar and Turbulent Free Convection from a Horizontal Cylinder. *Heat and Mass Transfer*, 18 (1975) 1049–1053

[12] Eckert, E. R. G.; Drake Jr., R. M.: *Analysis of Heat and Mass Transfer*. New York: McGraw Hill, 1962

[13] Erhardt, K.; Vix, U.: Direct Absorption of Concentrated Solar Radiation. In *Proc. 3rd Intl. Workshop on Solar Thermal Central Receiver Systems, Konstanz*, Becker, M. (Ed.), pp. 835–867, Berlin, Heidelberg, New York: Springer, 1986

[14] Etievant, C.; Amri, A.; Izygon, M.; Tedjiza, B.: *Central Receiver Plant Evaluation, Vol. 1-5*. Technical Report SAND86-8185, SAND87-8182, SAND88-8101, SAND88-8100, SAND88-102, Sandia National Laboratories, Albuquerque/NM, 1988

[15] Falcone, P. K.: *A Handbook for Solar Central Receiver Design*. Technical Report SAND86-8009, Sandia National Laboratories, Livermore/CA, 1986

[16] Fish, J. D. Private Communication

[17] Fricker, H. W.; Silva, M.; Winkler, C.; Chavez, J. M.: Design and Test Results of the Wire Receiver Experiment Almería. In *Solar Thermal Technology – Proc. 4th Intl. Symposium, Santa Fe/NM, 1988*, Gupta, B. P.; Traugott, W. H. (Ed.), pp. 265–277, New York: Hemisphere Publ. Co., 1990

[18] Geyer, M. Private communication, 1988

[19] Grasse, W.: *SSPS Results of Test and Operation 1981-1984*. Technical Report IEA-SSPS SR7, Köln (D), DLR, 1985

[20] Grasse, W.; Becker, M.; Finnstroem, B.: *Solar Energy for High Temperature Technology and Applications*. Köln (D): 1987

[21] Groeber, H.; Erk, S.; Grigull, U.: *Laws of Heat Transfer* (in German). Berlin, Heidelberg, New York: Springer, 1963

[22] Hell, F.: *Principles of Heat Transfer* (in German). Düsseldorf (D): VDI-Verlag, 1982
[23] Hunt, A. J.: New Approaches to Receiver Design: Prospects and Technology of Using Particle Suspensions as Direct Thermal Absorbers. In *Proc. 3rd Intl. Workshop on Solar Thermal Central Receiver Systems, Konstanz*, Becker, M. (Ed.), pp. 835–842, Berlin, Heidelberg, New York: Springer, 1986
[24] Kesselring, P.; Selvage, C. S. (Ed.): *The IEA/SSPS Solar Thermal Power Plants – Facts and Figures*. Volume , Berlin, Heidelberg, New York: Springer, 1986
[25] Lienhard, J. H.: *A Heat Transfer Textbook*. Englewood Cliffs: Prentice Hall, 1987
[26] Neuberger, A. W.; Becker, M.: Numerical Analysis of Gas Receiver Tube Flow. In *Colloques Internationaux du CNRS No. 306 Systémes Solaires Thermodynamic STS 80-28*, Lestienne, R. (Ed.), pp. 411–416, Paris (F): 1980
[27] Nikolai, U.; Harth, R.; Mueller, W. D.: Steam Reforming of Methane Driven by Solar Energy. In *Solar Thermal Technology – Proc. 4th Intl. Symposium, Santa Fe/NM, 1988*, Gupta, B. P.; Traugott, W. H. (Ed.), pp. 505–516, New York: Hemisphere Publ. Co., 1990
[28] Pritzkow, W.: The Volumetric Ceramic Receiver (VCR): Second Generation. In *Solar Thermal Technology – Proc. 4th Intl. Symposium, Santa Fe/NM, 1988*, Gupta, B. P.; Traugott, W. H. (Ed.), pp. 635–643, New York: Hemisphere Publ. Co., 1990
[29] Rabl, A.: Concentrating Solar Collectors. *Advances in Solar Energy*, (1985) 405–481
[30] Royere, C.: Solid Particles Solar Thermal Loop for Production of Heat at 1000 °C. In *Proc. 2nd Int. Workshop of the Design, Construction and Operation of Solar Central Receiver Projects*, Gretz, J. (Ed.), pp. 412–416, Varese (I): 1984
[31] Rozenmann, T.: Solar Reformer Pilot Plant. In *Proc. IEA-SSPS Experts Meeting on High Temperature Technology and Application, Atlanta/GA, IEA-SSPS TR 1/85*, Becker, M.; Skinrood, A. (Ed.), p. 159, Köln (D): DLR, 1985
[32] Schiel, W.; Geyer, M.; Carmona, R. (Ed.): *The IEA/SSPS High Flux Experiment*. Berlin, Heidelberg, New York: Springer, 1987
[33] Siebel, D. L.; Kraabel, J. S.: *Estimating Convective Energy Losses from Solar Central Receivers*. Technical Report SAND85-8250, Sandia National Laboratories, Livermore/CA, 1985
[34] Smith, D. C.; Chavez, J. M.: *A Final Report on the Phase I Testing of a Molten-Salt Cavity Receiver*. Technical Report SAND87-2290, Sandia National Laboratories, Albuquerque/NM, 1987
[35] Stoddard, M. C.: *Convective Loss Measurements at the 10 MW_e Thermal Central Receiver Pilot Plant*. Technical Report SAND85-8250, Sandia National Laboratories, Livermore/CA, 1985

6 Thermal Storage for Solar Power Plants

M. A. Geyer [1]

6.1 Impact of Storage on Solar Power Plants

Like with any other power plant, solar power plant (SPP) *output* must satisfy the demands of the utility market. During peak demand periods, kilowatt-hour prices are high and financial incentives are high for guaranteed supply. Solar plant *input* is limited by diurnal, seasonal and weather-related insolation changes. In order to cope with these fluctuations, the solar plant input may be backed up by fossil fuels, or the solar changes may be mitigated by a buffering storage system. The choice depends on demand, system and site conditions. In thermal SPPs, thermal storage and/or fossil backup act as:

- an output management tool to prolong operation after sunset, to shift energy sales from low revenue off-peak hours to high revenue peak demand hours, and to contribute to guaranteed output;
- an internal plant buffer, smoothing out insolation changes for steadying cycle operation, and for operational requirements such as blanketing steam production, component preheating and freeze protection.

In photovoltaic plants, which need no internal buffer, output management can be achieved with pumped hydro, battery or other electrochemical storage, and by Diesel-generator backup. The implications of battery storage are discussed in Chap. 8. In the following, for solar power plants, mainly thermal storage systems are considered.

6.1.1 Capacity Factor and Solar Multiple

A *storage hour* is defined as the storage capacity necessary to run a process connected to it at *rated output power* P_{out} for one hour. With the annual electric energy output W_a, the capacity factor CF of a SPP is given as

$$\mathrm{CF} = \frac{W_a}{P_{\mathrm{out}} \cdot 8{,}760} \qquad (6.1)$$

This capacity factor can also be applied to non-electrical processes. Figure 6.1 shows the ideal operating sequence of a solar plant with storage on a cloudless day, assuming negligible time delays between input and output. Between sunrise at t_1 and receiver start-up at t_2, incoming radiation is too low for receiver operation. During τ_R, receiver thermal output is still insufficient to run the conversion at cycle rated value. During τ_C, receiver output exceeds the required nominal input power, and surplus energy E_c charges the storage system with a *charging utilization factor* γ_c. During τ_D, the energy difference required to keep the conversion cycle running at rated power is retrieved from storage with a *discharge utilization factor* γ_d.

---

[1] Michael A. Geyer, Flachglas Solartechnik GmbH, Mühlengasse 7, D-5000 Köln 1

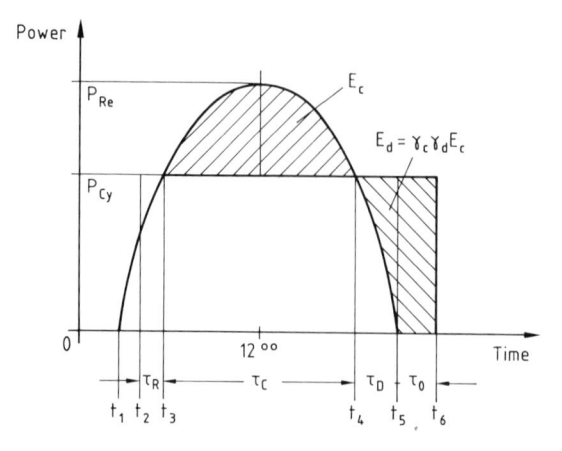

Fig. 6.1. Qualitative daily operating sequence of a thermal solar power plant with storage.

From sunset at t_5 through τ_O, the process runs at its rated power from storage only. Thus with storage, only the receiver needs to be designed for peak load P_{Re}, and the cycle may be P_{Cy}. Since charge and discharge utilization factors are < 1 due to storage and heat exchanger losses, the SPP transfers over the year less energy to the cycle via the storage system than it could transfer, if feasible, from the receiver directly. If δ_{St} is the fraction of annual absorbed receiver energy sent to storage, and γ_c/γ_d the average annual charge/discharge utilization factor, the annual plant system efficiency is reduced by the annual *storage loss factor* γ_{St}:

$$\gamma_{St} = 1 - \delta_{St} \cdot (1 - \gamma_c \cdot \gamma_d). \tag{6.2}$$

If, for example, 50% of receiver output is charged into a storage system with a 90% annual charge/discharge utilization factor, then thermal plant output is reduced to 95%. To compensate for this loss, the leveling effect of storage on cycle operation should improve annual cycle efficiency by 5%. Storage is already attractive, however, if, by shifting production from low to high tariff periods, increased revenues can pay for the additional storage costs. The ratio of receiver power at design point, P_{Re}, to nominal cycle inlet power P_{Cy}, is called the *solar multiple* (SM):

$$\text{SM} = \frac{P_{Re}}{P_{Cy}} \tag{6.3}$$

Figure 6.2 illustrates achievable rated power operating hours on a cloudless December 21st and June 21st for SM ranging from 1.0 to 2.8 in Almería, Spain. While the SM 2.8 is sufficient

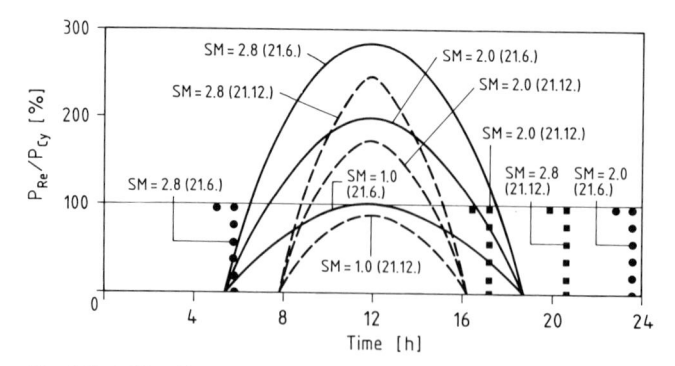

Fig. 6.2. Achievable operating hours for different solar multiples in summer and winter.

to operate 24 hours in summer, only 12 hours are possible with it in winter, and a minimum SM of 5 would be required to achieve 24-hour operation in winter.

6.1.2 Optimization of Solar Multiple and Storage Capacity

The results of the study of a 20 MW$_e$ gas-cooled central receiver SPP (GAST-20) is chosen to illustrate the optimization of storage. The GAST-20 study assumed a northern heliostat field, a gas-cooled metallic tube receiver operating between 500 and 800°C, and an *open Brayton cycle*. Fire-brick filled wind-heater modules, each with a usable thermal capacity of 100 MWh$_t$ (equivalent to 1.5 hours of rated power operation) were chosen for high temperature thermal storage. Annual SM of 1.0, 1.2, 1.4, 1.6, 1.8 and 2.0, were simulated by varying the storage capacity 1.5 h at a time in each case from 0 to 9 storage hours [8].

While field and receiver efficiency depend on SM only, cycle efficiency and resulting system efficiency depend on SM and storage capacity. Although added storage hours increase cycle efficiency by reducing partload operation, growing parasitic losses for charging/discharging of the storage diminish net electrical output. Thus a 5% improvement in cycle efficiency is already achievable merely by increasing the SM from 1.0 to 1.2 without adding storage, and a further 3% increase can be obtained by adding 1.5 h of storage. The gain from storage decreases again if storage capacity is greater than 4.5 hours, due to increased parasitics. The greatest cycle efficiency gain, 15%, is achieved with 1.5 storage hours with a SM of 1.4.

Figure 6.3 shows resulting annual system efficiencies. By reducing part-load operation, overall system efficiency increases from about 13% to almost 14% when SM increases from 1.0 to 1.2 (even without storage); system efficiency drops below design level with SM > 1.5 without storage, as unused summer surplus input exceeds the gains by improved partload performance; surplus which could be used, however, if storage were added. Highest system efficiency of 15.1% is obtained with 3 storage hours at SM = 1.4. Beyond that value, overall system efficiency again drops due to increased pumping parasitics. However, given sufficient storage capacity together with SM > 1.0, annual mean system efficiency is always greater than that at SM = 1.0.

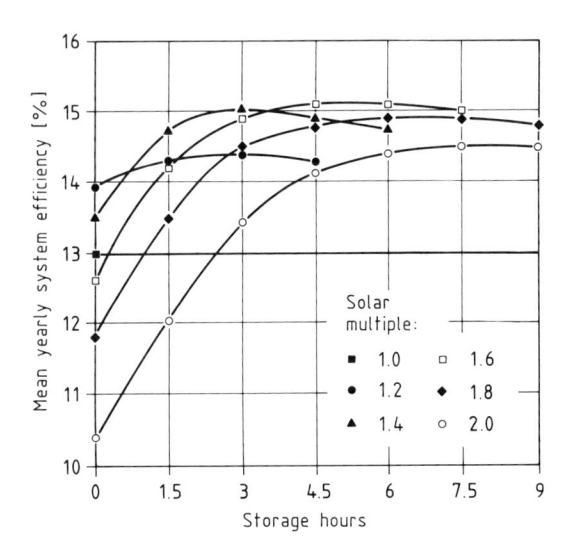

Fig. 6.3. Increase of annual system efficiency with addition of storage for different solar multiples.

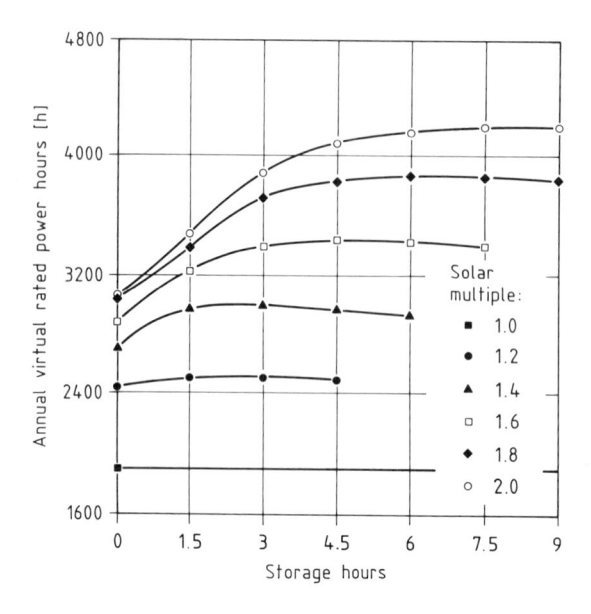

Fig. 6.4. Increase of annual rated power hours with addition of storage for different solar multiples.

In Fig. 6.4 total annual virtual rated power hours are plotted as function of storage capacity for different SM. Again, a considerable rise in rated power hours can be achieved without storage just by increasing SM to 1.2. With 4.5–9 hours of storage, over 4,000 virtual rated power hours can be achieved with a SM > 1.8, bringing such SPP into the range of conventional intermediate-load power plants. The real distribution of power levels achieved is illustrated in Fig. 6.5. At SM = 1.0, 1,510 rated power hours may be obtained with no storage. In reality, though, this SPP operates less than 800 hours at rated power and over 800 hours at less than 50%. At SM = 1.2, still without storage, rated power operation is almost doubled to 1,600 hours while operation with $\leq 50\%$ of rated power is reduced to 500 hours. With the addition of storage, rated power operation exceeds 80%, and with 4.5 storage hours at SM = 2.0, 3,700 hours of the total 4,080 operating hours run at rated output.

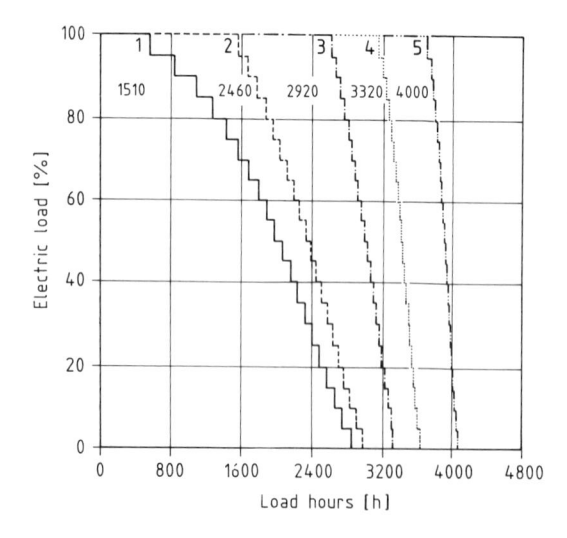

Fig. 6.5. Classified annual load distribution for various solar multiples and storage capacities.

Case	SM	Storage hours
1	1.0	–
2	1.2	–
3	1.4	1.5
4	1.6	3.0
5	2.0	4.5

Relative to the base case with the SM = 1.0 and no storage, cost analyses [8] showed that the specific electricity cost can be reduced by up to ≈ 15% if SM is increased to 1.4, even without adding storage (see also Chap. 10). With storage, a specific cost decrease of up to 25% is possible. The reason is obvious, looking at a tower SPP as example: With SM = 1.0 and no storage, specific costs are due 27% to the heliostat field, 28% to receiver and tower, and 25% to the conversion cycle. With SM = 2.0 and 6 storage hours, the heliostat field contributes 26%, receiver and tower 26%, conversion cycle 12% and storage 7% of costs.

6.2 Media for Thermal Storage

6.2.1 Sensible Heat Storage Media

The energetic densities of the thermal energy *storage media* now in use or proposed for SPP are compiled in Tab. 6.1.

Table 6.1. Thermal and cost data of storage media considered for solar thermal power plants. Volume specific heat capacity and media cost take into account the different operating temperatures for the storage media.

Storage Medium	Temperature Cold °C	Hot °C	Average density kg/m^3	Average heat conduct W/mK	Average heat capacity kJ/kgK	Volume specific heat capacity kWh$_t$/m^3	Media costs per kg $/kg	Media costs per kWh$_t$ $/kWh$_t$
Solid media								
Sand-rock-oil	200	300	1,700	1	1.30	60	0.15	14
Reinforced concrete	200	400	2,200	1.5	0.85	100	0.05	1
NaCl (solid)	200	500	2,160	7	0.85	150	0.15	1.5
Cast iron	200	400	7,200	37	0.56	160	1.00	32
Cast steel	200	700	7,800	40	0.60	450	5.00	60
Silica fire bricks	200	700	1,820	1.5	1.00	150	1.00	7
Magnesia fire bricks	200	1,200	3,000	5	1.15	600	2.00	6
Liquid media								
Mineral oil	200	300	770	0.12	2.6	55	0.30	4.2
Synthetic oil	250	350	900	0.11	2.3	57	3.00	43
Silicone oil	300	400	900	0.10	2.1	52	5.00	80
Nitrite salts	250	450	1,825	0.57	1.5	152	1.00	12
Nitrate salts	265	565	1,870	0.52	1.6	250	0.70	5.2
Carbonate salts	450	850	2,100	2	1.8	430	2.40	11
Liquid sodium	270	530	850	71	1.3	80	2.00	21
Phase change media								
NaNO$_3$		308	2,257	0.5	200	125	0.20	3.6
KNO$_3$		333	2,110	0.5	267	156	0.30	4.1
KOH		380	2,044	0.5	150	85	1.00	24
Salt-ceramics (Na$_2$CO$_3$-BaCO$_3$/MgO)		500-850	2,600	5	420	300	2.00	17
NaCl		802	2,160	5	520	280	0.15	1.2
Na$_2$CO$_3$		854	2,533	2	276	194	0.20	2.6
K$_2$CO$_3$		897	2,290	2	236	150	0.60	9.1

Although specific mass heat capacity is a more frequent reference, the number of thermal kilowatt-hours per cubic meter (kWh_t/m^3) is used here, since container volume and pumping power are the basic cost factors. The media costs (price base 1987) may not always reflect the lowest possible offer, but they provide orientation for the selection of a storage medium for a given temperature range. They do not include the often major costs of storage vessel and heat exchanger equipment. For temperatures of up to 300°C, thermal *mineral oil* can be stored at ambient pressure, and is the most economical and practical solution. *Synthetic* and *silicone oils*, available for up to 410°C, have to be pressurized and are expensive. *Molten salts* [5] and *sodium* can be used between 300 and 550°C at ambient pressure, but require parasitic energy to keep them liquid. For temperatures above 550°C, ceramic *fire bricks* become a competitive option.

6.2.2 Latent Heat Storage Media

Phase change media (PCM) provide a number of desirable features, e.g. high volumetric storage capacities and heat availability at rather constant temperatures [19]. Energy storage systems using the latent heat released on melting eutectic salts or metals have often been proposed, but never carried out on a large scale due to difficult and expensive internal heat exchange and cycling problems. Heat exchange between the heat transfer fluid (HTF) and the storage medium is seriously affected when the storage medium solidifies. Encapsulation of PCMs has been proposed to improve this [15]. Combining the advantages of direct-contact heat exchange and latent heat, hybrid *salt-ceramic* phase change storage media have recently been proposed in [16,18]. The salt is retained within the submicron pores of a solid ceramic matrix such as magnesium oxide by surface tension and capillary force. Heat storage is then accomplished in two modes: by the latent heat of the salt and also the sensible heat of the salt and the ceramic matrix.

6.2.3 Chemical Storage Media

Chemical storage system researchers hope to identify a suitable chemical or fuel with satisfactory energy densities. However, the high energy density of fossil fuel combustion can only be achieved in open combustion cycles, where neither the volume of oxygen required nor carbon dioxide emitted are an obstacle. Thermal storage systems using fuels, however, should only be closed conversion cycles without any exchange of oxygen or dioxide. In such closed systems, oxygen and carbon dioxide must be stored and transported in addition to the primary reactants, thus reducing considerably the volume-specific storage capacity. A further drawback is the inferior charge/discharge roundtrip efficiency.

6.2.4 Single Versus Dual Medium Concepts

In a *single medium storage* system, the HTF is at the same time the storage medium. The advantage is that no internal heat exchange between HTF and storage medium is necessary. If the liquid has low thermal conductivity and permits good thermal stratification, such as water and thermal oil, the one-tank *thermocline* concept requires the least tank volume since the hot and cold medium are contained in a single vessel. When thermal conductivity is higher, as in molten salts or sodium, a rapid balancing of the temperatures in the hot and cold regions takes place, making separate *hot* and *cold tanks* necessary. Since in that case twice as much tank volume as fluid content is required, a three-tank system in which there is only 1.5 times as much tank volume as fluid content has been investigated [17]. However, such systems, complicated to control, with extensive piping, and subject to increased heat losses from higher surface to volume ratios, are not in use today.

Dual medium concepts employ a storage medium that is different from the HTF because the storage medium – usually solid – is cheaper than the transfer fluid. The transfer medium exchanges its heat in direct or indirect contact with the storage medium. Apart from a drop in temperature between charging and discharging due to the intermediate heat exchange, the dual medium concept has the operational drawback that the outlet temperature of the heat transfer medium increases during charging and decreases during discharging, leading to non-usable storage capacities. In comparison, constant charging and discharging temperatures are maintained in a single medium hot/cold tank system until the tank in question is empty.

6.3 State-of-the-art of Thermal Storage for Solar Power Plants

Storage concepts will be presented and classified in the following in terms of the primary HTF and SPP type, starting with the low temperature oil-cooled solar farm systems, continuing with the intermediate temperature steam-, molten salt- and sodium tower SPP and ending with the high temperature gas-cooled tower SPP. The major experimental thermal storage systems for SPP are summarized in Tab. 6.2.

Table 6.2. Existing major solar thermal storage systems.

Facility / Location	Storage medium	Primary coolant	Design temperature cold °C	hot °C	Tank concept	Tank volume m^3	Thermal capacity MW$_t$
Irrigation pump Coolidge/AZ	Oil	Oil	200	288	1 tank Thermocline	114	3
IEA-SSPS (Farm) Almería, Spain	Oil	Oil	225	295	1 tank Thermocline	200	5
SEGS-1 Daggett/CA	Oil	Oil	240	307	Cold tank Hot tank	4,160 4,540	120
IEA-SSPS (Farm) Almería, Spain	Oil Cast iron	Oil	225	295	1 tank Dual medium	100	4
Solar One Barstow, CA	Oil Rock/sand	Steam	224	304	1 tank Dual medium	3,460	182
CESA-1 Almería, Spain	Molten salt	Steam	220	340	Cold tank Hot tank	200 200	12
Themis Targasonne, France	Molten salt	Molten salt	250	450	Cold tank Hot tank	310 310	40
CRTF Albuquerque/NM	Molten salt	Molten salt	288	566	Cold tank Hot tank	53 53	6.9
IEA-SSPS (tower) Almería, Spain	Sodium	Sodium	275	530	Cold tank Hot tank	70	5

6.3.1 Thermal Storage for Oil-Cooled Solar Plants

Thermal, synthetic and silicone oils, with operating temperatures from 300°C to over 400°C, are of particular interest as heat transfer media for thermal SPP. Unlike water/steam, oils do not require high-pressure piping, nor have they freezing problems as with sodium or molten salt.

Thermal oil as a storage medium has been tested in both the single tank thermocline and in hot/cold tank configurations. The 5 MWh$_t$ capacity one-tank thermocline storage system, operated with 114 m³ of thermal oil at temperatures between 225 and 295°C, was successfully tested at the IEA-SSPS project in Almería, Spain, and demonstrated a 92% roundtrip efficiency and excellent thermocline stratification.

Fig. 6.6. 120 MWh$_t$ two tank oil storage system at SEGS I, Daggett/CA, operating between 241°C and 307°C (courtesy of LUZ Intl. Ltd. Los Angeles/CA, USA)

Figure 6.6 shows the two-tank storage system at the SEGS I plant in Daggett/CA. The design storage capacity of 118.9 MWh$_t$ can support $2\frac{1}{2}$ turbine hours at 14.7 MW$_e$ rated power. The nominal heat loss rate of both carbon steel tanks is 205 kW$_t$. The tanks are protected by about 12 cm thick external fiberglass insulation, maintaining a minimum tank temperature of 220°C in the cold tank. In 1984, when the cost of oil was about 1.60 $ per gallon, specific investment costs amounted to 25 $/kWh$_t$, of which the tanks represented 24% and the oil 42% of the cost [11].

Since synthetic and silicone oil storage media are 9–16 times as expensive as mineral oils, dual medium concepts have been studied. The largest of such experimental systems was installed at the 10 MW$_e$ Solar One steam-cooled central receiver plant in Barstow/CA, a 182 MWh$_t$ capacity dual medium thermocline storage system, operating between 218°C and 304°C; such system would also be appropriate for an oil-cooled farm system. Its cylindrical steel tank (18 m ⌀, 14 m high) is filled with a compacted bed of rock and sand impregnated with heat transfer oil. Maximum heat loss was 3 MWh$_t$ in a 24-hour storage cycle, or less than 2% of maximum capacity. Roundtrip efficiency was about 70%. High amounts of alumina-silicate in the rock and sand acted as catalysts and led to accelerated thermal cracking of the oil [7]. Parasitic power needs were 3.5 kWh$_e$/MWh$_t$ for a complete charge/discharge cycle. It has been estimated that the investment costs of 62 $/kWh$_t$ for the pilot system would drop to 8 $/kWh$_t$ for a 1,000 MWh$_t$ system [9].

A dual medium storage concept was also tested at the IEA-SSPS project. Capacity of the one-tank storage module was 4 MWh$_t$. The 100 m³ volume of a steel vessel was filled with 115 pieces of 2.5 m⌀/0.23 m high-ribbed cast iron plates, weighing 358 tons. Operating temperature range was 225–295°C. Roundtrip efficiency, measured at around 70% [2], was similar to Solar One with no oil degradation. Although performance was most reliable, the 100 $/kWh$_t$ investment in the 4 MWh$_t$ prototype proved cast iron as too expensive for larger storage systems.

Only the first of the SEGS plants used thermal storage and mineral oil as storage medium. The use of synthetic oils as heat transfer oil in later SEGS plants made oil storage concepts prohibitively expensive. Instead, a parallel gas-fired boiler supplies steam, supplementing solar input and providing generating capacity when no solar energy is available [11]. While summer on-peak hours are covered mostly by afternoon direct field operation, winter on-peak periods start after the fields have gone into stow position. Legal provisions in the U.S. allow a maximum of 25% of total annual output from fossil sources. Under 1988 conditions of sale, thermal storage would be profitable at a investment cost of 25 $/kWh$_t$.

6.3.2 Thermal Storage for Steam-Cooled Solar Plants

At Solar One, CESA-1 and Eurelios, water/steam was used as HTF for the central receiver, allowing direct feeding of a steam-driven turbine. However, steam generated at 500°C and 100 bar cannot be stored directly and economically. Therefore, a separate oil loop transfers heat to and from rock/sand storage at Solar One, and a separate molten salt tank system was used at CESA-1. In both systems, plant performance was degraded by the high drop in temperature and pressure when steam was produced from storage. While steam is produced at about 500°C/100 bar in both receiver systems, steam from storage reached only 277°C/27 bar at Solar One, and 330°C/16 bar at CESA-1. This reduced turbine cycle efficiency by nearly 50%. While the storage temperature at Solar One was limited, by the mineral oil used, to ≈ 315°C; in comparison the molten salt of CESA-1 could well be operated up to 450°C. The real problem, however, is less the temperature limitations of the storage medium than the heat exchange between the liquid heat storage medium and a phase-changing HTF, i.e. condensing steam at charging and evaporating steam at discharging. When the 520°C/110 bar steam is desuperheated to its condensation temperature of 308°C, 704 kJ/kg (30%) are transferred to the salt. During condensation at 308°C, 1,259 kJ/kg (53%) are transferred, and at subcooling from 308°C to 240°C, 410 kJ/kg (17%). Since the molten salt can only absorb this energy as sensible heat through a rise in temperature (1.48 kJ/kgK at 290°C), 12.4 kg salt at 220°C must be provided for the condensation and subcooling of 1 kg steam. In the desuperheater, this kilo of steam yields only energy sufficient to heat the 12.4 kg of salt to 340°C. At discharge, the boiler temperature must be lower than the condensation temperature, which is only possible by lowering boiler pressure. Therefore, unless a latent heat storage system can be developed to match the steam/water phase change, fossil derived auxiliary input is the only economic alternative for large solar steam plants.

6.3.3 Thermal Storage for Molten Salt-Cooled Solar Plants

Molten salts are favoured central receiver coolants because of their high volume heat capacity, low vapor pressure, good heat transfer and low cost, which makes them economical enough to be used as a large bulk storage medium while their thermodynamic properties permit compact and efficient receivers. The storage tanks must be hydraulically separated from the receiver loop by an intermediate heat exchanger, however, in order to store storage medium at ambient pressure (the hydrostatic pressure of the more than 100 m high up- and downpipes would affect the storage tanks.)

Hitec$^©$, an eutectic mixture of 40% $NaNO_2$, 7% $NaNO_3$ and 53% KNO_3, and a 142°C freeze-point, has been used in the CESA-1 molten salt storage system (Fig. 6.7). The two

Fig. 6.7. 12.7 MWh$_t$ two tank molten salt storage system at CESA-1, Almería, Spain, operating between 220°C and 340°C (courtesy of CIEMAT-IER, Spain).

identical carbon steel, 4 m ⌀/16 m long tanks, with a 10 mm thick shell have a maximum storage capacity of 12.7 MWh$_t$. Full hot tank losses, measured over 24 hours were 535 kWh$_t$ or 4% capacity. Measured full cold tank losses were 271 kWh$_t$ or 2% capacity. The storage system is integrated into a steam/water cycle. During charging, 520°C/110 bar steam coming from the receiver is cooled down to 240°C after passing through three steam/salt heat exchangers simultaneously heating the salt from 220°C to 340°C. At discharge, 340°C molten salt is cooled down to 220°C in two heat exchangers, raising the steam temperature to 330°C. Complete charge/discharge cycle heat loss measured in the charging subsystem was 1,260 kWh$_t$. The discharging subsystem loss was 950 kWh$_t$ for a charging/discharging efficiency of 90%/91% [2]. Parasitic pumping power required for a complete charge/discharge cycle was 5 kWh$_e$/MWh$_t$.

At Themis, Hitec© salt is used at temperature from 250°C to 450°C. The hot and cold tanks were identical (5 m ⌀/17 m long, 310 m^3, 13 mm thick shell) with the exception of the steel used. They were designed for a maximum operating pressure of 3 bar plus hydrostatic pressure of the salt. The maximum storage capacity of the 537 ton salt inventory is 40 MWh$_t$. A total loss of 1.8 MWh$_t$ in the daily 40 MWh$_t$ charge/discharge cycle means that roundtrip efficiency is 95%. Hot tank temperature falls less than 3°C over-night when the tank is full at 450°C. The Themis storage system is integrated into a molten salt receiver loop which charges the tanks directly. At discharge, salt is cooled down from 450°C to 250°C in a steam generator and raises the steam temperature to 445°C. Salt costs were 0.67 $/kg in 1984, or 7.6 $/kW$_t$. Total storage costs were 1.8 $ million (1984), including salt, tanks, pumps, piping, valves, insulation and trace heating, whose initial capital investment was 45 $/kWh$_t$ [6].

At the Central Receiver Test Facility (CRTF) in Albuquerque/NM, a 7 MWh$_t$ molten salt thermal subscale storage research experiment to solve design, fabrication, operation and performance problems associated with a full-scale 1,200 MWh$_t$ system [20], used 79 tons of molten nitrate salt (60% NaNO$_3$ and 40% KNO$_3$ by weight) at temperatures between 280 and 560°C. The hot tank, 7.2 m high by 3.7 m carbon steel shell diameter, is insulated with a Technigaz© corrugated liner (Fig. 6.8) and refractory brick internal insulation, while the cold tank, 3.7 m high by 3.7 m carbon steel shell diameter, has external insulation only. Both tanks rest on a water-cooled concrete foundation. Roundtrip efficiency of over 93% was measured [1,14].

These three demonstration projects have shown that molten nitrate salt storage is technically feasible up to temperatures of 560°C. The economic potential in larger systems has been studied in a 1,200 MWh$_t$ preliminary storage system design in which, as illustrated in

Fig. 6.8. Corrugated hot tank of the molten salt thermal energy storage subsystem research experiment at CRTF, Albuquerque/NM, USA (courtesy of SNL, Albuquerque/NM, USA).

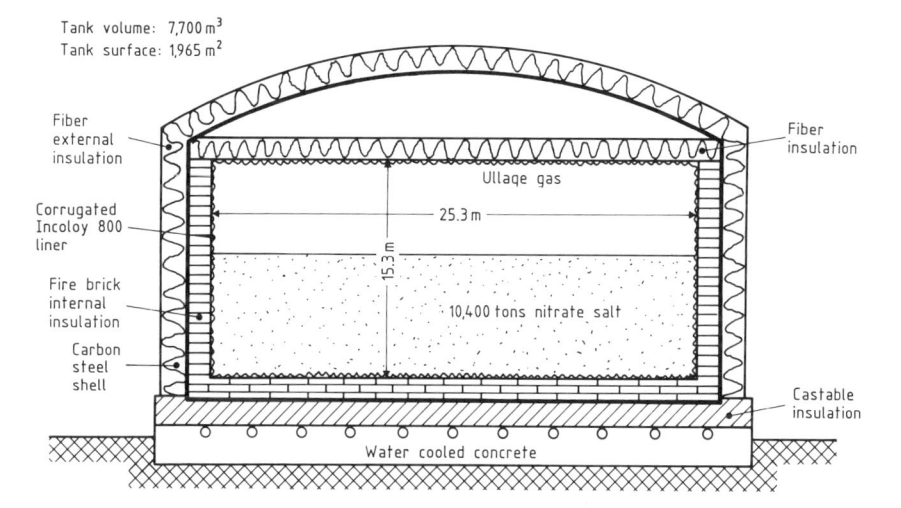

Tank volume: 7,700 m³
Tank surface: 1,965 m²

Fig. 6.9. 1,200 MWh$_t$ cylindrical hot tank concept for molten nitrate salt storage with corrugated liner (courtesy of SERI, Boulder/CO, USA).

Fig. 6.9, two cylindrical tanks (hot and cold), each 25 m in diameter and 18 m high, contain 10,500 tons of salt. The total initial investment, of which 46% corresponds to the salt and 34% to the tanks, was estimated to be 13 \$/kWh$_t$ [20]. Other estimates range from 8 \$ to 20 \$/kWh$_t$ for the same storage size [21].

A molten salt storage concept for temperatures of up to 900°C in which a molten ternary eutectic carbonate is stored in a below-grade, cone-shaped hot tank (Fig. 6.10) was investigated in a design study [10]. The cone shape of the inner thin-film lining allows relative thermal expansion and permits the use of low-cost structural and insulating materials for the outer layers. Three materials, Inconel 800, Nickel 201 and Inconel 600 were exposed to the highly corrosive salt vapor phase in the search for an appropriate lining material. It was

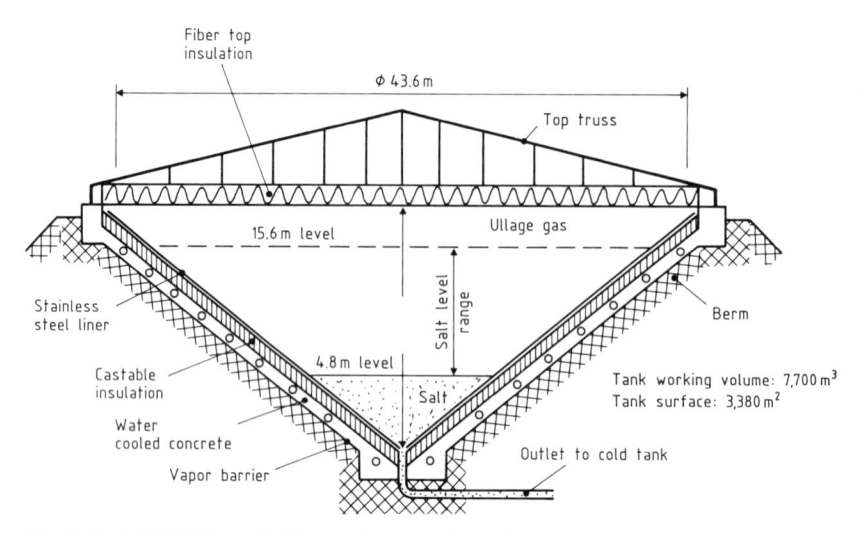

Fig. 6.10. 1,200 MWh$_t$ conical hot tank concept for molten nitrate salt storage with flat stainless steel liner (courtesy of Solar Energy Research Institute (SERI), Boulder/CO, USA).

found that a 1.27 cm thick liner of Inconel 600, which had the lowest corrosion rate of 1.3 mm/a, would have to be replaced every 10 years. Two 28 m diameter by 14 m high conical hot tanks with a working volume of 2,264 m^3 set on water-cooled reinforced concrete would be required for a 1,800 MWh$_t$ capacity. Total daily heat losses would amount to 35 MWh$_t$ or 2% of the total capacity, of which 66% are caused by foundation cooling. A preliminary estimate of the initial capital investment was $\approx$ 20 \$/kWh$_t$. Assuming a 30-year life with two liner replacements, 5% per year real cost of capital and energy at 0.06 \$/kWh$_e$, life-cycle costs would be 24 \$/kWh$_t$. It was recommended that the carbonate salt temperature be limited to 850°C, above which corrosion sharply accelerates. Molten salts may become attractive up to 850°C if the corrosion problem could be solved. But since no high temperature process can use molten salt at 850°C directly, an intermediate molten salt to gas heat exchange is still necessary. Above 600°C, direct-contact heat exchange was found to be more cost efficient than heat exchangers [4].

6.3.4 Thermal Storage for Sodium-Cooled Solar Plants

Due to its excellent thermodynamic properties, sodium has been found to be an efficient receiver coolant, for which nuclear state-of-the-art technology may be utilized. A 5 MWh$_t$, 280°C to 560°C sodium storage system, operated for four years at the IEA-SSPS project, consisted of a ferritic cold tank and an austenitic hot tank, both 3.3 m in diameter by 10 m in length with a volume of 70 m^3. Due to the small size of the tanks and the unfavorable surface-to-volume ratio, thermal losses were quite high: during a 12 h stand-still over night, the tanks would loose a total of 500 kWh$_t$ or 10% of the total capacity. The $\approx$ 1,000 m^3 sodium storage tank at the Superphenix fast breeder reactor at Creys-Malville, France, containing 700 tons of sodium, is the largest sodium tank ever built.

Although unit media costs of 12 \$/kWh$_t$ are three times greater than for molten salt, total system costs do not differ so much. Sodium storage employs an externally insulated stainless steel tank, while the much more expensive molten salt system container employs an internally and externally insulated carbon steel tank with an Incoloy liner. Cost analyses [21] even show a cost advantage for small sodium storage systems of up to 50 MWh$_t$, and at 3,000 MWh$_t$ sodium storage is still estimated to be 18 \$/kWh$_t$, in comparison to the 12 \$/kWh$_t$ of molten salt. For greater security, tanks should be installed in a weather proof building with safety berms in case of leakage.

6.3.5 Thermal Storage for Gas-Cooled Solar Plants

In gas-cooled receivers using ceramic materials, the gas to be fed to turbines or chemical processes could potentially reach temperatures above 1,000°C. For these temperatures, classical refractory material seems to be the only available storage medium. The most straight forward solution is the adaptation of the steel industry's existing wind heaters, known as *regenerators* or *Cowpers*, to solar requirements [12]. Such a heat exchanger, 50 m high by 11m diameter, with ceramic checkerwork, heats 200,000 m^3/h of ambient temperature air to 1,350°C

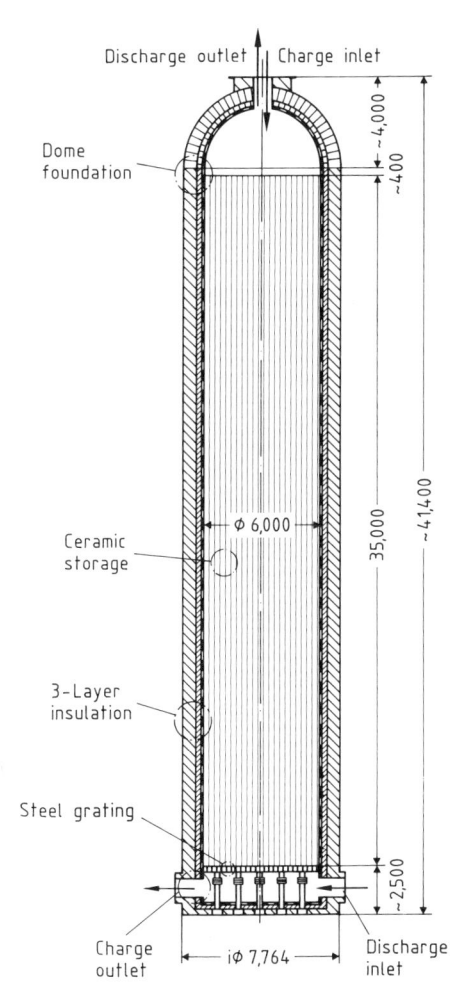

Fig. 6.11. Cowper type storage system for high temperature gas-cooled solar plants, operating up to 1,300°C. For a 300 K cycle temperature difference the capacity of the shown module is 100 MWh$_t$.

in a complete cycle lasting only 1–2 hours. Solar modifications consist mainly of a longer charge/discharge cycle, greater thermal capacity and reduced heat exchanging surfaces. A solar-adapted storage module, like the one shown in Fig. 6.11, could deliver 100 MWh$_t$ at an average cycle temperature difference of 300 K and maximum temperature of 1,350°C. 2,100 tons of MgO checkers store energy in a 1,000 m^3 internally and externally insulated, 7.7 m ⌀ and 41 m high steel container. The primary cost factor is the operating pressure: while the 2,100 tons of storage material have a constant cost of 1.5 Mio \$, container costs rise from 1 Mio \$ for ambient pressure operation, to over 2.5 Mio \$ for 4 bar, and to almost 20 Mio \$ for 40 bar. Cowper-type storage could also act as a pressure transformer between receiver and consumer: while one module is charged under receiver operating pressure, another module is discharged under output pressure [13].

Figure 6.12 shows the checker arrangement of typical hexagonal ceramic fire bricks in a commercial wind heater. If part of the conventional bricks were replaced by various layers of salt-ceramics with different melting points pressed in the same form, the total volume of such a storage module could be reduced by 30% [18].

Fig. 6.12. Fire brick arrangement in a high temperature industrial regenerator (hot blast stove). (Reproduced by courtesy of Didier-Werke AG, Wiesbaden, Germany).

6.4   Appendix: Frequently Used Symbols

CF Capacity factor

δ_{st} Fraction of annual thermal receiver output used for charging the storage system

γ_c Mean annual storage charging utilization factor

γ_d Mean annual storage discharging utilization factor

γ_{st} Mean annual storage charge/discharge roundtrip efficiency

P_{cy} Nominal conversion process thermal input power

P_{Re} Nominal receiver thermal input power at design conditions

t_3 Time when receiver thermal output power starts to exceed the required nominal thermal input power of the conversion cycle (receiver direct and storage charging operating mode)

t_4 Time when receiver thermal output power starts to become less than the required nominal input power of the conversion cycle and the difference must be extracted from storage (receiver direct and discharge operating mode)
t_6 Plant shutdown time
t_2 Time when receiver starts producing positive thermal output power
t_1 Sunrise
t_5 Sunset
τ_{DC} Period during which the receiver feeds the connected conversion process and charges storage (receiver direct and storage charge operating mode)
τ_D Period during which the connected conversion process is fed by the receiver and the storage system (receiver direct and storage discharge operating mode)
τ_O Period during which the process is fed only from storage (storage discharge only operating mode)
τ_R Period during which the connected conversion process is fed by the receiver only (Receiver direct operating mode)
SM   Solar multiple
W_a Annual design solar plant energy output

Bibliography

[1] *Molten Salt Thermal Energy Storage Subsystem Research Experiment.* Contractor Report SAND80-8192, Martin Marietta Corp., 1985

[2] Andujar, J. M.; Rosa, F.: *CESA-1 Thermal Storage System Evaluation.* Internal Report R-12/87, Almería, Spain, IEA-SSPS, 1987

[3] Becker, M.; Dunker, H.; Sharan, H.: *Technology Program "Gas-Cooled Solar Tower Power Station (GAST)", Analysis of its Potential* (in German). T 86-087, Bonn (D), BMFT, 1986

[4] Bohn, M.: Air Molten Salt Direct-Contact Heat Exchange. *Trans. ASME, J. Sol. Energy Engg.*, 107 (1985) 208–214

[5] Bradshaw, R. W.; Tyner, C. E.: *Chemical and Engineering Factors Affecting Solar Central Receiver Applications of Ternary Molten Salts.* Technical Report SAND88-8686, Sandia National Laboratories, Albuquerque/NM, 1988

[6] Etievant, C.; Amri, A.; Izygon, M.; Tedjiza, B.: *Central Receiver Plant Evaluation, Vol. 1–5.* Technical Report SAND86-8185, SAND87-8182, SAND88-8101, SAND88-8100, SAND88-102, Sandia National Laboratories, Albuquerque/NM, 1988

[7] Faas, S. E.: *10 MWe Solar Thermal Central Receiver Pilot Plant: Thermal Storage Subsystem Evaluation.* Technical Report SAND86-8212, Sandia National Laboratories, Livermore/CA, 1986

[8] Geyer, M.: *High-Temperature Storage Technology* (in German). Berlin, Heidelberg, New York: Springer, 1987

[9] Geyer, M.; Bitterlich, W.; Werner, K.: The Dual Medium Storage Tank at the IEA-SSPS Project in Almería(Spain), Part I: Experimental Validation of the Thermodynamic Design Model. *Trans. ASME, J. Sol. Energy Engg.*, 109 (1987) 192–198

[10] Ives, J. K.; Goodman, B.: High Temperature Molten Salt Storage Concept. In *Proc. 21st Intersoc. Energy Conv. Engg. Conf. (IECEC), San Diego/CA*, pp. 862–866, San Diego/CA: 1988

[11] Jaffe, D.; Friedlander, S.; Kearney, D.: The Luz Solar Electric Generating Systems in California. In *Proc. ISES Solar World Congress, Hamburg 1987*, Bloss, W. H.; Pfisterer, F. (Ed.), Oxford (UK): Pergamon Press, 1988

[12] Kainer, H.; Kalfa, H.; Palz, H.; Streuber, C.: High Temperature Storage for Gas Cooled Central Receiver Systems. In *Proc. 3rd Intl. Workshop on Solar Thermal Central Receiver Systems, Konstanz*, Becker, M. (Ed.), pp. 897–916, Berlin, Heidelberg, New York: Springer, 1986

[13] Kalfa, H.; Streuber, C.: Layout of High Temperature Solid Heat Storages. In *Solar Thermal Energy Utilization*, Becker, M. (Ed.), pp. 111–209, Berlin, Heidelberg, New York: Springer, 1987

[14] Kolb, G. J.; Nikolai, U.: *Performance Evaluation of Molten Salt Storage Concept*. Technical Report SAND87-2002, Sandia National Laboratories, Albuquerque/NM, 1987
[15] Olszewski, M. In *Proc. 1st IEA-SSPS Task IV Status Meeting on High Temperature Thermal Storage*, pp. 100–124, Köln (D): DLR, 1986
[16] Petri, R. J.; Ong, E. T.; Marianowski, L. G.: High Temperature Composite Thermal Energy Storage for Industrial Applications. In *Proc. 12th Energy Technology Conference and Exposition, Washington/DC*, (Ed.), 1985
[17] Stine, W. B.; Harrigan, R. W.: *Solar Energy Fundamentals and Design*. New York: John Wiley & Sons, 1985
[18] Tamme, R.; Allenspacher, R.; Geyer, M.: High Temperature Thermal Storage Using Salt/Ceramic Phase Change Materials. In *Proc. 21st Intersoc. Energy Conv. Engg. Conf. (IECEC), San Diego/CA*, pp. 846–849, San Diego/CA: 1986
[19] Tamme, R.; Geyer, M.: High Temperature Storage for Solar Thermal Applications. Solar Thermal Central Receiver Systems. In *Proc. 3rd Intl. Workshop on Solar Thermal Central Receiver Systems, Konstanz*, Becker, M. (Ed.), pp. 649–660, Berlin, Heidelberg, New York: Springer, 1986
[20] Tracey, T. R.; Scott, O. L.; Goodman, B.: Economical High Temperature Sensible Heat Storage Using Molten Nitrate Salt. In *Proc. 21st Intersoc. Energy Conv. Engg. Conf. (IECEC), San Diego/CA*, pp. 850–855, San Diego/CA: 1986
[21] William, T. A.; Dirks, J. A.; Brown, D. R.; Drost, M. K.; Antoniac, Z. A.; Ross, B. A.: *Characterization of Solar Thermal Concepts for Electricity Generation*. Technical Report PNL-6128, Richland/WA, Battelle Pacific Northwest Laboratories, 1987

7 Thermal Solar Power Plants Experience

W. Grasse, H. P. Hertlein and C.-J. Winter, [1]
with contributions by G. W. Braun

7.1 Introduction

In parallel with rising interest in solar power generation, several solar thermal facilities of different configuration and size were built, operated, and evaluated in the last decade and a half. Some of these facilities were of exploratory, first-of-a-kind or demonstration nature, in some cases designed merely as engineering experiments for the purpose of gaining performance and operating data at the subsystem and overall plant level. Most facilities were designed as modest-size experimental or prototype solar power plants (SPP) for producing electricity, in a few cases also for cogenerating thermal energy. Of all solar thermal technologies investigated, SPPs using parabolic trough concentrators were the first to reach sufficient maturity to be constructed on a commercial basis in a favorable regulatory environment. Table 7.1 provides an overview of the facilities built, their aggregate nominal capacity (MW_e), and their total collective/reflective area.

In this chapter, selected examples of the major technology lines of thermal SPPs are presented; also, major experience and lessons learned from experimenting with and operating such systems will be excerpted. This experience base is still fragmentary and, in some cases, preliminary – a fact not surprising considering the different approaches attempted and the first-generation technologies frequently involved. However, the data base is broad enough to identify major system operating characteristics, and to allow, with reasonable confidence, an extrapolation of future thermal SPP performance with mature technology under good solar resource conditions.

7.2 Farm Solar Power Plants with Line-Focussing Collectors

Using line-focussing parabolic troughs, a solar thermal power facility of about 35 kW_{mech} capacity was demonstrated successfully as early as 1913 in Egypt (Fig. 7.1). This facility had 1,233 m² of collector aperture and was designed for pumping water for field irrigation. Disturbances by World War I and the advent of the 'oil economy' stymied any subsequent development efforts.

Development activities started again in the mid 1970s in response to the sudden oil price increase. R&D programs financed by industry and governments spawned a multitude of al-

---

[1] see Chap. 4

Table 7.1. Construction status of concentrating thermal solar power plants (end of 1989).

Plant type	Developed since		Systems installed	Operational	Under construction	Comments
Parabolic trough (farm)	1973	number	≈ 15	8	1	
		MW_e	285	275	80	(a)
		m^2	1,799,000	1,789,000	464,000	
Central receiver (tower)	1973	number	6	–	1	
		MW_e	16	–	≈ 5	(b)
		m^2	117,000	–	≈ 40,000	
Parabolic dish (farm)	1977	number	4	3	–	
		MW_e	5.38	5.288	–	(c)
		m^2	35,672	34,547	–	
Parabolic dish (module with Stirling)	1977	number	≈ 15	≈ 3	–	
		MW_e	6.9	0.125	–	(d)
		m^2	50,200	≈ 540	–	

(a) 80 MW_e largest unit capacity in operation; about 600 MW_e total capacity contracted to be installed by 1993 in California/USA by one supplier (Luz International, Inc.) under provisions of the Public Utilities Regulatory Policy Act (PURPA); four 80 MW_e plants yet to be built. (b) A 30 MW_e demonstration facility in a preliminary design and component test phase undertaken in Europe; studies for 100–200 MW_e tower SPP pursued in the U.S. (c) A 4.88 MW_e dish-farm SPP installed in California under commercial terms; other dish farm SPP are operated for experimental purposes. (d) 50 kW_e largest unit tested; advance development of low-cost dishes and Stirling engines currently pursued in Germany and in the U.S.

ternate designs of collectors and SPP system approaches; several of these development steps are also highlighted in Figs. 7.1 and 7.2.

Technological progress of line-focussing collector technology can be illustrated by three significant examples (for technical data see Tabs. 7.2 and 7.13):

- the 150 kW_e facility at Coolidge/AZ, USA (1979), the first solar thermal full system experiment to demonstrate automated operation in an actual application environment;
- the 500 kW_e experimental Small Solar Power System plant in Almería, Spain (1981), designed, built and operated as a collaborative R&D project under the auspices of the International Energy Agency (IEA-SSPS); and
- the 30 MW_e Solar Electricity Generating Systems (SEGS II–VII; 1985–1989), developed commercially by a group of American, Israeli and German companies and marketed by Luz International Inc., Los Angeles/CA, USA.

7.2.1 Plant Configurations

System diagrams of the early 500 kW_e IEA-SSPS facility and of the SEGS VIII plant, illustrative of the most advanced commercial design, show the typical plant lay-out and the evolution in the design (Figs. 7.3 and 7.4); some observations can be highlighted:

- Each collector field consists of parallel loops of individual parabolic trough collectors in series. Heat transfer medium (HTF) is thermo-oil (suitable up to 300°C) or synthetic oil (stable up to 400°C; more expensive by a factor of 10). Water/steam as HTF is not yet used (advance development in progress).

7.2 Farm Solar Power Plants with Line-Focussing Collectors 217

Fig. 7.1. Evolution of parabolic trough collectors and systems (Part 1). (a) Earliest 50 hp trough collector system operated in Egypt for field irrigation in 1913 (F. Schuman); (b) 1-axis tracking parabolic troughs (Acurex; 1978); (c) fix-mounted parabolic troughs with moving receiver (test model, General Dynamics; 1978); (d) central receiver parabolic troughs (concept, FMC; 1978); (e) 2-axes tracking parabolic troughs (MAN; 1979); (f) 2-axes tracking parabolic troughs (Ansaldo; 1979).
ad (a) (courtesy of Roger-Viollet, Paris, France)
ad (b) (courtesy of SNL, Albuquerque/NM, USA)
ad (c) (courtesy of SNL, Albuquerque/NM, USA)
ad (e) (courtesy of MAN, Munich, Germany)
ad (f) (courtesy of Ansaldo, Genua, Italy)

Fig. 7.2. Evolution of parabolic trough collectors and systems (Part 2). (g) double-reflector parabolic troughs (Sunshine Project; 1981)*; (h) 1-axis tracking large-aperture parabolic trough collector (LUZ; 1988)**; (i) aerial view of SEGS III and SEGS IV 30 MW$_e$ parabolic trough plants, Kramer Junction/CA, USA***.
* (courtesy of NEDO, Tokyo, Japan)
** (courtesy of LUZ Intl. Ltd., Los Angeles/CA, USA)
*** (courtesy of SNL, Albuquerque/NM, USA)

The advantage of oil as primary HTF is a low vapor pressure, resulting in operating pressures < 5 bar. The disadvantage of oil is the low viscosity at low temperatures, which is critical at start-up after the plant has cooled down. By temperature stratification, oil offers the advantage of one-tank thermal energy storage of small to medium capacity (thermocline principle), but application is constrained for cost reasons, and by the limited temperature range of thermo-/synthetic oils.

- Small collector fields need some amount of storage to allow operation of the power conversion unit (PCU) independent from changes in oil temperature as a conseqence of irradiation transients. The oil inventory of large collector fields, particularly if in hybrid combination with one or more fossil-fueled water/steam heaters, provides sufficient operational flexibility without buffer storage.
- For maximizing annual generation, yet minimizing size and cost of collector fields, thermodynamic conversion must be as efficient as possible for the solar-induced broad range of operating conditions. Taking advantage of off-the-shelf PCUs for cost reasons, early small-capacity cycle designs tended to be rather straight-forward and not well adapted

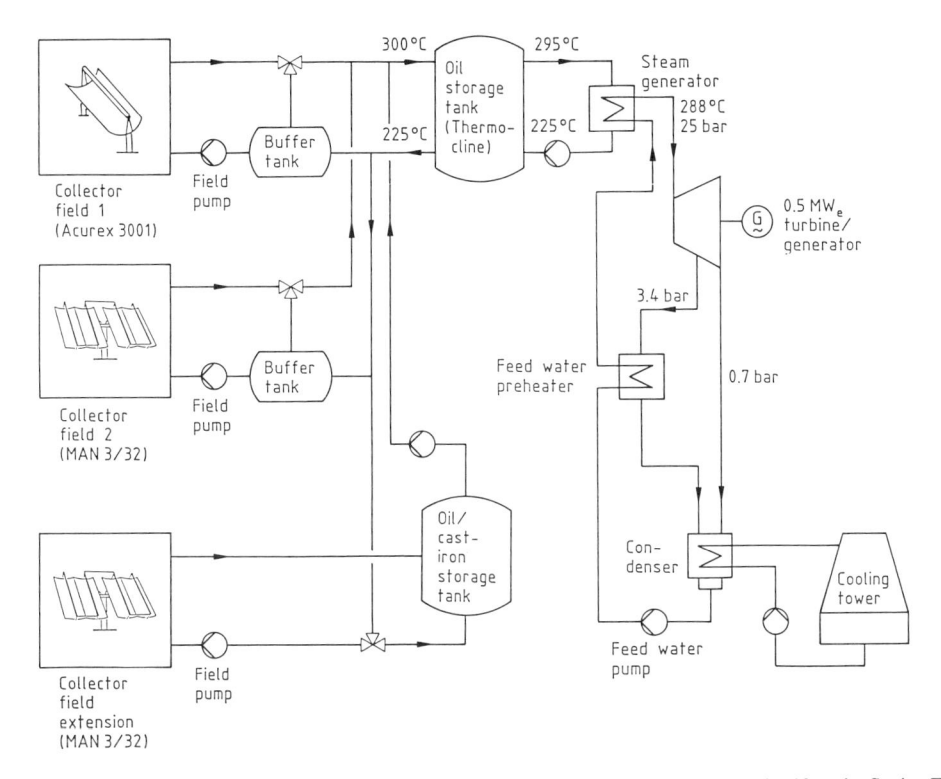

Fig. 7.3. Simplified system configuration of the IEA-SSPS 500 kW$_e$ thermal SPP in Almería, Spain. The experimental facility uses 1- and 2-axes tracking collectors in three different collector fields, and two storage tanks. The 500 kW$_e$ turbine/generator is an off-the-shelf, non-solarized unit.

to variable operating conditions. The large SEGS hybrid systems in use today incorporate highly sophisticated cycle configurations with (solar and/or fossil) superheating, and PCUs specifically adapted to solar operating conditions.

In this context, wet cooling is essential for best possible cycle efficiency. In sunny but arid regions, scarcity of water may necessitate that dry cooling be used for large thermal SPPs, affecting annual plant performance.

7.2.2 System Examples

(a) Coolidge solar thermal irrigation project. This 150 kW$_e$ irrigation facility, located at the Dalton Cole Farm in Coolidge/AZ, was designed for feeding electricity into a local grid from which an irrigation pump was operated. The system ran from late 1979 to late 1982 in a hybrid mode, and daily performance data (available irradiation, thermal energy collected, natural gas used, electrical energy generated) were recorded. In its last year, the plant functioned automatically with merely one staff technician who, for safety reasons, was needed to supervise PCU start-up (see Tab. 7.2 and 7.13 for plant technical data).

The Coolidge data show that performance of the solar plant (net generation, efficiency, collector field availability) improved over the three years of service (Tab. 7.3). Other major operating observations with relevance for future SPPs were [45,70]:

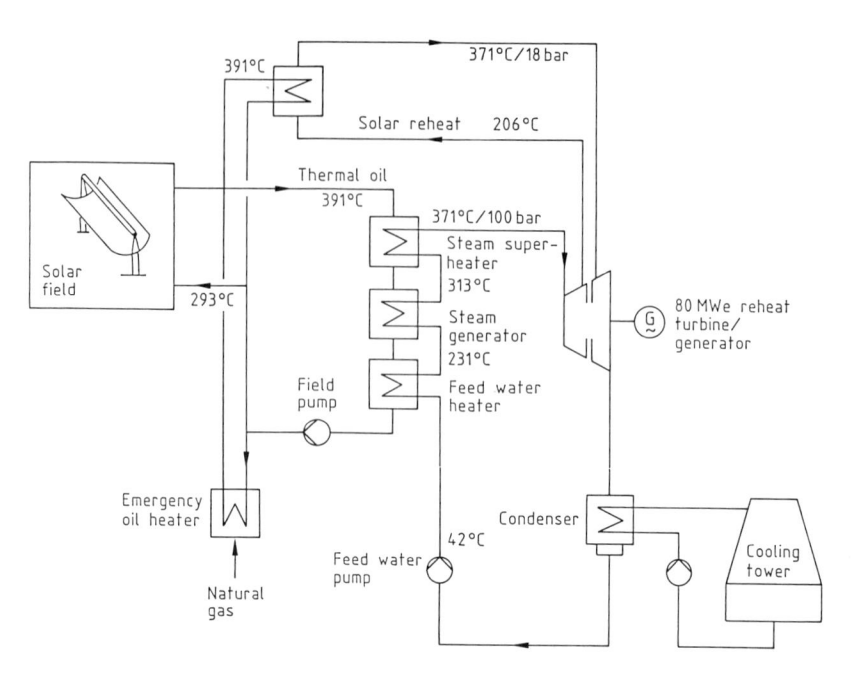

Fig. 7.4. Simplified system configuration of the 80 MW$_e$ solar electricity generating system (SEGS VIII), operational since 1989 at Harper Lake/CA, USA.

- the original Coilzak aluminum reflective surfaces deteriorated rapidly within one year; these surfaces were subsequently covered with a second-surface aluminized acrylic film (FEK-244) which proved optically effective (long-term performance and durability were not established)[2];
- demineralized water must be used for wet mirror cleaning (reflectivity of collector surfaces washed with hard water was lower than of those left dirty);
- flexhose and pump seal leaks were found to be safety/reliability hazards, causing two fires;
- mechanical motor drives have to be of adequate quality (many drive motors and pump seals failed).

(b) IEA-SSPS. The experimental parabolic trough IEA-SSPS farm plant was designed for 500 kW$_e$ net generation at 920 W/m^2 irradiation at equinox noon (Fig. 7.3). For side-by-side performance comparison, two different collector types were installed in three collector fields. For the same reason, two thermocline storage vessels, one with dual media, were incorporated and provided storage capacity equivalent to 0.8/0.37 MWh$_e$. A steam turbine generator was selected in preference to an ORC-based power conversion subsystem. One collector field consisted of one-axis tracking collectors with < 1 mm thin glass second-surface (S/S) silvered mirrors glued onto a flexible steel substrate. The other two fields were made up of two-axes tracking modules, each carrying four line-focussing troughs formed of sagged-glass S/S silvered mirrors. The sagged-glass concept was used later in the design of the Luz solar collector assemblies (see below).

---

[2] The idea of using front-silvered surfaces on polymer films is being pursued in heliostat development; environmental durability tests are currently carried out at SERI, Golden/CO, USA.

Table 7.2. Major characteristics of three trough SPPs.

	Coolidge	IEA-SSPS		SEGS III
Collector, type	Acurex	Acurex 3001	MAN 3/32	LUZ LS-2
Number	48 Assemblies	480 Units	140 Units	868 Assemblies
Aperture				
Unit m^2	44.6	5.58	32	235
Total m^2	2,140	2,688	4,480	203,980
Refl. surface	a) Aluminum	S/S thin	S/S sagged	S/S sagged
	b) FEK-244	glass/steel	glass	glass
		laminate		
Orientation	N-S	E-W	pole-mount	N-S
Parallel loops	8	10	24	62
HTF	Caloria HT-43	Santotherm 55		Monsanto VP-1
Op. temp. °C	200/288	255/295		248/349
Tracking	Shadow band	Sun sensor		μP control
Storage	Thermocline	a) Thermocline		None
Medium	Caloria	b) Oil/slab iron		
Capacity kWh$_t$	3,750	a) 5,000		
		b) 3,700		
PCU	ORC/1-stage	Steam/1-stage		Steam/2-inlet
Temp./press.	268°C/10.5bar	283°C/25bar		326°C/43.3bar (solar)
	41°C/0.1bar	39°C/0.07bar		510°C/104.5bar (gas)
				45°C/0.08 bar
Generator kVA	250	713		34,000
Backup Type	Gas boiler	None		Gas boiler
Capacity kW$_t$	1,035			56,000

S/S = second surface silvered; μP = microprocessor

Table 7.3. Average solar performance of the Coolidge irrigation system [38].

	Jan–Sep 80	Oct–Sep 81	Oct–Sep 82	Total
Solar output				
Gross MWh$_e$	97.1	133.2	170.0	400.3
Net MWh$_e$	56.0	75.1	119.1	250.2
Av. efficiency				
Gross %	2.45	2.47	3.5	2.9
Net %	1.41	1.39	2.45	1.81
Field availability %	89	93	98	

In 1982 and 1983, the plant was operated by utility personnel in two shifts, seven days per week, in grid-connected utility-like mode, insolation and availability permitting. Compensating for a low solar multiple (SM), the plant routinely was operated in storage charging mode for several hours before the steam generator and PCU were started. This operating strategy provided for a maximum of full-rated power production. Annual energy production performance was not representative of the capabilities of the plant, however. Energy production was curtailed by lower-than-expected local irradiation and low PCU efficiency, and by high thermal inertia and high irradiation threshold (> 350 W/m^2) for net generation. For a clear day, the plant demonstrated a 2.5% net efficiency.

Nonetheless, the plant operating experience provided valuable lessons for future trough SPPs; significant findings were, amongst others [17,30,43]:

- the expected performance advantage of a two-axes tracking, pedestal-mounted trough collector field could not be demonstrated. Additionally collected energy over one day, in comparison to the one-axis tracking collector, was compensated for by higher piping losses;
- maintenance of the one-axis tracking collector is considerably easier than for the two-axes tracking collector;
- the effect of thermal inertia is an important consideration to be included in plant sizing and performance analysis at the design stage;
- plant performance decreases sharply as compared to rated performance if irradiation is less than assumed for the design point;
- degradation of black-chrome absorptive coating on receiver tubes does not necessarily affect the output performance of collectors;
- flexhose, seal, joint and weld leaks leading to oil spills can be a significant maintenance factor (and environmental hazard).

(c) 30 MW$_e$ Solar electricity generating systems (SEGS). Taking the advantage of Federal/State tax benefits and of purchase agreements made possible by the Public Utility Regulatory Policies Act (PURPA), a series of plants based on one-axis trough collectors has been placed in operation in the service area of Southern California Edison Co. (SCE)/CA, USA (Figs. 7.1 and 7.3).

Each plant is structured as a third-party financial venture so as to maximize the value of tax benefits and the cash flow from electricity sales under negotiated or standard purchase agreements with SCE (Tab. 7.4). Plant annual performance is guaranteed by the manufacturer.

Table 7.4. The series of hybrid SEGS thermal SPPs.

Plant	Rating ($MW_{e,net}$)	Collector assembly	Start-Up	Power purchase agreement
I	13.8	LS-1	12/84	Special agreement
II	30	LS-1/LS-2	12/85	Special agreement
III	30	LS-2	12/86	Standard offer 4
IV	30	LS-2	12/86	Standard offer 4
V	30	LS-2	9/87	Standard offer 4
VI	30	LS-2	9/88	Standard offer 4
VII	30	LS-2/LS-3	12/88	Standard offer 4
VIII	80	LS-3	12/89	Standard offer 4

The solar collectors were developed by Luz Industries (Israel), building effectively on the experience accumulated in the U.S. and in Europe (e.g. IEA-SSPS) and forging them into a family of commercially marketable trough solar collector assemblies. These collectors progressed to ever larger trough apertures, higher concentration ratios and improved absorber emissivities (Tab. 7.5). Routine hybrid operation of the SEGS plants renders it difficult to determine their performance in solar-only operating mode from output statistics. One approach is to estimate coarsely the energetic value of the solar contribution in the hybrid input energy by the prorating of output according to the heat supplied from the fossil boiler and the solar field without consideration of the supply temperature. Table 7.5 shows the solar performance so calculated for the SEGS I–V plants in the first half of 1988 [38]. The solar performance improvement achieved in the more recent plants is apparent.

Although SEGS development started from an advanced state of trough collector development, a number of operational problems were encountered (and were corrected) in the first

Table 7.5. Solar performance of the SEGS I–V plants (Jan–Jun 1988) [38].

SEGS	I	II	III	IV	V
Net generation					
total GWh$_e$	6.11	23.50	35.74	36.00	36.91
from gas GWh$_e$	2.86	12.07	10.77	10.71	10.63
from solar GWh$_e$	3.25	11.43	24.97	25.29	26.28
Collector area (10^3 m^2)	82.96	165.38	203.98	203.98	233.12
Solar input GWh$_t$	103.7	206.7	255.0	255.0	291.4
Net efficiency (%)	3.1	5.5	9.8	9.9	9.0

two facilities. Staggered deployment of the series of SEGS plants carried out by one industrial supplier led to improved subsequent plants by applying the lessons learned, resulting in rapid and effective technology advance. Other key findings:

- leakage or failures of flexhoses, welds and pump/valve seals were the cause of a significant number of major oil leaks and fires, leading to subsequent design and component refinements;
- significant efficiency gains are attributable not only to improved solar collector assemblies, but also to the adaptation of the cycle configuration and the turbine-generator power block to the operating conditions of the SEGS plants;
- quality assurance during manufacture and field installation, and use of quality equipment, are more important than low investment cost.

7.2.3 Collector Subsystem

Several of the trough collectors illustrated in Fig. 7.1 underwent testing and were investigated at several R&D institutions, e.g. at Sandia National Laboratories, Albuquerque/NM, or were employed in thermal SPPs or experiments (see Tab. 7.6 for technical and design performance data of some representative collector assemblies).

The IEA-SSPS project in Almería tested and evaluated in detail the relative performance of fields of 1- and 2-axes tracking collectors with the objectives to

- compare their long-term performance;
- compare behavior of steel-sheet-laminated thin-glass mirrors (0.6 mm) with second-surface-silvered sagged-glass mirror reflectors;
- gather system-related experience with trough collectors using black-chromium or black-cobalt-based selective receiver coating, swivel-joint or flexible-hose pipe interconnections, and open- versus closed-loop subsystem control.

Key results [30] were, that

- a single 2-axes tracking collector unit, its aperture always oriented normal to incident irradiation, absorbs up to 30% more thermal energy than a 1-axis tracking, E-W oriented collector assembly (Fig. 7.5).
- when interconnected in a field set-up, the higher energy collection potential of two-axes tracking collectors shows up only at high irradiation levels; major reasons are higher thermal losses due to longer/more complex field piping and a large number of piping supports (Fig. 7.6);

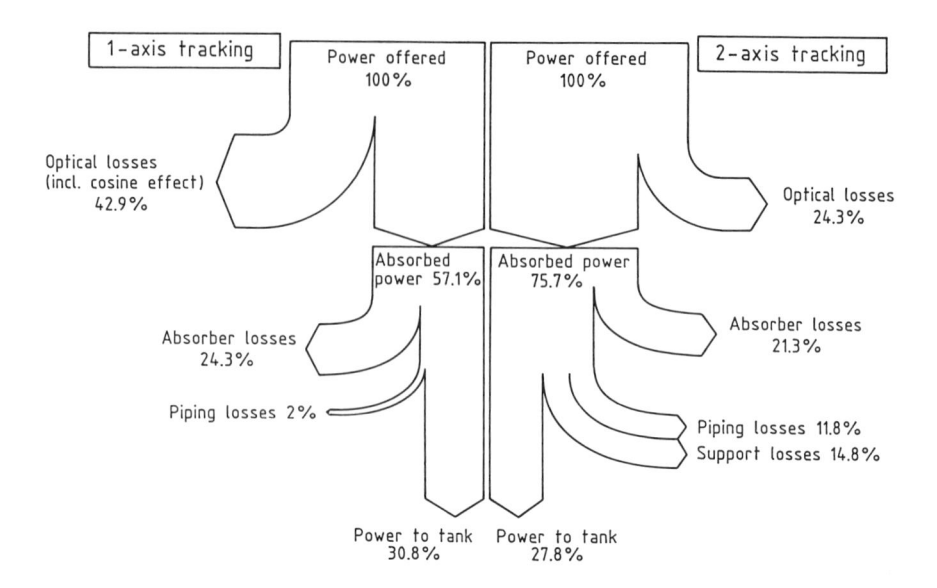

Fig. 7.5. 24-hour performance of 1-axis (Acurex 3001) and 2-axes (Helioman 3/32) tracking parabolic trough collectors over an arbitrary day with about 6 kWh/m²d of direct normal irradiation at Almería, Spain [30].

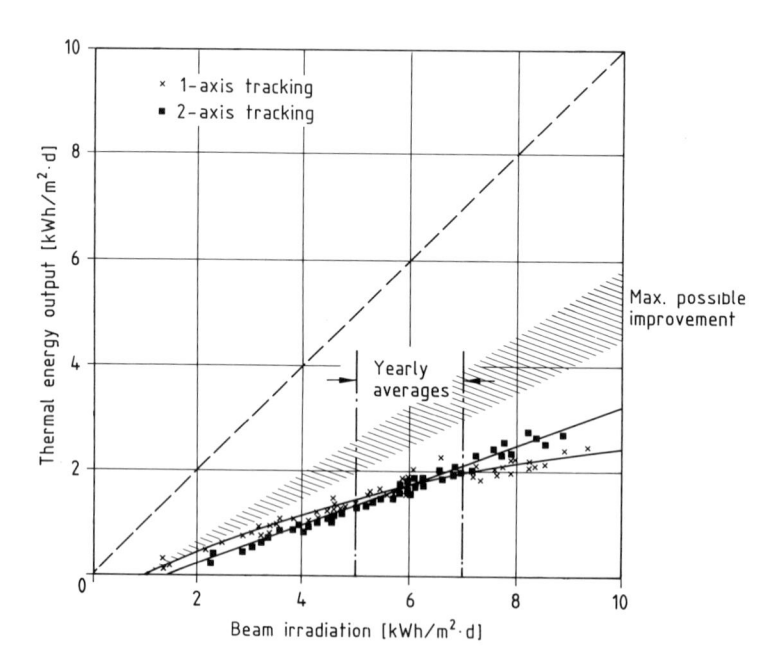

Fig. 7.6. Daily input-output performance of 1-axis (Acurex 3001) and 2-axes (Helioman 3/32) tracking parabolic trough collectors for different levels of total direct normal irradiation at Almería, Spain [30].

Table 7.6. Technical and design performance data of selected line-focussing parabolic trough collector assemblies (design values in brackets).

Manufacturer	Acurex	M.A.N.			LUZ		
Type	3001	3/32	M480	LS-1	LS-2		LS-3
Year	1981	1981	1984	1984	1986		1990
Trough width m	1.83	4 × 1.81	2.38	2.55	5.0		5.76
Collector length m	19.75	4.0	37.5	50.4	47.1		95.2
Aperture area m²	33.5	32.0	80.0	128	235		545
Rim angle	90	70.2	83.4	80	80		80
Pylon distance m	3.25	pedestal	6.0	6.3	8.0		12.0
Reflecting surfaces	thin glass/ steel-foil laminate	sagged glass S/S mirror			sagged glass S/S mirrors		
Mirror reflectivity	0.91(0.93)	0.85(0.89)	(0.93)		0.94		
Tracking	one-axis	two-axes	one-axis		one-axis		
Accuracy mrad	±0.25	±0.12	±0.10		±0.10		
Intercept factor	0.91 (0.92)	0.95 (0.96)	0.95		0.94		
Concentration	36	42	41	61	71		82
Optical efficiency	0.77	0.77		0.734	0.737		0.772
Absorber surface	black cobalt	black chromium			black chromium		
– Absorptivity	0.96	0.95	0.96	0.94	0.96 at 300°C		(0.97)
– Emissivity	0.27	0.25	0.17	0.30	0.24		0.15 at 350°C
Peak coll. eff.%	59(58)	48(57)	(65)	66	66(68)		68
Max. temp. °C	295	305	305	307	349		390
Spec. mass kg/m²	32.1	74.8	35				

- in terms of optical performance and physical ruggedness, thin-glass-on-metal reflecting surfaces proved as effective and robust as thick sagged-glass mirrors.
- periodic removal of air-transported dust/grime deposits from mirror surfaces is essential for maintaining reflectivity and collector field performance, cleaning intervals being dependent on local air quality and/or seasonal sandstorm occurrences, Fig. 7.7[3].

Polymer-based second-surface-silvered reflecting surfaces on steel substrates, which proved optically effective at Coolidge, are lightweight with the promise of lower cost; however, long-term performance and durability are not yet established.

7.2.4 Plant Performance Characteristics

A determination of plant performance using relationships of daily energy input to output was first attempted in the evaluation of the Japanese Nio and French Themis tower SPPs. Input-output performance of commercial plants is also increasingly published (SEGS) but, as stated earlier, solar-only performance of hybrid-operated plants is difficult to determine. In the absence of statistically relevant solar-only operating data, solar-only performance of hybrid SPPs is subject to interpretation. Nonetheless, a comparison of daily input-output relationships of the IEA-SSPS plant and of the SEGS III facility (based on linear regression, see Chap. 4) has been attempted [29]; as Fig. 7.8 indicates, considerable progress in trough

---
[3] This finding correlates with experience at other test locations; as a consequence, collectors of SEGS plants are cleaned at regular short intervals.

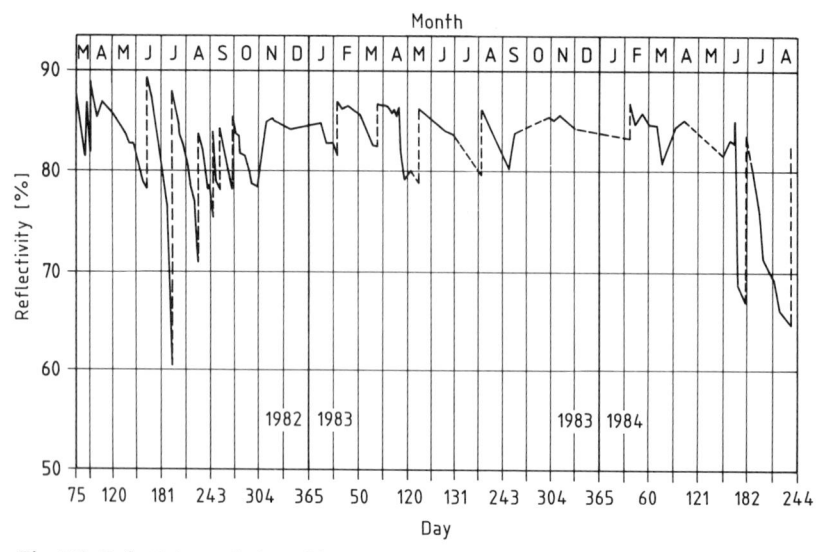

Fig. 7.7. Reflectivity variation with time of the Acurex Mod. 3001 parabolic trough mirror collectors at Almería, Spain [30]. Rapid improvements of reflectivity indicate occurrence of precipitation or heliostat washing.

collector SPP technology has been achieved in the past years. Several conclusions can be drawn:

- daily net energy performance on system level was improved from 4-6% to about 12%, primarily by minimizing non-active pipe length, higher CR, reduced thermal losses of receivers (vacuum insulation), and high collector field availability (in the 97–99% range for SEGS III–V [42]);
- minimum daily energy input needed before producing net output did not change significantly; it is argued that a threshold of 2.5–3.5 kWh/m²d of daily direct normal irradiation (at SM = 1.0) is a technology-specific constraint for trough SPPs irrespective of capacity[4].

Net output performance is improved (indicated by a steeper slope of the input-output curve) by yet higher efficiency of the thermodynamic cycle. Such refinement measures have in fact been undertaken for the SEGS plants. Starting with SEGS VI, a power conversion subsystem (power block) was installed which was specifically adapted to the SEGS operating conditions. As a consequence, the collector field size could be reduced by about 20% (affecting predicted yearly plant performance by only 1–2%).

7.2.5 Technical and Operational Potential

The considerable advance in trough SPP technology can be shown by a comparison of the efficiencies of the early 500 kW$_e$ IEA-SSPS facility (with SM = 1 and storage) with that of the hybrid 30 MW$_e$ SEGS III and future SEGS plants (Tab. 7.7). This comparison is also indicative of the annual performance which reasonably can be expected within the next few years, given high local irradiation availability and operating reliability.

---

[4] In the first half of 1988, solar-only performance was monitored at SEGS III [28]. These data indicate that slope of the input-output curve and threshold are dependent on the season (Fig. 7.9).

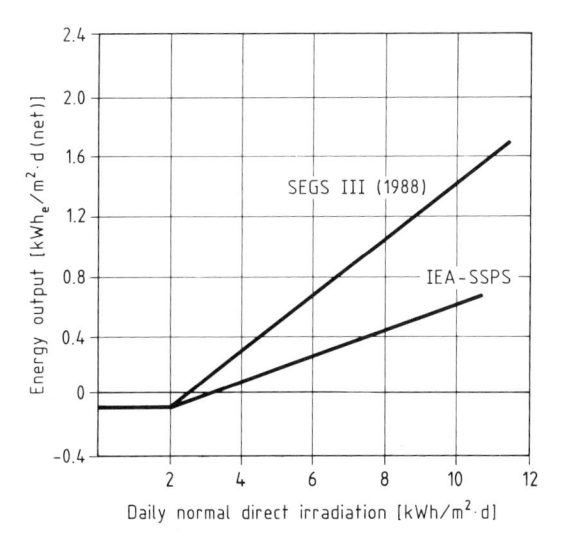

7.2 Farm Solar Power Plants with Line-Focussing Collectors

Fig. 7.8. Performance comparison of daily energy input to net output of the IEA-SSPS facility at Almería, Spain and of the 30 MW_e SEGS III plant at Kramer Junction/CA, USA in solar-only operating mode [29].

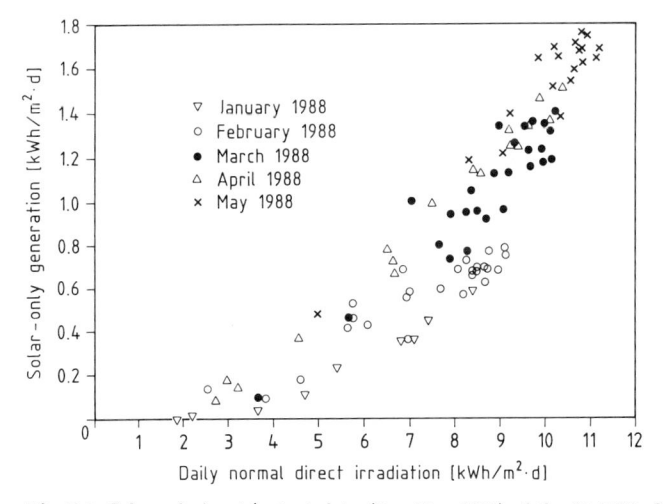

Fig. 7.9. Solar-only input/output data (Jan–May 1988) of the 30 MW_e SEGS III plant at Kramer Junction/CA, USA [28]

In summary, the following observations seem valid:
- early SPP facilities, lacking previous operating experience, were designed optimistically, actual performance falling short of expectations;
- collector and collector field performance improved significantly, from about 28% over a day with 'good' irradiation in the IEA-SSPS facility, to about 57% annually in the 30 MW_e SEGS III plant; this improvement is attributable as much to refined operating strategies as to better design, reliability and operational availability of today's collectors and collector fields;

Table 7.7. Improvement in efficiency stairsteps of parabolic trough SPPs (for solar-only operation in per cent of incident energy) [36,40,42].

Plant	IEA-SSPS			SEGS		
				III	VIII	XI
Capacity	(500 kW$_e$)			(30 MW$_e$)	(80 MW$_e$)	(80 MW$_e$)
Location	Almería/S.Spain			S.California		
Collector type		Acurex/M.A.N.		LS-2	LS-3	LS-4
	design point value	actual data	improvement potential	actual data	design values	design values
Time period	Equinox noon	1981–84 'good' 24h	'good' 24h	1988 year	1989 year	(1994) year
Incident solar	100.0	100.0	100.0	100.0	100.0	100.0
Absorbed by field	100.0	81.2	83.4	87.3	87.3	92.2
Thermal from field	52.0	22.7	41.9	49.4	–	–
Thermal to storage	51.5	21.2	41.1	–	–	–
Thermal to turbine	51.5	20.1	39.0	42.5	43.7	52.7
Gross electric	10.3	3.4	9.8	11.5	16.4	19.5
Net electric	10.2	2.5	8.8	10.1	14.3	18.0

- thermodynamic cycle efficiency was increased by virtue of higher operating temperatures, by longer periods of steady-state cycle operation, and by adaption of off-the-shelf PCU equipment to solar-specific operating conditions[5];
- internal parasitic energy requirements in operation and stand-by are critical for performance and must be minimized[6].

Commercial opportunities created by federal and state legislation in the U.S. provided the impetus for the rapid succession of the SEGS family of solar plants – and continue to do so. While the first plants (SEGS I and II) had difficulties in meeting performance targets/projections in the early years of operation, all subsequent plants (SEGS III and beyond) met targets and even exceeded projections in the first year. Instrumental for this was the fossil-fueled heater which was originally introduced mainly for meeting contractual performance guarantees independent of weather or time. The fossil heater, however, soon became a key element for improving plant performance (by superheating of solar-produced steam; by avoiding part-load operation in winter), and for maximizing annual revenues (45% of annual revenues but only 18% of annual energy are produced during summer on-peak periods [40]). For this reason, all SEGS plants are tuned for peak performance during summer on-peak periods, typically exceeding rated performance (30 MW$_e$) for most of the day from April to September (Fig. 7.10).

---

[5] IEA-SSPS had to use standard PCU equipment with paltry 16.9% daily efficiency, as compared to 29.4% for SEGS I/II, and 37.5% for SEGS III/VII; further improvements depend on even higher operating temperatures and on system capacities exceeding 80 MW$_e$.

[6] Parasitics, playing a dominant role in IEA-SSPS, were reduced to 12.2% annually in SEGS III (based on solar- and fossil-based generation); this is a level customary in conventional power plants.

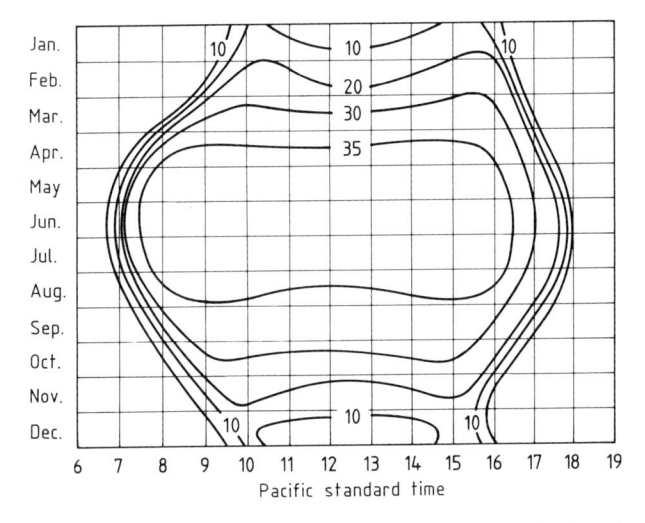

Fig. 7.10. Contour plot of gross solar-only electricity production of a typical 30 MW$_e$ SEGS power plant under irradiation conditions of the Mojave Desert/CA, USA [40]. Parameter is peak plant output (in MW$_e$) which can be achieved under average irradiaton conditions.

7.3 Farm Solar Power Plants with Point-Focussing Collectors

For a number of technical reasons, the higher performance expectation of two-axes tracking line-focussing collectors could not be achieved with a trough configuration of moderate concentration (IEA-SSPS). The argument has been raised that better results are obtainable with point-focussing collectors of higher concentration ratio. A few experimental solar thermal power plants were built using parabolic dishes in a distributed field arrangement as collectors of thermal energy (in contrast to individual parabolic dish units with individual power conversion units for each dish, see Sect. 7.5). Thermal energy was collected from the field of collectors, and standard turbine-generator equipment was used for central thermodynamic energy conversion. Information about operational experience with such plants is scarce, however.

7.3.1 Plant Configurations

The system configuration of solar farm power plants with dish collectors resembles those with trough collectors. Heat transfer fluid (usually thermal oil) passing the receiver of the first dish is routed through the receivers of several subsequent dishes, incrementally raising the temperature. Usually, several such strings of collectors (loops) operate in parallel.

System diagrams of two point-focussing solar farm facilities are shown in Figs. 7.11 and 7.12; for views of these and some other facilities see Figs. 7.13 and Fig. 7.14; technical data are compiled in Tab. 7.8 and Sect. 7.9.

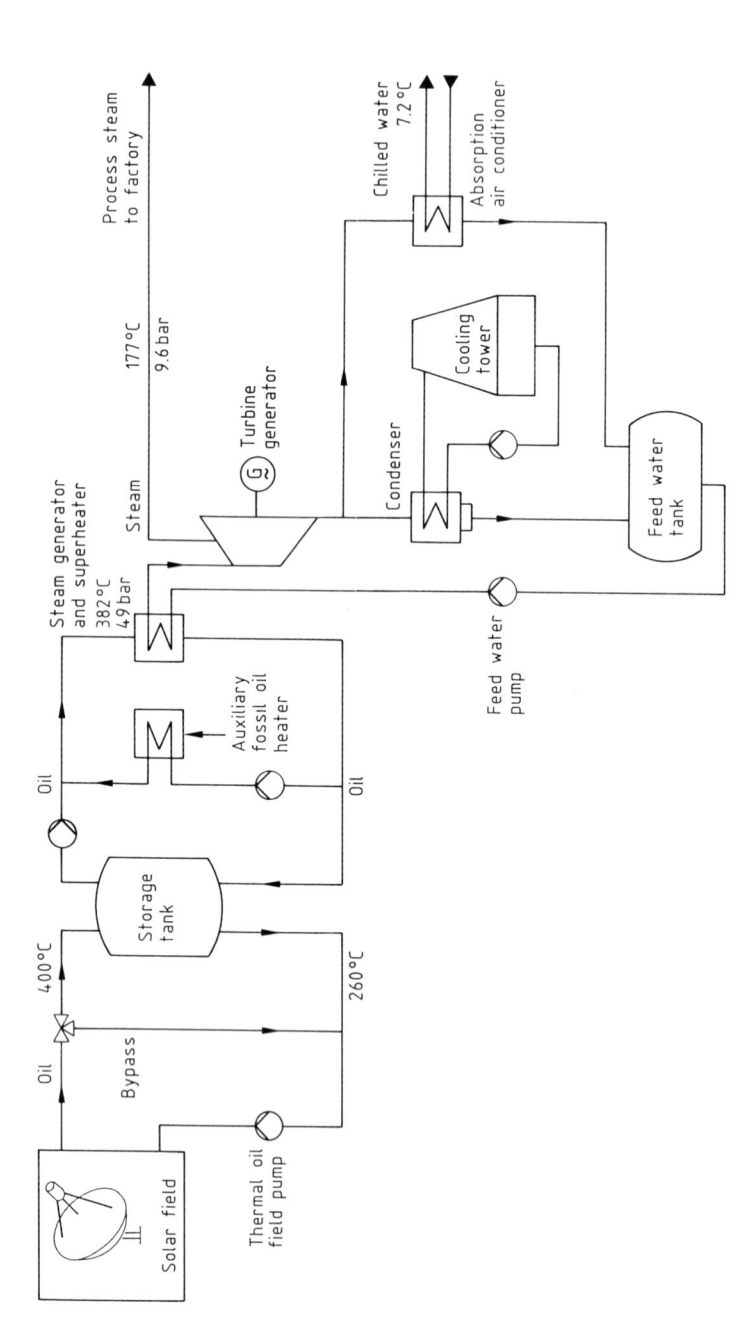

Fig. 7.11. Simplified system schematic of the cogenerating 400 kW$_e$/468 kW$_t$ point-focus Solar Total Energy Project (STEP) at Shenandoah/GA, USA (built and operated in partnership between U.S. Department of Energy and Georgia Power Company).

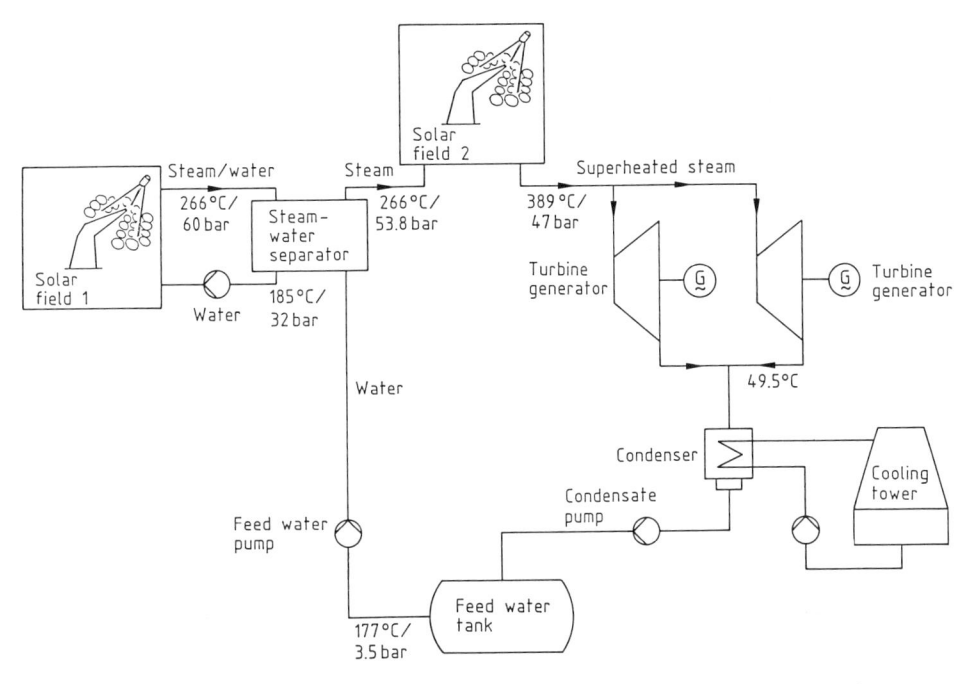

Fig. 7.12. Simplified system schematic of the 4.88 MW$_e$ Solarplant 1 facility at Warner Springs/CA, USA, using individually-facetted parabolic dishes (LaJet Energy Corp., Abilene/TX, USA).

7.3.2 System Examples

(a) Sulaibyah, Kuwait. The 100 kW$_e$/400 kW$_t$ Sulaibyah facility was designed and intended as an experiment for investigating the technical and operational performance and the viability of supplying the total energy needs of an agricultural research station by a solar power system (Fig. 7.14). The plant was configured as a hybrid system with a fossil-fired HTF heater and backup diesel generator for power supply. The plant provided electricity for irrigation pumping, and thermal energy for desalination and greenhouse climatization. Collector HTF was synthetic oil, and toluene was the working fluid in the power conversion subsystem (see Tab. 7.8).

The system was first operated in 1981 and served as experiment and test facility until 1987. Only few system performance data were published [76]. It was reported that the plant successfully demonstrated all operating modes, that it was capable of highly automated operation, and that about one half of the design value of power performance was attained on system level. Thermal inertia was high and morning start-up time longer than expected. At least 400 W/m^2 were needed for keeping the plant in operation. No data are available for assessing annual energy performance.

The system experienced oil leaks (absorber, tracking unit) and electronic malfunctions typical of a first-of-a-kind facility. Also, degradation of the absorptive coating of the receiver was observed.

(b) STEP, Shenandoah. This 400 kW$_e$/2,000 kW$_t$ cogenerating facility was intended as a system experiment using dish concentrators, with the objectives of producing engineering and development experience, and of determining the interaction of a total solar energy supply

Fig. 7.13. Point-focussing collectors and dish/farm solar power facilities (Part 1). (a) 5 m ⌀ parabolic dish with external receiver (MBB; 1980); (b) 400 kW$_e$/424 kW$_t$ cogenerating plant at Sulaibyah, Kuwait (MBB; 1981); (c) 7 m ⌀ parabolic dish with cavity receiver (SKI; 1981); (d) 400 kW$_e$/3 MW$_t$ cogeneration Solar Total Energy Plant (STEP) at Shenandoah/GA, USA (GE; 1982).
ad (a), (b) (courtesy of MBB, Munich, Germany)
ad (c), (d) (courtesy of SNL, Albuquerque/NM, USA)

system in an industrial user environment (Fig. 7.11). The plant was started up in 1982. The solar energy collected was experimentally used to determine to what degree the electrical air conditioning and process steam requirements of an industrial host could be met.

Oil as HTF is used to collect and to transport the absorbed energy from the dish collectors. The thermal energy coming from the field is supplemented with thermal energy from the gas-fired HTF heater. Superheated steam, produced in a steam generator, drives a conventional turbine/generator set. Steam extracted from the turbine provides process steam, and the low pressure exhaust steam is used as input for an absorption chiller which produces chilled water.

The plant is routinely started using the gas-fired boiler, which provides all thermal energy needed initially. The heater also is used to warm up the HTF which is circulated through the collector field until operating temperature is reached. Output from the field and the heater are then combined for generating steam. The facility operates in a hybrid mode from this point on. For meeting thermal and electric load demands, the heater output is adjusted so as to cover

Fig. 7.14. Point-focussing collectors and dish/farm solar power facilities (Part 2). (a) 4 m ⌀ multi-facetted parabolic dish with silvered polymer reflectors (LaJet; 1983); (b) 4.88 MW$_e$ dish-farm facility Solarplant 1 at Warner Springs/CA, USA (LaJet; 1984); (c) 7 m ⌀ glass-metal concentrator (ANU; 1985); (d) 25 kW$_e$ ANU experimental dish-farm system at White Cliffs, Australia.
ad (a), (b) (courtesy of LaJet Energy Corp., Abilene/TX, USA)
ad (c), (d) (courtesy of ANU, Canberra, Australia)

any deficit in thermal energy in case the collector field provides less than required. The plant can also operate in a solar-only mode, but a capacity mismatch between the collector field and the power conversion subsystem results in continuous part-load operation and associated low performance of the entire system.

Although STEP performance was below expectations, the experiment provided experience and data of value:

- auxiliary heaters are more efficient when placed in the steam loop rather than the primary heat transfer circuit;
- part-load efficiency of the heater is also an important system design issue;
- all weather-exposed components of the system must be qualified for local conditions (for example rain soaked the thermal insulation of the small cavity receivers and led to corrosion and leaks of the carbon-steel receiver tubes.).
- depending on temperature control capabilities, adequate margins must exist between the nominal system operating temperature and upper temperature limit of the HTF (STEP operating temperatures had to be reduced because of local HTF overheating);

- delamination of reflective polymer films from aluminum substrates in a moist environment remains a serious issue;
- adequate reliability of all components of the plant system is an important requirement for achieving adequate SPP availability.

(c) Solarplant 1, Warner Springs. This nominally 4.88 MW$_e$ SPP (Fig. 7.12) was privately financed and built, and is operated within PURPA provisions. The plant uses 700 of the LaJet LEC-460 dish concentrators with cavity receivers (for technical details see Tab. 7.8). Water is used as primary HTF, being evaporated in a field segment of 600 collectors and superheated from 276°C to 371°C in the remainder of the field. Power conversion is split into two turbines with 3.68 MW$_e$ and 1.24 MW$_e$ rating; the smaller PCU is used during start-up and shut-down, during periods when irradiation is too low to operate the main PCU, and whenever peak/excess energy becomes available. Annual (design) performance is 12 GWh$_e$/a.

The innovative LEC-460 dish collector is of lightweight construction and uses polymer-based stretched membrane reflector segments, with the provision to replace easily these membranes several times during the life of the plant [69].

Solarplant 1 went on line in 1985. It is claimed that the plant averaged 106% of projected output over a seven-day test period with all collectors on line, but performance data are not published. Significant equipment problems and operational problems with the steam loop were reported, associated particularly with daily cycling and start-up [52]. The start-up time is stated as 30–60 minutes for consecutive operating days, but up to half a day after an extended

Table 7.8. Dish-farm solar power plant data.

	Shenandoah	Sulaibyah	Solarplant 1
Rating kW$_e$	400	100	4,920
kW$_t$	2,000	400	n/a
Collector			
Type	SKI	MBB	LEC-460
Number	114	56	700
Aperture			
Single dish m^2	21	18	43
Total m^2	2,328	1,025	30,261
Surface	FEK-244 on aluminium	S/S glass/dish mold laminate	polymer stretched foils
Tracking	Polar	Polar	Polar
Receiver	conical cavity	external spherical	cavity
HTF	Syltherm-800	Synthetic oil	Water/steam
Op. temp.	260/363	235/345	277/371
Storage	thermocline	thermocline	none
Medium	Syltherm-800	Synthetic oil	none
Capacity kWh$_t$	1,600	700	none
PCU	steam/4-stage high speed/ condensing	ORC/1-stage radial/con- densing	steam/con- densing
Temp./Pressure	382°C/43.3bar 177°C(extr.)	320°C/15bar 242°C/0.4bar	360°C/41.4bar 110°C
Generator kVA	400	153	3,400 / 1,520
Backup	gas boiler	oil burner	none
Capacity kW$_t$	2,300	750	

S/S = second-surface silvered; PCU = power conversion unit

shut-down period. The value of Diesel generators for assuring a mimimum level of supply was experimentally investigated for the first time for a thermal SPP, using the recovered exhaust heat to keep headers and turbine warmed-up.

7.3.3 Plant Performance Characteristics

Of the farm-type dish solar systems, multi-day performance data were published only for the STEP system. However, as STEP represents a very early stage of SPP development, the performance is merely illustrative, but by no means conclusive, of the potential of a mature system of similar type at locations more favored with direct irradiation than Shenandoah/GA.

STEP operated continuously over ten and thirty consecutive-day periods during the summer of 1985 to determine solar contribution, capacity factor (CF), operations and maintenance (O&M) costs, and standby losses when keeping the system operational over several days [65]. During these test periods, the plant supplied the entire daily electrical and thermal energy needs of the industrial host, both from solar input (irradiation permitting) and from the gas heater. Despite 50% radiation-to-thermal conversion efficiency and > 95% operational availability of the collector field, significantly less solar energy than expected was collected and made available. This is attributed, mainly, to lower-than-average and highly transient irradiation during the test periods. Regression analysis of daily thermal energy input-output performance of the collector field (Fig. 7.15) suggests that about 2.5 kWh_t/m^2d of direct normal irradiation must be accumulated before achieving net output from the collector field. This value exceeds by far the amount of thermal energy needed for heating up the plant from ambient temperature; hence, the remainder must be attributed to thermal losses and parasitics [25,37,53,65,66].

The experimental attempt to satisfy all energy needs of an industrial user by a hybrid thermal SPP demonstrated some key facts of fundamental significance. During the tests, much fossil energy was expended to keep the solar facility operational so that the energy demand of the user could be satisfied without delay. The amount of natural gas consumed for that purpose was significantly higher than the energy needed if the industrial demands were covered conventionally. Keeping the solar system warmed up by the auxiliary heater, irrespective of the contribution ability of the solar field, is ineffective (Fig. 7.16).

If no flexibility exists to adjust user energy needs to local solar irradiation conditions, advance analysis is mandatory to determine whether demand can indeed be totally satisfied from solar in terms of energy (and cost). The effects of parasitic loads, plant availability, solar availability and integrated system performance are key considerations in this context.

7.3.4 Technological and Operational Potential

Operating experience with dish farm SPPs provides little basis for extrapolating technical and operational potential of future mature systems. Assessment of physical arguments is the only means for an evaluation.

The advantage of dish collectors (i.e. high concentration and temperature) does not really pay off in a farm configuration, mainly because the physical potential of high operating temperature cannot be exploited due to the low upper temperature limits of thermo-oils. As already shown with trough thermal SPPs, this deficit cannot be compensated for by the inherently higher energy yield with 2-axes tracking dishes when compared to 1-axis tracking trough collectors. This situation may have to be reassessed if the development efforts for using water/steam as a primary heat transfer medium prove successful (Solarplant 1). As water/steam is also considered as primary HTF in future trough collector SEGS plants,

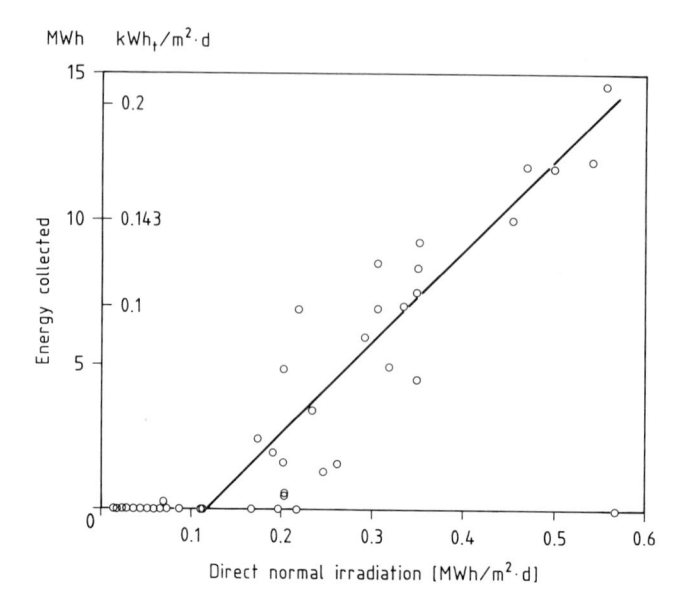

Fig. 7.15. Collector field input-output perfomance of the 400 kWe/483 kW$_t$ parabolic dish (C = 235) Solar Total Energy Project (STEP), Shenandoah/GA, USA. [67]

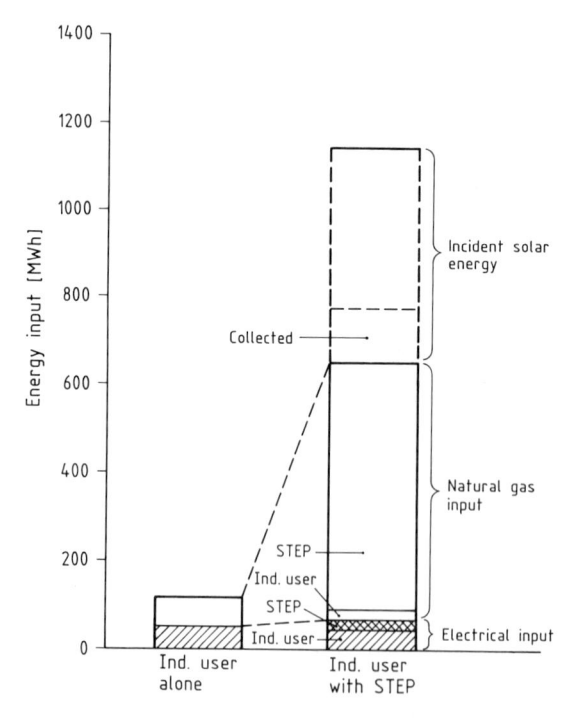

Fig. 7.16. Gas use when cogenerating electricity and thermal energy in response to supply needs of an industrial user; 400 kW$_e$/483 kW$_t$ Solar Total Energy Project (STEP), Shenandoah/GA, USA [67].

performance/reliability considerations and collector field cost are most likely the key issues in a comparative assessment of dish- and trough-based farm SPPs.

7.4 Central Receiver Solar Power Plants with Heliostat Fields

Development of central receiver SPPs (tower SPPs for short) was supported by funding authorities in several countries. The reason for this interest – aside from novelty of concept and engineering challenge – has been the possibility of collecting large amounts of concentrated solar irradiation without requiring a piping network for thermal energy collection, and the expectation of achieving economy-of-scale benefits in system sizes approaching those of utility power plants. The R&D interest evolved because the system design was complex, and prior experience with high irradiation flux conditions and associated material heat stress was lacking.

Development of solar tower systems began in the early sixties with pioneering work by G. Francia (Italy); until 1975 he operated a small facility with 135 m^2 of mechanically controlled mirrors and about 130 kW$_t$ capacity (but without thermodynamic cycle conversion) at San Ilario-Nervi near Genoa, Italy (Fig. 7.17). In 1978, a somewhat larger 400 kW$_t$ duplicate of this facility was installed as a high-temperature material R&D test bed at the Georgia Institute of Technology, Atlanta/GA, USA. In the early seventies, the 1,500 kW$_t$ French solar furnace at Odeillo, Pyrenees became operational, and was used in the mid seventies to demonstrate operational feasibility of an electricity producing cycle (64 kW$_e$).

Then, in rapid succession, six solar tower facilities were projected, built and operated in France, Italy, Japan, Spain and USA with strong financial involvement of respective governments, and a seventh plant reportedly began operation in the USSR, Crimea (Figs. 7.17 and 7.18). Technical data of the seven tower SPPs built are listed in Sect. 7.9.

7.4.1 Plant Configurations

Approaching utility power plants in nominal rating, the 10 MW$_e$ Solar One configuration at Barstow/CA, USA was of single-loop design, employing water/superheated steam as the primary heat transfer medium and as working fluid in the power conversion cycle (Fig 7.19).

Water is preheated in two steps, evaporated and superheated in a once-through external receiver (see Chap. 5). A dual-medium (oil/rocks) thermal storage can be charged/discharged via steam/oil heat exchangers. Heat supply for the steam turbine can come from the receiver directly, or from storage via a steam generator (at degraded steam conditions), or both simultaneously.

Two experimental tower SPPs employed dual-loop heat transport concepts using a liquid as primary coolant. Primary HTF was eutectic salt in the 2.3 MW$_e$ Themis plant at Targasonne/France, and sodium in the IEA-SSPS 500 kW$_e$ tower plant in Almería, Spain. Compared to single-loop and once-through water/steam configurations, the dual-loop concept allows higher receiver heat fluxes (see Chap. 5) yet reduces cycle fatigue stress of the receiver material, i.e. subjecting it to lower internal pressure and avoiding quenching effects by oscillating water columns.

Being good thermal conductors, hot and cold molten salts/metals must be stored in separate vessels if used as a heat transfer and storage medium (as exemplified by the Themis

Fig. 7.17. Views of experimental and demonstration type tower (central receiver) SPPs (Part 1). (a) 130 kW$_t$ experimental facility by G. Francia, water/steam, mechanically-tracked mirror reflectors (1968), San Ilario-Nervi, Italy;* (b) 1 MW$_e$ experimental plant, water/steam, eutectic salt storage (1981), Adrano, Sicily;** (c) IEA-SSPS 500 kWe demonstration plant, sodium; sodium storage (1981), Almería, Spain;*** (d) 1 MW$_e$ demonstration plant, water/saturated steam, steam storage (1981), Nio, Shikoku Island, Japan;**** (e) 2.3 MW$_e$ demonstration plant, salt, salt storage (1982), Targasonne, Pyrenees, France*****.

* (courtesy of Ansaldo, Genua, Italy)
** (courtesy of IRC, Ispra/Varese, Italy)
*** (courtesy of DLR, Cologne, Germany)
**** (courtesy of NEDO, Tokyo, Japan)
***** (courtesy of CNRS, Paris, France)

7.4 Central Receiver Solar Power Plants with Heliostat Fields

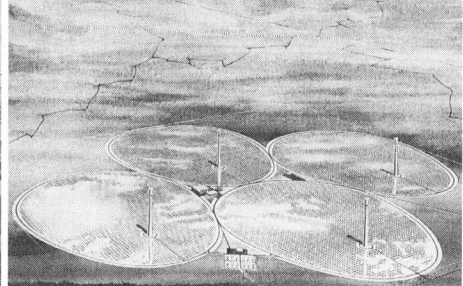

Fig. 7.18. Views of experimental and demonstration type tower (central receiver) SPPs (Part 2). (f) 1 MW_e demonstration plant, water/steam, salt storage (1983), Almería, Spain; (g) 10 MW_e pilot plant, water/steam, oil storage (1982), Barstow/CA, USA; (h) 5 MW_e demonstration plant, water/saturated steam (1988), Kertch/Crimea, USSR; (i) 20 MW_e GAST-20 study, air, salt storage (concept), Germany and Spain; (j) 200 MW_e Solar-200 study, different media (concept), USA.
ad (f) (courtesy of IER, Madrid, Spain)
ad (g) (courtesy of SNL, Albuquerque/NM, USA)
ad (h) (courtesy of Teploenergoprojekt, Moscow, USSR)
ad (i) (courtesy of Interatom, Bensberg, Germany)

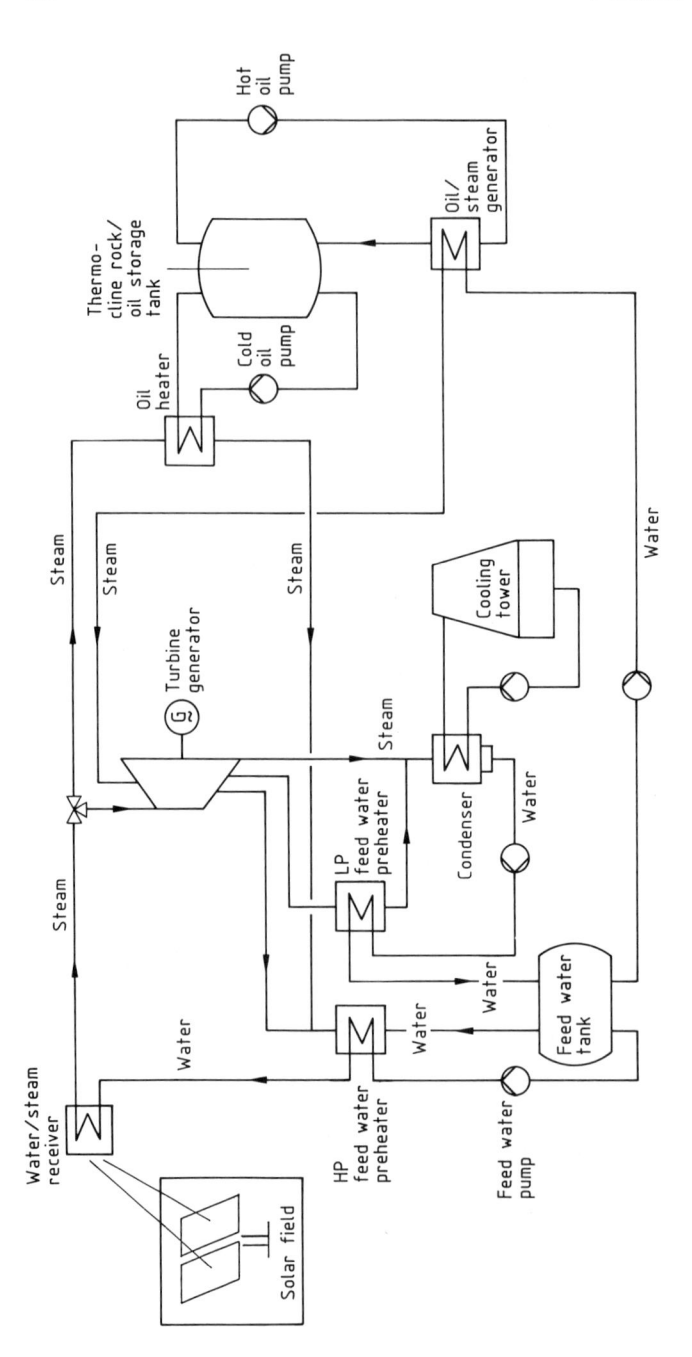

Fig. 7.19. System diagram of the Solar One 10 MW$_e$ pilot tower SPP at Barstow/CA, USA.

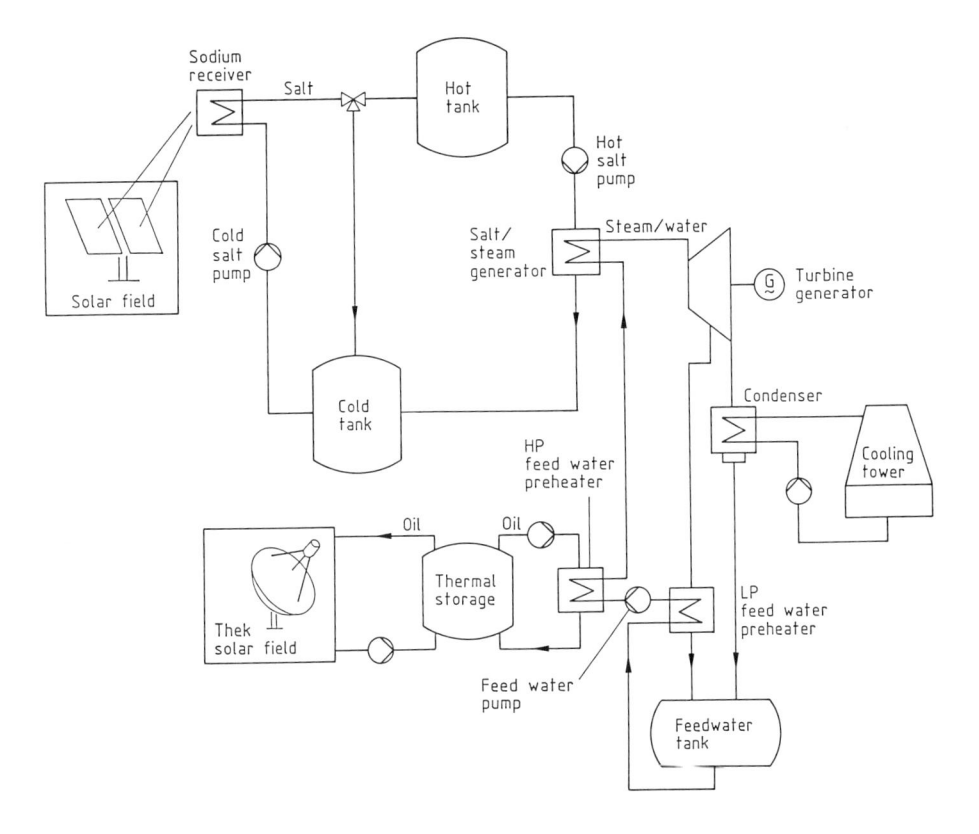

Fig. 7.20. System diagram of the Themis 2.5 MW$_e$ experimental tower solar power plant at Targasonne, Pyrenees/France.

system, Fig. 7.20). Using molten salt/metals as primary HTF and a dual-tank arrangement for storage, intermediate heat exchangers and associated losses are avoided, plant controllability is improved, and power-conversion is effectively decoupled from front-end solar energy input. The drawback is that salts and metals solidify well above 100°C, necessitating electric trace heating of all plant components in which liquids might freeze. Consequently, parasitic stand-by energy needs are increased. The choice of primary HTF becomes a key issue in design trade-offs.

Optimization of thermodynamic energy conversion is easier for large tower SPPs. While none of the operating plant systems (i.e. Solar One, Themis, IEA-SSPS) employed separate evaporators and superheaters, the phase-change and superheating steps are routinely separated in the steam cycle of large-capacity tower SPP designs [1,4]; this also simplifies incorporation of fossil-fired super-heaters into the plant design.

7.4.2 System Examples

Of the five European/U.S. tower SPPs, three used water/steam as primary HTF (10 MW$_e$ Solar One; 1.0 MW$_e$ CESA-1; 1.0 MW$_e$ Eurelios), and two used molten materials (2.4 MW$_e$ Themis; 0.5 MW$_e$ IEA-SSPS). The technical data is in Sect. 7.9 (Tabs. 7.12 to 7.16).

(a) Solar One. This project, located at Daggett (near Barstow) in Southern California, was a 10 MW$_e$ central receiver full system experiment, and was operated by SCE from early 1982 to late 1988 in pilot plant fashion (Figs. 7.18 and 7.19). Water/steam was both the heat transfer and working fluid for the thermodynamic cycle. One separate oil/rock thermal storage tank was coupled to the water/steam loop via heat exchangers, allowing operation of the power conversion subsystem at (reduced) steam conditions.

Although capable of operating in different modes, the plant was operated routinely using solar-generated steam without intermediate storage, thus improving annual energy efficiency. The storage subsystem was, for technical reasons, decommissioned after a fire incident.

Although annual energy production never reached design predictions, Solar One was the most successful tower SPP project so far. The long-term operation of the plant provided extensive data which were analyzed and evaluated, and which are most useful today for designing and assessing the performance of future tower SPPs. The size of Solar One and its utility-like operation rendered its performance less vulnerable to losses and parasitics which overshadowed the performance of the other smaller-sized experimental demonstration tower SPPs.

The wealth of information, experience and lessons learned from this pilot plant experiment has been comprehensively published [20,59,60]. Modelling and calculation codes for the design of tower SPPs were modified as a consequence of these data, providing the basis for all design studies currently undertaken (for instance the University of Houston Solar Central Receiver Code System, or the SOLERGY code developed by Sandia National Laboratories; see Sect. 7.8).

(b) CESA-1. The 1.0 MW$_e$ CESA-1 project was also a full system experiment, located near Almería in Spain; it was intended to demonstrate the feasibility of this type of plant, and to develop the specific technology and industrial base for tower SPP components. The plant started operation in early 1983 and was operated until the end of 1984. Water/steam was used as primary and secondary HTF, and molten salt as the storage medium in a two-tank configuration.

Designed to operate in six modes (direct, charging, discharging, direct and charging, direct and discharging, and buffered operation), the plant was operated only 324 h in grid-connected mode, producing about 130 MWh [7]. The short duration of operation provided useful information applicable mainly to the specific plant design [62].

(c) Eurelios. Eurelios was a 1.0 MW$_e$ full system experiment located in Adrano, Italy. Its objective was to demonstrate the grid-connected operation of a tower SPP, and to gain data for technical and economic evaluation (Fig. 7.17). Eurelios was the first tower SPP ever to be operated, being connected to the grid in the Spring of 1981. Operation continued through 1984.

The plant design incorporated a once-through water/steam receiver and a short-time buffer storage, using molten salt and a water/steam accumulator for steam superheating. Two types of heliostats of different size were used and were arranged in an East and West sector of a North field. A minimum irradiation of 450 W/m^2 of direct normal irradiation, a cloud cover < 25%, no haze, and > 75% heliostat availability were specified.

Due primarily to the extreme pipe length of the receiver, total start-up of the plant required typically about 2 hours. On the other hand, heliostats were moved into stand-by position three minutes after irradiation had dropped about 20% below nominal insolation. Also, local irradiation was more affected by cloud cover than expected. The total gross electricity production

---

[7] The CESA-1 facility was utilized later as a test bed for the development of air-cooled receiver components (GAST-20 program, see Chap. 5)

of the plant was only about 130 MWh$_e$, while parasitic power needs were higher (as compared to 14.5% efficiency prediction) [32,33].

(d) Themis. The Themis 2.4 MW$_e$ tower SPP, located at Targasonne in the French Pyrenees, was intended to conduct full system- and subsystem experiments, and to demonstrate the feasibility of this type of thermal SPP for deployment in sunbelt countries. The plant became operational in 1983 and was operated for 3 years (Fig. 7.18).

Themis used molten salt as primary HTF and as a storage medium in a two-tank configuration. A steam generator linked the secondary conversion cycle loop to the primary salt loop. This decoupled solar energy input and storage charging from power output generation, rendering plant operation more flexible. Thus, temporary drops in the collection of solar energy in the primary circuit hardly affected operation of the PCU at nominal cycle conditions.

Nominal operating procedure called for energy accumulation in the storage tanks until enough salt at sufficiently high temperature was available to sustain a rated output generation for 2-3 hours. Only at this point was the secondary circuit conditioned, the turbine started and power generated. Typically, three hours elapsed before adequate conditions for power production were achieved; about 45 minutes later, rated output power was attained, to be continued in the evening until storage was depleted to preset levels.

Themis performance was below prediction. Gross energy production was about 650 kWh$_e$/a on average, and net output was negative, due primarily to large parasitic loads (Tab. 7.9).

Table 7.9. Performance summary of the Themis 2.4 MW$_e$ tower SPP.

	Jul–Dec 1983	1984	1985	Jan–Jun 1986	Total
Plant Output					
Gross MWh$_e$	45.3	573.9	765.7	543.7	1929
MWh$_e$/m^{2}a	4.2	53.4	71.3	101.0	198
Net MWh$_e$	(743)	(2,697)	(2,182)	(1,131)	(6,752)
MWh$_e$/m^{2}a	(69.2)	(251)	(203)	(211)	(198)
On-line hours	51	470	526	388	1435

However, the Themis experience produced some significant findings and conclusions [24, 56]:

- the concept of separating solar thermal energy generation from power production by intermediate energy storage was successfully demonstrated;
- although heliostats were seriously damaged by the breaking of pedestals in two bad wind storms, 95% of heliostat availability was demonstrated; no corrosion was experienced with laminated glass mirrors;
- the design/layout of the primary (salt) loop can be simplified, trace heating concepts need to be improved, and parasitic loads in stand-by mode need to be drastically reduced to improve net power generation.

(e) IEA-SSPS. The 500 kW$_e$ IEA-SSPS tower plant was a full system experiment conducted in parallel with the IEA-SSPS parabolic trough project, both experiments under IEA auspices. The IEA-SSPS is the only tower SPP using sodium as a primary HTF and storage medium. The objective was to demonstrate the viability of this concept, to determine operational characteristics, and to compare performance of the two IEA-SSPS plants. The plant was operated from late 1981 until August 1986.

It was a two-tank storage configuration like Themis. A sodium steam generator decoupled the secondary water/steam loop from the primary sodium circuit. A steam motor was chosen as PCU.

The plant started operation after reaching levels > 300 W/m² of direct normal irradiance. After circulating through the receiver and reaching 500°C, sodium was accumulated/stored for about 2–4 hours in the hot tank. With hot storage sufficiently charged, the steam generator, the power circuit and steam motor were conditioned and output produced about 30 minutes later. Power generation continued until the hot tank was depleted to a minimum level.

Equipment outages and operating complications limited the ability to accumulate long-term system-level performance data. A combination of factors contributed to this situation such as high thermal inertia, SM = 1.0 despite storage, high-quality steam requirements for steam motor operation, and substantial parasitics. This situation was extensively analyzed and ways for plant improvement were determined. Design deficiencies were identified which, if avoided, would improve start-up and performance. Total gross production of the plant amounted to about 80 MWh_e. The project produced a number of important results and accomplishments:

- the technical feasibility and excellent component-level performance of a high-flux sodium receiver were demonstrated; predicted receiver performance were reached (see also Chap. 5);
- the reliability of trace heating elements is not sufficient to support reliable SPP operation when using HTFs with a freezing temperature > 100°C; repair/replacement of trace heating elements is difficult and tedious;
- high standards of quality control and assurance are essential to avoid hazards and costly O&M associated with sodium equipment[8].

7.4.3 Heliostat and Heliostat Field

Development history and outlook of heliostats (Fig. 7.21) indicate a trend from early rigid and heavy constructions with second-surface glass mirrors to lightweight low-cost constructions with front-surface-silvered polymer foil reflectors. Presently, two development lines are followed towards low-cost solutions, (a) the large-area glass-facetted configuration (150 m²) with correspondingly lower specific costs for support structures and drive train for a field of many heliostats, and (b) the so-called stressed membrane design (i.e. thin metal membranes stretched over front and rear of a circular supporting ring, the front surface being covered by reflective films or thin-glass mirror facets). Prototypes of such heliostats are currently being tested for performance at the CRTF and PSA test sites[9].

Via experimental tower SPPs, operational experience was gained with fields of glass-mirror heliostats, and with heliostats with as much as 65 m² of reflective surface (about 3,000 units in total with about 132,000 m² surface). This experience represents about 50 Mio million of heliostat operation.

Operational availability of the Solar One heliostat field remained above 96% in the yearly average (up to 99.7% per month), and averaged above 90% in all other tower SPPs operated routinely for extended time periods. However, due to inadequate grounding control, the electronics of heliostat units, subfields or the entire field proved vulnerable to damage by lightning effects (IEA-SSPS; Themis; Solar One), causing plant outages of several days (up to 15 operating days at Solar One).

---

[8] Sodium leak during valve replacement caused a serious fire in 1985 which destroyed the entire heat transfer, storage and plant control subsystems.

[9] CRTF = Central Receiver Test Facility; PSA = Plataforma Solar de Almería

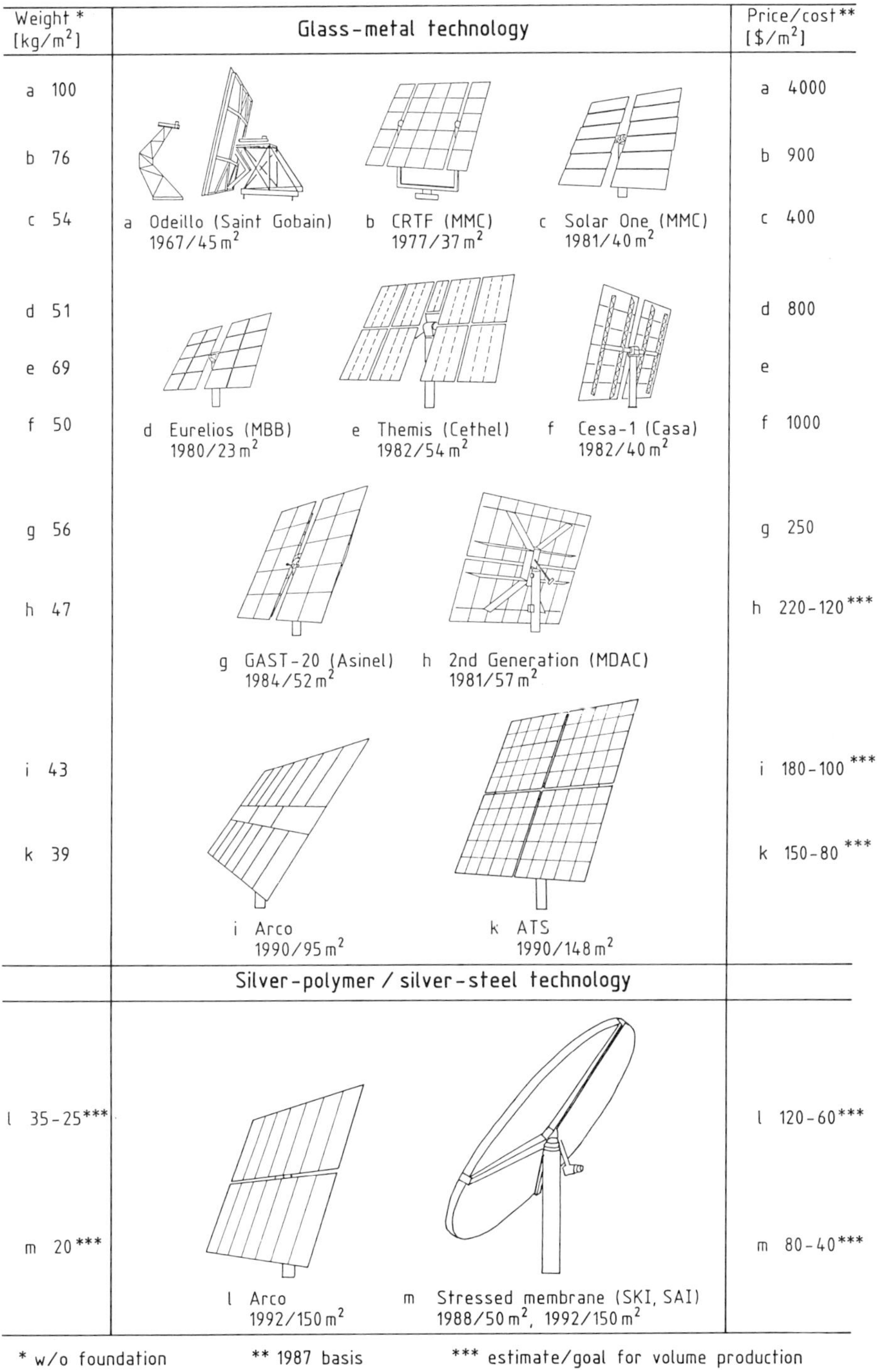

7.4 Central Receiver Solar Power Plants with Heliostat Fields 245

Weight* [kg/m²]	Glass-metal technology	Price/cost** [$/m²]
a 100		a 4000
b 76		b 900
c 54	a Odeillo (Saint Gobain) 1967/45 m² b CRTF (MMC) 1977/37 m² c Solar One (MMC) 1981/40 m²	c 400
d 51		d 800
e 69		e
f 50	d Eurelios (MBB) 1980/23 m² e Themis (Cethel) 1982/54 m² f Cesa-1 (Casa) 1982/40 m²	f 1000
g 56		g 250
h 47	g GAST-20 (Asinel) 1984/52 m² h 2nd Generation (MDAC) 1981/57 m²	h 220-120***
i 43		i 180-100***
k 39	i Arco 1990/95 m² k ATS 1990/148 m²	k 150-80***
	Silver-polymer / silver-steel technology	
l 35-25***		l 120-60***
m 20***	l Arco 1992/150 m² m Stressed membrane (SKI, SAI) 1988/50 m², 1992/150 m²	m 80-40***

* w/o foundation   ** 1987 basis   *** estimate/goal for volume production

Fig. 7.21. Status and trends in heliostat development (1988 status): configuration, technologies, specific weights and specific costs.

As high optical reflectivity is a key factor for optimum plant output performance, comprehensive reflectivity degradation measurements by soiling were carried out for the locations at Solar One (Barstow) and IEA-SSPS (Almería). Experience shows that mirror cleaning is unavoidable. Frequent washing can keep average reflectivities at 95%, requiring about 2% of heliostat investment cost annually for such maintenance [30]. Need and frequency for washing depends on local environmental conditions. Rain, if it occurs, effectively assists in rinsing off dust but is less effective in removing grime. Hence, cost-effective methods and procedures for the cleaning of reflective surfaces is an important issue when operating large tower SPPs.

Corrosion of the mirror reflective layer also was observed at Barstow and Almería. The corrosion growth rates and underlying causes were attributed mainly to moisture entering through protective paint layers and imperfections in the mirror edge seals. The number of affected mirror modules rose steadily but affected no more than about 1.5% of the total surface at Almería after five years (Fig. 7.22). Based on a limited but representative sampling, 0.061% of total reflective surface of the heliostat field was corroded at Solar One by mid-1986 (equivalent to about one heliostat). Hence, although mirror corrosion has had little effect on plant performance, the mirror corrosion history shows that protection of silvered mirror surfaces is an important issue for heliostat lifetime.

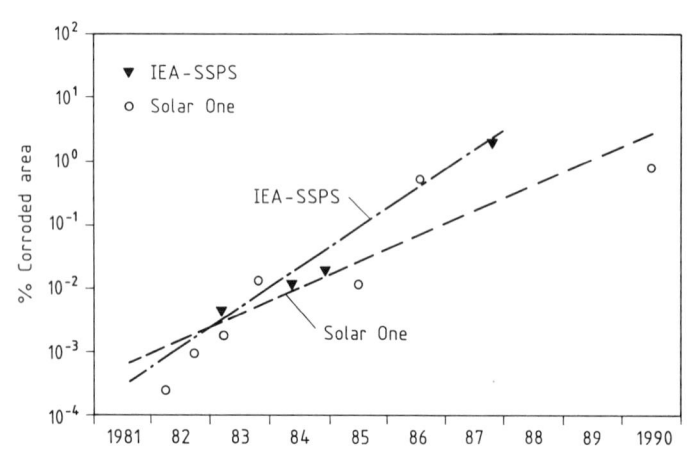

Fig. 7.22. Degradation experience of second-surface silvered glass mirrors at the Solar One and IEA-SSPS plant facilities

7.4.4 Plant Performance

At least a 2 year operation was achieved with nearly all tower SPPs, and much valuable experience has been gained from these activities. Although blurred at times as to what constitutes test, experiment or utility-like power production activity, much about characteristics and performance of the different tower SPPs and their subsystems and components was reported. But of all tower SPPs, only the 10 MW_e Solar One plant was operated for a sufficiently long time (> 6 years, 3 years of these in a power production mode), yielding a wealth of system-level experience, performance data, and lessons learned.

A plant availability of about 82% was demonstrated at Solar One, based on the aggregate of all time periods of operation or operational stand-by in the power production phase [60]. This is equivalent to 86% if lost hours are discounted during which output production was not possible due to bad weather (the design value was 90%). It is claimed that yearly overall availabilities up to 50% could have been achieved with the IEA-SSPS and the Themis plants with improved technology [24,29].

Maintenance activities were tracked at Solar One by computerized management systems used routinely by utilities. These records show that 60% of maintenance efforts was spent on preventive maintenance, and that 40% of maintenance costs was expended on solar-specific plant elements. By optimizing operation procedures, the original plant staff of 40 (for 7-day/3-shift operation) was pared down to about 20 at the end of the power production phase [60]; of this number, 8 persons would be needed for maintenance.

All tower SPPs were designed without the benefit of precursors, based solely on available irradiation data and knowledge about component and subsystem characteristics put to use in conventional power plants. It is not surprising, therefore, that original design and performance predictions were met only to a degree.

Solar One was designed with a SM = 1.0 using Barstow irradiation data for 1976 (8.0 kWh(DNI)/m^2d average). Assuming 100% equipment availability, design production was 26 GWh$_e$/a. Actual irradiation in the three years of power production not only was lower than in 1976 (by 16, 10 and 14%, respectively), but also remained lower than the 25-year irradiation average[10]. Accounting for actual irradiation and plant availability, the performance goal was adjusted to 15 GWh/a; about 10 GWh/a of net generation (1985/1986) were obtained in 7 days/week, 24-hour operation, equivalent to about 6% average annual energy efficiency (Tab. 7.10); highest monthly percentage of energy output to energy input was 9.8% (in August 1985) [59].

Highest capacity factors achieved in Solar One were 24% per month (in August 1985) and about 12% annually (Tab. 7.10), with little or no utilization of the thermal storage subsystem. Low and even negative capacity factors were observed in the low-irradiation months of December and January.

Table 7.10. Solar One power production performance, 1984-1987 (10 MW$_e$; 71,140m^2 total reflective area).

		1984–85	1985–86	1986–87
Average irradiation	kWh/m^2d	6.55	6.96	6.70
	MWh/m^2a	2.39	2.54	2.45
Gross generation	MWh$_e$	11,754	15,345	15,305
Parasitics	MWh$_e$	4,731	4,880	5,323
Net Generation	MWh$_e$	7,024	10,465	9,982
	kWh$_e$/m^2	98.74	147.10	140.32
Net annual eff.	%	4.13	5.79	5.73
Capacity factor		8.02	11.95	11.39

7.4.5 Plant Performance Characteristics

Relationships of daily energy input to net energy output were determined for Themis (2.4 MW$_e$), Solar One (10 MW$_e$), and IEA-SSPS (500 kW$_e$), based on observed performance as well as on performance estimates for improved plants (i.e. taking lessons learned from test and pilot plants into account) [24,29]. These input-output characteristics are contrasted

---
[10] Attributed to high volcanic activity in South America and strong South Pacific current anomalies in this time period

with the design performance of a hypothetical future 100 MW$_e$ solar tower plant of mature technology [1,4], assuming California irradiation conditions (Fig. 7.23).

Considering that performance can be improved when introducing storage and SM > 1, or that it is reduced as a consequence of transient output operation, extended operational stand-by periods or inclement weather, some general observations can be made:

- Solar One and (improved) Themis appear to have similar performance characteristics in spite of differences in capacity and primary heat transfer fluid. This experience contradicts recent study results [1,4] which indicate higher annual performance for large-capacity systems with liquefied metals/salts as primary HTF;
- daily energy input as high as 4 kWh(DNI)/m²d may be needed for net output generation using present technologies under SM = 1 conditions; energies collected below this level are consumed for covering (completely or in part) thermal losses and parasitics;
- with sufficiently large storage capacity, it is expected that as little as 2 kWh(DNI)/m²d may be needed as minimum input for large-capacity tower SPPs of advanced design, using molten-salt as primary heat transfer medium, SM = 1.6, and storage equivalent to 6 hour operation, provided cloud interference is not too high.

In essence, excellent direct irradiation conditions, and minimal thermal losses and parasitics both during operation and stand-by, are key design criteria for good net energy performance from solar input. Peak power performance, although of considerable technical interest, is an inadequate indicator for judging real system performance.

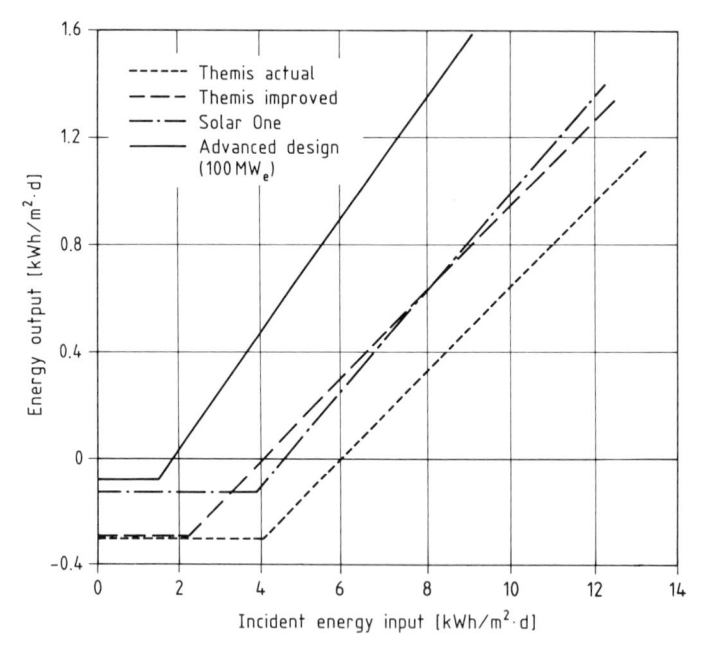

Fig. 7.23. Comparison of daily energy input-output characteristics for tower solar power plants (2.4 MW$_e$ Themis (actual and improved), 10 MW$_e$ Solar One (actual), 100 MW$_e$ advanced tower SPP design (SM = 1.6, 6 h storage) [1,4,24,29]).

7.4.6 Technological and Operational Potential

In addition to and mainly based on experience from operating experimental and pilot tower SPPs, several studies were undertaken to assess the performance of hypothetical systems of larger capacity and variances in plant configuration [1,4,31,73]. The improvements of performance expected to be achieved with mature technologies and optimal configuration is illustrated by comparing annual energy performances for the Solar One pilot plant and a future tower SPP with SM = 1.6 (Fig. 7.24). Expressed in terms of net annual energy per unit area of installed reflective area, a performance improvement from about 150 kWh$_e$/m²a to about 270–400 kWh$_e$/m²a is expected [29].

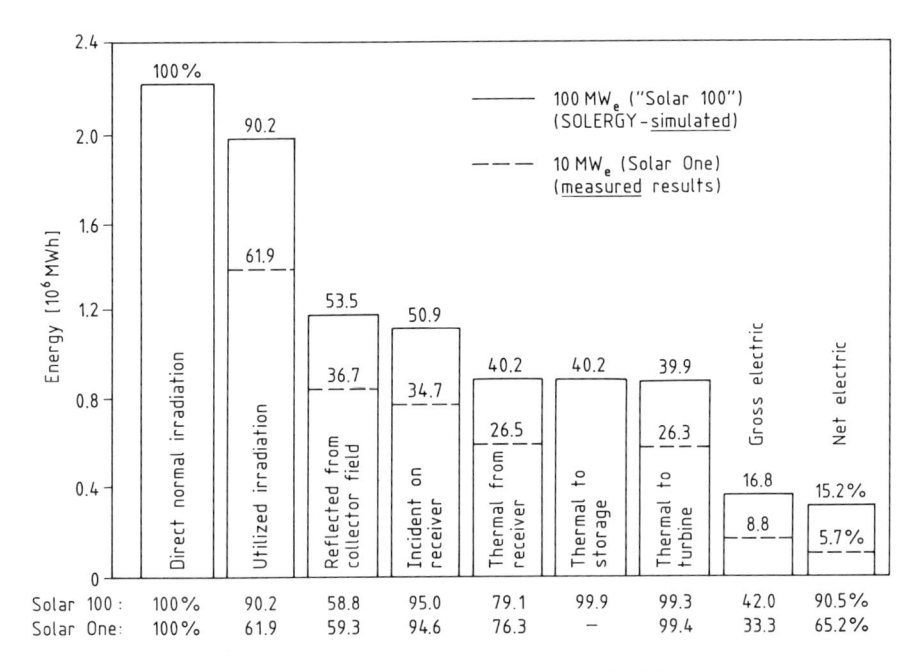

Fig. 7.24. Comparison of annual energy performance of the 10 MW$_e$ Solar One pilot plant and of a future 100 MW$_e$ tower plant with SM = 1.6 under the same irradiation condition (Solar-100) [29].

The following improvements on subsystem level are expected to contribute most to overall system performance:

- better utilization of available direct irradiation, attributable to higher availability of the heliostat field, better heliostat field performance and advanced receiver designs (mature technology; experienced maintenance);
- higher plant capacity factor by incorporating thermal storage and associated higher SMs;
- increase of thermodynamic cycle efficiency through optimal cycle parameters, a high degree of steady-state output operation, and larger power converters;
- reduction of internal losses and parasitic consumption.

Such performance improvements require continued efforts towards development of tower SPP technologies, and the accumulation of experience in operating plants in power production mode for extended time periods[11].

7.5 Individual Dish Solar Power Plants

Parabolic dishes can be designed to deliver electric energy directly by means of a PCU of appropriate size[12]. Each dish/converter assembly (or module) thus becomes a self-contained power producing unit. Several dish modules can be combined to form one SPP with their output collected electrically. The rating of such SPPs can be adapted to load needs and conditions of the local utility grid.

Inherent advantage of the individual dish/converter concept is that 2-axes tracking and high concentration/temperature offer the opportunity of using a high-efficiency power converter such as a Stirling engine. The constraints associated with heat transport over distances (thermal inventory, inertia and losses) are alleviated, but the capability of bulk thermal energy storage is lost. Need for precise collector contours, and for having to move sizeable masses (dish, PCU, support structures) when tracking the Sun, are further obstacles.

The unit size of individual dish/Stirling modules is defined by dish diameter, which is commonly adapted to available Stirling engine size. A limitation in unit capacity can be an advantage rather than a detriment: modularity results in high operational availabilities in multi-unit SPPs, economies of scale by volume manufacture, and fast feedback of operational experience from small-scale applications.

In any case, achieving optical precision with large, lightweight and non-rigid structures in dish/Stirling modules is a considerable engineering challenge and a requirement for cost reasons. This situation, together with the prototype Stirling engine development status, renders today's dish/Stirling modules still expensive in comparison to other solar thermal alternatives.

7.5.1 Configuration and Technology

Key element is the paraboloidal concentrator which is formed either by individual reflector elements held by a support structure, or by a continuous (but possibly subdivided) surface. The concave surface is covered by second-surface glass mirrors or by front-surface reflective (silvered or aluminized) films. Fig. 7.25 illustrates the diversity of construction approaches and technologies pursued in the past decade or yet under development (1987 status); Figs. 7.26 and 7.27 give views of historic, present and envisaged future dish configurations.

Positioning accuracy requirements for the power conversion subsystem (consisting of the receiver integral with thermodynamic converter, generator, and heat rejection device) increase with concentration. This may require engineering efforts to compensate gravity bending moments or thermal expansion of structural parts, or to sustain wind forces. The design challenges regarding structural integrity (particularly for large dishes of lightweight construction

---

[11] Performance demonstration with a tower SPP of > 10 MW$_e$ rating would be an essential intermediate step to ascertain the yet unverified performance prediction of > 400 kWh/m²a for tower plants of about 100 MW$_e$ and SM > 1.0. For this purpose, interested German, Swiss, Spanish and US industries formed a consortium in 1988 for pursuing the design, construction and operation of a 30 MW$_e$ solar tower facility (Phoebus project) [2].

[12] The fixed hemispheric bowl with a receiver moving in two axes for tracking a moving focal region belongs, strictly speaking, to the category of individual dish plants. Suffering from large cosine losses and extreme flux variations on the receiver (see Chap. 3), the hemispherical bowl concept has not been tested in an electricity-producing configuration, although a 10 m ⌀ prototype succesfully produced high-temperature steam.

7.5 Individual Dish Solar Power Plants

Weight * [kg/m²]	Glass-metal technology	Price/cost ** [$/m²]
1 162		1 1300
2 118		2 650
3 73	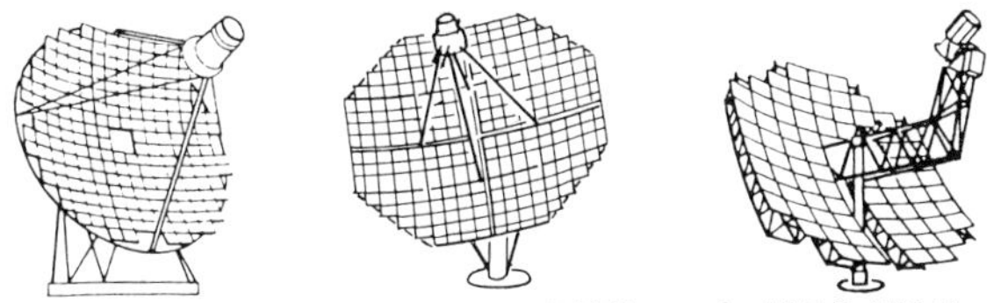 1 TBC (1977) ⌀ 11 m; C = 3000 2 Vanguard (1980) ⌀ 11 m; C = 2800 3 MDAC (1984) ⌀ 11 m; C = 2400	3 300–200
	Aluminized film technology	
4 74		4 ~1000 ***
5 56		5 ~300 ***
6 37	4 SKI (1980) ⌀ 7 m; C = 250 5 GE (future) ⌀ 12 m; C = 1000 6 LaJet (1986) ⌀ 7.4 m; C = 800	6 190–160
	Silver-polymer / silver-steel technology	
7 67		7 ~127 ***
8 55	7 Acurex (future) ⌀ 15 m; C = 1100 8 LaJet (future) ⌀ 15 m; C = 700	8 ~88
	Stretched-membrane technology	
9 52		9 <180 ***
10 30	9 SBP (1983) ⌀ 17 m; C = 600 10 SKI (future) ⌀ 15 m;	10 <140 ***

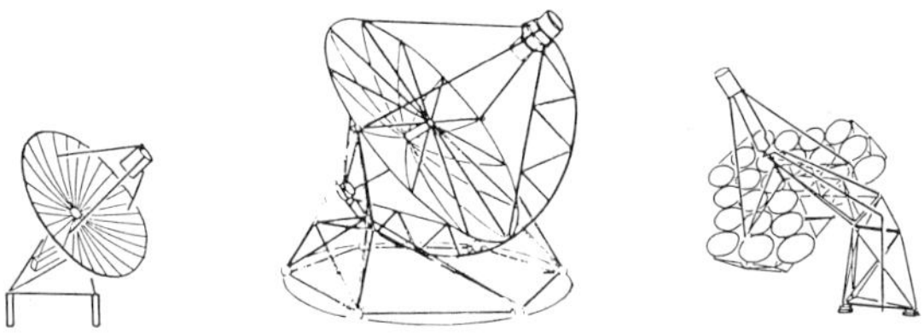

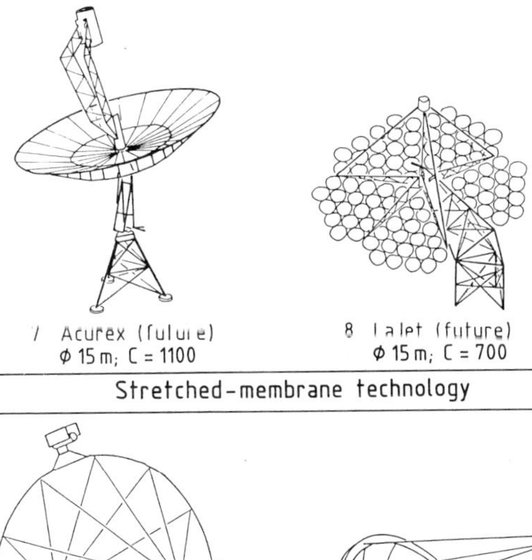

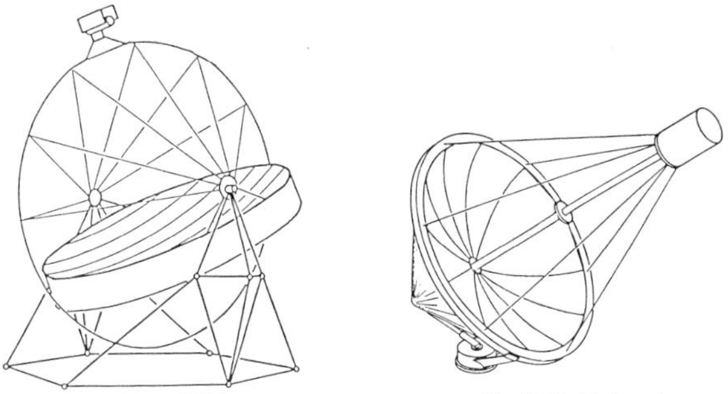

* w/o foundation   ** 1987 basis   *** estimate/goal for volume production

C = geometric concentration ratio

Fig. 7.25. Status and trends in parabolic dish development (1987 Status). Configuration, technologies, specific weights and specific costs.

Fig. 7.26. Views of historic, present and future dish configurations (Part 1). (a) parabolic dish, Mouchot/Paris (1878); (b) parabolic test bed concentrator, Edwards AFB/CA, USA (1977); (c) spherical bowl concentrator, Crosbyton/TX, USA (1980); (d) movable-slat point-focus SKI concentrator, Osage City, USA (1987); (e) individual-facet Vanguard parabolic dish, Rancho Mirage/CA, USA (1983).
ad (a) (courtesy of Olynthus-Verlag, Oberboezberg, Switzerland)
ad (b) (courtesy of SNL, Albuquerque/NM, USA)
ad (c) (courtesy of SNL, Albuquerque/NM, USA)
ad (d) (courtesy of SNL, Albuquerque/NM, USA)
ad (e) (courtesy of SNL, Albuquerque/NM, USA)

7.5 Individual Dish Solar Power Plants

Fig. 7.27. Views of historic, present and future dish configurations (Part 2). (f) individual-facet MDAC-25 concentrator, Los Angeles/CA, USA (1984); (g) individual stretched membrane facets concentrator, Warner Springs/CA, USA (1987); (h) large glass-metal membrane SBP concentrators, Solar Village, Saudi Arabia (1984); (i) fix-focus movable-segment Bomin-Solar concentrator, Freiburg, Germany; (j) 2nd generation 7.5 m ⌀ membrane SBP concentrator, Lampoldshausen, Germany.
ad (f) (courtesy of SNL, Albuquerque/NM, USA)
ad (g) (courtesy of SNL, Albuquerque/NM, USA)
ad (h) (courtesy of Schlaich, Bergermann und Partner, Stuttgart, Germany)
ad (i) (courtesy of Bomin Solar GmbH, Lörrach, Germany)
ad (j) (courtesy of DLR, Stuttgart, Germany)

and long focal length) may be exacerbated by engine vibrations (the interrelationship between focal length, rim angle and curvature of paraboloids is discussed in Chap. 3).

On the other hand, very high concentration is not necessarily superior; it has been shown [63] that dish/Stirling annual energy performance is not improved much beyond a concentration of about 1,500–2,000 (Fig. 7.28). Only a few Stirling engine designs exist and have been tested (see also Chap. 4), of which only a few engines are suited to operate in tilted positions (for reasons of lubrication)[13]. Brayton and Ranking cycle converters in dish applications have been assessed as alternatives to Stirling engines but proved inferior in terms of performance [8].

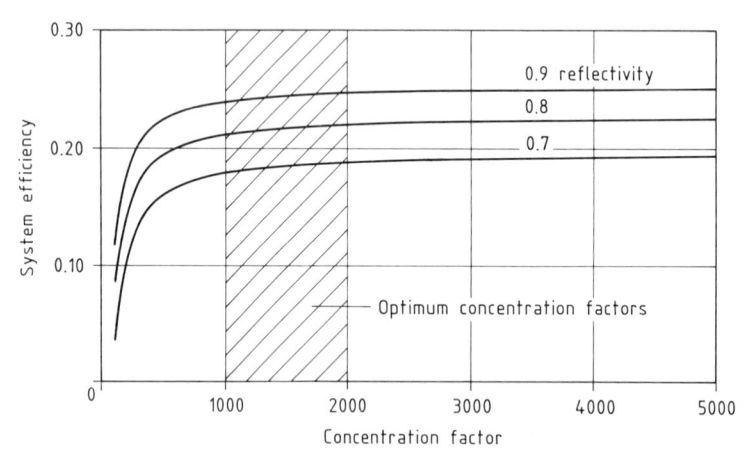

Fig. 7.28. Annual energy efficiency of dish/Stirling combinations as function of concentration ratio (35°C ambient; 600°C receiver; 0.95 intercept factor; 2,490 kWh(DNI)/m²a irradiation; 361 operating days) [63].

7.5.2 Dish/Stirling Examples

Vanguard 1 [14]. This 25 kW$_e$ prototype dish/Stirling module was operated at Rancho Mirage/CA in the Mojave desert in power production mode for an 18-month period (Feb 1984–Jul 1985, Fig. 7.26). The results of operational testing were analyzed and reported [23]. The 10.7 m ⌀, 86.7 m² dish with a 25 kW$_e$ Stirling/generator unit was designed for 20 kW$_e$ at 850 W(DNI)/m², 25 kW$_e$ at 1,000 kW$_e$(DNI)/m², and 200 W(DNI)/m² operating threshold (see Sect. 7.9 for further technical data). With over 30% peak power efficiency, this module achieved the best performance ever measured for any solar thermal power system.

MDAC 25 [15]. This 10.5 m ⌀, 91.5 m², 25 kW$_e$ dish/Stirling system was commercially developed (Fig, 7.27). Several units were built and operated by utilities at different locations in the U.S. In 1985, commercial development activities were terminated; operating experience was reported to a limited degree [19,41].

---

[13] A stationary Stirling engine placed on a tower and its receiver illuminated by heliostats in tower-SPP-like manner has recently been proposed within the collaborative Renewable Energy R&D program of the IEA (SSPS project), and was preliminarily tested using the IEA-SSPS tower facility in Almería, Spain.

[14] manufactured by Advanco Corp.

[15] McDonnell Douglas Astronautics Corp., Huntington Beach/CA, USA.

SBP 50 [16]. Two units with 17 m $\oslash$, 227 m^2 reflective stretched membranes with 50 kW$_e$ Stirling engines were built and operated near Riyad/Saudi Arabia and were in operation from 1986 to 1989 (Fig. 7.27). Peak net electric output to the grid exceeded 34 kW$_e$ under optimum conditions. The systems suffered from reliability problems but generally demonstrated the same operational characteristics as the other dish/Stirling systems. Total operating hours for both systems were over 4,000 hours. Cleaning of the thin-glass mirrors glued to the metal membranes proved particularly effective. As the collecting surface could be walked on, two men with dry brushes could improve reflectivity from the low 70% to over 90% in less than an hour. This demonstrated a key advantage of the glass-metal membrane reflector technique [7,54]).

7.5.3 Plant Performance

The Vanguard 1 experiment provides well documented and representative operating experience for single dish/Stirling units; supporting performance data were also provided by the SBP 50 tests in Saudi Arabia. Following statements and conclusions are based mainly on this information.

- Excluding times for planned maintenance during daylight hours, Vanguard 1 achieved 64% operational system availability based on total operating hours during the 18-month operating period (72% if based on operating days with irradiation sufficiently high for operation). To maintain this availability, continuous on-site presence of technical personnel was required. 38% of downtime was caused by Stirling engine malfunction, and 20% each by the need for dish, receiver and control system repairs.
- Almost immediate response (in the order of 1 minute) of electric output to thermal energy input was recorded, caused by the small thermal mass which is characteristic of compact PCUs such as a 25/50 kW$_e$ Stirling engine with a generator. As a consequence, rapid irradiation-following capability was verified in the Vanguard 1 experiment (Fig. 7.29). Stirling engine output is usually controlled by H$_2$-pressure variations. However, thermal lag is evident after the gas temperature drops below the set level; in this case time is needed to reach the temperature set level again, after which pressure control can take over again.
- Decrease in mirror reflectivity by soiling immediately impacts on the output. This becomes evident by slope differences in gross electrical power output as a function of direct irradiation input (Vanguard 1; Fig. 7.30). Reflectivity values have been found to differ as a function of location on the dish, rendering it difficult to determine overall mirror reflectivity accurately. It has been suggested [23] that the power slope method can serve not only as a detector for reflectivity changes but also as a tool for best characterizing the system performance, irradiation threshold for operation (x-intercept), and electrical power lost as a result of receiver, wind and mechanical losses (y-intercept).
- Both Vanguard 1 and SBP 50 provided information on the importance of structural strength and the interplay between optical accuracy and performance. Output of the SBP 50 module degraded under windy conditions (Fig. 7.31) due to flux spillage (observed) and increased receiver convection losses (not measured). Annual average wind speed at the SBP 50 test location was 3.9 m/s, leading to a calculated annual energy performance reduction of 15% as compared to calm conditions. For Vanguard 1 with specified beam accuracy of ± 1.25 cm (C = 2,500), a worst case deflection of 1.0 cm, due to the weight of the Stirling PCU, was calculated for a dish movement from 10° to 75° of elevation,

---

[16] SBP = Schleich, Bergermann und Partner, Stuttgart, Germany.

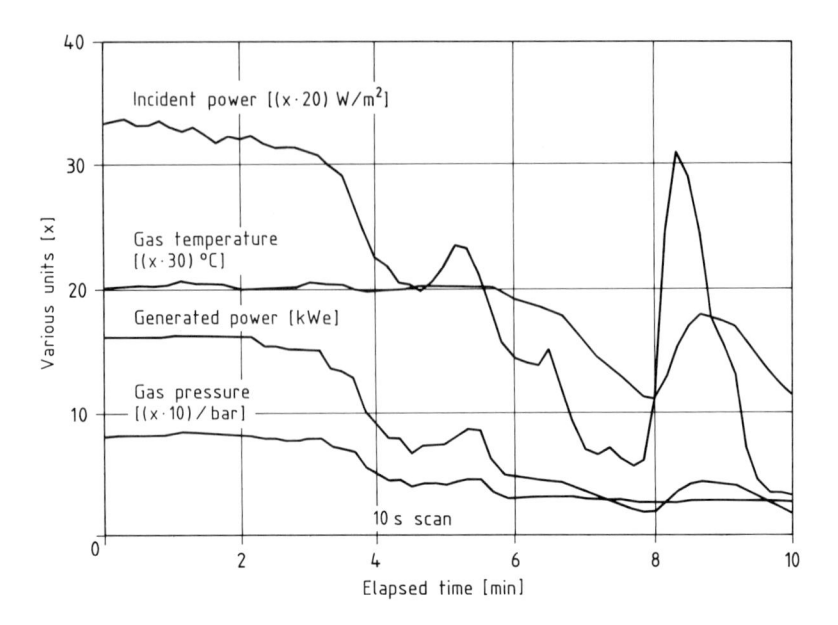

Fig. 7.29. Capability to respond to irradiation transients 25 kW$_e$ Stirling/generator power conversion block, Vanguard 1 module (relative scales) [23].

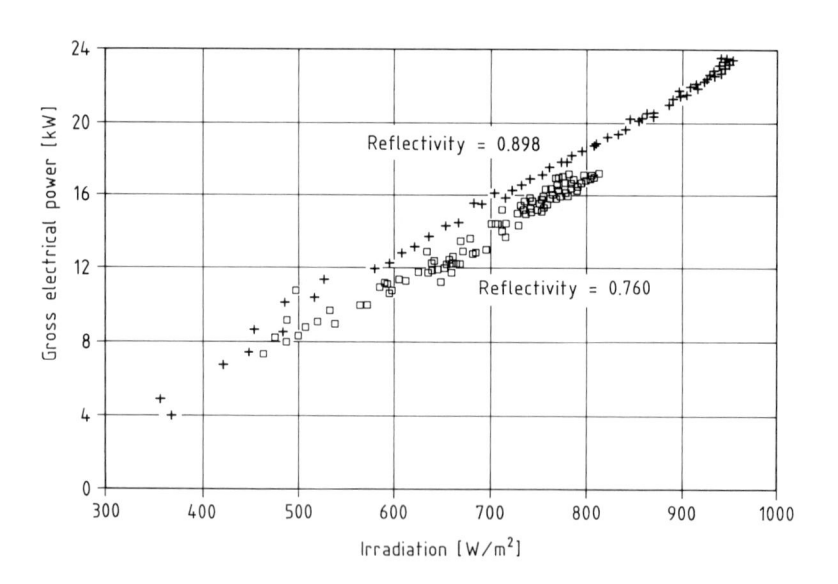

Fig. 7.30. Effect of dish reflectivity on electrical output as a function of direct irradiation, Vanguard 1 module [23].

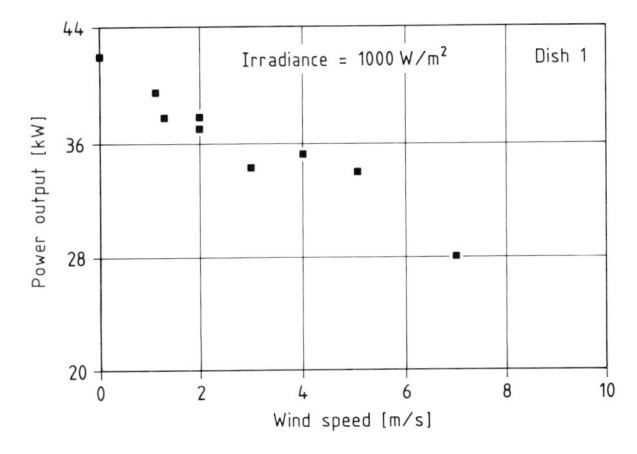

Fig. 7.31. Influence of wind speed on the performance of the SBP 50 kWe stretched-membrane dish/Stirling module [35].

although the engine block was mounted on a stiff tripod and was decoupled from the dish structure.
- Image deflection and convective effects (varying with dish elevation) contributed to uneven performance of the cavity receiver of Vanguard 1. However, energy flow to the four pistons is a prerequisite for mechanically smooth operation and for optimal conversion efficiency. Temperature differences over 100°C were measured between the four receiver quadrants, equivalent to a 15% difference in upper cycle temperature between the four pistons of the Stirling engine (Fig. 7.32)[17].

7.5.4 Plant Characteristics

Characteristics of daily energy input to net energy output have been determined using data of the Vanguard 1, MDAC 25 and SBP 50 dish/Stirling units [35,41,63] (Fig. 7.33). At all test locations, clouds interfered little with irradiation (i.e. with the direct radiation portion) during operating days during which data for these characteristics were accumulated. Observations regarding the development status achieved are:

1. at irradiation levels of 8 kWh(DNI)/m²d and 4 kWh(DNI)/m²d, daily net energy efficiencies have been achieved in the order of
   - 24% and 20%, with glass-mirror facetted dish constructions, concentration of 2,000–2,500, and 25 kW_e Stirling engine;
   - 14% and 8%, with stretched-membrane glass-mirror dish construction, concentration ratio of 600–800, and 50 kW_e Stirling engine;
2. 1.5–2 kWh(DNI)/m²d of direct irradiation must be available and must be collected daily before net electricity output can be attained with Stirling engines of 25–50 kW_e rating.

---

[17] Stiffness/rigidity problems and the need for air-cooled heat rejection is avoided in a tower-mounted, heliostat-illuminated "dish"/Stirling configuration, at the expense of cosine-losses of the heliostat field. However, thermal energy storage and SM > 1 become more readily feasible.

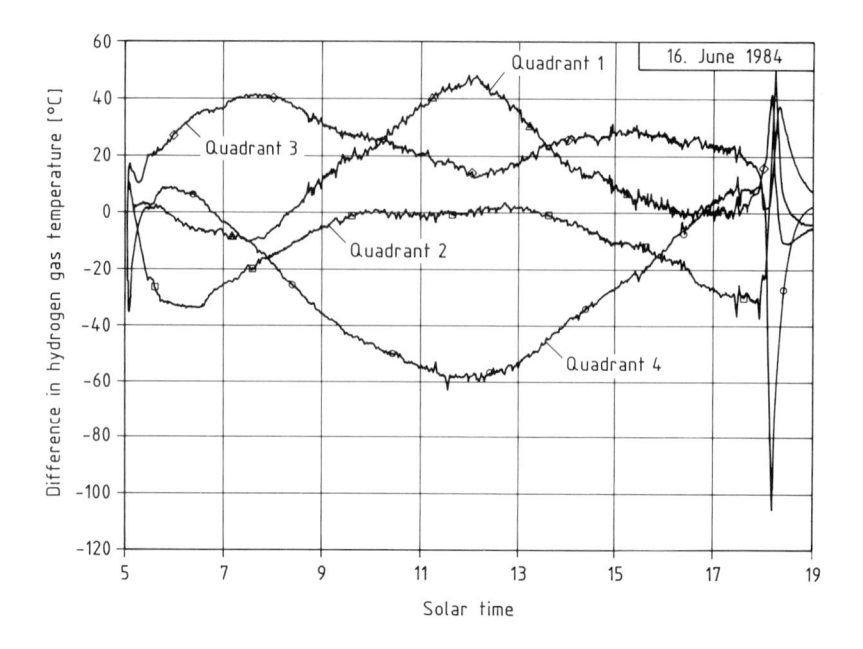

Fig. 7.32. Temperature variation between the cavity receiver quadrants of the Vanguard 1 dish/Stirling module [23].

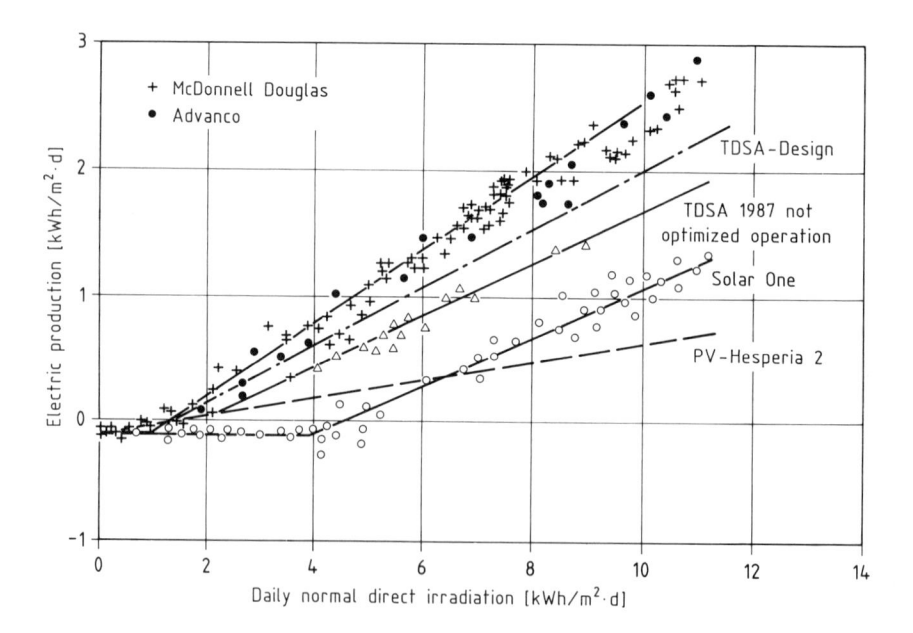

Fig. 7.33. Measured daily energy input-output performance of different dish/Stirling units [63].

7.5.5 Technological and Operational Potential

Although not yet representative of cost-optimized designs, performance results with the Vanguard 1 and MDAC 25 dish/Stirling modules have set standards with respect to energy efficiencies and operational thresholds to be matched by future advances in dish/Stirling technology, as well as by other concentrating solar thermal technology alternatives. Given adequate direct irradiation, it is nonetheless expected that dish/Stirling availability can still be further improved. Annual energy performance, if calculated on the basis of the input-output curves of Fig. 7.33 and Barstow irradiation, could reach 23–27% [63] (Tab. 7.11).

Table 7.11. Measured efficiencies and calculated limits of performance for dish/Stirling modules [63].

	Measured efficiencies (%)				Projected efficiency (%)
	peak power	net daily energy	net monthly energy	net yearly energy	net yearly energy
Vanguard 1	28–32	28–29	20–23	9–10	23–24
MDAC 25	30	28	12–20	–	25
SBP 50	17–18	9–11	–	–	14–15

7.6 Comparison of Thermal Solar Power Plants

Although an appreciable number of experimental, demonstration or even (emerging) commercial thermal SPPs has been installed, the performance assessment of SPPs is largely based – with the exception for the SEGS family of parabolic trough plants – on information which yet carries little statistical significance. Hence the following conditions prevail under which any comparison must be carried out:

- not all thermal SPPs were operated under similar, or even comparable, conditions (operating philosophy; irradiation; environment);
- the thermal SPP technologies are of different maturity, each technology – again with the exceptions of the SEGS plants – representing only the 1st or 2nd generation development status reached after 10–15 years of development efforts.

Hence, any comparison on the basis of actual performance cannot be but preliminary at present and, without doubt, is subject to future modification and refinements. Interpretation and conclusions of results should therefore be undertaken with appropriate caution and judgment.

7.6.1 Performance Comparison

Performance in terms of energy produced, and ultimately in terms of cost for production or of revenue achieved, are the bottom-line criteria for a comparison of power plant economics (see also Chap. 10). Lacking reliable information of that type, substitute criteria are helpful to convey technical merit (or superiority). In the past, peak power efficiencies of systems, subsystems and components were frequently employed for this purpose.

With caution, the transfer function based on aggregate daily net energy output to daily solar direct irradiation input may be used for an efficiency-based performance comparison (see Chap. 4). Such transfer functions can make technology-specific distinctions and relative performance differences apparent. Transfer functions do not convey a complete picture, however, as effects such as reliability, availability, durability, operating complexity, controllability, or O&M requirements of the technology involved, are not fully incorporated.

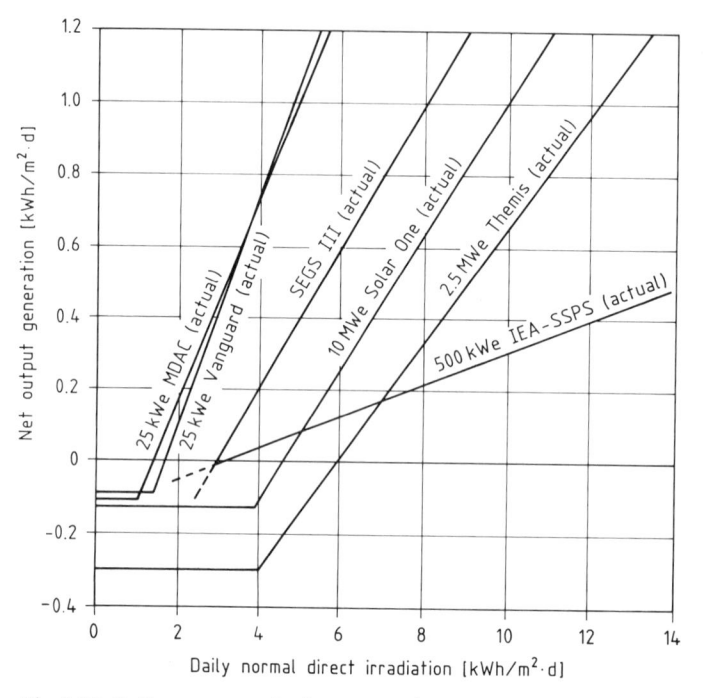

Fig. 7.34. Daily energy transfer (input-output) functions of thermal SPP alternatives (based on actual and experimental data).

The comparison of trough, tower and dish thermal SPPs of present-day and (expected) future mature design (Fig. 7.34) suggests the following conclusions:

- energies below technology-specific limits of accumulated daily direct irradiation cannot be utilized for output generation, but may contribute towards covering thermal losses and/or parasitic power needs;
- tower and farm SPP have higher thresholds for net output generation than dish/Stirling systems;
- slope of the transfer function above threshold is best for dish/Stirling systems, the slope being indicative of the quality of thermal energy transport, conversion efficiency, and thermal losses;
- by virtue of size, compactness and character of the conversion cycle, dish/Stirling modules exhibit the best performance in comparison to other alternatives, but are of relatively small single unit capacity.

7.6 Comparison of Thermal Solar Power Plants

A comparison of power production performance of the 10 MW$_e$ Solar One plant and the 25 kW$_e$ MDAC-25 dish/Stirling module, carried out over a period of 6 months of operation in 1986 at Barstow/CA, supports these findings (Fig. 7.35); with about 2,500 kWh(DNI)/m²a available irradiation, Solar One achieved the equivalent of 71 kWh/m²a, and MDAC-25 about 185 kWh/m²a.

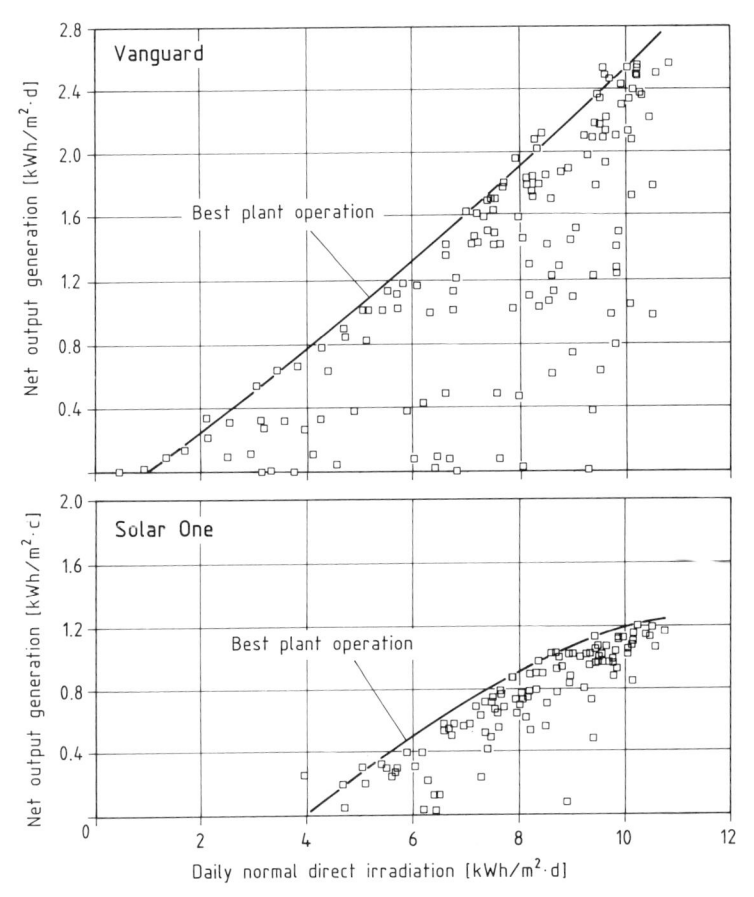

Fig. 7.35. Performance envelope of the Solar One tower plant and the MDAC-25 dish/Stirling module, both operating for six months in 1986 under identical irradiation conditions at Barstow/CA [38].

7.6.2 Long-term Operating Histories

Some thermal SPPs have been operated for extended periods under utility-like conditions: the commercial 13.4 MW$_e$/30 MW$_e$ SEGSs plants and the 10 MW$_e$ Solar One facility. By plotting the annual cumulative output generated in these periods, it is shown that consistent output performance could be achieved already with these first generation facilities approaching utility scale (Fig. 7.36). For a fair comparison, the contribution of natural gas to the output of the SEGS systems must be taken into account. On this basis, net annual performance of early trough and tower SPPs is comparable in terms of MWh$_e$ generated per kW$_e$ installed. Technical improvements are evident, looking at the performance of the more recent SEGS systems which took up operation only one or a couple of years after SEGS I. This fact demonstrates the significant influence of learning-curve effects in the rapid and continuous SEGS systems development. Other thermal SPP technologies so far did not benefit from such continuity. The maturity level of the thermal SPP technologies differs therefore at the present time (for further discussion on SPP technology comparison see also Chap. 10).

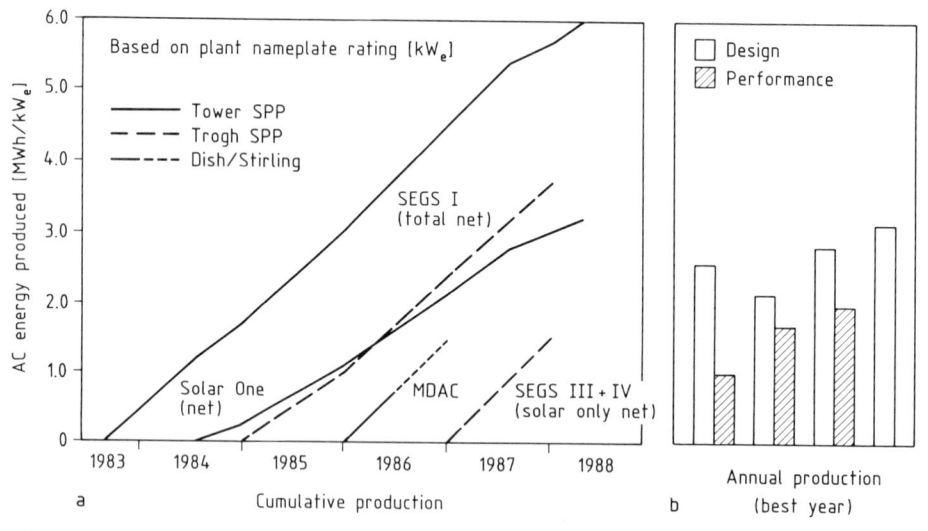

Fig. 7.36. Cumulative and annual specific performance (MWh$_e$ generated annually per kW$_e$ installed of thermal SPPs in continuous production operation.

7.7 Test Sites for Solar Thermal R&D

During 1970–1975, solar test facilities were established for high-temperature materials R&D in France (Solar Furnace, Odeillo) and in the U.S. (White Sands Proving Grounds). Between 1975 and 1985, additional test facilities were established in the U.S. for academic research (Advanced Component Test Facility (ACTF), Atlanta/GA) or in support of government-sponsored development of solar thermal technologies (Central Receiver Test Facility (CRTF) and Distributed Receiver Test Facility (DRTF), both at Sandia National Laboratories, Albuquerque/NM). The latest and most versatile entry (1986) is the Plataforma Solar de Almería (PSA), Almería, Spain, a German/Spanish collaboration utilizing hardware elements which were constructed in the framework of national (CESA-1) and multilateral activities (IEA-SSPS). Plans exist to convert the Solar One facility at Barstow/CA into a high-capacity test center for the advancement of tower SPP technology (see Tab. 7.14).

7.8 Thermal Solar Power Plant Modelling and Calculation Codes

Numerous models and calculation codes for thermal SPPs or for plant components/subsystems were developed by universities, research laboratories, and industry. The following list of short code descriptions is but a brief excerpt of the computation codes existing.

7.8.1 Performance Models

(a) Heliostats/Heliostat Fields

MIRVAL [47]

A Monte Carlo ray-tracing program that simulates individual heliostats and a portion of the receiver as it calculates optical performance of well-defined solar thermal central receiver systems. Created for detailed evaluation and comparison of fixed heliostat, field, and receiver designs, it accounts for shadowing, blocking, heliostat tracking, and random errors in tracking and in the conformation of the reflective surface, irradiation, angular distribution of incoming solar rays to account for limb darkening and scattering, attenuation between the heliostats and the receiver, reflectivity of the mirror surface, and aiming strategy. Three receiver types (external cylinder, cylindrical cavity with downface aperture, north facing cavity) and four heliostat types are included in the code (other heliostats or receivers can be evaluated by changing a small number of subroutines). Information needed includes a file that groups the heliostats in the heliostat field in a regular way; sunshape information; heliostat type, configuration, dimensions, and performance; receiver type and dimensions; zoning options; irradiation; attenuation; and start/stop times for a calculation.

        Availability: Model available to the public.
        Used by: Engineers.
        Language: Fortran.
        Machine: Mainframe.

HELIOS [71,72]

Evaluates the performance of central receiver heliostats, parabolic dish, and other reflecting collector systems. Calculations are made with up to 559 individual concentrators, or up to 559 cells containing multiple concentrators, and a single target surface. Uses cone optics to evaluate flux density. Safety considerations with respect to abnormal heliostat tracking can be evaluated. Effects included in detail in HELIOS are declination of the Sun, Earth orbit eccentricity, molecular and aerosol scattering in several standard clear atmospheres, atmospheric refraction, angular distribution of sunlight, reflectivity of the facet surface, shapes of focused facets, and distribution of errors in the surface curvature, aiming, facet orientation, and shadowing and blocking. HELIOS is used where a detailed description of the heliostat is available and an extremely accurate evaluation of flux density is desired. Input includes problem and output data, Sun parameters, receiver parameters, facet parameters, heliostat parameters, time parameters, and atmospheric parameters.

        Availability: Model available to the public.
        Language: Fortran.
        Machine: Mainframe.

DELSOL2 [21]

A performance and design optimization code that uses an analytical Hermite polynomial expansion/convolution-of-moments method for predicting images from heliostats. Performance is evaluated on the basis of zones that are formed by sectioning the heliostat field radially and azimuthally or on the basis of individual heliostats. Time-varying effects of irradiation, cosine, shadowing and blocking, and spillage are calculated, as are the time-independent effects attributable to atmospheric attenuation, mirror reflectivity, receiver reflectivity, receiver irradiation and convection, and piping losses. Optimization runs use a data base created by a performance run in order to determine field boundaries and a system configuration that is based on the lowest levelized energy cost for the total system. Many system sizes can be optimized in a single run. Input data include such parameters as site location, irradiation, sunshape; heliostat field information; heliostat information; receiver-related information; flux-related information; efficiency reference values for power, irradiation and convection, hot and cold piping losses; optimization-related input on heliostat density, receiver width and height; cost data; and economic analysis information.

>Availability: Model available to the public.
>Used by: Engineers.
>Language: Fortran IV.
>Machine: Mainframe.

NS cellwise performance code [48]

Evaluates the optical performance of a specified central receiver field to produce flux maps for external surround, cavity, and flat plate types of receivers. The flux map algorithm is based upon an analytical Hermite polynomial expansion/convolution-of-moments method for predicting images from heliostats. Receiver panel powers and gradients can be printed for each instant. Also, diurnal and annual performance data are generated.

Special timing sequences provide sunrise startup data and cloud passage data. In addition, 'Drift studies' provide for multiple flux maps with the heliostats either fixed (Sun drift) or slewing on either axis. A typical flux calculation takes less than 30 seconds on a VAX 780. Inputs are entered via three data modules: (1) heliostat, (2) collector field, and (3) receiver data modules. Default values based on Solar One are provided for many of the input requirements.

>Availability: Model available to the public for a nominal fee.
>Used by: Mechanical engineers.
>Language: Fortran.
>Machine: Mainframe.

HFLCAL [44]

Calculates optionally the annual performance of a specific configuration of heliostat field, tower, receiver and conversion cycle. The program also calculates, for a fixed tower height and specified design point, the heliostat field and receiver geometry yielding the maximal annual energy per m^2 of reflective area. By repeating the optimization procedure for different tower heights and weighting the plant data with given cost functions for plant components, the economic optimum can be determined. The flux reflected by a single heliostat is assumed to have a Gaussian distribution within a circular cone. The half width is computed from the sun-shape, the beam quality, the tracking error and the astigmatic aberration on root-mean-square (RMS) basis at each instant considered. Receiver and cycle are simulated by steady-state part load figures which provide the operation at constant mass flow or at constant temperature of the cooling media. One or two modular cavity receivers with elliptical or rectangular apertures, external flat plate or cylindrical receivers are optionally available. The HFLCAL output provides graphical representations of the heliostat field and of the receiver aperture/absorber in terms of isocontours of the efficiencies and of the flux distribution, respectively.

>Availability: Model available to the public at a fee.
>Used by: System engineers.
>Language: Fortran.
>Machine: Mainframe.

b) Trough Collectors

Optical analysis and optimization of parabolic trough collectors [13]

The results of a detailed optical analysis of parabolic trough solar collectors are summarized by universal graphs and curve fits. The graphs enable the designer of parabolic trough collectors to calculate the performance and optimize the design with a hand calculator. The method is illustrated by specific examples that are typical of practical applications. The sensitivity of the optimization to changes in collector parameters and operation conditions is evaluated. The variables affecting collector efficiency fall into several groups: (1) operating conditions (irradiation, tracking mode, operating temperature, flow rate), (2) properties of materials (reflectance, absorptance), (3) receiver type (absorber shape, evacuated or non-evacuated), and (4) concentrator geometry (concentration ratio and rim angle).

>Availability: Model available to the public.
>Used by: Parabolic trough designers.
>Machine: Programmable calculator.

7.8 Thermal Solar Power Plant Modelling and Calculation Codes

(c) Power Conversion

CYCLE3 [6]

Models and analyzes various Rankine-cycle energy conversion systems including supercritical cycles with or without regeneration and/or superheat. The program will accommodate a variety of working fluids, generating the required thermodynamic properties internally. Properties of 15 power-conversion working fluids are presently included in the program. CYCLE3 is an extension of an earlier program and has the additional capability to model cycles that include expansions in the two-phase ('wet-vapor') region. CYCLE3 is interactive and predicts cycle efficiency. Required input includes upper and lower cycle temperatures, upper cycle pressure, and efficiencies of system components.

> Availability: Model available to the public.
> Used by: Energy conversion engineers; mechanical, chemical, thermodynamic engineers.
> Language: Fortran.
> Machine: Mainframe.

DYNAG [34,39]

This code simulates the non-steady state operational behaviour of the 20 MW_e GAST reference central receiver solar power plant; it models the open air gas turbine cycle including compressor, receiver, pipes and mixing chambers/points, combustion chambers, gas turbine and waste heat boiler (air pressure losses). The bottoming water/steam turbine cycle does not influence the gas turbine cycle and hence is not part of the simulation. In addition, the dynamic process parameters, being objects of interest of the non-steady state analysis, occur primarily in the gas cycle. The gas is considered as a stationary medium for most of the operation modes in question. Subroutine REGEL represents the flow characteristics of valves for simulating control actions.

The code is useful primarily for the analysis of transient operation over very short time periods. Longtime performance (e.g. a complete day run) can also be calculated, but at the expense of considerable computing time.

Operating characteristics of components and boundary conditions of the operating regime are of fundamental influence on results. Using the code, start-up procedures, cloud transients, shutdown, cooling down at night, and load rejection can be analyzed (for pure solar, fossil support, and solar/fossil hybrid operation).

> Availability: Available to the public at a fee.
> Used by: System engineers.
> Language: Fortran.
> Machine: Mainframe.

(d) Tower solar power plant systems

RC cellwise optimization code [46,57]

Provides data for 'waterfall' charts, receiver flux maps, annual energy, design point power, and capital, operations and maintenance cost data for optimized systems. This optics model is designed so that input is entered through a main program and three data modules: (1) heliostat, (2) collector field, and (3) receiver data modules. Default values based on Solar One are provided for many of the input requirements.

A tool for use in design studies and performance assessment of central receiver heliostat fields and their interaction with the receiver. It provides the capabilities to analyze and optimize the collector-receiver system and to specify the coordinates of the individual heliostat field and boundaries of the resulting cost-optimized solar plant. Rather than to carry out a Monte Carlo analysis of the entire system in one step, the analysis proceeds in a step-by-step fashion through cost and performance of each major system element.

> Availability: Model available to the public for a nominal fee.
> Used by: Engineers.
> Language: Fortran 77.
> Machine: VAX 11/785, IBM.

ROSET (Representation of solar electric-thermal) [55]

A set of five computer programs designed to provide an estimate of the probable energy output of a solar thermal power system. The first four programs use hourly weather data for a year to calculate hourly electric energy output for a solar thermal system. The last program uses one or more electric energy output files created by the first four programs to provide an energy distribution for each hour of a typical day (one typical day per month).

Several types of collector systems can be simulated: point-focus central receiver, line-focus central receiver, point-focus distributed receiver, fixed-mirror distributed focus, line-focus distributed receiver-tracking collector, line-focus distributed receiver-tracking receiver, low-concentration nontracking collector, Fresnel lens, and shallow solar ponds.

This code was designed to function as part of a larger program package that can be used to determine the value of solar energy power systems (SEPS) to electric utilities. The results from ROSET, estimates of the electric production derived from the solar thermal power system, are provided so that the utility load forecast can be modified to incorporate the SEPS generation. The final program determines the break-even cost of each SEPS penetration and the SEPS marginal value, where value is the utility's present worth savings of reduced operating costs and modified capital additions. These values can be combined with total SEPS cost to determine the maximum amount of SEPS capacity that can be economically justified as an addition to a utility system.

>   Availability: Documentation available from the National
>   Technical Information Service (NTIS), Springfield/VA.
>   Used by: Utilities, small energy cooperatives.
>   Language: Fortran IV.
>   Machine: Mainframe.

GASBIE [27]

This code was designed to simulate the operation of the GAST-20 gas-cooled tower plant. Since this system has no solar storage, the optional use of fossil support is provided. The program requires as input clear irradiation data, the distribution of clear and cloudy periods day by day, and the heliostat field efficiencies as calculated by HFLCAL. The code determines, in user-specified time steps, the annual performance of the plant and the required fossil energy dependent on the required output profile. In addition, the thermal parasitics due to start-up and shut-down of the receiver, the fossil burner and the thermodynamic cycle are computed, whereas electric parasitics are determined in terms of gross electric power produced at every instant. The subsystems are specified by their part load curves including the dependence on the relevant fluid-mechanical and thermodynamical parameters under steady-state conditions. The program is structured as a constellation of routines with logical separate functions. System components, strategies of operation, etc., may be modified without fundamental structural changes to the program.

>   Availability: Available to the public at a fee.
>   Used by: System engineers, plant designers.
>   Language: Fortran.
>   Machine: Mainframe.

(e) Parbolic trough solar power plant systems

SIMSOL (Simulation of solar farm plants) [10,12]

A code for performance and design optimization of solar power plants with distributed receiver systems for process heat or electric power production. It is a quasi-steady state computer model that calculates both the daily and annual performance as well as special operational modes of solar farms with parabolic troughs. The time-step, usually 15 minutes, can be selected. For calculation purposes, the plant is divided into the collector field and the power conversion cycle which may be coupled by a storage system. Input data is direct irradiation data and other meteorological data, such as cloudiness, ambient temperature and wind speed. By means of a power balance calculation, the mass flow, temperature and energy of the thermal oilv can be obtained. Input data for the power conversion cycle is the industrial or urban load profile as well as the energy balance of the collector field with respect to the storage system. For every time-step which may be integrated over periods of a day, a month or a year, the following operational data, amongst others, may be obtained: irradiation, shading losses, heat losses from the collector, the piping and storage, internal consumption.

By means of SIMSOL it is possible to carry out an overall system optimization regarding among others the following issues: number of collectors, collector field geometry, mass flow and temperature spread of thermal oil, storage size. The operational behaviour of the plant may be modelled for characteristic days, as for example for solstices and equinox.

7.8 Thermal Solar Power Plant Modelling and Calculation Codes

Availability: Available to the public at a fee.
Used by: Engineers.
Language: BASIC.
Machine: HP 9845.

(f) Dish solar power plant systems

DBS (Dish-Brayton-system) [50]

From a design point, evaluates performance of a dish collector, Brayton-cycle engine, and receiver. Input includes air properties; design irradiation levels; receiver temperature; receiver irradiation properties; collector diameter, properties, surface errors, reflective properties; and gas turbine component efficiencies. Output for the dish is reflected energy; for the engine: efficiency, output in kilowatts, temperature of the components; for the receiver: efficiency, optimum receiver aperture and diameter, heat losses; and for the entire system: efficiency and power output.

Availability: Model available to the public.
Used by: Mechanical engineers.
Language: Fortran.
Machine: Mainframe.

ENERGY

This model predicts the annual performance of point-focusing concentrators, including the energy conversion subsystem (Organic Rankine, Brayton, or Stirling). The model uses as input the direct normal irradiation and ambient temperature for the site; size of plant; number of dishes; and properties of the concentrator, receiver, and energy conversion subsystems. The outputs produced include the monthly and annual energy produced, system efficiency, and cost per kilowatt-hour. The model has also been used to predict the performance of the buffer storage subsystem.

Availability: Model available to the public.
Used by: Engineers, economists.
Language: Fortran.
Machine: Mainframe.

(g) System comparison

Yearly average performance of the principal solar collector types [58]

Estimates the yearly total energy deliverable by the principal collector types: flat plate, evacuated tubes, CPC, collectors that track about one axis, collectors that track about two axes, and central receivers. The method is recommended for rating collectors of different types or different manufacturers on the basis of yearly average performance. The method is also useful for evaluating the effects of collector degradation, the benefits of collector cleaning, and the gains from collector improvements (due to enhanced optical efficiency or decreased heat loss per absorber surface). Only three variables are needed: (1) the operating threshold (for thermal collectors this is the average heat loss divided by the optical efficiency), (2) the geographical latitude, and (3) the yearly average direct normal irradiation. The rms deviation between the correlation and the exact results is about 2% for flat plates and 2% to 4% for concentrators.

Availability: Document available from NTIS.
Used by: Engineers, consultants, utilities, policy analysts.
Machine: Programmable calculator.

7.8.2 Economic Analysis Models

(a) Tower solar power plants

HELCAT (Heliostat cost analysis tool) [15]

A heliostat cost analysis tool that processes manufacturing, transportation, and installation cost data to provide a consistent structure for cost analyses. HELCAT calculates a representative product price based on direct input data including purchased and raw materials; consumables; direct labor hours per heliostat; land size; building size; equipment and tooling cost; quantity and other costs such as subcontracts, site-retained capital, and direct transportation charges; and various economic, financial, and accounting assumptions. A set of nominal economic and financial parameters that can be changed to describe a specific manufacturer's business practice is included for the user.

    Availability: Model available to the public.
    Used by: Utilities, economists, engineers.
    Language: Fortran.
    Machine: Mainframe.

SOLTES (Simulator of large thermal energy systems) [26]

A modular, heat-and-mass balance code that simulates the quasitransient or steady-state response of thermal energy systems to time-varying data such as weather and loads. The modular form of SOLTES allows complex systems to be modeled. Component performance is described by thermodynamic models. A pre-processor is used to construct and edit system models and to generate the main SOLTES program. It is possible to realistically simulate a wide variety of thermal energy systems such as solar, fossil, or nuclear power plants; solar heating and cooling; geothermal energy; and solar hot water.

    Four categories of input data are required to run SOLTES: (1) component/information routine data, (2) executive routine control data, (3) fluid loop definition data, and (4) fluid property data. Weather data and load data are optional categories.

    Availability: Model available to the public.
    Used by: Engineers.
    Language: Fortran.
    Machine: Mainframe.

SOLERGY [68]

Simulates the operation and power output of a user-defined solar central receiver power plant for a time period of up to one year.

    Recorded or simulated weather data and plant component models are utilized to calculate the power flowing through each part of the solar plant and account for thermal and electric parasitics both during the day and overnight. A plant control subroutine monitors these powers and determines when to start-up/shut-down all the major plant subsystems. There are two different plant control subroutines available – output maximizing and value maximizing. The user selects one of them to control plant operation for the entire simulation run. Parasitic electrical power is computed on a 24-hour basis.

    The program is based on an improved and expanded version of a solar central receiver annual energy simulation code called STEAEC, and was renamed SOLERGY.

    Availability: Model available to the public.
    Used by: System engineers, economists.
    Language: Fortran IV.
    Machine: Mainframe.

ASPOC (A solar plant optimization code) [61]

This program was developed in the framework of the GAST-20 Technology Program. Its main objective is to perform a fast and economic calculation to optimize all parameters of a solar power plant. Typical applications are parametric analysis of heliostat characteristics, field lay-out, tower, receivers, thermal cycles, etc. The task consists of finding the combination of parameters that maximize the annual thermal production per unit of the reflecting surface, or minimize the annual energy levelized cost for solar plants with cavity or cylindrical receivers, and with thermal storage or fuel support. The method consists on a stepping procedure to maximize the absolute value of the desired function. It uses a selective directed search of a surrounding n-dimensional grid of points in order to find the increasing direction, given an initial estimation. The procedure is repeated until the improvement is small enough. The code is organized in modules to facilitate software maintenance and the updating of cost and efficiency subroutines.

    Availability: On request (Asinel, P.O. Box 233, 28030 Mostoles, Spain).
    Used by: System Engineers.
    Language: Fortran 77.
    Machine: Mainframe.

7.8 Thermal Solar Power Plant Modelling and Calculation Codes

(b) Cogenerating solar power plants

STESEP (Solar total energy system evaluation program) [51]

Provides trade off evaluations relating to cascading thermal power conversion systems, determination of optimal collector sizes and operating conditions, and comparison of solar total energy concepts in various types of commercial buildings in different parts of the country. The program can be used either for a deterministic approximization of weather conditions or hourly weather data. It determines component demand loads based on steady-state energy balances to meet the load requirements for electrical, heating, ventilation, and air-conditioning, and process heat demand loads for average day conditions. Component sizes and use are estimated to determine yearly energy use. System capital and return costs are evaluated to determine the life-cycle cost. The program provides yearly energy use and life-cycle cost of the system. The user must provide electrical, heating, ventilation and air-conditioning, and process heat demand loads; system capital and recurring costs; component parametric data; and some meteorological information.

    Availability: Limited to members of the National Energy Software Center.
    Fees vary.
    Used by: Systems designers, engineers.
    Language: Fortran IV.
    Machine: Mainframe.

(c) Trough solar power plants

ECOSOL (Economy of solar power plants) [11]

A code for evaluating the economic performance of renewable energy plants such as solar thermal plants. Inputs include investment costs, operating and maintenance costs, technical performance (rated energy output, life time of plant), costs for competitive fuel, financing arrangements and economic conditions. Results are the following economic criteria: payback period, internal rate of return, life cycle costs, investible expenditure and solar energy costs. For the calculation a discount method is used. All calculations may be made for different investment years and using various assumptions for development of plant, operating and fuel prices such that predictions on the 'break-even point' of the solar investment can be made.

    Availability: Available to the public at a fee.
    Used by: Engineers, economists.
    Language: BASIC.
    Machine: HP 9845me.

(d) Plant comparison

SOLSTEP (Simulation of localized solar thermal electricity production) [14]

A tool for evaluating the thermodynamic and economic performance of alternative solar thermal power plants. Inputs include meteorological tapes and irradiation data (ranging from daily to every 15 minutes), thermodynamic performance parameters (optical efficiency, heat transport efficiency, energy conversion efficiency, and energy storage efficiency), and solar system costs (capital, operating, and maintenance). Output is annual power production estimates, levelized energy cost for point estimates, and a variety of capacity factors for optimized systems.

    Availability: Model available to the public.
    Used by: Large electric utilities, researchers.
    Language: Fortran.
    Machine: Mainframe.

(e) Economics

BUCKS – Economic analysis model for solar electric power plants [16]

Developed for the economic analysis of solar thermal central receiver technology in utility networks and for comparative evaluation of alternate plant designs. The model calculates power production costs for a single solar thermal central receiver plant. Specifically, it calculates levelized busbar energy costs; it does not include transmission and distribution costs or other indirect utility costs.

Two types of data are required: (1) plant performance and size information, and (2) cost and economic data. Data are supplied for a reference plant where detailed costs are known. A plant similar in design but different in size is then described with input variables that include net electrical generation from the plant: Collector, receiver, turbine, and storage capacities; parasitics required; storage charging/discharging rates; and operational hours during the year. Cost and economic information is provided in FORTRAN data statements as follows: escalation rates, years of escalation, tax rate, construction years, capital form and rates, year of commercial operation, and indicator to use or charge the referenced plant parameters. The referenced design cost and economic data which are contained in the subroutines for as many as three designs of six subsystems include cost estimates of each of the 33 solar-related items; cost estimates of each of the 27 nonsolar-related items; costs of each of the seven O&M tasks; estimated plant lifetime; contingency, insurance, and property tax; O&M escalation rate; and storage media replacement.

>Availability: Model available to the public.
>Used by: Planners, economic analysts.
>Language: Fortran.
>Machine: Mainframe.

CASSOL (Cash flow analysis for solar plants) [9]

A code for evaluating the economic performance of renewable energy plants such as solar thermal plants. Inputs include investment and capital costs, revenues, operating and maintenance expenses, financing arrangements, economic conditions and various income tax effects, as for example depreciation and tax credits. Results of the year-by-year cash flow analysis are the net cash flow as well as the discounted cash flow. Both internal rate of return and payback period are calculated. These economic criteria are helpful for investment based decisions.

>Availability: Available to the public at a fee.
>Used by: System engineers, economists.
>Language: BASIC.
>Machine: HP 9845.

Levelized energy cost model [22]

An indicator of average cost of energy (standardized to a levelized busbar equation) over the life of a system that enables the user to compare energy systems on a cost basis. Inputs are five separate capital costs, operations and maintenance costs on an annual basis, capital replacement costs, tax rate for the state in question, insurance rate, discount rate, and general inflation rate. Final output is the levelized cost of energy in dollars per kilowatt hour.

>Availability: Model available to the public.
>Used by: Utilities, solar system contractors, economists, solar researchers.
>Language: Fortran (with BASIC and PI versions).
>Machine: Mainframe, microcomputer, and programmable calculator (HP and TI).

The Model [74]

Equally applicable to large and small systems, this model is designed to assess the economic attractiveness of investing in solar thermal power plants. It has the capacity to analyze a variety of alternative financing methods. Inputs include capital costs, loan interest and amount, lease payments, federal and state tax rates and credits, and a depreciation schedule for solar and non-solar equipment. Output is a year-by-year statement of net cash flow from the project and a yearly and total calculation of the effective net present value.

>Availability: Model available to the public.
>Used by: Researchers, utilities, entrepreneurs.
>Language: Visicalc, SuperCalc.
>Machine: IBM PC.

7.9 Appendix: Solar Thermal Facility Data

Table 7.12. Central receiver (tower) solar power plants (electricity generation).

Name		Solar One	Themis	CESA-1	IEA-SSPS	Sunshine	Eurelios	SES-5
Location		Barstow	Targasonne	Almeria	Almeria	Nio	Adrano	Kertsch
Country		USA	France	Spain	Spain	Japan (Shikoku)	Italy (Sicily)	USSR (Crimea)
Design Conditions								
Coordinates	deg	35 N	43 N	37 N	37 N	34 N	38 N	45 N
	deg	117 W	2 E	2 E	2 E	134 E	15 E	36 E
Altitude	m	593	1,660	500	500	6	215	.
Design								
Time	Sol.Time	12:00	Equinox	W. Solstice	Equinox	L. Solstice	Equinox	.
			12:00	10:00	12:00	14:00	12:00	
Irradiance	W(DNI)/m^2	950	1,040	700	920	750	850	800
Solar Multiple		1.00		1.93 (S.S.)			0.75	.
Output	MW_e	10.0 (S.S.)	2.4	1.0	0.5	1.0 (gross)	1.0	5.0 (?)
	MWh_e/a	26,000	3.0	3.988	.	517	.	5–7,000
Efficiency	% net	17.4	20.5	12.9	16.5	10.3	14.5	.
Plant area	ha	52.6	.	.	.	.	.	24
Project Data								
Operation Start		1982	1982	1983	1981	1981	1981	1985
Construction Time	months	30	36	36	18	32	.	.
Cost	Mio$(US)	141	37	18	18	25.0	8.2	.
O&M Staff	Manyear	36	37	32	27.5	(10)	20	.
O&M Cost	Mio$(US)/a	2.97	2.53	.	.	2.1	.	.
Status (end of 1989)		concluded	concluded	test fac.	test fac.	dismantled	concluded	test fac.
Technical Information								
Heliostat Field								
Manufacturer		MMC	Cethel	Casa/Sener	MMC/MBB/Asinel	Mitsubishi	MBB/Cethel	.
# of Units		1,818	201	300	93	807	112/70	1,600
Refl. Area	m^2	71,095	10,740	11,880	3,655	12,912	2,576 + 3,640	40,000
Field Area	ha	29.1	.	7.7	.	.	3.5	15
Target Height	m	90.8	106	60	43	69	55	80
Receiver		External	Cavity	Cavity	Cavity/Extern.	Cavity	Cavity	External
Manufacturer		Rockedyne	CNIM	TR/Babcock	IA/Sulzer	Mitsubishi	Ansaldo	.
Rating	MW_t	43.4	8.9	7.7	2.5	56.7	.	5.0
Aperture	m^2	302	16.0	11.6	9.7/8.3	.	15.9	154

7.9 Appendix: Solar Thermal Facility Data

Heat Transfer								
Medium		2-loop	2-loop	1-loop	2-loop	1-loop	1-loop	.
		Water/Steam	Salt	Water/Steam	Sodium	Water/Steam	Water/Steam	Water/Steam
Temperature	°C	104/516	250/450	190/525	270/530	115/249	37/512	256
Pressure	bar	105	1.0–2.0	132/108°C	6.0	40.0	62.0	40
Power Conversion								
Engine		2-inlet	.	2-inlet	6-pist/5-stage	Impulse Sat.	2-inlet	.
		Condens. Turb.	.	Condens. Turb.	Steam Motor	Steam Turbine	Condens. Turb.	Condens. Turb.
Manufacturer		GE	Alstrom	Siemens/Bazan	Spilling	.	Ansaldo	.
Medium		Water/Steam	Water/Steam	Water/Steam	Water/Steam	Water/Steam	Water/Steam	Sat. Steam
Temperature	°C	45/510	89/410	55/520(330)	193/500	115/187	36/510	256
Pressure	bar	0.09/100	0.19/40	0.15/98	0.3/100	12.0	64.0	40
Generator Rating	kVA	12,500	2,300	1,500	617	.	1,100	.
Storage								
		Oil/Rocks	2-tank Salt	2-tank Salt	2-tank Sodium	Water/Steam	Water/Salt	Press. Water
Capacity	MWh_e	28	12.5	3.5	1.0	3.0	.	140
	MWh_t	182	40.0	16.0	5.5	17.8	0.036	.
Temp. Range	°C	218/302	250/450	220/340	275/530	197/249	275/430	.
Performance								
Average								
Irradiation	kWh/m²d	6.97 ('85/'86)	4.51 ('84)	.	5.37	.	.	.
Threshold	W(DNI)/m²	450	300	.	300	250	450	.
Production	MWh_e/a	15,350 (gross)	574 (gross)	.	39 (gross; '84)	517 (calc.)	130 (MWh; gross)	.
Yearly Eff.	% gross	8.49	12.4 (calc.)	.	.	.	.	.
	% net	5.78	10.4 (calc.)	.	.	.	.	.
Peak								
Output	MW_t	37.0	.	.	.	.	.	.
	MW_e	11.7	.	.	.	0.8	0.75	.
Production	MWh_e/d	87.0	.	.	.	2.4 (gross)	.	.
Efficiency	% gross	.	19.0	27.0	9.5	9.2	.	.
	% net	8.7	17.0	.	8.1	.	.	.
Monthly CF	%	24	.	.	.	.	.	.

* achievable

S.S. = Summer Solstice

Table 7.13. Trough farm-, dish farm- and individual dish solar power plants (electricity generation).

Name		STIP 150	IEA-SSPS	SEGS 1	SEGS 3/4	SEGS 5	TDSA	Solar Plant 1
Location		Coolidge	Almeria	Barstow	Kramer Jct.	Kramer Jct.	Al-Jubaylah	Warner Springs
Country		USA	Spain	USA	USA	USA	Saudi Arabia	USA
Design Conditions								
Coordinates	deg	33 N	37 N	35 N	35 N	35 N	25 N	33.3 N
	deg	112 W	2 W	117 W	118 W	118 W	47 E	116.7 W
Altitude	m	434	500	600	.	.	685	823
Design								
Time	Sol.Time	.	Equinox	.	.	.	12:00	.
Irradiance	W(DNI)/m²	599	920	.	.	.	1,000	.
Solar Multiple		0.7	1.0	.	.	.	.	.
Output	kW$_e$	103 (170)	500	13,800	30,000	30,000	2 × 50.0	4,880
	MWh$_e$/a	.	.	30,630	85,052	92,400	.	12.37
Efficiency	% gross	.	10.3	11.0	11.0	11.0	.	16.4
	% net	4.0	10.1	.	.	80.5	23.1	14.8
Plant area	ha	.	.	30	73.8	.	.	13.5
Project Data								
Operation Start		1979	1981	1984	1986	1987	1986	1984
Construction Time	months	.	.	15	10	10 (est.)	5	12
Cost (Year)	Mio$(US)	5.5 ('78)	12.2 ('81)	72.0 ('84)**	104.0 ('86)**	123.7 (est.)**	1.5 ('84)	18.6
O&M Staff	Manyear	0.6	14	25	15	15 (est.)	3	5
O&M Cost	k$(US)/a	14.940	.	.	3,200	2,000 (est.)	.	.
Status (end of 1989)		inoperative	test fac.	operating	operating	operating	inoperative	test operation
Technical Information								
Collector Field		Throughs	Throughs	Throughs	Throughs	Throughs	Dishes	Dishes
		1-axis/E-W	1- + 2-axes/E-W	1-axis/N-S	1-axis/N-S	1-axis/N-S	2-axes	2-axes/polar
Manufacturer		Acurex	Acurex + MAN	LUZ	LUZ	LUZ	SBP	LaJet
# of Units		384	40 + 84	740	868	992	2	700
Refl. Area	m²	2,140	2,674 + 4,928	82,960	203,980	233,120	454.0	30,261
Field Area	ha	.	2.80	.	.	.	0.12	13.5
Receiver		line foc.	line foc.	line foc.	line foc.	line foc.	external	cavity
Manufacturer		Acurex	Acurex/MAN	LUZ	LUZ	LUZ	USAB	LaJet
Rating	MW$_t$	.	3.66	.	.	.	2 × 0.144	0.04

7.9 Appendix: Solar Thermal Facility Data

Heat Transfer							
Medium		Caloria HT-43	Santotherm 55	Esso 500	Monsanto VP-1	Monsanto VP-1	Hydrogen	Water/Steam
Temp. max/range	°C	288/200	295/225	307	349/248	349/248	700	277/371
Pressure	bar	9	25				25–180	
Power Conversion								
Engine		single-inlet	single-inlet	single-inlet	dual-inlet	dual-inlet	Kinematic	single-inlet
		ORC Turbine	Condens. Turb.	Condens. Turb.	Condens. Turb.	Condens. Turb.	Stirling	Condens. Turb.
Manufacturer			Stal-Laval	Mitsubishi	Mitsubishi	Mitsubishi	United Stirl.	Stal-Laval
Medium		Toluene	Water/Steam	Water/Steam	Water/Steam	Water/Steam	Hydrogen	Water/Steam
Temp. HP/LP	°C	268/41	280/39	415***	510/327	510/327	700	360
Press. HP/LP	bar	10.3/0.1	25/0.07	36	104.5/43.4	104.5/43.4	25–180 (range)	41.4
Generator Rating	kVA	250	713	14,700	38,000	38,000	50.0	3,680 + 1,240
Fossil Backup		1.035 (gas)	None	8.5 (gas)	14.6 (gas)	14.6 (gas)	None	None
Thermal Storage	MW$_t$	Caloria HT-43	Santotherm 55	Esso 500	None	None	None	None
Capacity	MW$_e$	0.9	0.8	37.0				
	MW$_t$	3.75 (MWh$_t$)	2.5					
Temp. Range	°C	200/288	225/295					
Performance								
Average								
Irradiation	kWh/m²d		6.2	6.9 ('85)	6.79 ('86)	6.79 ('86)	6.3	7.02
Threshold	W(DNI)/m²	300	300	300	300	300	300–400	300
Production	MWh$_e$/d	1.05 (netpeak)	153.4					
	MWh$_e$/a	178 (gross)		13,980 ('88)*	82,714 ('88)*	77,405 ('88)*		
Yearly Eff.	% gross	3.5	4.3					
	% net	2.45	2.5	3.1*	11.1*			
Peak								
Output	kW$_t$							20,330
	kW$_e$			15,000	103,800			4,800
Efficiency	% gross	5.4						
	% net	4.3					18.0	
Monthly CF	%							
Fossil Use	GWh$_t$/a	none	none			63.2 ('88)	none	
					63.9 ('88)			

* hybrid input
** includes financing/cost of private investment
*** with gas superheater

Table 7.14. Trough farm-, dish farm- and central receiver (tower) solar power plants/projects (cogeneration and process heat production).

		STEP 100	Sulaibyah	STEP	Sonntlan	Soleras	Al-Ain	Fairfield
Name		Meekathara	Sulaibyah	Shenandoah	Las Barrancas	Yanbu	Dubai	Fairfield
Location		Australia	Kuwait	USA	Mexico	Saudi Arabia	Un. Emirates	USA
Country								
Design Conditions								
Coordinates	deg	27 S	29 N	33 N	26 N	24 N	23.75 N	40 N
	deg	110 E	48 E	85 W	112 W	38 E	55.5 E	112 E
Altitude	m		95	280	6			
Purpose		Cogeneration	Cogeneration	Cogeneration	Cogeneration	Desalination	Desalination	Enh. Oil Rec.
Design			Equinox					every day
Time	Sol.Time		12:00	12:00				9:00–15:00
Irradiance	$W(DNI)/m^2$		970	630	700			950
Solar Multiple						./		N/A
Output	kW_e	100	100	400	117		500***	N/A
	kW_t	470	424	3,000			300	1.0
Efficiency	% net	7.7	10.2	14.5	6.5			
Plant area	ha							1.6
Project Data								
Operation Start		1982	1981	1982	1983	1983	1984	1982
Construction Time	months		16	12		18		10
Cost (Year)	Mio$(US)	2.5	14.8		6.8			
O&M Staff	Manyear	3	2					0.5
O&M Cost	k$(US)/a							
Status (end of 1987)		inoperative	test fac.	test fac.	test fac.	inoperative	operating	inoperative
Technical Information								
Heliostat Field		Trough/2-axes	Dish	Dish	Trough/2-axes	Dish	Trough/1-axis	Heliostats
Manufacturer		MAN	MBB	Sol. Kinetics	MAN	Pwr. Kinetics	MAN	Arco Solar
# of Units		30	56	114	108	18	30	30
Total Aperture	m^2	960	1025	4386	3305	359	640	1620
Field Area	ha							1.2
Target Height	m	N/A	N/A	N/A	N/A	N/A	N/A	19.8
Receiver		line foc.	spherical	cavity	line foc.		line foc.	cavity
Aperture	m^2	N/A	N/A	24.0 (tot.)	N/A	N/A	N/A	9.3 (tot.)

7.9 Appendix: Solar Thermal Facility Data

		Facility 1	Facility 2	Facility 3	Facility 4	Facility 5	Facility 6
Heat Transfer							
Medium		2-loop	2-loop	2-loop	2-loop		1-loop
Medium		Oil	Oil	Syltherm 800	Oil	Oil	Water/Steam
Temperature	°C	175/275	235/345	180/400	160/300	140/230	285
Pressure	bar	20	15	48			68.9
Power Conversion							
Manufacturer		Screw Exp. MAN	ORC–Turbine Linde	MTI			none
Medium		Water/Steam	Toluene	Water/Steam	Water/Steam		
Temp. Range	°C	./250	242/320	./380	70/250		
Pressure	bar	20	0.4/15	48			
Generator Rating	kVA		153 (kW$_e$)				
Fossil Backup	MW$_t$		0.75 (oil)				none
Storage							
Medium		Oil	Oil	Syltherm 800	Oil	Oil	none
Capacity	MWh$_t$	1.73	0.7		0.8	0.215	
	MWh$_e$	0.6	0.13	2.9		N/A	none
Temp. Range	°C	235/345		./390	./300		
Performance							
Average							
Irradiation	kWh/m²d		4.6	3.95 (June '85)			
Threshold	W(DNI)/m²		430	130 (for field)		600	
Production	MWh$_e$/d		14 (meas.)				
	MWh$_t$/d						
	MWh$_e$/a	15.22 (net '84)	1.6 (est.)	294 (sim.)		8.6 (peak)	7.3 (peak)
	MWh$_t$/a			1,580 (sim.)			N/A
Yearly Eff.	% gross						
	% net						705.3 ('83)
Peak							
Output	MW$_t$			–/136			
	MW$_e$		10.6	154**/429			0.85
Efficiency	% gross		8.6	5.0			N/A
	% net			1.2			
Monthly CF	%	not determined	not determined	not determined			none
Fossil Use	MWh$_t$/a	none	none	6.333 (sim)	none	none	none

** Electricity production only
*** cbm/d of drinking water

Table 7.15. Solar thermal test facilities

Name		CNRS	CRTF	GIRT/ACTF	DRTF	MSSTF	DLR
Location		Odeillo	Albuquerque	Atlanta	Albuquerque	Albuquerque	Lampoldshausen
Country		France	USA	USA	USA	USA	Germany
General Conditions							
Coordinates	deg	43 N	35 N	33 N	35 N	35 N	.
	deg	.	107 W	84 W	107 W	107 W	.
Altitude	m	1,600	.	.	.	.	.
Testing purpose		Metallurgy[1]	Component testing[2]	Component testing[3]	System evaluation[4]	Thermal loop tests[5]	Component testing[1]
DNI Irradiation							
Average	kWh/m²d	.	.	.	.	.	.
Peak	kWh/m²a	.	.	900	.	.	.
Rating	kW$_t$	1,500	5,500	325	.	2 × 80	140
	kW$_e$	N/A	N/A	N/A	.	.	.
Facility Area	ha	.	.	.	.	.	.
Operation Start		1970	1978	1977	.	.	1986
Staff	Manyear	.	.	.	.	.	5
Status (end of 1989)		test fac.	operating	test fac.	operating	test fac.	operating
Technical Information							
Concentrator		Tracking heliostats + fixed paraboloid	Individually tracking heliostats	Mechanically controlled heliostats	Two dishes	Parabolic troughs	Dish
Manufacturer		63	MMC	550	.	.	SBP
# of Units		63	222	550	2	.	1
Total Aperture	m²	2,835	8,680	532	95 (11m ⊘)	.	227 (17m ⊘)
Focal Length	m	.	.	.	6.6	.	13.6
Concentration		.	.	.	.	.	1,000
Target							
Height	m	120/140/160	61	22.8	N/A	N/A	N/A
Type		Furnace	Centr. rec.	Centr. rec.	Centr. rec.	Line abs.	Centr. rec.
Image Size	cm²	0.28–38.5	30,000	1,900–7,800	1,800	N/A	.
Peak Flux	W/cm²	500–1,5000	260	125	1,500	.	.
Peak Temp.	°C	3,200	2,000	2,100	.	.	.
Power Conversion							
Engine		N/A	N/A	N/A	Stirling	.	.
Manufacturer		N/A	N/A	N/A	USAB	.	.

[1] High-temperature material research
[2] Thermodynamic engine testing
[3] High-temperature research
[4] Thermal component testing
[5] Performance testing of trough fields

7.9 Appendix: Solar Thermal Facility Data

Table 7.16. Technical data on parabolic dish collectors

Manufacturer		MBB	MDAC	LEC 460	LaJet LEC 1700	SKI	ANU	Acurex Innovative	SBP	Advanco Vanguard	PKI /35
Type											
Plant		Kuweit		Solarplant 1		Shenandoah	White Cliffs		Al-Jubaylah	Rancho Mirage	Molokai
Gross Aperture	m^2	19.6	91.0	43.2	120.6–171.0	38.2	20.0	.	227	86.7	306
Unshaded Ap.	m^2	18.3	91.0	41.3	120.6–164.0	.	19.8	177.5	212	79.8	304
Focal Length	m	2.0	.	5.55–5.7	11.7	.	.	7.5	13.6	6.2	15.0
Diameter	m	5.0	.	9.42	19.6	7.0	5.06	15.0	17.0	10.7	17.5 × 17.5
Concentration		200	2,000	225 (1,700)	1,000	235	410	1,925	600	2,500	350–500
Mirror Element Area	m^2	N/A	.	1.8	1.8	.	0.011	0.5	0.0004	0.27	0.72
Mirror Element Number		N/A	82	24	67 to 95	21	≈2,300	40 + 24	1,420	336	392
Design Reflectivity		0.94	.	0.80 (.822)	0.90	0.86 (.82)	0.80	0.91	0.88	0.92	0.87
Reflective Surface		SS glass mirrors	.	PMMA foil	PET foil	FEK 244 foil on substrate	SS glass mirrors	PMMA foil on steel membrane[7]	SS glass mirror on steel membrane	SS glass mirror	SS glass mirror
Dish Structure		reinforced plastic	steel tubing	steel tubing	steel tubing	stamped Al-gores	fiber glass substrate moulded on ring	.	double membrane evacuated drum	steel tubing racks	structural steel
Weight	kg/m^2	.	73	36.0[2]	69.3[2]	37[2]	80[2]	55	52	118.5	100
Peak Collection Efficiency		0.94	0.88	0.77	0.75	0.56 (.3)	0.70	0.92	0.85	0.83	0.86
Rating at 1,000 W/m^2	kW$_t$	13.5	76	30.2	105.0–147.5	16.3[4]	14.0	161.0	178.6	71.8	243.5
Receiver Aperture	m^2	0.023	0.034	0.20	1.0	0.21	0.13	0.7	0.38	0.03	1.8
Drive Power	W	.	.	2 × 93	145 (3,012 stow)	.	20	.	2,000	700	.
Max Op. Wind Speed	km/h	.	56	56	.	.	80	.	.	50	65
Cost	$(US)/m^2	3,500	300[1]	11.1[2]	72[1]	1,135[6]	320[5]	144[3]	200[1]	650	.

[1] estimate
[2] without receiver
[3] 10k production volume
[4] at 950 W/m^2
[5] incl. absorber and controls
[6] installed
[7] structurally integrated sheet metal panels
( ) measured
⟨ ⟩ peak value
SS = second surface

Bibliography

[1] *Arizona Public Service: Utility Solar Central Receiver Study, Vols. 1 & 2.* Technical Report DOE/AL/38741-1, Springfield/VA, Arizona Public Service Co., Black & Veatch Engineers-Architects, Babcock & Wilcox, Pitt-Des Moines, Inc., Solar Power Engineering Co., University of Houston, 1988

[2] *International 30 MW$_e$ Solar Tower Plant, Feasibility Study – Phase I: Presentation of Results.* Technical Report, Madrid, Miner/DLR, 1987

[3] *Phoebus Executive Summary of Phase IA Work.* Technical Report, Köln (D), DLR, 1988

[4] *Solar Central Receiver Technology Advancement for Electric Utility Applications: Phase I Topical Report, Vols. 1 & 2.* Technical Report DOE/AL/38741-3, Springfield/VA, Pacific Gas & Electric Co., Bechtel National Inc., 1988

[5] Tower Stations, Thermodynamic Conversion of Solar Energy (in French). *Etropien*, 103 (1982)

[6] Abbin, J. P.; Leuenberger, W. R.: *Program CYCLE – A Rankine Cycle Analysis Routine.* SAND74-0099 (Revised), Sandia National Laboratories, Albuquerque/NM, 1977

[7] Al-Rubaian, A.; Hansen, J.: 50kW$_e$ Solar Concentrators with Stirling Engine – Results of the Six Months of Operation in Saudi Arabia. In *Proc. ISES Solar World Congress, Hamburg 1987*, Bloss, W. H.; Pfisterer, F. (Ed.), Oxford (UK): Pergamon Press, 1988

[8] Alvis, R. L.: *Some Solar Dish / Heat Engine Design Consideration.* Technical Report SAND84-1698, Sandia National Laboratories, Albuquerque/NM, 1984

[9] Amannsberger, K.: *Cash Flow Analysis for Solar Plants (CASSOL).* Internal Report, München (D), MAN Technologie GmbH, 1985

[10] Amannsberger, K.; Bittner, I.: System Optimization, Simulation and Comparison with First Experimental Results of Solar Thermal Plants (Distributed Collector Systems). In *Proc. ASME Winter Annual Meeting, Phoenix/AZ*, American Society of Mechanical Engineers, New York: 1982

[11] Amannsberger, K.; Schoelkopf, M.: Economic Assessment of the Economics of Regenerative Energies (in German). In *ISES/BSE Tagungsbericht, Karlsruhe (D)*, 1982

[12] Amannsberger, K.; Wiedmann, U.: System Optimization and Operational Simulation of Solar Farm Plants Applied to a Plant with Diesel Waste Heat Utilization. In *Conf. Proc. Systemes Solaires Thermodynamiques*, p. 137, Marseilles (F): 1980

[13] Bendt, P.; Rabl, A.; Gaul, H. W.: Optimization of Parabolic Trough Solar Collectors. *Solar Energy*, 29 (1982) 407–417

[14] Bird, S. P.: *Assessment of Solar Operations for Small Power Systems Applications, Volume 5, SOLSTEP: A Computer Model for Solar Plant System Simulation.* PNL-4000, Pacific Northwest Laboratories, Richland/WA, 1980

[15] Brandt, L. D.; Chang, R. E.: *Heliostat Cost Analysis Tool.* SAND81-8031, Sandia National Laboratories, Livermore/CA, 1981

[16] Brune, J. M.: *BUCKS – Economic Analysis Model of Solar Electric Power Plants.* SAND77-8279, Sandia National Laboratories, Livermore/CA, 1979

[17] Casal, F. G.: *Solar Thermal Power Plants.* Berlin, Heidelberg, New York: Springer, 1987

[18] Castro, M.; Peire, J.; Martinez, P.: Five-Year Cesa-1 Simulation Program Review. *Solar Energy*, 38 (1987) 415–424

[19] Coleman, G. C.; Raetz, J. E.: *Field Performance of Dish/Stirling Solar Electric Systems.* Technical Report, Orange/CA, McDonnell Douglas Energy Systems, 1986

[20] Criner, D. E.; Gould, G. L.; Soderstrum, M. G.; Ege, H. D.; Wolfs, K. E.; Bigger, J. E.: *10 MW$_e$ Solar-Thermal Central-Receiver Pilot Plant, Volume 1: Report on Lessons Learned.* Technical Report AP-3285, Electric Power Research Institute, Palo Alto/CA, 1983

[21] Dellin, T.; Fisk, M. J.; Yang, C. L.: *A User's Manual for DELSOL2 – A Computer Code for Calculating the Optical System Design for Solar Thermal Central Receiver Plants.* SAND81-837, Sandia National Laboratories, Livermore/CA, 1981

[22] Doane, J. W.; O'Toole, R. P.; Chamberlain, R. G.; Bos, P. B.; Maycock, P. D.: *The Cost of Energy from Utility-Owned Solar Electric Systems.* JPL 5040-29, Pasadena/CA, Jet Propulsion Laboratory, 1976

[23] Droher, J. J.: *Performance of the Vanguard Solar Dish-Stirling Engine Module.* Technical Report EPRI AP-4608, Electric Power Research Institute, Palo Alto/CA, 1986

[24] Etievant, C.; Amri, A.; Izygon, M.; Tedjiza, B.: *Central Receiver Plant Evaluation, Vol. 1–5.* Technical Report SAND86-8185, SAND87-8182, SAND88-8101, SAND88-8100, SAND88-102, Sandia National Laboratories, Albuquerque/NM, 1988

[25] Fair, D.: *Solar Total Energy Project – Test Report for Thirty Consecutive Day Test.* Technical Report, Atlanta/GA, Georgia Power Co., 1985

[26] Fewell, M. E.; Grandjean, N. R.: *User's Manual for Computer Code SOLTES-1B (Simulator of Large Thermal Energy Systems)*. SAND78-1315, Sandia National Laboratories, Albuquerque/NM, 1980
[27] Fsadni, M.: *Description of the Program GASBIE for the Simulation of the GAST Plant Operation*. GAST-IAS-BT-100200-057, Bergisch-Gladbach (D), Interatom GmbH, 1983
[28] Geyer, M.; Klaiss, H.: 194 MW of Solar Electricity with Trough Collectors (in German). *Brennstoff-Wärme-Kraft*, (1989) 288–295
[29] Grasse, W.: *Design Basics for Solar-Thermal Power Plants – Results and Experiences from Operating Experimental Facilities* (in German). Volume 704 of *VDI-Berichte*, Düsseldorf (D): VDI-Verlag, 1988
[30] Grasse, W.: *SSPS Results of Test and Operation 1981–1984*. Technical Report IEA-SSPS SR7, Köln (D), DLR, 1985
[31] Grasse, W.; Klaiss, W.: *Solar Tower Power Plants – Analysis of Their Development Status* (in German). Technical Report Internal Report, Stuttgart (D), DLR, 1988
[32] Gretz, J.: Concept and Operation Experiences with EURELIOS. In *3rd International Workshop on Solar Thermal Central Receiver Systems*, Becker, M. (Ed.), pp. 65–80, Berlin, Heidelberg, New York: Springer, 1986
[33] Gretz, J.; Strub, A.; Palz, W.: *Thermo-Mechanical Solar Power Plants – Eurelios, the 1 MW_e Experimental Solar Thermal Electric Power Plant*. Dordrecht: Reidel, 1984
[34] Guillen, J.: System Dynamic Behavior. In *GAST – The Gas-Cooled Solar Tower Technology Program*, Becker, M.; Boehmer, M. (Ed.), pp. 37–50, Berlin, Heidelberg, New York: Springer, 1989
[35] Hansen, J.: *TDSA Project Joint Test and Operation (August 1986–1989 / 1986–December 1989)*. Technical Report, Köln (D), DLR, 1990
[36] Harats, Y.; Kearney, D.: Advances in Parabolic Trough Technology in the SEGS Plants. In *ASME Intl. Solar Energy Conference*, San Diego/CA: 1989
[37] Hicks, T.: *Solar Total Energy Project – Test Report for Continous Fourteen Day Commercial Operations*. Technical Report, Atlanta/GA, Georgia Power Co., 1985
[38] Holl, R.: *Status of Solar-Thermal Electric Technology*. Technical Report EPRI GS-6573, Electric Power Research Institute, Palo Alto/CA, 1989
[39] Jansen, K. H.: *DYNAG Code for the Investigation of Non-Steady State Operation Modes of the GAST Reference Plant* (in German). IAS-BT-100200-063, Bergisch-Gladbach (D), Interatom GmbH, 1983
[40] Jensen, C.; Price, H.; Kearney, D.: The SEGS Power Plants: 1988 Performance. In *ASME Intl. Solar Energy Conference*, San Diego/CA: 1989
[41] Johansson, L.: *Daily Performance Data of the McDonnell Douglas Dish-Stirling Module*. Technical Report, Louisville/KY, Phoenix Holdings Inc., 1987
[42] Kearney, D.; Gilon, Y.: Design and Operation of the Luz Parabolic Trough Solar Electicity Plants. p. 53, Düsseldorf (D): VDI-Verlag, 1988
[43] Kesseling, P.; Selvage, C. S. (Ed.): *The IEA/SSPS Solar Thermal Power Plants – Facts and Figures*. Volume , Berlin, Heidelberg, New York: Springer, 1986
[44] Kiera, M.: *Description of the Computing Code System HFLCAL* (in German). GAST-IAS-BT-200000-075, Bergisch-Gladbach (D), Interatom GmbH, 1986
[45] Larson, D. L.: Operational Evaluation of the Grid-connected Coolidge Solar Thermal Electric Power Plant. *Solar Energy*, 38 (1987) 11–24
[46] Laurence, C. L.; Lipps, F. W.: *A User's Manual for the University of Houston Computer Code – RC: Cellwise Option for the Central Receiver Project*. SAN/0763-3, Houston/TX, University of Houston, 1980
[47] Leary, P. L.; Hankins, J. D.: *User's Guide for MIRVAL – A Computer Code for Comparing Designs of Heliostat-Receiver Optics for Central Receiver Solar Power Plants*. SAND77-8280, Sandia National Laboratories, Livermore/CA, 1979
[48] Lipps, F. W.; Vant-Hull, L. L.: *A User's Manual for the University of Houston Solar Central Receiver System – Cellwise Performance Model*. SAN/0763-4-1/2-2/2, Houston/TX, University of Houston, 1980
[49] Mancini, T. R.: *Point-Focus Concentrating Collector Technology Development*. Technical Report SAND87-1258, Sandia National Laboratories, Albuquerque/NM, 1987
[50] Maynard, D. P.; Gajanana, B. C.: *Analytical Foundation/Computer Model for Dish-Brayton Power System*. JPL 5105-9, Pasadena/CA, Jet Propulsion Laboratory, 1980
[51] McFarland, B. L.: *Manual for the Solar Total Energy System Evaluation Program*. SAND78-7045, Canoga Park/CA, Rockwell International Energy Systems Group, 1979
[52] McGlaun, M. A.: LaJet Energy Company Update of Solar Plant 1. In *Solar Thermal Technology Conference*, 1987
[53] Ney, E. J.: *Solar Total Energy Project Summary Report*. Contractor Report SAND87-7108, Shenandoah/GA, Georgia Power Co., 1988

[54] Noyes, G.: *TDSA – Results of Weather and Dish-Stirling Analysis*. Technical Report Technical Note 001/88, Stuttgart (D), DLR, 1988
[55] O'Doherty, R.; Finegold, J.; Herlevich, A.: ROSET: A Solar Thermal Electric Power Simulation User's Guide. 1982
[56] Pharabod, F.; Bezian, J. J.; Bonduelle, B.; Rivoire, B.; Guillard, J.: Themis Evaluation Report. In *Proc. 3rd Intl. Workshop on Solar Thermal Central Receiver Systems, Konstanz*, Becker, M. (Ed.), pp. 91–104, Berlin, Heidelberg, New York: Springer, 1986
[57] Pitman, C. L.; Vant-Hull, L. L.: *The University of Houston Solar Central Receiver Code System: Concepts, Updates and Start-Up Kits*. SAND88-7029, Houston/TX, University of Houston, 1989
[58] Rabl, A.: Yearly Average Performance of the Principal Solar Collector Types. *Solar Energy*, 27 (1981) 215–233
[59] Radosevich, L. G.: *Final Report on the Experimental Test and Evaluation Phase of the 10 MW_e Solar Thermal Central Receiver Pilot Plant*. Technical Report SAND85-8015, Sandia National Laboratories, Livermore/CA, 1985
[60] Radosevich, L. G.: *Final Report on the Power Production Phase of the 10 MW_e Solar Thermal Central Receiver Pilot Plant*. Technical Report SAND87-8022, Sandia National Laboratories, Livermore/CA, 1987
[61] Ramos, F.; Mateos, J.; de Marcos, J.: Optimization of a Central Receiver Solar Electric Power Plant by the ASPOC Program. In *Solar Thermal Technology – Proc. 4th Intl. Symposium, Santa Fe/NM, 1988*, Gupta, B. P.; Traugott, W. H. (Ed.), p. 61, New York: Hemisphere Publ. Co., 1990
[62] Sanchez, F.: Results of Cesa-1 Plant. In *Proc. 3rd Intl. Workshop on Solar Thermal Central Receiver Systems, Konstanz*, Becker, M. (Ed.), pp. 46–64, Berlin, Heidelberg, New York: Springer, 1986
[63] Schiel, W.: Dish/Stirling Systems – Technical Design, Operation Experience and Development Trends (in German). In *Solarthermal Power Plants for Heat and Electricity Generation*, p. 117, Düsseldorf (D): VDI-Verlag, 1988
[64] Stine, W.: *Power from the Sun – Priciples of High-Temperature Solar Thermal Technology*. Technical Report SERI/SP-273-3054, Solar Energy Research Institute, Golden/CO, 1987
[65] Stine, W. B.; Heckes, A. A.: Energetics of Extended Operation of a Hybrid Solar Total Energy System. In *Proc. ASME-JSME-ISES Solar Energy Conference, Honolulu/Hawaii*, 1987. SAND86-17733
[66] Stine, W. B.; Heckes, A. A.: Energy and Availability Transport Losses in a Point-Focus Solar Concentrator Field. In *Proc. 21st Intersoc. Energy Conv. Engg. Conf. (IECEC), San Diego/CA*, 1986. SAND86-0004
[67] Stoddard, M. C.: *Convective Loss Measurements at the 10 MW_e Thermal Central Receiver Pilot Plant*. Technical Report SAND85-8250, Sandia National Laboratories, Livermore/CA, 1985
[68] Stoddard, M. C.; Faas, S. E.; Chiang, C. J.; Dirks, J. A.: *SOLERGY – A Computer Code for Calculating the Annual Energy from Central Receiver Power Plants*. SAND86-8060, Sandia National Laboratories, Livermore/CA, 1987
[69] Strachan, J. W.: *An Evaluation of the LEC-460 Solar Collector*. Technical Report SAND87-0852, Sandia National Laboratories, Albuquerque/NM, 1987
[70] Torkelson, L.; Larson, D. L.: *1981 Annual Report of the Coolidge Solar Irrigation Project*. Technical Report SAND82-0521, Sandia National Laboratories, Albuquerque/NM, 1982
[71] Vittitoe, C. N.; Biggs, F.: *A User's Guide to HELIOS – A Computer Program for Modeling the Optical Behavior of Reflecting Solar Concentrators, Appendices Concerning HELIOS-Code Details*. SAND81-1562 (Part III) and SAND81-1180 (Part I), Sandia National Laboratories, Albuquerque/NM, 1981
[72] Vittitoe, C. N.; Biggs, F.; Lighthill, R. E.: *HELIOS: A Computer Code for Modeling the Solar Thermal Test Facility – A User's Guide*. SAND76-0346, Sandia National Laboratories, Albuquerque/NM, 1978
[73] William, T. A.; Dirks, J. A.; Brown, D. R.; Drost, M. K.; Antoniac, Z. A.; Ross, B. A.: *Characterization of Solar Thermal Concepts for Electricity Generation*. Technical Report PNL-6128, Richland/WA, Battelle Pacific Northwest Laboratories, 1987
[74] Williams, T. A.; Cole, R. J.; Brown, D. R.; Dirks, J. A.; Edelhertz, H.; Holmlund, I.; Malhotra, S.; Smith, S. A.; Sommers, P.; Wilke, T. L.: *Solar Thermal Financing Guidebook*. PNL-4745, Richland/WA, Pacific Northwest Laboratories, 1983
[75] Winter, C.-J.; Nitsch, J. (Ed.): *Hydrogen as an Energy Carrier – Technologies, Systems, Economy*. Berlin, Heidelberg, New York: Springer, 1988
[76] Zewen, H.; Schmidt, G.; Moustafa, S.: The Kuwait Solar Thermal Power Station: Operational Experiences with the Station and its Agricultural Application. In *Proc. 8th Biannual ISES Solar World Congress, Perth/W.Australia 1983*, Szokolay, S. V. (Ed.), p. 1527, Oxford (UK): Pergamon Press, 1984

8 Photovoltaic Power Stations

W. H. Bloss, H. P. Hertlein, W. Knaupp, S. Nann and F. Pfisterer [1]

8.1 Introduction

Photovoltaic (PV) power generation is gaining increasing importance as a renewable energy supply. From small-scale space applications in the 1960s, production of photovoltaic modules has increased to about 45 MW peak power per year worldwide. This development was accompanied by a cost reduction for solar modules to as low as 4–5 US \$/W_p in 1989[2].

The principle of photovoltaic energy conversion was outlined in Chap. 2. The basic device and the smallest independent operational unit of all PV systems is the solar cell. Its size can be chosen according to the planned application; in practice it varies from several mm^2 for *consumer electronics* applications such as pocket calculators, wrist watches, etc. to the present standard cell size of $10 \cdot 10$ cm^2.

To achieve higher-power PV modules, a large number of solar cells has to be interconnected. This property of photovoltaics – its modularity – allows construction of mW-size PV energy supplies as well as multi-MW PV solar power plants, in contrast to most other solar energy-providing systems. Performance and reliability of photovoltaic systems have been demonstrated in a large variety of small- and medium-scale stand-alone applications, as well as in MW-size grid-connected power stations.

Besides modularity, the main advantages of photovoltaic energy conversion are:

- the irradiance threshold for start of output operation is very low, practically no energy inventory is needed within the system for operation;
- non-concentrating photovoltaic supply systems can be constructed without involving moving parts. PV systems operate noiselessly, without emissions into the environment, and with minimal need for maintenance;
- non-concentrating PV systems are capable of converting diffuse irradiation with about the same efficiency as direct irradiation.

The main obstacle for using multi-MW-scale photovoltaic systems at present is the cost, caused primarily by the cost of PV modules.

This chapter dealing with photovoltaic solar power plants (SPP) will outline key technical aspects of solar cells and the state-of-the-art in solar cell development, aspects and technological status of PV modules and PV arrays, power conditioning elements and systems, and

---

[1] Werner H. Bloss, Werner Knaupp, Stefan Nann and Fritz Pfisterer, Institut für Physikalische Elektronik, Universität Stuttgart, Pfaffenwaldring 47, D-7000 Stuttgart 80
Hansmartin P. Hertlein, Deutsche Forschungsanstalt für Luft- und Raumfahrt (DLR), Linder Höhe, D-5000 Köln 90

[2] W_p denotes the power output at Standard Test Conditions (STC), i.e. air mass (AM) = 1.5, 25°C cell temperature, 1,000 W/m^2 for flat-plate modules, and 850 W/m^2 of direct irradiance for concentrating modules.

supporting structures, tracking and concentrating devices. Then aspects of PV power plant design will be discussed, and the performance of and operational experience with PV power plants will be reported.

8.2 Technical Aspects of Solar Cells

8.2.1 IV-Characteristic of Solar Cells [3]

The IV-characteristic of a solar cell, as already cited in Sect. 2.6, is given by

$$I = I_0 \left\{ \exp\left(\frac{qV}{AkT}\right) - 1 \right\} - I_{ph} \tag{8.1}$$

with $I_0 =$ reverse saturation current, $q =$ electronic charge, $A =$ diode quality factor (A-factor), and $I_{ph} =$ photocurrent. In principle this equation also holds for pin-junctions, pn-heterojunctions, Schottky-barriers, and MIS-junctions. These devices can all be used for solar cells; with respect to (8.1) they only may differ in the value and the physical origin of I_0 and A [28,33]. From (8.1) the short circuit current I_{sc} and the open circuit voltage V_{oc} can be calculated:

$$V = 0: \quad I = I_{sc} = -I_{ph} \tag{8.2}$$

$$I = 0: \quad V = V_{oc} = \frac{AkT}{q} \cdot \ln\left(\frac{I_{ph}}{I_0} + 1\right). \tag{8.3}$$

According to the diffusion theory of Shockley [92], I_0 is given by

$$I_0 = qN_vN_c \left[\exp(-E_g/kT)\right] \left(\frac{L_n}{n_n\tau_n} + \frac{L_p}{p_p\tau_p}\right), \tag{8.4}$$

N_v, N_c are the effective densities of states in the valence and conduction band and L_n, L_p, n_n, p_p, τ_n, τ_p are the diffusion lengths, the densities, and the lifetimes of electrons and holes, respectively. With (8.4) we obtain from (8.3), assuming $A = 1$ and $I_{ph} \gg I_0$,

$$V_{oc} = \frac{E_g}{q} - \frac{kT}{q} \cdot \ln\left[\frac{1}{I_{ph}} \cdot qN_vN_c\right] \left(\frac{L_n}{n_n\tau_n} + \frac{L_p}{p_p\tau_p}\right). \tag{8.5}$$

(8.5) demonstrates that the bandgap E_g of the semiconductor used is the dominant parameter determining V_{oc}.

Graphs of IV-characteristics of a solar cell are shown in Fig. 8.1 for various irradiation levels. To the first approximation I_{ph} and I_{sc} increase linearly with radiation flux density, whereas V_{oc} according to (8.3) increases logarithmically.

The maximum power point at V_{max} and I_{max} (see also Sect. 8.5.1) is obtained from (8.1) by calculating $d(I \cdot V)/dV = 0$:

$$V_{max} + \frac{AkT}{q} \cdot \ln\left(1 + \frac{qV_{max}}{AkT}\right) = V_{oc}, \tag{8.6}$$

$$I_{max} = -\frac{qV_{max}}{AkT} \cdot I_0 \cdot \exp\frac{qV_{max}}{AkT}. \tag{8.7}$$

---

[3] The balance of the chapter can be understood without concern for the detailed derivation of this section.

8.2 Technical Aspects of Solar Cells

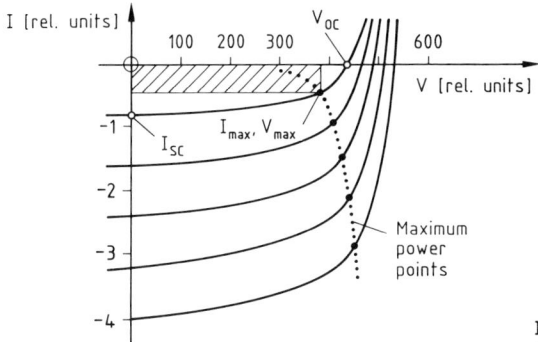

Fig. 8.1. IV-characteristics of a solar cell at different radiation flux densities.

The numerical solution of these equations together with (8.5) shows that

- $V_{\max}$ increases with increasing I_{ph},
- $V_{\max}$ approaches V_{oc} with increasing V_{oc}.

This behavior is indicated in Fig. 8.1 by the dotted line.

The fill factor of a IV-characterisitic is defined as

$$FF = \frac{I_{\max} V_{\max}}{I_{sc} V_{oc}} \qquad (8.8)$$

With $I_{sc} = -I_{ph}$ and using (8.6) and (8.7), FF can be calculated:

$$FF = \frac{V_{\max}^2 \cdot (1 - I_0/I_{ph})}{(V_{\max} + AkT/q) \cdot V_{oc}} \qquad (8.9)$$

The numerical solution of (8.9) shows that FF increases when the photocurrent I_{ph} increases and/or the reverse saturation current I_0 is reduced.

These properties of solar cells derived from the basic theory of pn-junctions are reasons for

- the possibility to increase solar cell efficiencies by applicating concentration of irradiation (i.e. increase of I_{ph}), and
- the success in developing high-efficiency monocrystalline Si cells ($\eta > 20\%$) by using silicon of outstanding quality (i.e. reduction of I_0); see below.

8.2.2 Temperature Effects

The IV-characteristic of a solar cell is sensitively affected by temperature variations. At constant illumination an increase of cell temperature reduces V_{oc} according to (8.5). On the other hand, I_{ph} is observed to increase slightly with increasing temperature. This is, among other effects, due to increasing minority carrier diffusion lengths L_n and L_p with increasing temperature [31]. This behavior is illustrated in Fig. 8.2, which also demonstrates that V_{oc} as well as $V_{\max}$ decrease with increasing temperature.

Practical outdoor operation of photovoltaic cells results in increased solar cell temperatures with increasing insolation. Consequently, the effects shown in Fig. 8.1 and 8.2 overlap:

- increasing irradiation increases $V_{\max}$ according to (8.6) and (8.7), and
- increasing irradiation increases T and, according to Fig. 8.2, decreases $V_{\max}$.

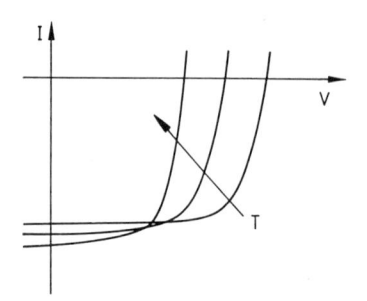

Fig. 8.2. Influence of operating temperature T on the IV-characteristic of a photovoltaic generator at constant illumination.

An example is given in Fig. 8.3. It indicates that with widely varying insolation the output voltage (V_{max}) varies only in a small range (shaded area in Fig. 8.3). This behavior argues for the application of so-called *constant voltage systems* for power conditioning (see below).

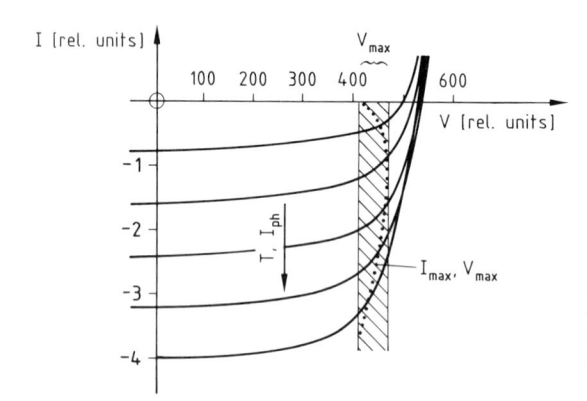

Fig. 8.3. IV-characteristics of a solar cell operating under practical outdoor conditions (increasing temperature with increasing insolation, practical V_{max}-range is shaded).

8.2.3 Radiation Absorption and Material Selection

The interaction of radiation with different materials is characterized by the absorption coefficient α. The variation of α with photon energy is shown in Fig. 8.4 for different semiconductors which have been used in photovoltaics. The absorption edge is determined by the energy bandgap of the material. The slope of the absorption curves (in a logarithmic scale) is extremely steep for semiconductors with direct transitions, *direct semiconductors*, whereas the so-called *indirect semiconductors* show absorption curves with much smaller slopes.

Due to Beer's law describing the penetration of radiation through matter,

$$I(z) = I_1(1-r)e^{-\alpha z}, \tag{8.10}$$

where I_1 is the irradiance, r the reflection coefficient at the surface, and z the direction of penetration; the absorption coefficient α determines the required thickness of photovoltaic layers. Let αz be in the order of 5, which expresses that within the thickness

$$z_s = L_\alpha = 5/\alpha \tag{8.11}$$

more than 99% of the radiation is absorbed, then this absorption depth describes the minimum thickness of the absorber material for photovoltaic generators. It can be reduced by the use of reflecting back surfaces. Whereas crystalline Si cells require thicknesses ranging between 50

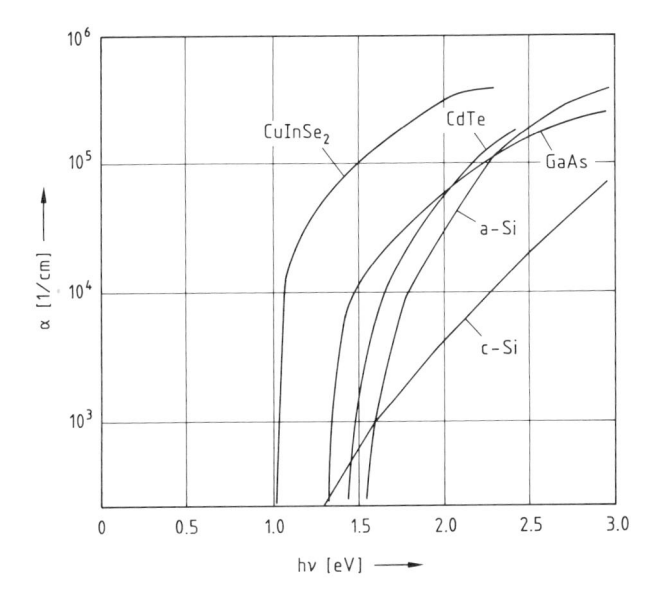

Fig. 8.4. Absorption coefficient of semiconducting materials used in photovoltaics.

and 100 μm, cells consisting of direct semiconductors or amorphous Si can be realized with thicknesses below 1 μm. The use of direct semiconductors implies a substantial reduction of material and leads to the concept of thin film solar cells.

8.2.4 Tandem Systems

The performances of tandem cells especially is affected by variations in the solar spectral irradiance. The fraction of actually converted energy (cell sensitivity) for cells with differing spectral responses is found by integrating the given solar cell response with the present solar spectrum divided by the total broadband irradiance. Some major trends are evident from these calculations [78]:

- Monocrystalline silicon solar cells are less sensitive to variations in the solar spectrum than amorphous silicon. The narrow response range of amorphous silicon ($\approx$ 340–800 nm) makes the cell more sensitive to variations in solar spectral irradiance than cells made of crystalline silicon ($\approx$ 380–1,100 nm).
- Cell sensitivity is less influenced by atmospheric conditions if global rather than direct irradiance is utilized. Therefore, the sensitivity increases for modules tilted towards the sun, and for tracking and concentrating devices.
- Still more pronounced are the effects for multijunction solar cells, especially with two terminals only, when different photocurrent densities originating in top and bottom cells cause mismatch losses.

Therefore, if in the future different types of solar cell materials are available, statistical information about the spectral distribution related to distinct geographical regions and seasons are required for forecasting and optimizing the output of a PV power plant. Figure 8.5 illustrates typical spectral irradiances measured at similar solar zenith but under different cloud conditions: a partly cloudy sky with strong direct beam and an overcast sky with no direct beam present. Each measurement is normalized by the respective broadband (280–4,000 nm) pyranometer observation. The influences of clouds on the spectrum can clearly be seen:

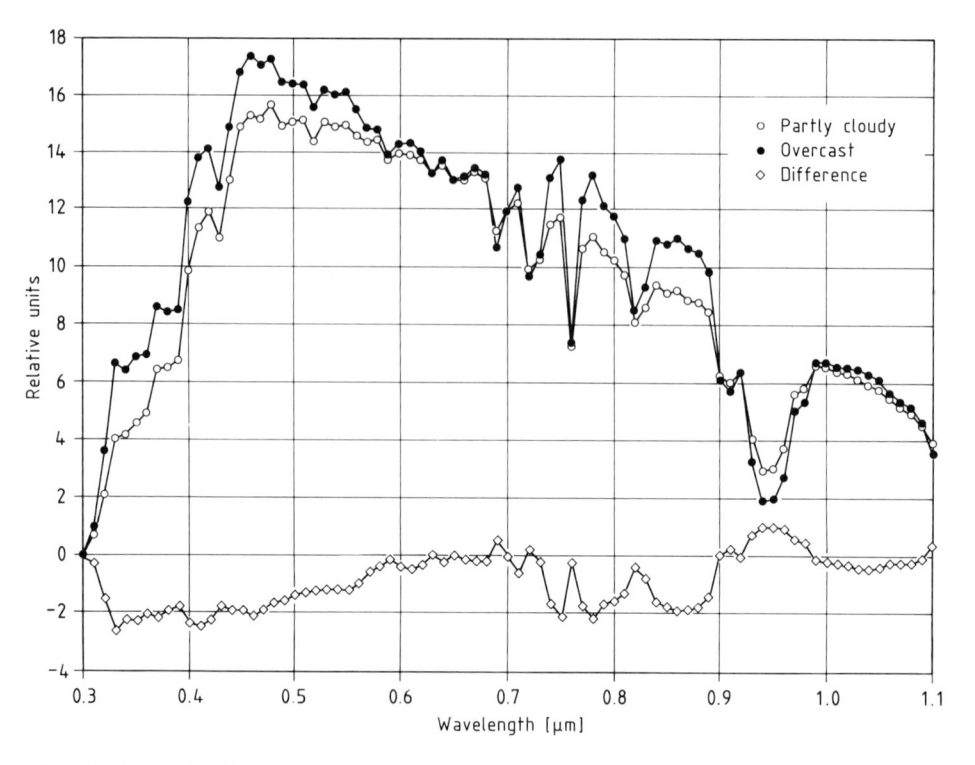

Fig. 8.5. Spectral irradiance, measured with a latitude tilted (48°) Spectroradiometer, under two different cloud conditions and a relative air mass of 1.5 (normalized at 630 nm [76].

shift of the spectrum to UV/blue wavelengths, enhancement by surface albedo beyond the chlorophyll absorption band near 650 nm, stronger water vapor and droplet absorption in the red/IR.

8.3 Status of Solar Cell Development

Since its early beginning, the use of photovoltaics in the power sector has been dominated by solar cells based on crystalline silicon. The worldwide PV installations – estimated peak power 200 MW – are exclusively built with single- and multicrystalline[4] silicon solar cells. The consumer-electronics market, however, is dominated by amorphous silicon cells. The total shipments of amorphous Si cells amount to about 60 MW peak power up to now (end of 1989).

---

[4] The term *multicrystalline* has been introduced for describing cast silicon with crystallite sizes of several mm (e.g. *SILSO* from Wacker, Germany) in order to avoid confusion with *polycrystalline silicon*. Poly-Si is the *electronic-grade* Si (EG-Si) final product of the *Siemens process*; it results from pyrolysis of SiHCl$_3$. Poly-Si with its grain size in the μm-range is the starting material for single crystal growth and for casting of *multicrystalline* blocks. Instead of multicrystalline, the term *semicrystalline* is also used (Solarex, Crystal Systems, USA).

Crystalline Si cells and CdTe thin film cells are the only alternatives that have slightly entered this particular market.

Thin film cells based on compound semiconductors have been investigated since the early beginning of photovoltaics in 1954. Only recently, two types of compound semiconductor cells crossed the 10%-efficiency barrier and show promising stability: cells based on CuInSe$_2$ and on CdTe. Concentrating PV systems require a special solar cell development. Here single-crystalline Si and single-crystalline III-V-compounds (GaAs) are the most advanced materials. A promising concept for the future are tandem structures. This principle may be applied for high-efficiency structures, for concentrating systems, and for medium-efficiency low-cost thin film structures.

8.3.1 Crystalline Silicon Solar Cells

Crystalline silicon with its bandgap of 1.1 eV and its indirect band structure is a material not optimized for solar energy conversion. Nevertheless, due to the lack of successful competitors, crystalline silicon dominates the power market as outlined above.

Present Technology. The classical technology of crystalline Si solar cell production starts from electronic-grade silicon (EG-Si) with an impurity content $< 10^{-9}$. The wafers used for cell fabrication are obtained from single-crystalline rods produced according to Czochralski (CZ) or by float-zoning (FZ). Photovoltaic cells are then produced by vapor phase diffusion techniques for pn-junction formation and by vacuum deposition of front and back contact and antireflection coating. Techniques with reduced cost, like diffusion from the solid phase and screen printing and sintering, are also applied.

The state-of-the-art structure of an industrial Si solar cell is outlined in Fig. 8.6. A small surface-near region of a p-type wafer is n$^+$-doped for establishing the pn-junction.

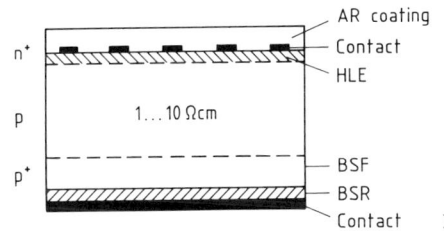

Fig. 8.6. Schematic structure of a crystalline Si solar cell.

The BSF (back surface field), a drift field implemented through a gradient in the p-doping concentration, acts as a photocarrier mirror at the rear side of the cell. An optical mirror, the BSR (Back Surface Reflector), provides increased absorption probability for incident light and allows the use of thin Si wafers (down to 200 μm or, with additional measures, even to 50 μm). Finally, the *high-low-emitter* (HLE) concept realizes a drift field at the front surface by implementation of a gradient of the doping concentration in the n$^+$-layer. The action of the HLE is comparable to that of BSF.

A successful modification of this technology is the substitution of single crystal pulling by casting techniques [95][5]. Worldwide more than one third of crystalline Si solar cell production is based on multicrystalline wafers. The shipments in 1988 were 13.1 MW$_p$ of single crystalline Si and 7.9 MW$_p$ of multicrystalline solar cells [7].

---

[5] Multi- or semicrystalline ingot casting and solidification is done by Wacker (*SILSO*), Solarex, Crystal Systems, Osaka Titanium, and Photowatt.

Grain boundary effects and effects caused by further structural imperfections in multicrystalline wafers reduce the solar cell efficiency. By hydrogen treatment and application of getters, the efficiency reduction can be kept low. Multicrystalline cells with average efficiencies arround 12–13% are obtained in industrial production; monocrystalline cells approached 14–15%.

Cost reduction. Since the early seventies strong efforts have been made for cost reduction:

- A number of low-cost processes for base material production resulting in *solar grade* Si (SoG-Si) have been investigated [11,64,95]; however, the relevance of such processes has decreased in recent years, since the main efforts in silicon solar cell technology are more directed towards economic wafer production and to high-efficiency cells. As long as the material consumption of the photovoltaic industry is less than 10% of the total silicon production, the photovoltaic industry can make use of low-cost off-specification *electronic grade* Si (EG-Si), and there is no need for large efforts in SoG-Si production. Furthermore, in order to obtain efficiencies in the range of 15%, the SoG-Si production processes would have to be improved to approach the Siemens process.
- In order to get rid of sawing techniques which always cause material losses and require a number of production steps, various types of ribbon pulling and casting and foil preparation techniques (altogether about 20 different methods) have been investigated [4]. Today, edge-defined film-fed growth (EFG; Mobil Solar, USA) is the only process used for the industrial production of ribbon-type Si solar cells [26].
- In order to realize junction formation and contacting, a number of simplifications such as solid phase diffusion with dopants deposited by screen printing or spin-on, combined with BSR and contact formation, have been introduced and successfully applied (see above).

Increase in efficiency. The second main strategy for improving the overall applicability of photovoltaics is to increase solar cell efficiency. The latest developments are based mainly on the use of highest-quality FZ-silicon and on the application of passivation: any surface of Si crystals or any interface with other materials causes recombination centers and recombination losses. SiO_2-coating of free Si-surfaces and thin SiO_2-interlayers at contacts have been proven to suppress these losses. It was the *passivated emitter solar cell* (PESC) that resulted in demonstrated efficiencies surpassing 20% [42]. Simultaneously, the *interdigitated back contact* (IBC) concept and the *point contact* (PC) solar cell were developed [101]. The structure of the PC-cell is shown in Fig. 8.7. The PC-cell is characterized by:

- Si-wafers of outstanding quality to minimize volume recombination losses;
- the IBC concept to avoid contacts at the illuminated surface and, consequently, shading losses;
- the PE-concept (SiO_2-layers) to minimize surface and interface recombination losses. It is realized throughout the whole surface of the Si wafer, except at the miniature point contacts.

All these structures can be produced by means of well-proven technologies of integrated circuit production. A further modification, the PERC-cell (*passivated emitter and rear contact cell*), holds the efficiency record of 24% at the present time[6].

---

[6] announced by M. A. Green/UNSW at the 21st IEEE Photovoltaic Specialist Conference, Orlando/FL, USA, May 1990.

Fig. 8.7. Point contact silicon solar cell [101].

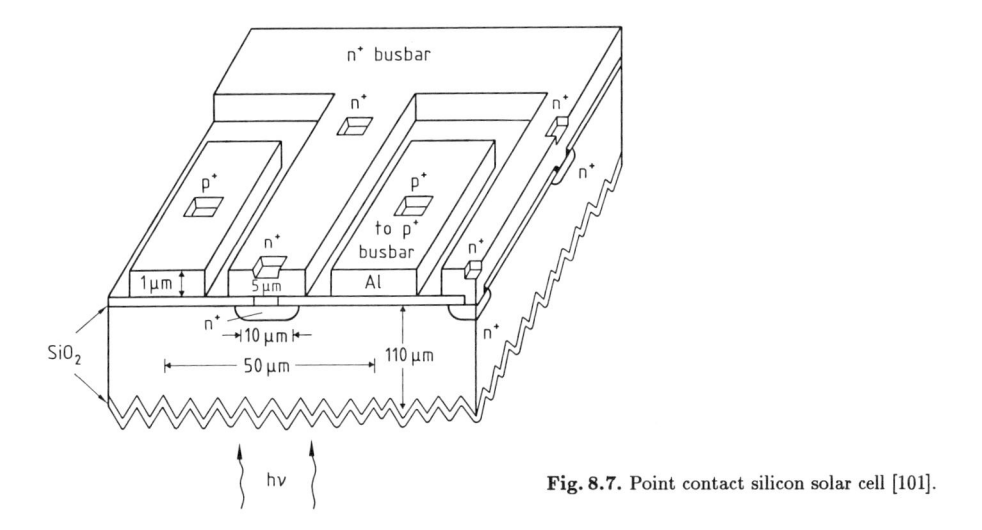

8.3.2 Amorphous Silicon Thin Film Solar Cells

During the last years by far the greatest research efforts in the field of thin film solar cells have been made with films and solar cells based on amorphous silicon (a-Si). It is true that the amorphous state does not provide a band structure like a crystalline semiconductor and that dopability of the material and diffusion length of photocarriers are still limited, but the extensive research in this field and the development of particular film deposition technologies resulted in efficiencies of a-Si based solar cells of up to 12% [61].

The ease of fabrication of small-area integrated modules with thin film cells founded the economic success of a-Si solar cells in the consumer electronics market. A worldwide shipment of 13.9 MW_p in 1987 is reported, 9.7 MW_p of which were produced in Japan [7].

The optical bandgap of a-Si:H is near 1.7 eV and thus somewhat higher than the optimum bandgap for photovoltaic conversion (1.5 eV). However, changes of the bandgap can be realized by preparation of amorphous alloys: the substitution of Si atoms in the amorphous network by C or N atoms as well as by Ge or Sn results in a shift of E_g to larger, and for Ge and Sn to lower values. These properties not only allow the bandgap to be tuned to optimum for single-cell- and tandem-PV systems, but also for special applications, e.g. cells acting as semitransparent coatings of automobile sunroofs powering cooling fans.

Still a serious problem concerning a-Si solar cells is degradation of efficiency with time (*Staebler-Wronski effect*). It can be attributed to an increase of the number of *dangling bonds* in the bulk material acting as recombination centers, which may be caused by weak bond breaking (Si-H bonds, Si-Si bonds) [98] or breaking of paired bonds [9]. These mechanisms have to be considered to be system-inherent. Up to now, outdoor stability tests of a-Si cells and generators always resulted in a decrease of efficiency not tolerable for the power sector. Research efforts are going on with the aim of minimizing degradation by applying stacked *quasi-tandem* structures and by applying advanced thin-film deposition techniques (various types of Chemical Vapor Deposition (CVD) techniques).

8.3.3 Polycrystalline Thin Film Solar Cells

Thin film solar cells based on compound semiconductors have attracted increasing attention in recent years, because rapid progress has been made in the fabrication of efficient devices. Furthermore, an inherent stability problem like in a-Si seems not to exist. The most advanced candidates at present are cells based on $CuInSe_2$ and CdTe. A second group can be mentioned which also has given promising results: $CuGaSe_2$, $Cu(In,Ga)Se_2$ and CdSe.

Most of these materials cannot be made n-type by doping, they show a preferential type of conductivity. For solar cells, therefore, pn-heterojunctions with their particular advantages and disadvantages [28,31] are required.

The characteristic features of these polycrystalline cells are:

- single-cell efficiencies of more than 10% were achieved at several labs and with a considerable number of production methods; even large-area modules of $CuInSe_2$ from Arco Solar/Siemens Solar exceed the 10%-bound [73];
- among the production technologies the low-cost process of screen printing and sintering is applied successfully for CdTe-modules [51];
- low-cost substrates such as glass or metal foils can be used;
- the amount of ecologically harmful materials (Cd-compounds) can be minimized.

For more details see [19,23,82]. In conclusion, CdTe- and $CuInSe_2$-cells have good prospects for application in the power sector.

8.3.4 Concentrator Cells

For large-scale stand-alone or grid-connected PV systems the application of concentration is still in discussion, even for Southern or Central Europe (see Sect. 8.7).

Silicon concentrator cells achieved efficiencies of 28% at an optical concentration $C = 150$ [94]. The production technology of these cells is in principle based on that of the high efficiency cells mentioned above (PESC cell, PC cell).

III-V concentrator cells offer a great potential for high concentration systems. 28.1% under concentrated radiation were obtained by Varian Associates[7].

8.3.5 Tandem Solar Cells

In the case of tandem solar cells, two different development lines have to be distinguished: high-efficiency tandem cells and low cost tandem structures based on thin film cells. The efficiency of a tandem structure already passed the 30% mark [3]. With a mechanical stack composed of a GaAs top cell from Varian Ass. and a crystalline Si bottom cell from Stanford an efficiency of 31% at a concentration of 350–500 suns was measured at Sandia National Laboratories. The development of tandem configurations of this type continues: efficiencies of 35% with 2-cell tandems and of 40% with 3-cell tandems are expected[8].

From the theoretical point of view, low-cost 2-cell tandem solar cells can be made with efficiencies slightly above 15% with subsystems showing efficiency values of about 10%. The most successful effort in developing low-cost thin film-based tandems today (1989) is an a-Si/$CuInSe_2$ tandem fabricated at Arco Solar/Siemens Solar, with an efficiency of 15% [73].

---

[7] A pilot production of GaAs cells is being installed at Spectrolab, Sylmar/CA, USA.
[8] In July 1989 Boeing announced 37% efficiency with a GaAs/GaSb tandem concentrator solar cell with prismatic covers at a concentration of 100 suns.

A production line for modules using this type of tandem solar cells is planned. For detailed information about further efforts see [19,23,82].

8.4 Photovoltaic Modules

The smallest electrical unit formed with solar cells is a module. This section deals with the status of non-concentrator module technology and interconnection schemes of solar cells within modules.

8.4.1 Status of Non-Concentrator Module Technology

The main functions of a module are:

- electrical interconnection of solar cells;
- mechanical and environmental protection of the cells and their interconnections;
- structural support.

The construction elements of modules with discrete or thin film solar cells are outlined in Fig. 8.8 [8]. The requirements for the front cover are: low reflectance, good optical transmission, chemical durability, abrasion resistance, mechanical stability, and simple processing.

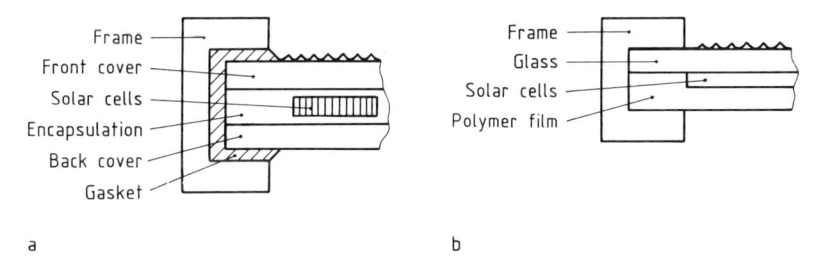

Fig. 8.8. Present constructions of non-concentrator modules with (a) discrete and (b) thin film solar cells.

The transmittance through the front cover varies due to reflection losses of radiation at the surface as a function of the angle of incidence. A reduction of reflection losses can be achieved by shaped surfaces (see Figs. 8.8 and 8.9) through multiple reflections at the module surface. The shape with lowest reflection loss is the *Gothic Vee*. Specific disadvantages of structured surfaces are their cost and their sensitivity to dust deposition.

A commonly used compromise covering all requirements is a low-iron, tempered glass with a typical thickness of 2–3 mm. Alternative front covers like low cost synthetic materials are mainly used for consumer electronics and indoor PV applications. At present, disadvantages like poor mechanical and ultraviolet stability and increased soiling makes an outdoor application of synthetic materials not advisable [39].

The frame of a module is formed by aluminium or stainless steel profiles with a sealant gasket to minimize stress concentration caused by differences in thermal expansion coefficients. Additionally, some manufacturers also apply a weatherproof seal between glass and gasket. These materials are typically Butyl tape, Silicone rubber, Neoprene tape, Polyvinyl Chloride, Polyethylene tape or Kapton tape. Advantages of these materials are low cost and

60° Vee   90° Vee   Gothic Vee

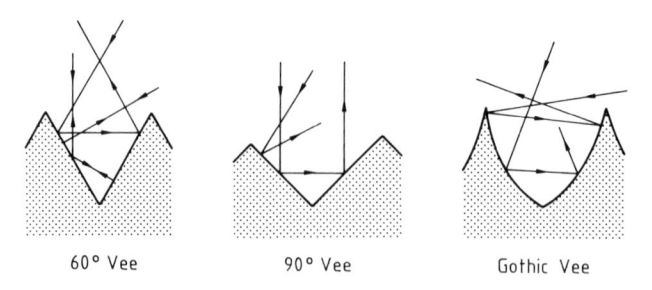

Fig. 8.9. Shaped surfaces for reduced net reflection losses through multiple reflections [70].

easy application. A disadvantage is the poor fire resistance [99]. Current developments apply injection molding techniques with synthetics for the frame of thin film modules or pure laminates without any frame. Requirements for encapsulation materials are UV-stability, resistance against microorganisms and electrochemical corrosion, good adhesion to solar cells and covers, easy processing, low sensitivity to water sorption, no outgassing of volatile substances, thermal stability, and low cost. Several encapsulation materials have been proven suitable for the photovoltaic industry [35,74], e.g.:

- PVB, Polyvinyl Butyral, a thermoplastic material. It is very hygroscopic and requires careful temperature and humidity control during storage, conditioning and sheeting operations. Water vapor blocking at the back cover is realized by glass or a metal foil. The excellent adhesion of PVB is the main advantage.
- EVA, Ethylene Vinyl Acetate, a thermoplastic material with a better light transmission characteristic than PVB. With some additives a good adhesion can be realized. A peroxide-catalysed cure to guarantee creep resistance and high lifetime is required.

The back cover of a photovoltaic module is realized either by a glass plate or by a back sheet system of Tedlar/aluminium foil/Tedlar for protecting the solar cells from environmental influences.

For further reduction of module specific cost, integration of the module in the support structure (e.g. structural frame profiles or direct gluing of the laminates) along with the diversification of module types for different applications (power plants, small end-user systems) have to be considered.

8.4.2 Module Design and Interconnection of Cells

The efficiency of solar modules depends, among other factors, on the solar cell temperature and the interconnection scheme. In the following some design aspects will be outlined.

Operating temperature. As indicated in Sect. 8.2, an increase of the solar cell temperature causes a significant decrease of efficiency. Therefore, it is necessary to keep the solar cell temperature as low as possible. The solar cell temperature rises to a level at which the heat removal by convection, radiation and conduction via array support structure plus the electrical output balances the solar energy inflow. This temperature level is known as the equilibrium or operating temperature [85].

Several thermal models were developed to calculate the operating temperature in a simplified analytical way [10,34,90]. Other models describe the solar cell temperature for special

8.4 Photovoltaic Modules

sites and cases with the help of a regression analysis of photovoltaic system performance data [86].

Possible measures to reduce heating in PV modules are to increase heat dissipation from the back cover through convection and emission (e.g. a significant decrease in solar cell temperature by use of a black metal plate with fins, Fig. 8.10 is reported [30]). Another approach uses interference filters to reject the part of the spectral solar irradiance unusable for solar energy conversion.

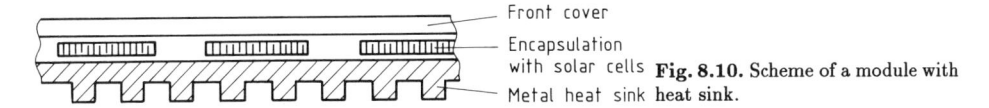

Fig. 8.10. Scheme of a module with heat sink.

Interconnection of solar cells. Solar cells are not exactly identical, even if carefully sorted, and a mismatch of the *IV*-characteristics always occurs. Secondly, shading can occur accidentally through fallen leaves or through other rows of modules, bordering roofs, trees and wires. As a result two effects must be considered:

1. Loss of output power from the module.
2. Power dissipation in the shaded cells (*hot spots*, see below).

The second point is more important, since heat dissipation in shaded cells can lead to the destruction of cell or encapsulation. The output power of partially shaded modules and/or mismatched solar cells depends on factors like time variance of shadow, non-uniformity of irradiance, lateral thermal gradients, and *IV*-characteristics of the relevant cells. Several modelling approaches were reported [24,32].

The consequence of partial shading will be illustrated by two simple examples, one for series connection and one for parallel connection:

In Fig. 8.11 two solar cells with identical *IV*-characteristics are connected in series. One cell is partially shaded leading to a short circuit current lower than that of the unshaded cell. The resulting IV-characteristic is graphically constructed by adding the voltage values at particular current values. Without protection measures the shaded cell becomes reverse biased. Depending on the operating point P, the reverse bias can reach high negative voltages, and forces the shaded cell to act as a load. The dissipated power in this load heats the cell and the encapsulation material. This so-called *hot spot* can damage a module irreversibly. If the reverse bias voltage exceeds the negative breakdown voltage of the cell, it might even be destroyed.

The *IV*-characteristic of the discussed cell combination depends strongly on the reverse characteristic of the shaded cell. It follows from Fig. 8.11 that the output power is increased,

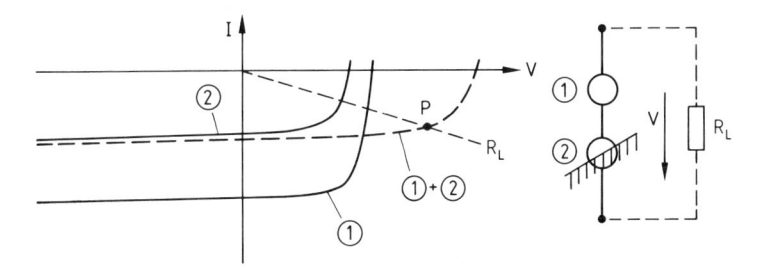

Fig. 8.11. *IV*-characteristic of series-connected solar cells with one shaded cell.

if cells with low breakdown voltages are used. This leads to the use of shunt- or bypass-diodes connected across shaded cells or strings.

By addition of shunt diodes across shaded strings a low breakdown voltage is artificially produced. If a current greater than the short circuit current of the shaded cell is forced through the shaded cell, this cell is switched to reverse bias but the shunt diode is switched to forward conduction, thus limiting the dissipated power in the cell. Figure 8.12 illustrates the effect of a bypass diode in the same configuration as Fig. 8.11.

In Fig. 8.13 two solar cells with identical IV-characteristics are connected in parallel. One cell is partially shaded leading to an open circuit voltage lower than the value of the unshaded cell. The resulting IV-characteristic is graphically constructed by adding the current values at particular voltage values. Without protection measures the terminal voltage at the shaded cell could reach values higher than the cell's own open circuit voltage, depending on the operating point P, and could force the shaded cell to act as a load. The dissipated power in this load heats the cell and the encapsulation material. The resulting IV-characteristic depends on the characteristic of the shaded cell in forward direction.

A reduction of power losses can be achieved through blocking diodes. If the operating voltage of the system exceeds the open circuit voltage of the shaded cell, the blocking diode prevents forward current flow and thus power dissipation in this cell. Figure 8.14 illustrates the effect of a blocking diode in the same configuration as in Fig. 8.13.

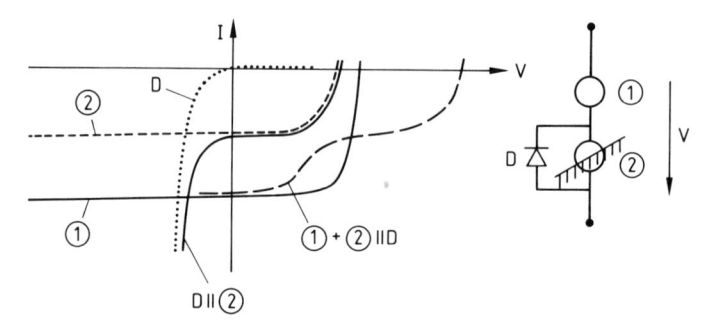

Fig. 8.12. The effect of a bypass diode at a partially shaded cell in series-connection with unshaded solar cells.

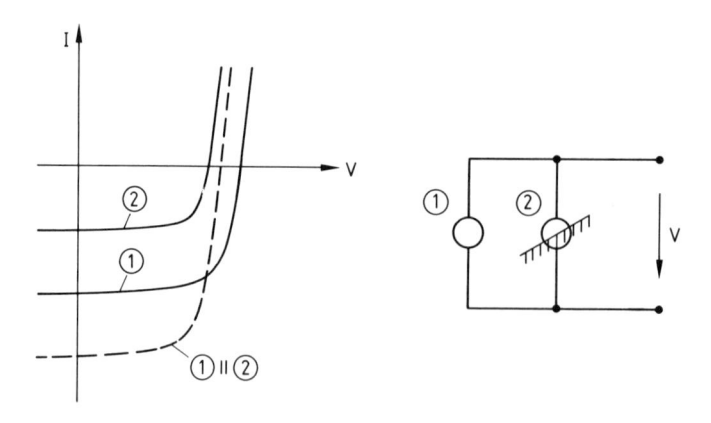

Fig. 8.13. IV-characteristic of parallel-connected solar cells, one cell shaded.

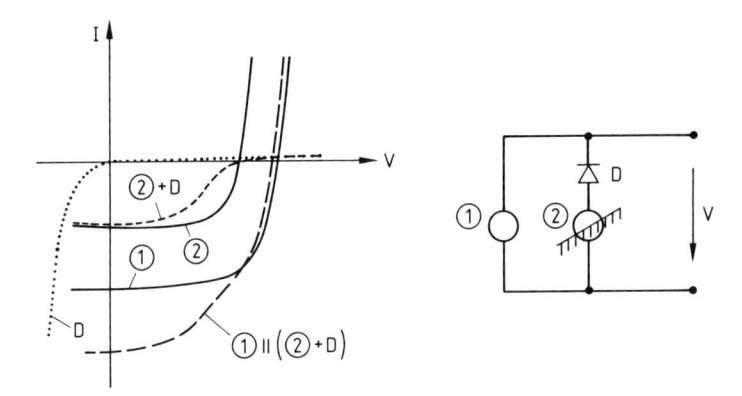

Fig. 8.14. The effect of a blocking diode at a partially shaded cell in parallel-connection with unshaded solar cells.

Regarding the use of blocking diodes, a comparison is necessary between permanent power losses through these diodes and possible power gains by allowing operation despite the occurrence of shadow or mismatch conditions (see Sect. 8.8.2).

8.5 Power Conditioning Systems

The varying solar irradiance requires an electronic adaptation of the fluctuating current/voltage to the load. A general photovoltaic system is outlined in Fig. 8.15.

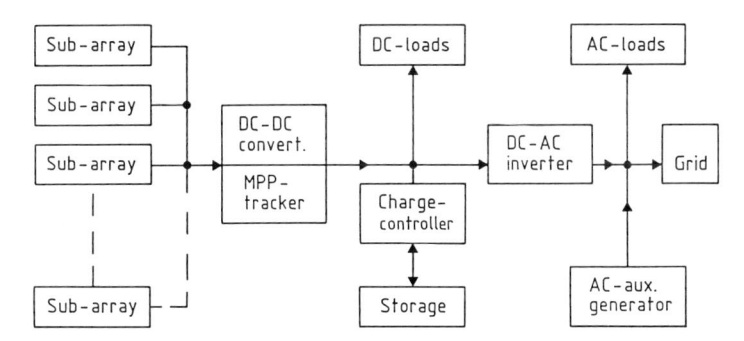

Fig. 8.15. General photovoltaic system.

8.5.1 DC-DC Converter, Maximum-Power-Point Tracking

DC-DC converter. The device to change direct-current voltage levels is a DC-DC converter. Several types of DC-DC converters for different power ranges are known [5,27]. With switching elements like transistors or thyristors a direct voltage is switched on and off. The average value of the output voltage is controlled for example by variation of the duty cycle or the frequency of the switch-on/switch-off period.

Applying the flyback converter principle [27] (see Fig. 8.16), energy from the input (e.g. the solar generator) is only transferred to the transformer as a storage element when the switch is closed, and not to the load. During the off-period of the switch, energy is transferred to the load. The input capacitor C_1 smoothes the input voltage and delivers the intermittent input current. The control logic for the switch is not shown. Disadvantages are the need for large transformers with large core sections and air gaps to store energy, and the limitation of accessible power to several hundred Watts [5]. The possibility of multiple outputs and the simplicity, on the other hand, are attractive.

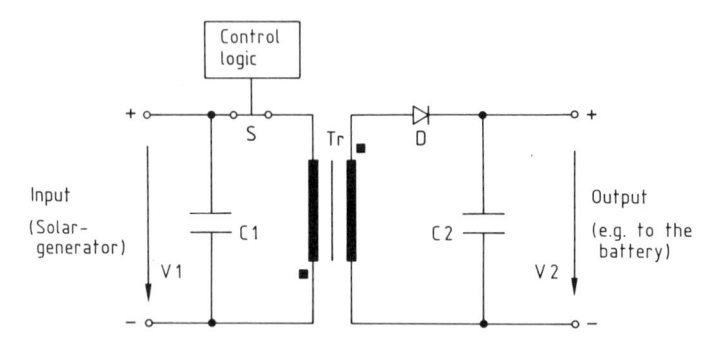

Fig. 8.16. Flyback converter with transformer isolation, (buck-boost converter).

Maximum-power-point tracking. For operation at the maximum power point (MPP) a controlled DC-DC converter providing an impedance matching between the dynamic resistance of the solar generator and the load is necessary. The gain in energy output achieved by applying an MPP-tracker instead of simpler systems has to be counterbalanced with the increased cost.

Considering a *constant-voltage* load, like a battery for example (see Sect. 8.2.2), a well-adapted solar generator or a simple DC-DC converter with fixed output voltage is usually more cost-effective than using an MPP-tracker. For grid connection MPP-tracking is more attractive, because a DC-AC inverter can be made to perform MPP-tracking at low additional cost.

A simple method to calculate the maximum power point of a PV generator is to put the derivative power $(v \cdot i)$ to zero:

$$dP = \frac{\partial(v \cdot i)}{\partial i} \cdot di + \frac{\partial(v \cdot i)}{\partial v} \cdot dv = 0 \qquad (8.12)$$

$$\frac{dv}{di} = -\frac{v}{i} = R_m \qquad (8.13)$$

Equation (8.13) is fulfilled by an impedance matching of the dynamic resistance R_m of the solar generator and the resistance of the load. The control logic is realized e.g. with a microcontroller as a part of a power controller or with the help of a graphical procedure using operating characteristics [20].

8.5.2 DC-AC Inverter

The transformation of DC-power into AC-power is accomplished by a DC-AC inverter. The basic principle is shown in Fig. 8.17. A filter for the higher harmonics follows the AC output of the square wave generator.

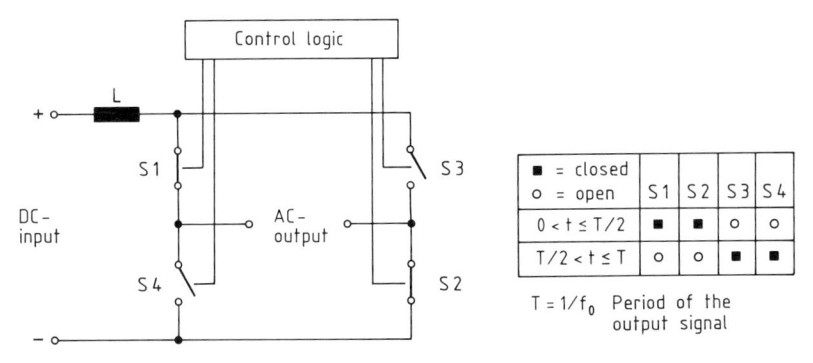

	S1	S2	S3	S4
■ = closed, ○ = open				
$0 < t \leq T/2$	■	■	○	○
$T/2 < t \leq T$	○	○	■	■

$T = 1/f_0$ Period of the output signal

Fig. 8.17. Basic DC-AC inverter circuit (square wave generator).

A new method is the step-wave approximation of the sinoidal wave form with several DC sources at defined voltage ratios. Good efficiencies with low harmonic distortion are reported [15]. Another high-efficiency DC-AC conversion technique is to modulate a high frequency square wave generator with a reference sine wave signal. A sine wave is obtained after filtering of the pulse pattern.

The DC-AC inverters are subdivided into two categories:

1. Externally commutated inverters; they require an external AC source for the commutation voltage, they are simple, low-cost approaches for grid connection, yet with a high distortion factor.
2. Self commutated inverters; they do not require external AC sources. Commutation is carried out through internal energy storage elements. They are in principle suitable for stand-alone applications only.

For realizing power switches up to a power range of $\approx$ 50 kVA, transistors and thyristors are more suitable; above 50 kVA, thyristors are used almost exclusively.

8.5.3 Batteries and Charge Regulators

Storage is a requirement in many cases of terrestrial applications of solar energy to match the supply with the load demand. Storage elements are already used in conventional power plants for instant power reserve, peak-power levelling, and grid frequency control [62,104]. Various methods of storing electrical energy are known [45,63,84], especially electrochemical storage in batteries.

The predominant choices for storage are lead-acid and nickel-cadmium batteries, because of their reliability, availability, and flexible sizing. Main disadvantages are their high cost, low energy density, and the need for a careful treatment [44]. In many cases charge regulators are necessary to control the charge/discharge cycles, charge condition, and battery temperature. The number of possible cycles during lifetime decreases drastically with the depth of discharge (see Fig. 8.18). A correct charge and discharge control requires sophisticated regulators com-

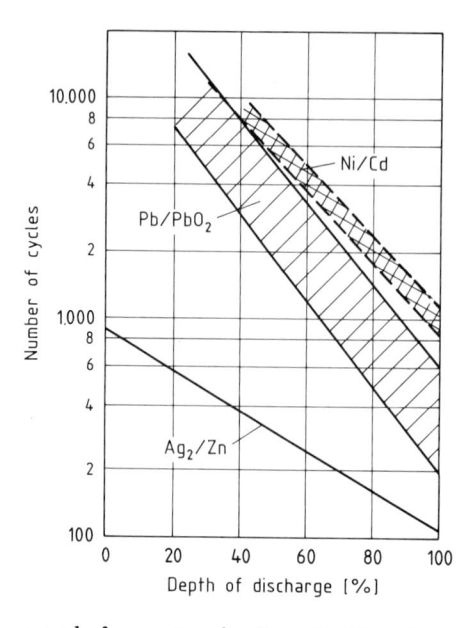

Fig. 8.18. Number of cycles versus depth of discharge for several electrochemical cells [44].

posed of current and voltage limiting elements [60]. The optimum control strategy for charge regulation is uncertain because reliable battery models are missing.

At present several advanced electrochemical cells are being developed to improve energy density, self discharge rate, and cycle life. Two promising systems are $Na/\beta Al_2O_3/S$ and $Zn/ZnBr_2/Br_2$. The sodium/sulphur battery uses molten reactants with an operating temperature of 330°C and βAl_2O_3. It meets the requirements of low maintenance, high energy density and efficiency. Disadvantages are limited cycle life and number of thermal cycles. The zinc/bromine battery, mainly developed in the U.S., Japan and Austria, seems to be very suitable for solar energy applications due to the high cycle number of more than 2,000 [103]. Table 8.1 summarizes theoretical and practical energy density values of several electrochemical systems.

Table 8.1. Characteristic values of several electrochemical batteries [103].

	$Pb/H_2SO_4/PbO_2$	$Na/\beta Al_2O_3/S$	$Zn/ZnBr_2/Br_2$
Cell voltage	2.05 V	2.1 V	1.83 V
Energy density			
(theoretical)	161 Wh/kg	792 Wh/kg	438 Wh/kg
(practical)	40 Wh/kg	90 Wh/kg	160 Wh/kg

8.6 Supporting Structures

For an economic design of a photovoltaic power plant the choice of a suitable support structure is an important cost factor. In Europe, through the implementation of fifteen PV pilot projects supported by the Commission of the European Communities (CEC), different solutions for exclusively fixed tilted structures were realized. In the following a review of some realized low-cost support structures is given with a summary of design criteria, parameter variations, and approaches proposed by several studies.

8.6.1 Basic Design Considerations

The development of a low-cost support structure means the optimization of material, fabrication, installation, versatility, durability, and environmental demands. Criteria are:

1. Material.
   - acceptable performance under load conditions;
   - availability in the open market;
   - lifetime with respect to system lifetime.
2. Fabrication.
   - potential for assembly line production techniques;
   - ease of transport;
   - tolerances required by common fabrication techniques.
3. Installation.
   - ease of array connection to support structure;
   - requirements for special tools and training;
   - ease of attachment of structure to its foundation;
   - constraints by structure design on site preparation;
   - protection against theft.
4. Versatility.
   - adaptability of the structures to multiple uses;
   - adaptability for changing module / array size;
   - possibility of replacing individual modules.
5. Durability.
   - level of maintenance;
   - structural behavior under design loads;
   - corrosion and abrasion resistance.
6. Environment.
   - aesthetics of support structure;
   - influence of irradiation on the complete array;
   - thermal effects of arrays on the support structure;
   - resistance to fire and vandalism.

8.6.2 Review of Selected Support Structures

As a result of several European studies [36,40,41], two support structures were proposed. Figure 8.19 shows a low-cost support structure realized in the 100 kW_p PV plant on the island of Kythnos, Greece. The second proposed support structure is unconventional. It was derived from viniculture structures (Fig. 8.20).

Production and installation costs, including site preparation, mounting of modules and wiring of array field, of all support structures realized in EC pilot projects vary between 70 $/m^2 and 240 $/m^2.

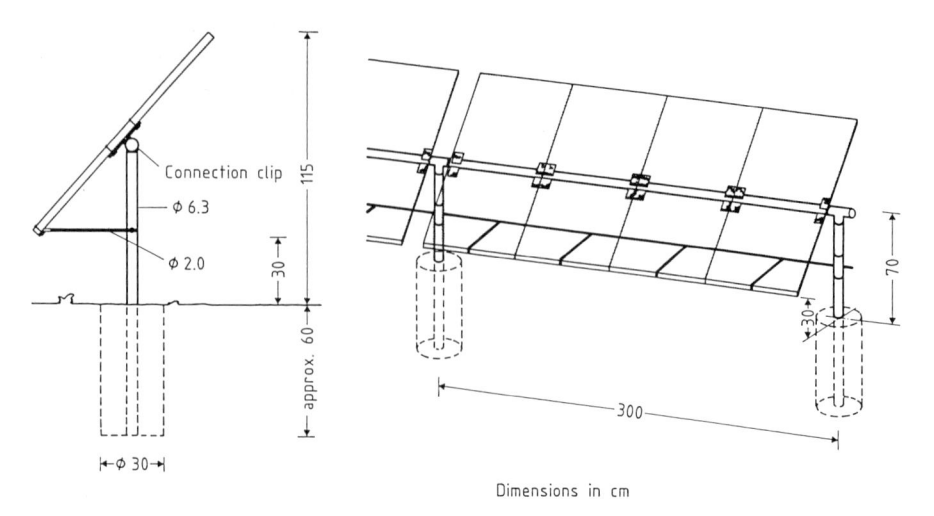

Fig. 8.19. Low-cost support structure on Kythnos [36].

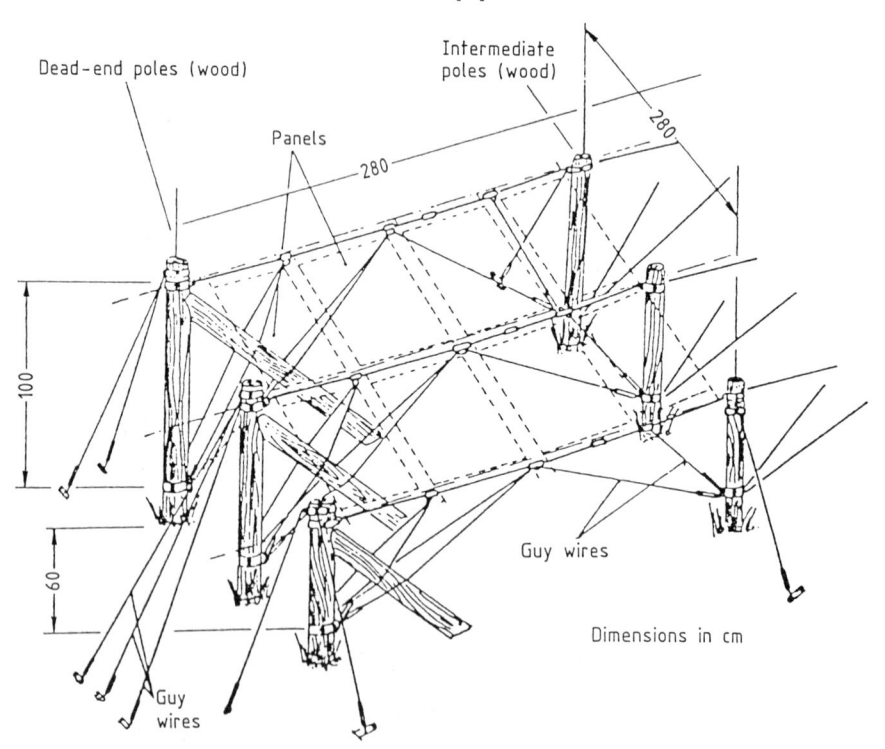

Fig. 8.20. Low-cost support structure [41].

The subject of further studies was to develop a support structure for various soil conditions [13], and to develop a standard building block design for low-cost flat plate arrays [25]. Figure 8.21 shows the final solution.

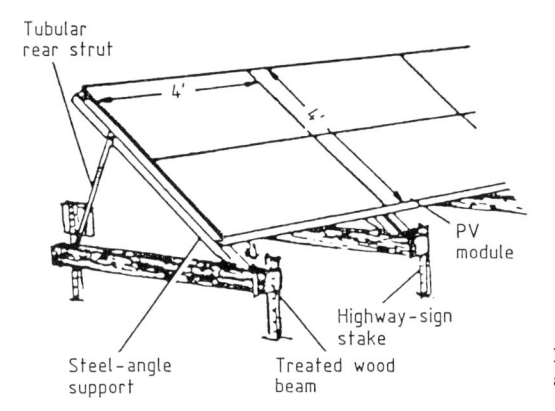

Fig. 8.21. Structure and foundation design for a standard building-block array field [13].

For future cost reduction the following design parameters should be taken into account:

- Examination of soil conditions referring to the possibility of ramming pedestals directly into the earth and avoiding concrete foundations.
- Use of structural shapes designed to the required stresses rather than selection of standard available shapes which may be heavier than required.
- Integration of the structural design for the module- and support frame. It may reduce material requirements for the overall system.
- Wind and snow load assumptions imposed on structures in the interior of the collector field. Adequate consideration of wind shielding or blocking either by the peripheral array structures or by the addition of separate windbreak structures could affect structural design and construction cost of the majority of supports inside collector fields.

8.6.3 Support Structures for Tracking Arrays

At present (1989), there are only two small test sites in Europe for tracking photovoltaics R&D; Adrano, Italy, and Widderstall/Merklingen, Germany[9]. Several large tracking PV power plants are operational in the U.S. (see Sect. 8.9).

A first approach was to adopt two-axes tracking heliostat structures to tracking PV needs [67]. Double axes tracking heliostats from the 10 MW Solar One plant at Barstow/CA, USA, cost 685 \$/m^2 [59]; the estimate for heliostats of tomorrow's SPPs (Phoebus project) is 250 \$/m^2 (with eventual costs under 100 \$/m^2 projected [6]).

Another, probably more viable approach for sun-trackers in photovoltaic systems is the passive thermohydraulic system. It operates for example by liquid flowing between sealed frame parts or by vapor pressure difference in a double-acting hydraulic actuator (Fig. 8.22). The inaccuracy in tracking up to 10° is tolerable for non-concentrating devices. Good performance of one-axis tracking thermohydraulic systems are reported [88]. The entire thermohydraulic tracking system at the 300 kW PV power plant in Austin costs 105 \$/m^2, including foundations but without modules, installation, and site preparation [49].

---

[9] Widderstall is a test facility of the Centre for Solar Energy and Hydrogen Research (ZSW), Stuttgart, Germany.

Fig. 8.22. Robbins' passive thermohydraulic tracker at the PV test site Widderstall/Merklingen, Germany (courtesy of ZSW, Stuttgart, Germany).

8.7 Tracking and Concentrating Systems

The main impediment to the widespread use of photovoltaic conversion is the cost of the solar cells. A possible approach to overcome this limitation is the use of low-cost optical concentrators, which focus solar radiation onto cells with an area much smaller than the concentrator aperture. In addition, concentration leads in principle to an increase in cell efficiencies.

Concentrators are mainly used in thermal SPP. A review of typical solar concentrators is given in Chaps. 4 and 7. Some of them are applicable to photovoltaic power conversion. The main difference between thermal and photovoltaic concentrator application is that higher temperatures under concentrated sunlight lead to an increase of efficiency for thermal collection, but to a decrease of efficiency for photovoltaic conversion (typically 0.45% of power per degree Kelvin).

Typical concentration ratios are small, often less than 25. A low concentration ratio means a large field of view. There are four basic reasons why the field of view is an important characteristic of a PV concentrator module:

1. Since production of concentrators and trackers should be cheap, only a limited fabrication accuracy is feasible.
2. Photovoltaic systems are very sensitive to an inhomogeneous illumination of modules and arrays (see Sect. 8.4). This can not only cause power losses but also serious damages to the modules and arrays.
3. The larger the field of view the more diffuse radiation is utilized. This is especially important for regions with high diffuse portions.
4. If the angular acceptance of the concentrator is too small, the circumsolar radiation will be lost.

The operation of solar cells under high illumination, e.g. removal of heat from these cells, demonstration of efficient low-cost concentrating optics and design of reliable cost-effective tracking structures have been in the centre of attention for 15 years [12]. Today the Fresnel concentrator is mostly applied and still investigated to improve this technology. A second system used is the V-trough concentrator, applied in the world's largest photovoltaic power plant at Carrisa Plains.

In the following sections these systems will be discussed and some remarks will be given concerning other concentrators like the CPC (compound parabolic concentrator), concentrators with parabolic geometries and concentrators using dispersive elements.

8.7.1 Fresnel Modules

As an example the principle of a point-focus Fresnel module is shown in Fig. 8.23. The Fresnel lens is made from polymers and is covered by glass. To achieve higher tracking tolerances a terminal concentrator (see Sect. 3.4) optical element is applied. A key feature of the design is the material selection and installation of an isolator between cell and heat sink, which withstands the high temperature changes, but permits the highest possible heat flux to the module housing. The cell package consists of the cell mounted on a substrate of BeO, with appropriate accessories for thermal cooling, electrical isolation, photoelectric energy removal, geometric positioning, and support of the secondary optical element.

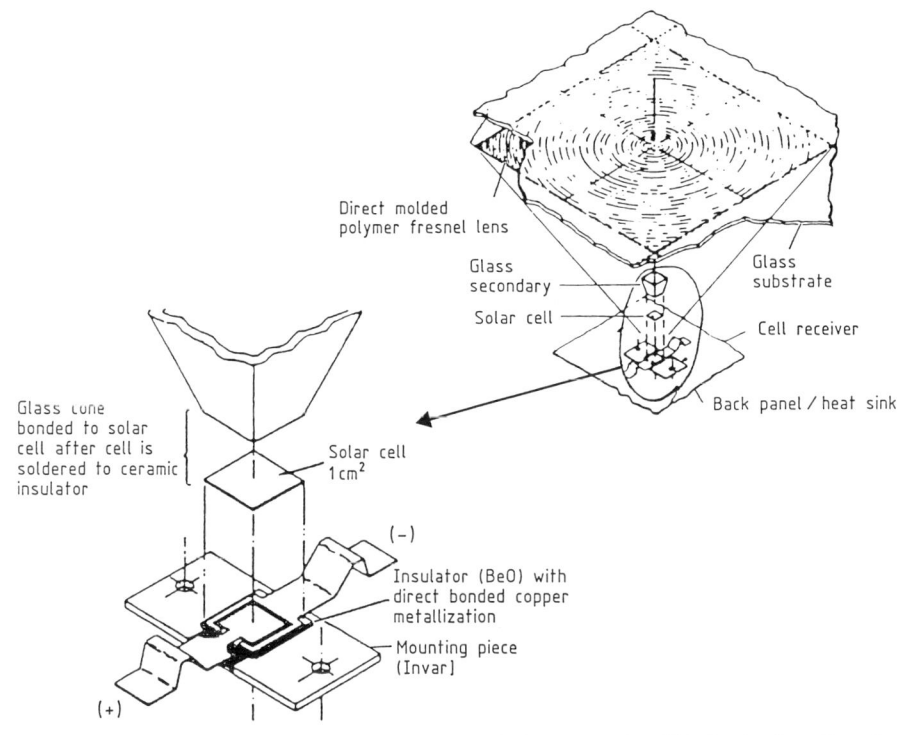

Fig. 8.23. Possible cell mount and design approach for a high concentration (500×) point-focus Fresnel module [65].

Refractive Fresnel lens technology is the optic choice of today. Four major Fresnel concentrator PV systems were built and operated successfully:

- Soleras 350 kW$_p$ at Riyadh, using two-axes tracking 40×-point-focus Fresnel lenses (by Martin Marietta),
- Dallas-Fort Worth (DFW) Airport 27 kW$_p$ plus 140 kW$_t$ at Dallas, using one-axis tracking 25×-linear Fresnel lenses (by Entech),
- Sky Harbour Airport 225 kW$_p$ at Phoenix, using two-axes tracking point-focus Fresnel lenses (by Martin Marietta). This plant has been decommissioned.
- The fourth major PV concentrator project has recently been installed in Austin/TX and is rated at 300 kW$_p$. The collector field consists of laminated, arched linear Fresnel lens

PV concentrators (by Entech) with two-axes tracking modules and 22.5× concentration. The prototypes of these modules achieved more than 16% peak efficiency [80].

The most important recent advances made in linear Fresnel collector technology are prism covers which effectively eliminate front grid shadowing, even with metallization fractions as high as 20–40% (by Entech). These large metallization fractions in turn permit use of cells fabricated with low-cost 1-sun technology [21].

There are other designs of PV concentrator modules currently being pursued (by Sandia National Laboratories, Black & Veatch, Electric Power Research Institute, Intersol, Alpha Solarco and Watsun) reaching concentrations up to 500×.

8.7.2 V-Trough Concentrator

The V-trough concentrator technology as exemplified by Carrisa Plains utilizes not only one-sun cells, but also ordinary one-sun module technology in connection with simple plane mirrors. The main idea behind this concept is that both technologies are available as mass products. A low concentration ratio does not require high accuracy in fabrication and tracking.

Figure 8.24 shows a double-axes tracked device with a geometrical concentration of two. This value may not be the optimum value, but is the upper limit, if one utilizes one-sun modules. In principle, the optimum concentration is a function of the fractions module/mirror costs and diffuse/direct irradiance.

Fig. 8.24. Example of a V-trough concentrator, double-axes tracked (courtesy of ZSW, Stuttgart, Germany).

The modules in a V-trough with an opening angle of 60° receive approximately the same amount of diffuse radiation as those without the mirrors, assuming a reflectivity of 0.8 (under environmental conditions). This was calculated [77] using an anisotropic diffuse sky radiation model [81]. This characteristic of the low concentration device makes the concentrator attractive even for areas with high diffuse radiation portions. It should be emphasized, however, that the aperture area of the whole system with reflectors is twice that of the flat-plate module.

But in case of the Carrisa Plains photovoltaic SPP this approach failed. The higher module operating temperatures caused by concentration and inadequate convective cooling (hindered by the mirrors) caused discoloration of the encapsulation material (EVA) by thermal effects. Module temperatures approached 90°C for extended periods of time during summer with high ambient temperatures experienced at Carrisa Plains. This discoloration effect reduced the plant efficiency from as high as 10.3% in 1984 to 8.5% in 1987 [100].

8.7.3 Parabolic Geometries

More sophisticated concentrator designs like CPC (Compound Parabolic Contentrator) and parabolic geometries are introduced in Chap. 3. But there is a serious problem with these higher concentrating devices: the solar cells are no longer homogeneously illuminated, unless highly sophisticated materials and high fabrication and tracking accuracy are guaranteed (see Sect. 3.5.4). It turned out that these demands could not be achieved at cost competitive with conventional arrays [22].

Parabolic troughs were tested in connection with active cooling systems for heat and electricity cogeneration. These photovoltaic-thermal hybrid plants have a limited number of applications where they can be effectively used. In general co-generation plants produce 5 to 7 times more low-temperature ($< 100°C$) thermal energy than electrical energy. Since loads requiring large amounts of low-temperature thermal energy are difficult to identify, and since thermal transport of low-temperature heat is very expensive, this is a rather specialized collector option.

8.7.4 Further Concentrator Concepts

There are two further concentrator concepts which have not had any importance for SPP applications up to now. However, they demonstrate the different principles of concentration of solar radiation that may be possible. The two developments are the FLUKO (Fluorescent Concentrator) and the DISCO (Dispersive Concentrator). The idea behind the FLUKO is as follows: Incoming solar irradiance (diffuse or direct) is absorbed by a dye dispersed in a polymer plate and re-emitted with a certain Stokes shift to lower frequencies. Thus, it cannot be reabsorbed (ideal assumption) by the other dye molecules. Because of total internal reflection at the surfaces of the polymer plate most of the light is guided within the plate and therefore concentrated onto solar cells, which are fixed at the edges of the plate.

Nearly every geometric-optical concentration demands tracking. The FLUKO does not. But the loss mechanisms involved in the different steps of absorption, fluorescence and reflection result in concentrator efficiencies of less than 10% [38].

The DISCO is a holographic beamsplitter for multijunction devices. It can be combined with a Fresnel lens [50], or spherical and cylindrical dispersive holographic lenses can be produced, which allow spectral splitting and concentration by the same optical element [43]. The use of multiply exposed holograms allows a separation of different wavelength regions. The wavelength can be adjusted and optimized for various combinations of solar (tandem) cells. A dichromated gelatin plate is used to record the hologram.

DISCO module efficiencies achieved today are not higher than those of conventional tandem structures using the same solar cells without concentration and without beam splitting. This is because satisfactory diffraction efficiencies are only realized for small spectral bandwidths today.

8.7.5 Perspectives of Tracking and Concentrating Systems

The question addressed is which PV system option is most cost effective for a future generation of multi-MW plants. To answer this question, a complex analysis is needed if concentrating devices like Fresnel lenses are to be compared with tracking or fixed-tilted flat-plate systems. The most important parameters which impact an appropriate life-cycle cost analysis are unit costs of concentrator and flat-plate modules, and of trackers and their components. The predominant insolation is the most important site specific parameter.

In two studies [77,83] six options including fixed, one-axis and two-axes tracking flat-plate collectors as well as V-troughs and other concentrators utilizing linear and point-focus Fresnel optics, were studied. Results were:

- One should not expect that tracker or concentrator technologies are the key to an essential reduction of PV energy costs.
- None of the six competing options is a clear winner over the others for tomorrow's technologies.
- For present technology, Fresnel-concentrator options show a small cost advantage at high irradiation sites. For low to medium irradiation, performance of the tracking flat-plate and the V-trough options is superior.

The comparisons are dominated by the manufacturing costs of solar cells. As long as ordinary flat-plate technology is expensive, it may be advantageous to *refine* the cells by Fresnel lenses and trackers. If the flat-plate module costs decrease, due to improvements in thin film technologies for example, the area-related costs for tracking and concentrating technologies become more and more important. This trend is demonstrated in Fig. 8.25, which compares the costs of one-axis, two-axes tracked and V-trough systems relative to fixed flat-plate systems as a function of module costs. The calculation is appropriate at sites with moderate climates (50% diffuse irradiance on a horizontal surface). The area-related balance-of-sytem (BOS) costs of the tracked arrays are assumed to be 50% and 100% above the area-related BOS costs of the fixed arrays (70 \$/m^2). The cost ratio Q is the fraction of

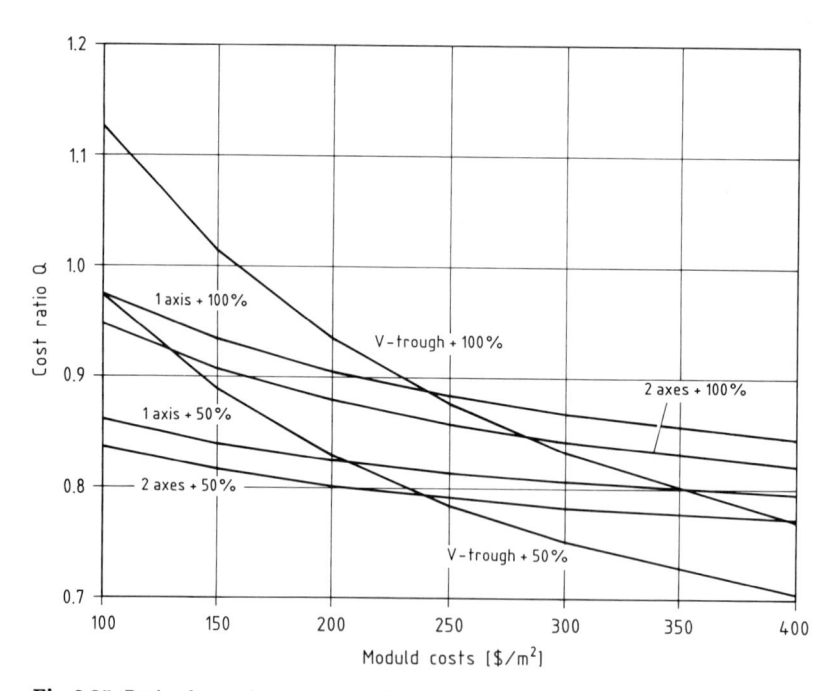

Fig. 8.25. Ratio of annual energy costs of tracked relative to fixed flat-plate systems in moderate climates as a function of area-related module costs. The parameter varied is the percentage increase above area-related BOS costs of a fixed flat-plate system (70 \$/m^2) due to tracking provisions. For $Q > 1.0$, annual energy costs of tracking systems are higher than for fixed systems [77].

total energy costs of the tracked system relative to the fixed flat-plate. As long as the cost ratio Q is less than one, trackers and V-troughs are calculated to be more economic.

Today, after 15 years of research, not one of the three general categories of PV system technologies – flat-plate crystalline silicon and thin films, tracking and concentrating systems – emerged as clear cut winner. Only parabolic trough concentrators have been ruled out as technology in PV systems.

8.8 Design Considerations for Grid-Connected Power Plants

The design of grid-connected photovoltaic SPPs has been the subject of many years of research, which can now be supplemented by data from numerous operating systems in the U.S., the European Community and in Japan (see Sect. 8.9). However, not all of these demonstration plants operated successfully, often because of inappropriate design of the electrical circuit.

In the following some design considerations for large grid-connected photovoltaic power plants without storage will be reviewed.

8.8.1 Site and System Selection

The conception of a photovoltaic SPP should be to meet demands taking into consideration applicable restrictions. The most important demand is the power or energy desired per year, day and hour by the plant's users. The most important restrictions are the availability of the resources – capital, land area and solar irradiation – as well as any environmental restrictions.

Certain site characteristics are favorable to the installation of a photovoltaic SPP of any design [55]. Favorable climate characteristics are, for example:

- high insolation,
- regular distribution of irradiation patterns during the year,
- low average ambient temperatures,
- low probability for hail, storm, sand and snow drift,
- low soiling rate.

The following aspects are advantageous to minimize shadowing and foundation costs:

- no large boulders or trees,
- low latitudes,
- cheap land,
- sloping site, tilted south;
- flat site,
- no big rocks.

In addition some infrastructure demands are of significance, such as:

- traffic roads, skilled labor,
- vicinity to grid transformer station and consumers.

Some characteristics listed above are not clear for all cases. For example low, continuous structures such as long fixed or horizontally tracked rows (e.g. SMUD, Pellworm) require relative flat fields, but, because of their low profile, are suitable for areas with high wind speed regimes.

The area requirements of photovoltaic SPPs using present technology range from 6 ha/MW$_p$ (Pellworm), 8 ha/MW$_p$ (Lugo) to 11 ha/MW$_p$ (Carrisa Plains) depending on collector technology (fixed, tracked or low-concentrating) and latitude. Because of shadowing, the highest power density (W$_p$ per m^2 of land) is achieved for non-concentrating and non-tracking devices. For latitudes lower than 40° tracking systems require about twice the land area of fixed tilted ones [37]. Regarding land use, tracking has a small advantage only at very high latitudes because tracking systems also receive irradiance coming from north during summer. In all cases, higher photovoltaic efficiencies would universally decrease the land area requirement.

The same considerations concerning efficiency of land use apply to systems with high concentration and tracking. In addition, concentrator modules cannot receive diffuse radiation or only a fraction of it. Not utilizing diffuse radiation is equivalent to a loss of 20–50 % of the annual land-area-related efficiency, expressed as the fraction of utilized to incoming irradiation. Therefore, the performance of concentrator systems depends strongly on the climate of the site. On the other hand, concentration leads to an increase in cell efficiency (see Sect. 8.2).

The additional annual solar irradiance received by tracked arrays in moderate climates (e.g. central Europe) is at least 29% for one-axis tracked and 33% for two-axes tracked arrays relative to an optimum tilted fixed array [77]. All calculations were performed using long-term weather data and under the assumption of a non-isotropic distribution of the diffuse sky radiation according to the Perez model [81]. In addition, reflection losses as a function of incidence angles at the module surface were taken into account.

The additional irradiance received by a tracker with east-west oriented horizontal axis (tracking the Sun's elevation) ranges between 4% and 7% only. But tracking of the Sun's azimuth with horizontal north-south oriented axis yields 15 to 30% more, depending on the latitude of the site. As the site approaches the equator, this horizontal tracker approaches more and more the performance of the vertical and polar-axis azimuth trackers. For higher latitudes vertical- or polar-axis trackers show the better performance with a surplus of more than 29%. Double-axes tracking yields only a few percent more.

However, 38% surplus due to two-axes tracking at sites like Albuquerque/NM translates into 800 kWh per m^2 and year, whereas the 33% surplus at Stuttgart/Germany results in only 400 kWh per m^2 and year [77].

The value of the energy not only depends on the total energy delivered, but also on the profile of the daily and seasonal output most valuable to the system. Most utilities have to satisfy a demand which results in daily and seasonal load profiles with variations to the value of the energy. For example, European utilities might wish to maximize photovoltaic system output in the winter evenings. With such objective, tilt angles steeper than those for maximum annual irradiance collection, or even arrays not south-oriented, may be more effective.

In the U.S., however, south-western utilities find that the photovoltaic system output would hit its highest revenue in the summer early afternoon, because their loads are dominated by air-conditioning (one of several reasons why the highest installed PV capacity is found in that region).

The integration of a high percentage of PV generation into a conventional grid (above about 10% of installed capacity) may cause a technical problem because fluctuations in PV energy output due to migrating clouds must be compensated by control action of the conventional plants of the grid. These fluctuations can be leveled out by distributive siting of photovoltaic SPPs.

8.8.2 Electrical Circuit Design Aspects

Electrical circuit design in a photovoltaic power plant has to fulfill two essential demands:

1. to transmit produced electricity with good overall efficiency and low cost;
2. to guarantee reliability and safety.

Interconnection, wiring, module arrangement. A critical plant design parameter is the nominal voltage level. It depends e.g. on load requirements, I^2R-losses, the inverter available, and safety codes. A desired voltage level can be realized by connecting a number of modules in series/parallel. As discussed above (see Sect. 8.4), dissimilar modules or subarrays cause mismatch losses resulting in reduced output power and possible damage to cells.

A second array loss mechanism arises due to shadowing e.g. from other subarray rows. Depending on the application blocking- and bypass-diodes and the mode of shadowing of arrays among each other, the interconnection scheme of modules within subarrays, subarrays within arrays, and arrays within the total PV generator can be optimized. Since the application of blocking- and bypass-diodes causes losses even in ideally matched interconnected modules, the optimization problem is complex.

Concerning the output of a PV system, a high nominal voltage is advantageous to keep current and I^2R-losses low. In contrast, high voltage levels lead to hardware availability problems (e.g. transistors), increase the cost for high voltage DC hardware (e.g. switches, fuses), and demand additional safety measures.

Module and array wiring must be designed to be water- and UV-resistant. Copper wire is advantageous to ensure minimum line resistance. In humid environments it is advisable to treat each connection with anti-corrosive materials to prevent metal-to-metal corrosion and subsequent increased line-resistance. In harsh environments such as coastal areas with salt spray, pressure-type connectors should be replaced by soldered connections.

Power conditioning unit (PCU). Because of decreasing inverter efficiency at partial loads it may be useful to split the DC-AC inverter into several units with lower nominal power. This kind of solution was realized at the 300 kW$_p$ photovoltaic SPP at Pellworm, Germany. Three inverters are operating in parallel according to the actual load demand. Figure 8.26 shows the efficiency versus power output of this coupled inverter system.

Table 8.2 shows typical electrical losses for the 6.5 MW$_p$ PV power plant at Carrisa Plains/CA, USA [93].

Table 8.2. Carrisa Plains/CA, solar power plant loss factors at rated output [93].

Power conditioning unit loss	3.0%
Module mismatch loss	0.2%
Array mismatch loss	1.0%
Module resistive loss	0.3%
Wiring resistive loss	1.0%
Switchgear loss	0.1%
Blocking diode loss	0.3%
Tare loads	1.1%

Grounding, fault and surge protection. Grounding of an electrical system provides a well defined path of low resistance from various points in the system to earth ground. This path i designed to carry current during fault conditions. In conventional electrical power systems lik

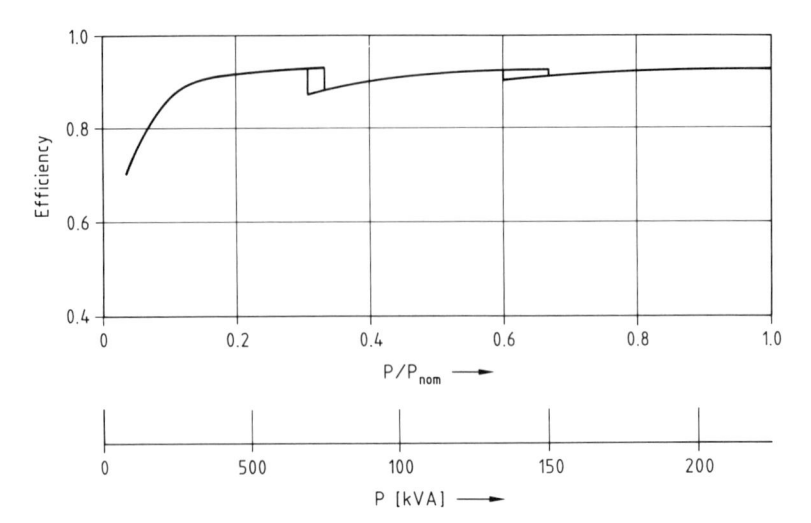

Fig. 8.26. Efficiency of a coupled inverter system versus power output and load demand (Pellworm, Germany) [66].

a utility grid this fault current can be very high and therefore may quickly trip overcurrent protective devices. In contrast, during a fault a photovoltaic SPP can only provide a small increase in current starting from the optimum operating point. This difference in operating characteristic causes difficulties in protecting both equipment and personnel because faults within a photovoltaic SPP are often difficult to detect and hence may persist for an indefinite period [58].

Fault protection measures depend upon application and upon priority of fire protection, safety of personnel and energy production. As photovoltaic SPP in a commercial setting are usually unmanned and automatically monitored, the primary concern is energy production. Therefore, faults are allowed and a repair of the system may not be warranted until the power output is noticeably degraded.

Protection measures against either direct lightning strokes or electromagnetic induced currents into the power system, e.g. by nearby lightning discharge, have to be considered during plant design [58].

Modular building block. The basic idea of the modular building block concept consists of independent array fields each with its own power conditioning unit and control logic. The great advantage is the high reliability of the whole PV power plant, because a single failure does not paralyse the whole plant. Another advantage is the better over-all efficiency of the DC-AC inverters, due to this modularity. So far only the Carrisa Plains photovoltaic SPP uses a building block approach. The building block size in this case is approximately 700 kW, each with its own DC-AC inverter.

8.8.3 Plant Monitoring

Plant modelling is a necessary instrument for both forecasting the energy output of the plant as well as understanding and improving the plant performance by comparing the modelled output with the one actually measured. Therefore, a special strategy for plant monitoring is also a matter of plant design.

8.8 Design Considerations for Grid-Connected Power Plants

The primary aims and reasons for monitoring are:

- to find out whether the system is operating properly,
- to prevent damages because of unsafe conditions,
- to determine the best operating method,
- to detect fault or degradation,
- to permit plant comparisons and validation of models.

The most important data required are energy and other cumulative and average parameters on a daily basis such as:

- plane-of-array irradiance,
- PV array actual energy output,
- inverter output,
- ambient temperature,
- energy to and from grid.

The most important value is the irradiance upon the array, which is diminished by shadowing, soiling, and reflection losses, even before the sun's energy is converted by the photovoltaic cell with its inherent loss mechanisms.

Additional losses in the system occur: resistive and array mismatch losses, losses by departure from maximum power point tracking, and DC-AC conversion losses. Further degradation

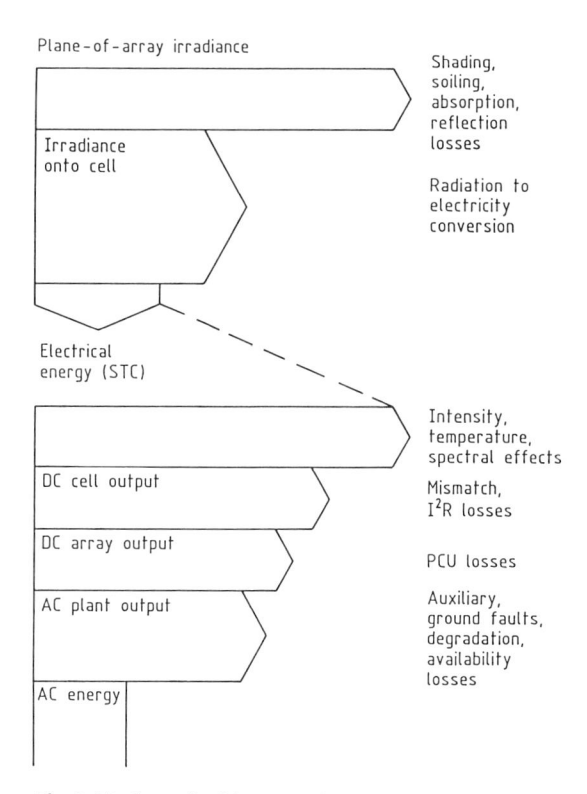

Fig. 8.27. Generalized loss tree of a grid-connected PV plant without storage (not to scale!).

of output occurs reducing stand-by parasitics, and ground-faults reducing system-, subsystem- and component-availability.

These effects are important for monitoring and modelling and are summarized in the loss-tree of a grid-connected plant (Fig. 8.27). The actual performance in terms of annual energy yield is in the range of 6–12% of annual plane-of-array irradiance for today's photovoltaic SPP.

8.9 PV Plant Operating Experience

8.9.1 The Experience Base

Several multi-year application and demonstration tests of larger-size PV systems in grid-connected mode have been carried out by public, industrial, and private organizations. These facilities are located mainly in the U.S. and in Europe. Their performance and operational reliability provide information about the level of maturity attained with PV components and systems today [2,29,75].

The construction and operating status of the larger grid-connected PV test facilities (Tab. 8.3) shows that about 13 MW_p of capacity are operational today, comprising different array mounting configurations, collection principles and cell technologies. Views of representative systems are shown in Fig. 8.28, technical features of some plants are compiled in Sect. 8.12.

8.9.2 Operating Experience

Detailed performance evaluations of grid-connected PV plants have been carried out in the U.S., sponsored by government and the utility industry [47,52,53,54,68,72,79,87,96]. Monitoring of the European PV demonstration plants is in progress and will augment materially the documented experience once comparable results become available [14,16,17,56,105]. The following discussion about plant performance will be drawn on the entire experience base.

Array rating. Array DC output power rating under average operating conditions, or even those close to standard test conditions (STC), were observed to be universally lower than manufacturer-supplied nameplate DC power rating; normalized to nameplate rating, some results from US field test [93,96,97] are shown in Tab. 8.4.

Array efficiency. PV module characteristics determine PV array performance. In multi-year field tests, performance of crystalline-cell flat-plate PV modules degraded less than 1%/a in efficiency, excepting early generation modules. Also, less than 0.1%/a annual failure rate due to breakage, electrical shorts or open circuits were observed for modules of present production. At such rates of failure, array fields with 10% initial efficiency would still have about 8% efficiency after 20 years of operation [1,46,48].

The in-situ array input-output relationship and the derating influence of temperature can be inferred from U.S. data. Taking the SMUD, Lugo, and Phoenix facilities as examples, the decrease of array DC power efficiency with temperature was verified to lie within expected bounds; in terms of annual frequency distribution of hourly averaged DC energy efficiencies, the array power efficiencies also exhibit dependencies on irradiance levels (see Fig. 8.29).

Array operating efficiencies in newer plants like SMUD are higher than in earlier facilities like Lugo or Phoenix. The highest power efficiencies observed in these plants occur at irradi-

8.9 PV Plant Operating Experience

Fig. 8.28. Views of representative grid-connected photovoltaic solar power plants. (a) 2,350 kW$_p$ Sacramento Municipal Utility District (SMUD) field test facilities (PV 1 and PV 2) near Sacramento/CA, USA (204 flat-plate 1-axis tracking arrays); (b) 350 kW$_p$ SOLERAS PV field test facility near Riyadh/Saudi Arabia (160 fresnel lens point-focusing 2-axes tracking arrays); (c) 100 kW$_p$ Kythnos PV field test facility on Kythnos Island, Greece (40 latitude-tilt fixed flat-plate arrays); (d) 6,450 kW$_p$ power plant at Carrisa Plains near Bakersfield/CA, USA (756 mirror-enhanced V-trough flat-plate 2-axes tracking arrays)
ad (a) (courtesy of SMUD, Sacramento/CA, USA)
ad (b) (courtesy of MRI, Kansas City/MI, USA)
ad (c) (courtesy of Siemens Solar GmbH, Munich, Germany)
ad (d) (courtesy of Siemens Solar Inc., Chatsworth/CA, USA).

Table 8.3. Major grid-connected PV power plants > 30 kW (construction and operating status as of 1989).

Date operational	Location	Country	Nameplate rating (kW$_{DC}$)	Status	Last year of performance evaluation
	Flat-Plate, Fixed, Latitude-Tilt				
1979	Mt. Laguna/CA	USA	60	D	1981
1981	Beverly/MA	USA	2 × 50	O; M	1984
	Lovington/NM	USA	2 × 50	O; M	1984
1982	Oklahoma/OK	USA	135	O; M	1984
	San Bernadino/CA	USA	60	D	1982
	Aghia Roumeli/Crete	Greece	50	O; M	in progress
	Epcot Center/FL	USA	73	O	
1983	Mt. Bouquet	France	50	O; M	in progress
	Nice Airport	France	50	D	
	Fota Island	Ireland	50	O; M	in progress
	Terschelling Island	Netherlands	50	O; M	in progress
	Kythnos Island	Greece	50	O; M	in progress
	Pellworm Island	Germany	300	O; M	in progress
	Marchwood	UK	300	O; M	in progress
	Chevetogne	B	63	O; M	in progress
1984	Georgetown/DC	USA	300	O; M	1984
	Vulcano Island	Italy	80	O; M	in progress
1985	CSFE	Spain	100	O; M	in progress
1986	Delphos	Italy	300	O; M	in progress
	Alabama/AL	USA	100	O; M	in progress
1988	Kobern-Gondorf	Germany	390	O; M	1989
1991	Neurather See	Germany	340	C	in progress
	Bochum	Germany	30	C	in progress
1992	ENEL	Italy	10 × 330	C	in progress
	Flat-Plate, Tracking				
1982	Lugo/CA (2-axes)	USA	1,000	D	1987
1983	Carrisa/CA (2-axes)	USA	6,450	D	1986
1984	SMUD PV 1/CA (1-axis)	USA	1,180	O; M	1986
1986	SMUD PV 2/CA (1-axis)	USA	1,170	O; M	1988
	PV 300/TX (1-axis)	USA	326	O; M	1988
	Vista/VA (1/2-axes)	USA	75	O; M	1988
	Fresnel Concentrator, Tracking				
1981	Dallas-Ft. Worth/TX	USA	27/140	O; M	1984
	Phoenix/AZ	USA	225	D; M	1984
1982	Solar Village	Saudi Arabia	350	O	1985
1989	Austin II/TX	USA	300	C	in progress
	Trough Concentrator, Tracking				
1981	Blytheville/AR	USA	240	D	1982
1982	Albuquerque/NM	USA	47	D	1982
	Hawaii	USA	35	D	1982

O = operating; D = operation discontinued; C = under construction; M = monitored

ance levels above 600 W/m². This seems intuitively inconsistent as rising operating module temperatures with increasing irradiance level usually degrade power efficiency. These observations are not yet consistently explained. The decrease in efficiency at Phoenix and Lugo at highest irradiance levels is temperature induced, however, the slight efficiency increase at SMUD is assumed to be due to simultaneously occurring low ambient temperatures. It has

8.9 PV Plant Operating Experience

Table 8.4. PV array DC power rating at different operating conditions. Nameplate: manufacturer supplied; Condition a: 25°C cell temperature; Condition b: 20°C ambient temperature; insolation 1,000 W/m² for flat-plate modules; 850 W/m² for concentrator modules; mirror enhancement at Carrisa is equivalent to a concentration of about 2.

Name Location	Sky Harbor Phoenix	SMUD Sacramento	Lugo Hesperia	Georgetown Washington	Carrisa Carrisa Plains
Type	Fresn.conc. 2-axes	Flat-plate 1-axis	Flat-plate 2-axes	Flat-plate lat.tilt	FP/mirr.enh. 2-axes
Nameplate	100%	100%	100%	100%	100%
at Cond. a:	68%	88%	79%	81%	91%
at Cond. b:	62%	79%	73%	78%	81%

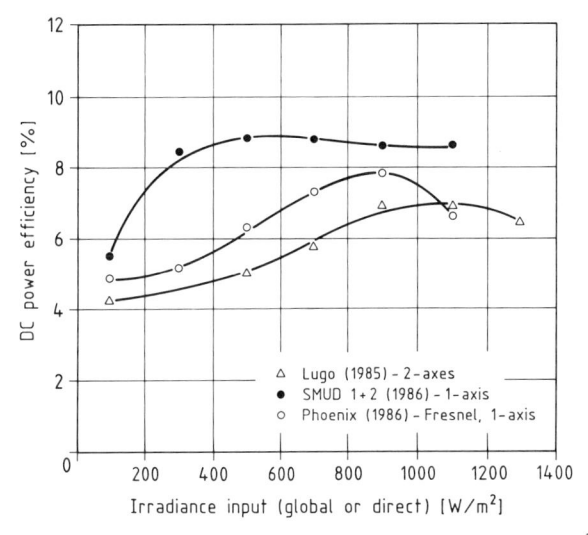

Fig. 8.29. Distribution of annual array DC power efficiencies as a function of plane-of-array (POA) irradiance and influence of temperature [48,97].

Facility	Array efficiency depending on temperature (%/°C)
SMUD1	0.034
SMUD2	0.024
Lugo Subfield 1	0.026
Lugo Subfield 2	0.008
Phoenix	0.026

been demonstrated, though, that lower efficiencies during low irradiance operation are due to panel shading and early morning panel frost [96].

Inverter performance. Efficiency, operating reliability, and safety are the principal issues characterizing inverter performance. High inverter efficiency, although of interest at nameplate rating, becomes of critical importance at part-load. Differences in observed efficiencies are exemplified in Fig. 8.30. Efficiency is high above 50% of nominal load, but differs markedly

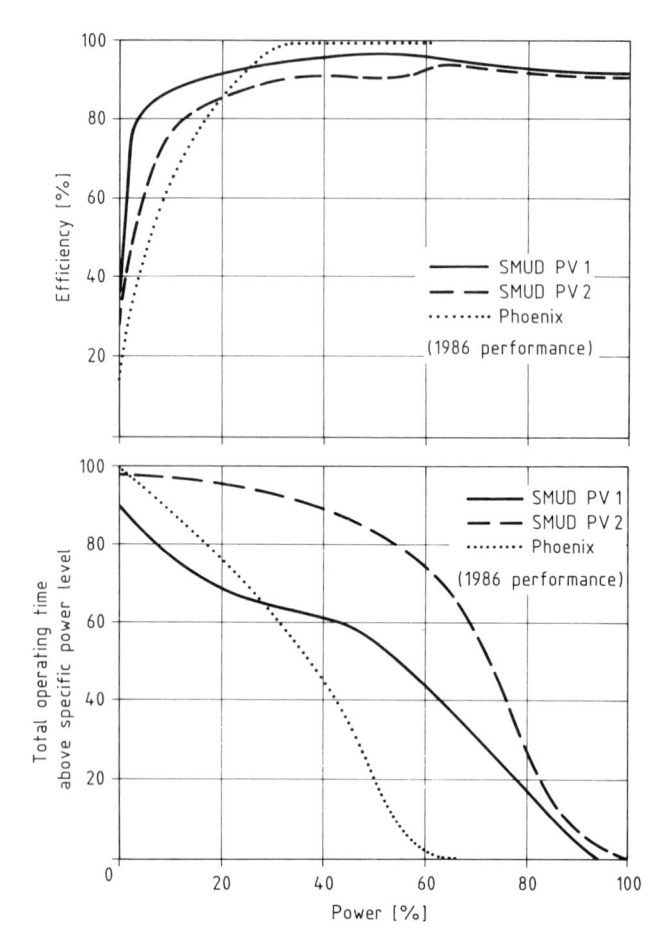

Fig. 8.30. Representative performance of DC-AC converters. (SMUD PV1: 1,000 kVA, 12 month operation, 96% ave. annual efficiency; SMUD PV2: 1,000 kVA 9 month operation, 93.1% ave. annual efficiency; Phoenix: 300 kVA, 12 month operation, 96.7% ave. annual efficiency).

at 10% load. The significance of high part-load efficiency, particularly at sites which are less insolation-favored, is underlined when considering efficiency in conjunction with the yearly percentage of time the inverter operates below rated power level. Inverters of present design offer significantly improved part-load efficiency (> 90% efficiency at 10% load), but field performance experience has not yet been reported.

Failure of an inverter leads to the shut-down of the array area portion connected to this inverter. Early inverters tended to be the plant component with lowest reliability; in the past, inverter malfunction caused up to 85% of overall yearly plant downtime, or caused complete plant shutdowns (e.g. SMUD PV2).

Plant performance. A minimum irradiance is needed to reach sufficient array voltage and power for inverter start-up. The total irradiance thresholds are

- about 50 W/m^2 of global irradiance for non-concentrating flat-plate array system;

8.9 PV Plant Operating Experience   319

- about 100 W/m² of direct irradiance for parabolic trough concentrator systems;
- about 100–200 W/m² of direct irradiance for point-focus concentrator systems.

Once irradiance threshold is exceeded, output is delivered synchronous to irradiation input; an irradiation threshold in the sense as for thermal SPP does not exist.

Using annual average array efficiency as a basis and disregarding possible effects of temperature, characteristics of DC-output to radiation-input may be approximated by straight lines (see Fig. 8.31). These curves can be used to estimate yearly DC energy output performance of these plants under alternate site-specific insolation conditions. Such analysis does not have to deal with energy throughput, as differences in short- and long- efficiencies lie within tolerance limits of data gathering and evaluation.

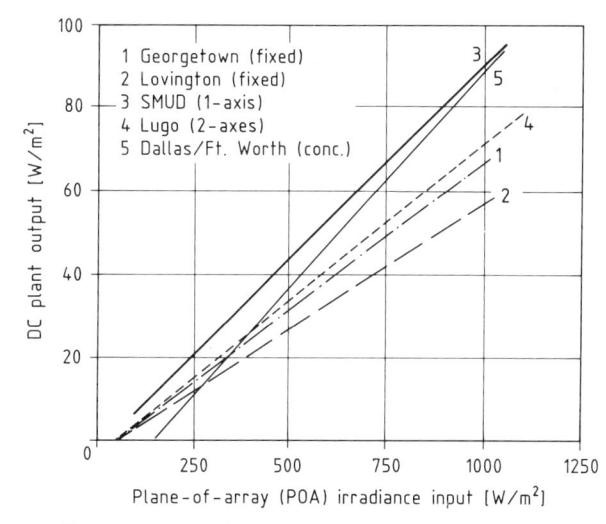

Fig. 8.31. Irradiance input to DC-output characteristics of representative PV solar power plants [13,52,53,54].

Temperature and temperature-induced aging effects with time may become significant for concentrating systems. For example, performance of the Soleras point-concentrating air-cooled system with Si-cells depends not only on ambient temperature, but also showed performance degradation with time at high irradiance levels (see Fig. 8.32). Accelerated aging of Si-cells in a high temperature environment and its effect on lifetime may favor GaAs cells in future concentrator systems.

Some PV plant input-output characteristics in terms of average daily energy densities (kWh/m²d) may be inferred from published data [93,96,97]. The curves shown in Fig. 8.33 are approximations derived from monthly performance averages of DC or AC energy output, plane-of-array irradiation, and plant availability.

A decrease of energy output at higher daily irradiation levels can be observed in all installations, although to a different degree. The monthly temperature averages associated with the data points for Carrisa indicate that temperature effects indeed contribute to the reduction in specific output at high irradiation levels.

Plant capacity factor (CF). Continuous production at minimal operation and maintenance (O&M) cost has been the operating philosophy in most PV field tests. This allowed to determine actual CFs under prevailing irradiation conditions (Tab. 8.5). The average CF of Carrisa in the years 1984–1986 during the peak demand periods (as defined by the local utility) was almost 80% (see Fig. 8.34). These figures show that substantial capacity credit during peak demand periods are achieved, given favorable circumstances, and that yearly and monthly CF

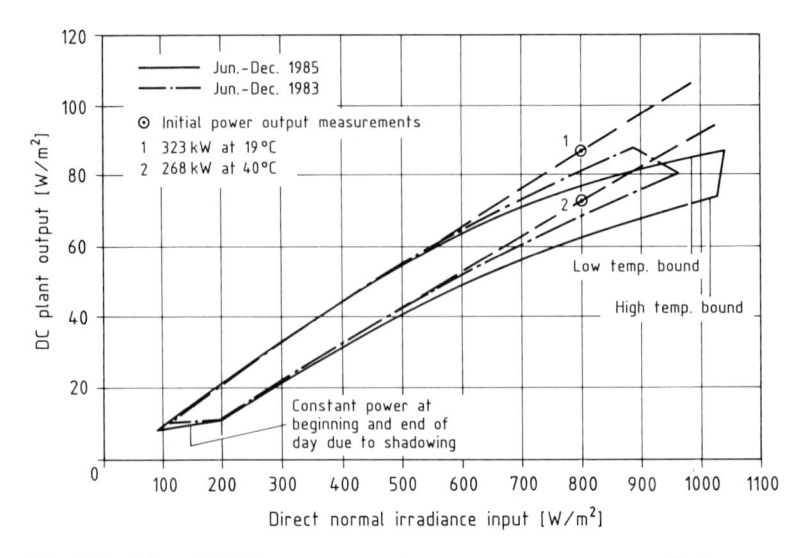

Fig. 8.32. Soleras 350 kW$_p$ concentrator plant input-output characteristic [87].

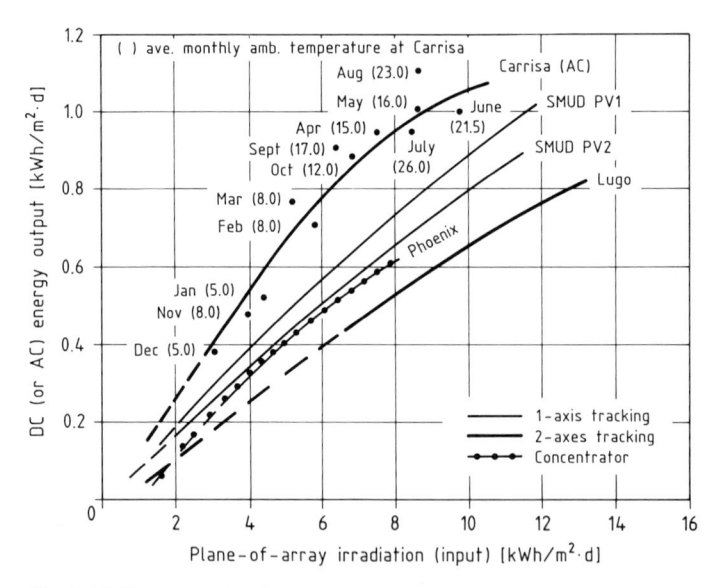

Fig. 8.33. Representative characteristics of PV power plant energy input-output performance [93,96,97].

are rather consistently repeated [47]. Such performance, however, is contingent upon irradiation conditions remaining consistent from year to year, and upon high levels of irradiation coinciding with periods of high end-use demand.

Plant reliability and availability. Although site-specific experiences differ in detail, power conditioner and inverter failures together with electrical problems (ground faults), emerged as the

Table 8.5. PV plant capacity factors (CFs) observed at some US southwest locations (based on actually measured DC array output, determined by regression analysis under standard operating conditions of 20°C ambient temperature and 1,000 W/m^2 insolation; low CFs would result if based on nameplate array peak rating)

Facility		Yearly average	Monthly High	Monthly Low
Phoenix	(1986)	23.3	36.3	14.9
SMUD 1	(1986)	27.9	48.1	8.8
SMUD 2	(1986)	26.2	42.0	5.9
Lugo	(1986)	35.2	45.7	23.6
Carrisa	(1986)	28.9	39.2	18.0
	(1984)	28.7	40.6	11.4

major reasons for plant downtime. In two-axes tracking systems, mechanical and electronic control problems were major factors degrading yearly performance [102].

For these reasons, initially recorded availabilities in early-generation US PV field test facilities has been as low as 50 or 60%; in contrast, availability consistently in the 95% range has been observed in one early facility in the U.S. without a PCU. Availability of the two-axes tracking Phoenix system changed from 53 to 85% after replacement of the PCU. The more recent systems (Lugo, SMUD 1) demonstrated availabilities consistently well above 90%, with monthly averages ranging between 75 to 99% (discounting non-operating times of SMUD 1 due to catastrophic PCU failure).

By way of example, representative plant downtime reasons, and output lost due to downtime, are shown in Tab. 8.6. The strong influence on plant availability by inverter downtime,

Table 8.6. Representative plant downtime causes and output lost due to downtime [96].

Causes	conc./1-axis	Phoenix 1-axis (1986)	SMUD 2-axes (1986)	Lugo (1985)
Array	%	–	–	–
Tracker	%	23	1	55 [a]
Tracker control	%	–	13	–
Electrical	%	18	1	–
Inverter/PCS	%	32	85	9
Utility Interconnect	%	–	–	6
Array storage	%	N/A	N/A	30
Others	%	27	–	–
Total downtime	%	100	100	100
DC output	MWh/a	258	2,014/1,544	2,244 [b]
DC output lost	MWh/a	44.1	69.9/53.3	106.7
Monthly energy lost	%	14.6	3.5/3.5	4.8
Annual availability	%	85	91/92	96

[a] Repair of wind damage accounts for 96% of this figure
[b] Reduced by 2.3% due to shading

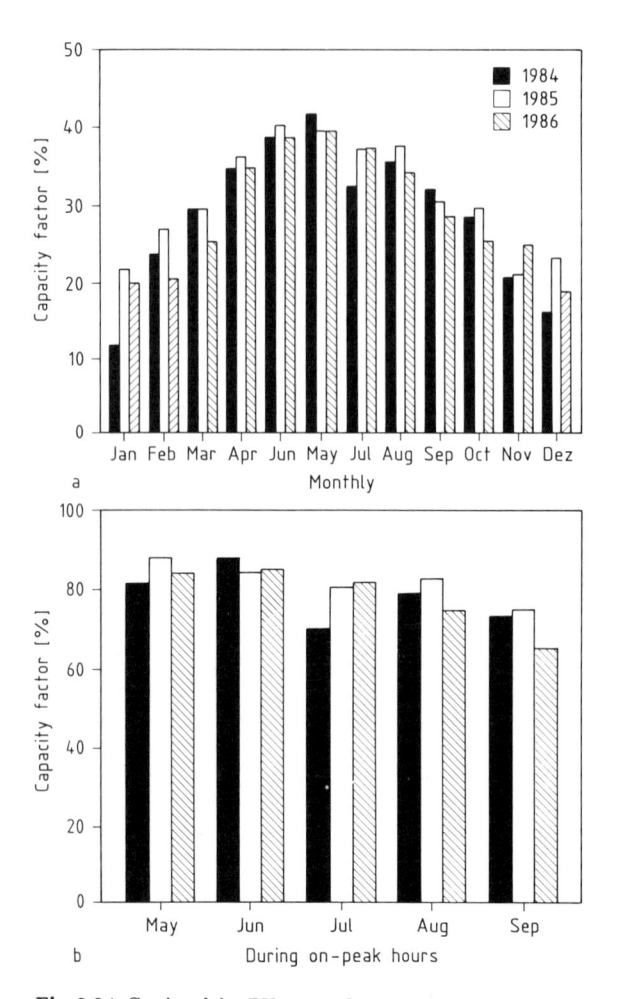

Fig. 8.34. Carrisa plains PV power plant capacity factors, (a) monthly, (b) during on-peak hours (12:30–18:30 hours, pacific daylight time) in peak season (May through September) [47].

tracker reliability, and the need for tracker stowage (to avoid damage by wind) becomes obvious. For example, two-third of the losses in solar energy collection at SMUD in 1986 were caused by inverter malfunctions, while tracker drive problems and wind-enforced array stowage were the significant loss factors at Lugo.

Energy production. The production performance histories of some US field test systems (see Fig. 8.35) show that PV plants can be consistent producers. Energy production per kW installed is a function of local irradiation and of the reliability of primarily non-photovoltaic components. Production of some plants was even halted completely for major replacement or overhaul of such components. For example, operation at Georgetown had to be discontinued in early 1986 for analyzing and correcting inherent installation deficiencies. SMUD PV2 was forced to be taken off-line in May 1987 after a double-contingency failure (e.g. ground fault in conjunction with blocking diode failure) caused a major fire which destroyed the PCU

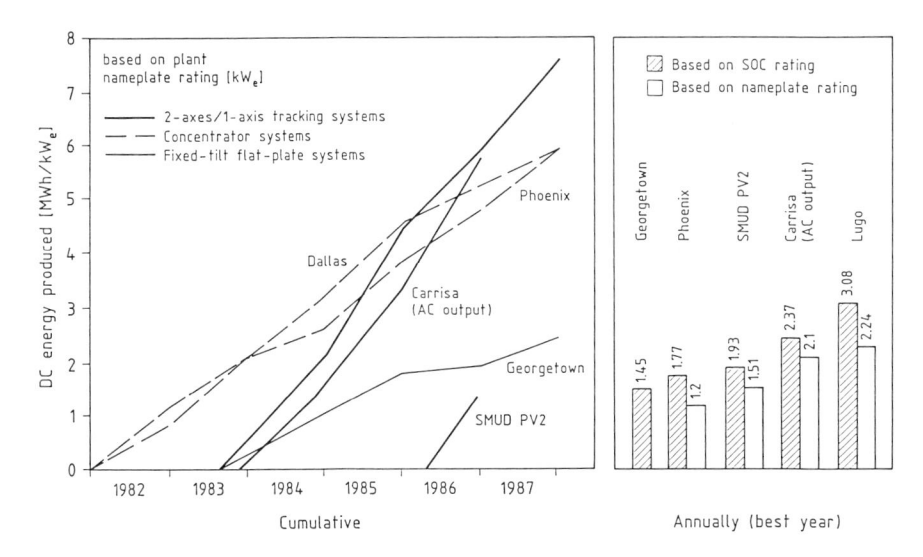

Fig. 8.35. Cumulative and annual (best year) specific generation performance (MWh produced per kW installed) of representative PV solar power plants in continuous production operation. SOC = standard operating conditions: 20 °C ambient temperature, in conjunction with 1,000 W/m² for flat-plate systems and 850 W/m² for concentrator systems

beyond repair. The PV modules remained unaffected and, after PCU replacement, were put into service again. As a consequence of such experiences, new guidelines for the electrical design of PV systems were incorporated in electrical safety codes.

As a footnote on field performance experience, obtaining good data from and achieving reliability in on-site data acquisition systems is a challenge sometimes difficult to achieve. Operational reliability of data acquisition systems (DAS) frequently proved lower than the availability of the PV system to be monitored.

Operations and maintenance. O&M needs were monitored and analyzed in several PV test systems [89,97]. Excluding time spent for modifications or for correcting initial system deficiencies, average routine maintenance time has been reported from as low as 11 hours/month in 1986 for the 1.0 MW$_p$ SMUD facilities of more recent design, to about 34 hours/month for the early 211 kW$_p$ Phoenix tracker facility.

Using 25 \$/manhour as a basis and considering yearly energy production, the O&M costs correspondingly are in the range of 0.14 cent/kWh (SMUD PV2; 1986) to 4.7 cent/kWh (Phoenix; 1986).

8.9.3 Summary and Conclusions

Several conclusions can be drawn from the accumulated experience in designing and operating PV power systems:

- modularity, wide range of possible system sizes, electronic-only energy conversion, low O&M needs, and the potential for high operational reliability/availability are the most characteristic attributes of photovoltaic SPPs;
- failure rates of state-of-the-art flat-plate crystalline PV modules are below expectations and R&D program goals, with typical performance degradation < 1%/a;

- high reliability and stable performance was demonstrated for Fresnel point-focussing modules, but performance is sensitive to tracking errors and depends on availability of direct irradiation;
- operational reliability and availability of PV systems is strongly influenced by inverter and electromechanical component reliability, as well as by the electrical integrity of the system and the electrical fault protection philosophy;
- PV plant output rating under field operating conditions is lower than the nominal plant capacity based on PV module rating; unrealistically high plant ratings penalize key performance parameters like CF and plant capacity credit;
- production of over 2 MWh_e/a per kW installed plant capacity under California insolation conditions ($\approx$ 2,550 kWh (DNI)/m^{2}a) have been demonstrated;
- unattended operation, little need for maintenance (e.g. as low as 10 hours per year per MW_e capacity), and minimal operating cost (e.g. as low as 0.2 c/kWh$_e$) for plants up to 6 MW_p have been demonstrated;
- flat-plate modules become effectively cleaned by natural precipitation in most field tests; under such conditions, washing PV modules as routine maintenance activity appears not to be cost-effective;
- environmental impact of PV power plants is minimal; wherever competition for land use is low land utilization at up to 8–11 ha/MW_e of capacity for two-axes tracking systems may pose no obstacle;
- the design process of PV power plants is well understood; a modular unit design is replicable until total system capacity is reached (land area permitting);
- modular nature of PV power plants permits staggered plant construction and reduction of investment risk (one module can already produce while the next one is being constructed);
- construction time for PV plants is short (8 to 12 months).

The past field test experience proved that today's PV technology and associated power conditioning equipment are technically viable even for large-capacity PV power plants [93]; however, PV array field costs are still too high for economically competitive bulk power generation (for discussion on cost see Chap. 10).

8.10 Photovoltaic Solar Systems Modelling and Calculation Codes

About fifteen to twenty design and performance analysis models for photovoltaic power supply systems have been developed in the past five years and are in the process of being refined (against actual system performance data). Stand-alone and grid-connected configurations with electrochemical battery storage, occasionally including additional renewable or backup energy technologies, have been modelled. Approaches and details differ; not all calculation codes are in wide use or are well documented. The following list is therefore a limited selection of available calculation codes[10].

SOMES

A model developed for the evaluation and optimization of autonomous energy systems which may comprise solar modules, wind turbine(s), battery storage and a backup energy source. The PV model treats the module efficiencies as a function of incident irradiance. The battery representation is based on Shepherd's model, incorporating variable charge efficiencies for both charge and discharge regimes. The inverter model relates

---

[10] Descriptions of several of these codes were provided by Liam Keating, National Microlelectronics Research Center, Ireland [57].

output to input power using the $P_{(out)} = a(P_{(in)} - b)$ relationship (a,b are constants). The power management model is predefined; sheddable and switchable loads can be incorporated. SOMES is accepted by Dutch organizations as an official PV system design tool.

    Availability: Available to the public for a fee.
    Used by: Electrical engineers, project planners.
    Language: Turbo PASCAL.
    Machine: Cyber Mainframe; IBM-PC.
    Documentation: see [18].

PVSS

A model to simulate the operation of a PV plant and to optimize both the plant and subsystem size. The PV model is based on the standard solar cell equation. The I-V relationship is given as a function of a set of specialized parameters unique to the solar module or array being modelled. A linear variation of state-of-charge with given charge/discharge rates is assumed for the battery (constant series resistance and charge efficiency). Both inverters and DC voltage regulators are described by a power and efficiency that is a quadratic function of load. The user is provided with the option of selecting any one of sixteen different PV Plant configurations. Model used to optimize the 80 KW$_e$ Vulcono Plant in Italy.

    Availability: Not available to the public.
    Used by: Electrical/Utility engineers.
    Language: FORTRAN IV-Plus.
    Machine: VAX 11/780.
    Developed by: N.N., ENEL-Centro Ricerca Elettrica, Italy.

PVFORM

This model was developed as a computer simulation program for system design. It also facilitates the evaluation of the system economic potential. The PV array model is based on stipulated solar module power and efficiency characteristics; these must be determined by the user. The array tilt angle can be changed by the user for successive simulation runs. The battery model uses data provided by the user on the battery capacity, minimum state-of-charge, equalization state-of-charge/schedule, etc., to calculate the current state-of-charge. The inverter model is based on two parameters, the input rating of the PCU and it's efficiency at full power. The load model requires the user to enter the expected load in watts, for each hour of a simulated day. The battery management is modelled, the remainder of the power management is assumed to be optimized. Performance predictions are claimed to be accurate to within 5% of actual experience.

    Availability: Available free to the public.
    Used by: Engineers, PV plant analysts and project planners.
    Language: ANSI FORTRAN 77.
    Machine: Mainframe, Minimainframe, IBM-PC.
    Documentation: see [71].

LOSADIS.PV

While this model was developed as an evaluation tool for PV designers, its function is also to give experience in the field and laboratory testing of PV arrays. The PV array model is based on the standard cell equation; the parameters of the model for many different types of solar modules are stored in the program's database. A single, simple equation is used for modelling battery charge and discharge operation; battery parameters needed for the model are available for both lead-acid and NiCd batteries. End-user electrical requirements are modelled in detail. Battery state-of-charge management is modelled, the remainder of the power management is assumed to be optimized. The model is used by thirty user groups in France.

    Availability: available in the research domain.
    Used by: Engineers and PV plant researchers.
    Language: Microsoft BASIC.
    Machine: IBM-PC and Compatible.
    Developed by: N.N., Centre d'Etudes Nucleaires, Cadarache, France

HERMINES

A model designed to simulate the transient performance of PV systems and hence to evaluate different design methods. The standard solar cell equation is used to model the PV array; eight parameters are used in the model and must be provided for each type of cell used. Battery charge and discharge are described by separate equations which relate current and voltage by means of specially defined battery parameters. Battery management is modelled, the remainder of the power management is assumed to be optimized. Model was used to design the Paomina Rondilinu (Corsica) installation. The model is in use at AFME/France.

> Availability: Available to the public via Minitel, free of charge.
> Used by: Engineers, PV plant analysts/researchers.
> Language: FORTRAN 77.
> Machine: HP 1000 PC or Compatible.
> Developed by: D. Mayer, Ecole des Mines de Paris, Mayer/Nolay, France.

ASHLING

Developed by PV plant designers, researchers, managers and university specialists within a concerted European action, the code is intended for the simulation and optimization of stand-alone or grid-connected PV power plants with or without battery storage. The mathematical model for the module is based on the standard solar cell equation. The battery representation uses the Shepherd equation for battery charge and discharge. A 3rd-order polynomial approximation to the input/output power curve is used to model the inverter. The power management strategy is based on considerations of the various ways in which each component within the PV array system can be controlled or regulated. The software provides a menu of components (i.e. modules/subarrays, batteries, inverters) which allow the user to build a PV plant, and to select specific data files from the program-provided databases (currently for 8 different modules and for about 10 different inverters, the data coming from the European pilot and demonstration plants). Other components such as battery chargers, MPP trackers, DC/DC converters, etc. are treated as part of the system control software. The model has been extensively validated on component level.

> Availability: Available to the research domain, free of charge.
> Used by: PV plant researchers, designers and operators.
> Language: Turbo Pascal 4.0.
> Machine: IBM-PC or Compatible.
> Documentation: see [69].

INSEL

A general purpose simulation code for electric component and system analysis, stand-alone system design and energy flow analysis, with special features for renewable energy applications. The modular, block diagram-oriented program includes models for PV elements, wind energy converters, motor/generator sets, batteries, hydrogen storage systems (consisting of electrolysis, compressor, pressure storage and fuel cell), regulators and inverters. These blocks represent realistic system components. Also, blocks for reading arbitrary data files, blocks for writing data to files and plotting, integrators and other mathematical utilities are included. Using a simple simulation language, the components can be 'connected' to form a descriptive system model. The model can be changed interactively within the working stage so that model structure and parameters can be varied quickly.

> Availability: Intended to be made available to the public later.
> Used by: Engineers, physicists and PV system researchers.
> Language: ANSI FORTRAN 77.
> Machine: Mainframe, IBM-PC of Compatible
> Documentation: see [91].

8.11 Appendix: Frequently Used Symbols

A	diode quality factor
E_g	bandgap
FF	fill factor
$I_{\max}$	current at maximum power point
I_0	reverse saturation current
I_{ph}	photocurrent
I_1	irradiance
I_{sc}	short-circuit current
L_n	electron diffusion length
L_p	hole diffusion length
N_v	effective density of states, valence band
N_c	effective density of states, conduction band
$P_{\max}$	maximum power
T	temperature
$V_{\max}$	voltage at maximum power point
V_{oc}	open circuit voltage
z	penetration depth, ξ-direction
α	absorption coefficient
τ_r	electron lifetime
τ_p	hole lifetime

8.12 Appendix: Photovoltaic Facility Data

Table 8.7. Some demonstration photovoltaic power plants > 30 kWe (Part 1).

Name		Carrisa	Lugo	Soleras	SMUDPV1	SMUDPV2
Location		Carrizo Plain	Hesperia	Al-Jubaylah	Rancho Seco	Rancho Seco
Country		USA	USA	Saudi Arabia	USA	USA
Design Conditions						
Site						
Latitude	deg N	35 N	34 N	25 N	38 N	38 N
Longitude	deg E or W	120 W	117 W	47 E	121 W	121 W
Altitude	m	615	1,140	685	17	17
Design						
Array Output	kW_{DC}	5,700*/759*	1,000*	350	1,180*	1,170*
Plant Output	kW_{AC}	5,100*/707*	850*	.	1,000***	1,000***
Irradiance	W/m^2	.	.	800 (AM 1.5)	1,000 (AM 1.5)	1,000 (AM 1.5)
Wind Speed	m/s	.	.	.	1.5	1.5
Plant Area	ha	70.8	8.1	5.2	4	4
Project Data						
Start of Operation		12/1983	12/1982	1981	8/1984	3/1986
Construction Time	months	16 + 4	9.0	.	.	.
Total Cost (Year)	Mio$(US)	60 (est.)	13.5 (est.)	32.0	18 (est.)	10.4 (est.)
O&M Staff	Manyear	2.0	0.2	4.0 (est.)	.	.
O&M Cost	k$(US)/a	48	.	.	.	.
Status (as of 1987)		operating	operating	operating	operating	operating
Technical Information						
PV-Field						
Cell	Mat/Dim.	ASI	ASI	Applied Solar	ASI	ASI/Sol./Mob.
	cm	mono k-Si; 10 × 10	mono k-Si; ⊘	mono k-Si; 5.7 ⊘	mono k-Si; 10 × 10	mono/poly/mono
Module	Type	Flat Plate/Flat Plate	Flat Plate	Fresnel (C=33)	Flat Plate	Flat Plate
		M51	M51	8 cell/module	M51	M51/.../...
	Dimension cm	30 × 120/30 × 120	30 × 120	Lens: 30.5 × 122	30 × 120	30 × 120/.../...
	Rating $W_p/(\%)$	11.4/12.5	10.4	.	11.4	12.5/.../...
	Number	96,768/17,200	27,648	5,120	28,672	19,584/1,600/2
	Total Area m^2	34,836/6,192	9,950	3,810	10,654	10,358
Array	Tracking	2-axes; pole-mounted	2-axes	2-axes	1-axis (N-S)	1-axis (N-S)
	Mounting	mirror enh./no mirr.	pole-mounted	pole-mounted	row-tilt	row-tilt
	Dimension m	9.5 × 10/12 × 12	9.8 × 9.8	12.1 × 2.7	.	.
	Number	367/43	108	160	112	68/.../...
	Aperture m^2	95/144	97.0	23.8	92	104/.../...

8.12 Appendix: Photovoltaic Facility Data

Bus Voltage	V_DC	550*/575*	525*	280	600	600
Field Area	ha	64.0/8.0	8.0	4.0	.	.
Power Conditioning		Helionetics/Toshiba	Helion./Garrett	Helionetics	Windworks	Toshiba
INV Rating	KVA	9 × 650/1 × 750	2 × 500/1 × 1,000	300	1,000	1,000
Commutation		self/self	self/line	line + self	line	self
Input Voltage	V_DC (Range)	.	.	.	.	.
Eff. at 100%	%	96/98	96	94.5	.	98.5
Eff. at 50%	%	97/98	97	92	.	.
MPP Tracking		yes/yes	yes	yes	no	yes
Storage, Lead Acid	Ah	none	none	1,700	none	none
Performance						
Average						
Irradiation	kWh/m²d	5.6	5.9 ('85)	6.1	6.3 (total)	6.03 (total)
	kWh/m²a	1,940 (global)	2,145 ('85)	150	1,757 (global)	1,757 (global)
Amb. Temp.	°C(ave//max/min)	.	.	40//46/0	.	.
Production	kWh_DC/d	.	5,970 ('85)	.	.	.
	MWh_DC/a	38,400	4,377 ('84-'85)	1,600	2,023 ('86-'87)	1,776 ('86-'87)
	kWh_AC/d	.	5,967 ('83-'84)	625 ('85)	.	.
	MWh_AC/a	13,500 ('86)	2,184 ('83-'84)	>100	1,921 ('86-'87)	1,712 ('86-'87)
Operations Range	W/m²	.	>75	.	>50	>50
Array Field Eff.	%/a	.	6.2 ('85)	.	9.16	8.55
System Eff.	%/a	.	.	.	8.71 (gross, AC)	8.03 (gross, AC)
Peak						
Power	kW_DC	5,700* + 759*	793*	375 (...)	784	.
	kW_AC	4,600 + 635	753*	.	735	.
Array Field Eff.	%	.	7.7*	10.9 (design)	.	.
Capacity Factor	%	29 ('86); 22 ('89)	35 ('85); 35 ('89)	.	25 ('86)	23 ('87)

* at Standard Test Conditions (1,000 W/m²; 25°C cell temp.)
** at Standard Operating Conditions (1,000 W/m²; AM = 1.5; 25°C ambient temp.; 1 m/s wind speed)

330 8 Photovoltaic Power Stations

Table 8.8. Some demonstration photovoltaic power plants > 30 kWe (Part 2)

Name			PV 300	Pellworm	Kythnos	Marchwood	Alabama
Location			Austin	Pellworm Island	Kythnos Island	Marchwood	
Country			USA	Germany	Greece	U.K.	USA
Design Conditions							
Site							
Latitude		deg N	30 N	55 N	38 N	50–56 N	31–35 N
Longitude		deg E or W	98 W	9 E	24 E	6 W–2 E	85–88 W
Altitude		m	.	.	.	.	.
Design							
Array Output		kW_{DC}	326*	300	100	30	100
Plant Output		kW_{AC}	300	.	.	.	.
Irradiance		W/m^2	.	.	.	.	.
Wind Speed		m/s	.	.	.	.	.
Plant Area		ha	.	2.6	1.6	.	.
Project Data							
Start of Operation			12/1986	7/1983	7/1983	6/1983	6/1986
Construction Time		months	5.0	12	.	.	.
Total Cost (Year)		Mio$(US)	3.12	5.6	.	.	.
O&M Staff		Manyear	.	.	.	.	.
O&M Cost		k$(US)/a	.	.	.	.	.
Status (as of 1987)			operating	operating	operating (?)	operating	operating
Technical Information							
PV-Field							
Cell	Mat/Dim.	cm	ASI	AEG	Siemens	Lucas BP	Chronar
			mono k-Si	poly k-Si; 10 × 10	mono k-Si; ...0		a-Si
Module	Type		Flat Plate	Flat Plate	Flat Plate	Flat Plate	Flat Plate
				Glass-Glass-Al	Glass-Glass-Al	Glass-Glass-Al	
	Dimension	cm	.	56 × 46	146 × 110	.	.
	Rating	W_p/(%)	> 2.067 (12.4)*	19.2	125	33	.
	Number		154	17,568	800	960.5	11,254
	Area	m^2	17.13 (gross)	4,514	.	426	.
Subarray	Tracking		1-axis	fixed	fixed	fixed	fixed
	Mounting		pole-mounted	.	.	.	.
	Dimension	m	.	.	.	.	.
	Number		154	359	40	59	.
	Total Area	m^2	2.638 (gross)	1.726	.	.	.

8.12 Appendix: Photovoltaic Facility Data

Bus Voltage	V_{DC}	516 (max)	346	250	220	300
Field Area	ha	1.2	1.65	.	.	.
Power Conditioning						
INV Rating	KVA	Toshiba 300	AEG $2 \times 75 + 1 \times 300$	Siemens $4 \times 25/3 \times 50$	1×10	.
Commutation		self	self + line	–/self	.	.
Input Voltage	V_{DC} (Range)	300–550	.	.	.	.
Eff. at 100%	%	98.6	.	97/93.5	.	89
Eff. at 50%	%	.	.	./91.0	.	85
MPP Tracking		.	.	yes/–	.	.
Storage, Lead Acid	Ah	none	6.000	1.500	370	.
Performance						
Average						
Irradiation	kWh/m²d	.	.	.	.	.
	kWh/m²a	.	.	.	.	.
Amb. Temp.	°C	.	.	.	.	.
Production	kWh$_{DC}$/d	.	.	.	.	.
	MWh$_{DC}$/a	.	.	.	.	.
	kWh$_{AC}$/d	.	.	.	.	.
	MWh$_{AC}$/a	.	145 (214 peak)	170	25	.
Operations Range	W/m²	.	.	.	.	.
Array Field Eff.	%/a	.	.	.	.	.
System Eff.	%/a	.	.	.	.	.
Peak						
Power	kW$_{DC}$	.	286.0	86.3	29.4	54 (4/89)
	kW$_{AC}$	.	.	.	.	9.5 (4/89)
Array Field Eff.	%	25 ('88)	6.7	6.8	7.1	.
Capacity Factor	%	.	.	.	.	.

* at Standard Test Conditions (1,000 W/m²; 25°C cell temp.)

Bibliography

[1] *Annual Progress Report: Photovoltaics FY 1986.* Technical Report DOE/CH/10093-H3, U.S. Department of Energy, 1987
[2] *EPRI/CRIEPI Workshop Memorandum,* 1986
[3] Leading Edge: Solar-to-Electric Conversion Efficiency of New Concentrator Cell Reaches Record 31%. *PV Insider's Report,* Curry, R. (ed.), 7 (1988)
[4] Shaped Crystal Growth 1986. *J. Crystal Growth,* 82 (1987)
[5] *Switching Networks, Technology and Components* (in German). Fürth (D): Siemens AG, 1985
[6] *Utility Solar Central Receiver Study.* Midterm Review April 1987; Final Review, August 1987; Summary Report (Draft), 1987
[7] World PV Module Production up 23%; US Share 30%. *PV News,* Maycock, P. D. (ed.), 8 (1989)
[8] Adcock, J. P.; Knecht, R. D.: *Flat-Plate Solar Array Project: Government and Industry Responding to National Needs.* Technical Report SERI/SP-320-3382, Golden/CO, Solar Energy Research Institute, 1988
[9] Adler, D.; et al. *Noncryst. Sol.,* 66 (1984) 273
[10] Anis, W. R.; Mertens, R. P.; v. Overstraeten, R.: Calculation of Solar Cell Operating Temperature in a Flat-Plate PV Array. In *Proc. 5th E.C. PV Solar Energy Conference, Athens 1983,* Palz, W.; Fittipaldi, F. (Ed.), p. 520, Dordrecht (NL): D. Riedl, 1984
[11] Aulich, H. A.: Progress in Solar-Grade Silicon Production. In *Proc. 17th IEEE PV Specialist Conference, Orlando,* p. 390, New York: IEEE, 1984
[12] Backus, C. E.: *Terrestrial Photovoltaic Power Systems with Sunlight Concentrators.* Tuscon/AZ, Arizona State University, 1974
[13] Bechtel National Inc.: *Design of Low-Cost Structures for Photovoltaic Arrays – Final Report.* Technical Report SAND79-7002, Sandia National Laboratories, Albuquerque/NM, 1979
[14] Belli, G.; Iliceto, A.: The Photovoltaic Plants of Orbetello and Zannone. In *Proc. 7th E.C. PV Solar Energy Conference, Sevilla,* Goetzberger, A.; Palz, W.; Willeke, G. (Ed.), p. 245, D. Reidel, 1987
[15] Beyer, U.: Experimental Comparison Between Different DC/AC Inverters for Small Grid-Connected Photovoltaic Applications. In *Proc. 6th E.C. PV Solar Energy Conference, London,* Palz, W.; Treble, F. C. (Ed.), p. 430, Dordrecht (NL): D. Reidel, 1985
[16] Blaesser, G.; Krebs, K.: Summary of PV Pilot Monitoring Data 1984–1985. In *Proc. 7th E.C. PV Solar Energy Conference, Sevilla,* Goetzberger, A.; Palz, W.; Willeke, G. (Ed.), p. 84, D. Reidel, 1987
[17] Blaesser, G.; Krebs, K.; Ossenbrink, H.; Rossi, E.: Acceptance, Testing and Monitoring the CEC Photovoltaic Plants. In *Proc. 6th E.C. PV Solar Energy Conference, London,* Palz, W.; Treble, F. C. (Ed.), p. 481, Dordrecht (NL): D. Reidel, 1985
[18] Blok, C.; Terhorst, E.: *SOMES – A Simulation and Optimization Model for Autonomous Energy Systems, User Guide.* Technical Report, Utrecht (NL), University of Utrecht, 1987
[19] Bloss, W. H.; Schock, H. W.: Advances in Polycrystalline Thin-Film Solar Cells. In *Proc. 8th E.C. PV Solar Energy Conference, Florence,* Solomon, I.; Equer, B.; Helm, P. (Ed.), p. 1571, Dordrecht (NL): Kluwer, 1988
[20] Boehringer, A.: Configuration and Control of Satellite Energy Supply Systems (in German). *etz-A,* 92 (1971) 114
[21] Boes, E. C.: Photovoltaic Concentrator Technology: Recent Results. In *Proc. 8th E.C. PV Solar Energy Conference, Florence,* Solomon, I.; Equer, B.; Helm, P. (Ed.), p. 673, Dordrecht (NL): Kluwer, 1988
[22] Boes, E. C.: Photovoltaic Concentrators: Recent Developments. In *Proc. 5th E.C. PV Solar Energy Conference, Athens 1983,* Palz, W.; Fittipaldi, F. (Ed.), p. 257, Dordrecht (NL): D. Riedl, 1984
[23] Bube, R. H.: Thin-Film Polycrystalline Solar Cells. In *Proc. ISES Solar World Congress, Hamburg 1987,* Bloss, W. H.; Pfisterer, F. (Ed.), p. 64, Oxford (UK): Pergamon Press, 1988
[24] Bucciarelli, L. L.: Power Loss in Photovoltaic Arrays Due to Mismatch in Cell Characteristics. *Solar Energy,* 23 (1979) 277
[25] Carmichael, D. C.; Alexander, G.; Castle, J. A.; Post, H. N.: Low-Cost Modular Array-Field Designs for Flat-Panel and Concentrator Photovoltaic Systems. In *Conf. Rec. 16th IEEE PV Specialist Conference, San Diego,* p. 378, New York: 1982
[26] Chao, C.; Bell, R. O.: Effect of Solar Cell Processing on the Quality of EFG Nonagon Growth. In *Conf. Rec. 19th IEEE PV Specialist Conference, New Orleans,* p. 366, New York: IEEE, 1987
[27] Chryssis, G.: *High Frequency Switching Power Supplies: Theory and Design.* New York: McGraw Hill, 1984
[28] Coutts, T. J.; Meakin, J. D. (Ed.): *Current Topics in Photovoltaics.* Orlando: Academic Press, 1985

[29] DeMeo, E. A.: Photovoltaics for Electric Utility Applications. In *Tech. Digest Int. 3rd PVSEC, Tokyo*, Takahashi, K. (Ed.), p. 777, Tokyo: 1987
[30] Diguet, D.; Anguet, J.: Comparative Study of the Thermal Characteristics of Various Solar Cell Module Structures. In *Proc. 3rd E.C. PV Solar Energy Conference, Cannes*, Palz, W. (Ed.), p. 726, Dordrecht (NL): D. Reidel, 1980
[31] Fahrenbruch, A. L.; Bube, R. H.: *Fundamentals of Solar Cells*. Orlando: Academic Press, 1983
[32] Feldmann, J.; Singer, S.; Braunstein, A.: Solar Cell Interconnections and the Shadow Problem. *Solar Energy*, 26 (1981) 419
[33] Fonash, S. J.: *Solar Cell Device Physics*. Orlando: Academic Press, 1981
[34] Fuentes, M. K.: *A Simplified Thermal Model for Flat-Plate Photovoltaic Arrays*. Technical Report SAND85-0330, Sandia National Laboratories, Albuquerque/NM, 1987
[35] Georgia, K.; Samuelson, G.: Properties of a New Photovoltaic Encapsulant. In *Conf. Rec. 18th IEEE PV Specialist Conference, Las Vegas*, p. 1036, New York: IEEE, 1985
[36] Gloeckl, J.; Puccetti, P.; Helm, P.; Traeder, K.: Design and Costs of Support Structures. In *Proc. Final Design Review Meeting on E.C. PV Pilot Proj., Brussels 1981*, Palz, W. (Ed.), p. 2, Dordrecht (NL): D. Reidel, 1982
[37] Goetzberger, A.; Stahl, W.; Voss, B.: Comparison of Yearly Efficiency and Cost of Energy for Stationary, Tracking and Concentrating PV Systems. In *Proc. 7th E.C. PV Solar Energy Conference, Sevilla*, Goetzberger, A.; Palz, W.; Willeke, G. (Ed.), p. 250, D. Reidel, 1987
[38] Goetzberger, A.; Wittwer, V.: *Solar Energy* (in German). Stuttgart (D): Teubner, 1986
[39] Gonzalez, C.: Photovoltaic Array Loss Mechanisms. *Solar Cells*, 18 (1986) 373
[40] Grassi, G.: Flexible and Continous Low Cost PV Structures. In *Tech. Digest Int. 3rd PVSEC, Tokyo*, Takahashi, K. (Ed.), p. 1672, Tokyo: The Japan Times, Ltd., 1987
[41] Grassi, G.; Paoli, P.; Leonardini, L.; Vitale, E.; Conti, P.; Colpizzi, E.: Low-Cost Support Structure for LArge Photovoltaic Generators. In *Proc. 5th E.C. PV Solar Energy Conference, Athens 1983*, Palz, W.; Fittipaldi, F. (Ed.), p. 450, Dordrecht (NL): D. Riedl, 1984
[42] Green, M. A.; Wenham, S. R.; Blakers, A. W.: Recent Advances in High Efficiency Silicon Solar Cells. In *Conf. Rec. 19th IEEE PV Specialist Conference, New Orleans*, p. 6, New York: IEEE, 1987
[43] Griesinger, M.; Bloss, W. H.; Reinhardt, E. R.: Optimization of Tandem Solar Cell Systems Based on Dispersive Concentrating Elements. In *Proc. 5th E.C. PV Solar Energy Conference, Athens 1983*, Palz, W.; Fittipaldi, F. (Ed.), p. 240, Dordrecht (NL): D. Riedl, 1984
[44] Gross, F.: Electrical Energy Technology (in German). p. 595, Köln (D): TÜV Rheinland, 1987
[45] Hartmann, E.; Baer, G.: Hydraulic Pump Storage Power Plants – Proven Effective in Over 50 Years of Operation (in German). p. 55, Düsseldorf (D): VDI-Verlag, 1987
[46] Hester, S.; Hoff, T.: Long-term PV Module Performance. In *Conf. Rec. 18th IEEE PV Specialist Conference, Las Vegas*, p. 198, New York: IEEE, 1985
[47] Hoff, T.: The Value of Photovoltaics: A Utility Perspective. In *Conf. Rec. 19th IEEE PV Specialist Conference, New Orleans*, p. 1145, New York: IEEE, 1987
[48] Hoff, T. E.; Matsuda, K. M.; Betts, E. G.: *PG&E Photovolatic Module Performance Assessment*. Technical Report EPRI AP-4464, Electric Power Research Institute, Palo Alto/CA, 1986
[49] Hoffner, J. E.; Yenamandra, R.: Construction Experience with a 300-Kilowatt Photovoltaic Plant in Austin, Texas. In *Proc. 1987 Ann. Meeting American Solar Energy Soc., Portland*, p. 121, 1987
[50] Hughes Aircraft Company: *Development of a Holographic Beam Splitter for Use with a Conventional Fresnel Lens*. Technical Report SAND86-7049, Sandia National Laboratories, Albuquerque/NM, 1986
[51] Ikegami, S.: II-IV Compound Semiconductor Solar Cells. In *Tech. Digest Int. 3rd PVSEC, Tokyo*, Takahashi, K. (Ed.), p. 677, Tokyo: 1987
[52] Inglis, D. J.: *Photovoltaic Field-Test Performance Assessment, Technology Status Report Number 1*. Technical Report EPRI-AP-2544, Electric Power Research Institute, Palo Alto/CA, 1982
[53] Inglis, D. J.: *Photovoltaic Field-Test Performance Assessment, Technology Status Report Number 2*. Technical Report EPRI-AP-3244, Electric Power Research Institute, Palo Alto/CA, 1983
[54] Inglis, D. J.: *Photovoltaic Field-Test Performance Assessment, Technology Status Report Number 3*. Technical Report EPRI-AP-3792, Electric Power Research Institute, Palo Alto/CA, 1984
[55] Jones, G. J.; Post, H. N.; Stephens, J. W.; Key, T. S.: Design Consideration for Large Photovoltaic Systems. In *Conf. Rec. 18th IEEE PV Specialist Conference, Las Vegas*, p. 1307, New York: IEEE, 1985
[56] Kaut, W.; Gillett, W. B.: Preliminary Experience from the CEC Photovoltaic Demonstration Programme and Future Prospects. In *Proc. 7th E.C. PV Solar Energy Conference, Sevilla*, Goetzberger, A.; Palz, W.; Willeke, G. (Ed.), p. 103, D. Reidel, 1987
[57] Keating, L.: *Computer Simulation and Expert Systems for Photovoltaic Applications*. PhD. Thesis, NMRC University College, Cork (IR). to be published

[58] Key, T.; Menicucci, D.: Photovoltaic Electrical System Design Practice: Issues and Recommendations. In *Conf. Rec. 19th IEEE PV Specialist Conference, New Orleans*, p. 1128, New York: IEEE, 1987
[59] Klaiss, H.; Geyer, M.: Economic Comparison of Solar Power Electricity Generating Systems. In *GAST – The Gas-Cooled Solar Tower Technology Program*, Becker, M.; Boehmer, M. (Ed.), Berlin, Heidelberg, New York: Springer, 1989
[60] Koethe, H. K.: *Practical Aspects of Solar- and Wind-Electric Energy Supplies* (in German). Düsseldorf (D): VDI-Verlag, 1982
[61] Konagai, M.: Technical Status of Amorphous Silicon Solar Cells. In *Proc. ISES Solar World Congress, Hamburg 1987*, Bloss, W. H.; Pfisterer, F. (Ed.), p. 56, Oxford (UK): Pergamon Press, 1988
[62] Kraemer, K.: Development State of Lead Acid and Alkaline Batteries (in German). p. 191, Düsseldorf (D): VDI-Verlag, 1987
[63] Ledjeff, K.: Unconventional Energy Storage Systems. In *Proc. 8th E.C. PV Solar Energy Conference, Florence*, Solomon, I.; Equer, B.; Helm, P. (Ed.), p. 26, Dordrecht (NL): Kluwer, 1988
[64] Leipold, M. H.: Critical Technology Limits to Silicon Material and Sheet Production. In *Proc. 4th E.C. PV Solar Energy Conference, Stresa*, Bloss, W. H.; Grassi, G. (Ed.), p. 985, Dordrecht (NL): D. Reidel Publishing Co., 1982
[65] Levy, S. L.: *Conceptual Design for a High-Concentration (500X) Photovoltaic Array*. Technical Report EPRI/AP 3263, Electric Power Research Institute, Palo Alto/CA, 1984
[66] Lowalt, H. J.: 300 kW Photovoltaic Pilot Plant Pellworm. In *Proc. Final Design Review Meeting on E.C. PV Pilot Proj., Brussels 1981*, Palz, W. (Ed.), p. 179, Dordrecht (NL): D. Reidel, 1982
[67] Martin Marietta Aerospace: *Design and Fabrication of a Low-Cost Two-Axes Solar Tracking Structure for Photovoltaic Concentrator Arrays*. Technical Report SAND82-7150, 1983
[68] Mayorga, H.; Hostetter, R.; Inglis, D. J.; Davis, B. L.: *Photovoltaic Field-Test Performance Assessment, Technology Status Report Number 4*. Technical Report EPRI-AP-4466, Electric Power Research Institute, Palo Alto/CA, 1986
[69] McCarthy, S.; Keating, L.; Wrixon, G. T.: *Development of ASHLING, a PV Plant Simulation Program*. Technical Report, Cork, Ireland, NMRC University College, 1990
[70] Meinel, A. B.; Meinel, M. P.: *Applied Solar Energy*. Reading/Mass.: Addison-Wesley, 1977
[71] Menicucci, D. F.; Fernandez, J. P.: *User's Manual for PVFORM: A Photovoltaic System Simulation Program for Stand-Alone and Grid-Interactive Applications*. Technical Report SAND85-0376, Sandia National Laboratories, Albuquerque/NM, 1988
[72] Menicucci, D. F.; Poore, A. V.: *Today's Photovoltaic Systems: An Evaluation of Their Performance*. Technical Report SAND87-2585, Sandia National Laboratories, Albuquerque/NM, 1987
[73] Mitchell, K. W.; Eberspacher, C.; Ermer, J.; Pier, D.; Milla, P.: Copper Indium Diselenide Photovoltaic Technology. In *Proc. 8th E.C. PV Solar Energy Conference, Florence*, Solomon, I.; Equer, B.; Helm, P. (Ed.), p. 1578, Dordrecht (NL): Kluwer, 1988
[74] Mon, G.; Wen, L.; Ross, R.: Encapsulant Free-Surfaces and Interfaces: Critical Parameters in Controlling Cell Corrosion. In *Conf. Rec. 19th IEEE PV Specialist Conference, New Orleans*, p. 1215, New York: IEEE, 1987
[75] Moore, T.; DeMeo, E.; Cummings, J.: Opening the Door for Utility Photovoltaics. *EPRI Journal*, (January/February 1987) 5
[76] Nann, S.: Cloud Cover Modifier for Solar Spectral Irradiance Modelling. In *Proc. ISES Solar World Congress, Kobe 1989*, p. 422, Oxford (UK): Pergamon Press, 1990
[77] Nann, S.: Potentials for Tracking Photovoltaic Systems and V-Troughs in Moderate Climates. *Solar Energy*, 45 (1990) 385
[78] Nann, S.: Uncertainties in Determination of Short-Circuit Current from Measured and Modelled Spectral Solar Irradiance. In *Proc. 9th E.C. PV Solar Energy Conference, Freiburg*, Palz, W.; Wrixon, G. T.; Helm, P. (Ed.), Dordrecht (NL): Kluwer, 1989
[79] O'Neill, M. J.; Hudson, S. L. In *Proc. 1978 Annual Meeting, Denver/CO: Solar Diversification*, Boer, K. W.; Franta, G. E. (Ed.), pp. 855–859, Newark/DE: American Section of ISES, 1978
[80] O'Neill, M. J.; Walters, R. R.; Perry, J. L.; McDanal, A. J.; Jackson, M. C.; Hesse, W. J.: Fabrication, Installation and Initial Operation of the 2000 SQ.M. Linear Fresnel Lens Photovoltaic Concentrator System at 3M, Austin/TX. In *Conf. Rec. 21st IEEE PV Spec. Conference, Kissimmee, 1990, to be published*
[81] Perez, R.; Seals, R.; Ineichen, P.; Stewart, R.; Menicucci, D.: A New Simplified Version of the Perez Diffuse Irradiance Model for Tilted Surfaces. *Solar Energy*, 39 (1987) 221
[82] Pfisterer, F.; Bloss, W. H.: Polycrystalline Thin Film Solar Cells, State-of-the-Art. In *Proc. 2nd Int. PVSEC, Beijing*, p. 561, Tianjin, China: Tianjin Inst. of Power Sources, 1986

[83] Post, H. N.; Thomas, M. G.: Photovoltaic Systems for Current and Future Applications. *Solar Energy*, 41 (1988) 465
[84] Quast, P.: Compressed Air Storage (in German). p. 89, Düsseldorf (D): VDI-Verlag, 1987
[85] Rauschenbach, H. S.: *Solar Cell Array Design Handbook*. New York: Van Nostrand, 1980
[86] Risser, V. V.; Fuentes, M. K.: Linear Regression Analysis of Flat-Plate Photovoltaic Systems Performance Data. In *Proc. 5th E.C. PV Solar Energy Conference, Athens 1983*, Palz, W.; Fittipaldi, F. (Ed.), p. 623, Dordrecht (NL): D. Reidel, 1984
[87] Salim, A. A.; Huraib, F. S.; Eugenio, N. N.: Performance Comparison of Two Similar Concentrating PV Systems Operating in the U.S. and Saudi Arabia. New York: IEEE, 1987
[88] Salim, A. A.; Huraib, F. S.; Eugenio, N. N.: PV Power – Study of System Options and Optimization. In *Proc. 8th E.C. PV Solar Energy Conference, Florence*, Solomon, I.; Equer, B.; Helm, P. (Ed.), p. 688, Dordrecht (NL): Kluwer, 1988
[89] Schaefer, J.: Photovoltaic Operating Experience. *EPRI Journal*, (March 1988) 40
[90] Schott, T.: Operation temperatures of PV Modules – A Theoretical and Experimental Approach. In *Proc. 6th E.C. PV Solar Energy Conference, London*, Palz, W.; Treble, F. C. (Ed.), p. 392, Dordrecht (NL): D. Reidel, 1985
[91] Schumacher-Groehn, J.: *Documentation of the Simulationprogram INSEL* (in German). Technical Report, Oldenburg (D), University of Oldenburg, to be published
[92] Shockley, W.: *Electrons and Holes in Semiconductors*. New York: van Nostrand, 1950
[93] Shushnar, G. J.; Schlueter, L. E.: Operational Experience with the Carrisa Plain PV Plant. In *1st PV Annual Systems Symposium*, 1987
[94] Sinton, R. A. PhD. Thesis, Stanford University, 1987
[95] Sirtl, E.: Production Methods for Mono- and Multicrystalline Silicon. In *Proc. ISES Solar World Congress, Hamburg 1987*, Bloss, W. H.; Pfisterer, F. (Ed.), p. 34, Oxford (UK): Pergamon Press, 1988
[96] Stokes, K. W.: *Hesperia Photovoltaic Power Plant: 1985 Performance Assessment*. Technical Report EPRI-AP-5229, Electric Power Research Institute, Palo Alto/CA, 1987
[97] Stokes, K. W.; Risser, V. V.: *Photovoltaic Field Test Performance Assessment: 1986*. Technical Report EPRI AP-5762, Electric Power Research Institute, Palo Alto/CA, 1988
[98] Stutzmann, M.; Jackson, W. B.; Tsai, C. C.: Light Induced Metastable States in Hydrogenated Amorphous Silicon: A Systematic Study. *Phys. Rev.*, B 32 (1985) 23
[99] Sugimura, R. S.; Otth, D. H.; Ross, R. G.; Arnett, J. C.; Samuelson, G.: Candidate Materials for Advanced Fire Resistant Photovoltaic Modules. In *Conf. Rec. 18th IEEE PV Specialist Conference, Las Vegas*, p. 1164, New York: IEEE, 1985
[100] Sumner, D. D.; Whitaker, C. M.; Schlueter, L. E.: Carrisa Plains Photovoltaic Power Plants 1984–1987 Performance. In *Conf. Rec. 20th IEEE PV Specialist Conference, Las Vegas*, p. 1289, New York: IEEE, 1988
[101] Swanson, R. M.: Point-contact solar cells: Modeling and Experiment. *Solar Cells*, 17 (1986) 85
[102] Thomas, M. G.; Fuentes, M. K.; Lashway, C.; Black, B. D.: Reliability of Photovoltaic Systems: A Field Report. In *Conf. Rec. 18th IEEE PV Specialist Conference, Las Vegas*, p. 1336, New York: IEEE, 1985
[103] Voss, E.: Secondary Electrochemical Cells for Solar Energy Storage. In *Proc. 8th E.C. PV Solar Energy Conference, Florence*, Solomon, I.; Equer, B.; Helm, P. (Ed.), pp. 19–25, Dordrecht (NL): Kluwer, 1988
[104] Wagner, U.: Power Control by Energy Storage (in German). p. 177, Düsseldorf (D): VDI-Verlag, 1987
[105] Wrixon, G. T.; McCarthy, S.: Field Experience of a 50 kWp Photovoltaic Array at Fota Island. In *Proc. 6th E.C. PV Solar Energy Conference, London*, Palz, W.; Treble, F. C. (Ed.), p. 491, Dordrecht (NL): D. Reidel, 1985

9 Solar Fuels and Chemicals, Solar Hydrogen

M. Fischer and R. Tamme [1]

9.1 Introduction

The term *solar power plant* commonly refers to electricity production by photothermal or photovoltaic conversion. Within this book, these main conversion techniques are called the *solar thermoelectric path* and the *solar photoelectric path*. In addition, a third possibility exists, which, following the above used nomenclature, is called the *solar fuels path* [46]. This path represents the conversion of solar energy into chemical energy, and is important due to its potential to overcome the problems of long term storage and transport of solar energy, as well as for the intrinsic value of the chemicals themselves. With respect to the present discussion, endergonic reactions (with a positive change of the Gibbs free energy of reaction ΔG) are especially suited so that useful energy (exergy) of the solar radiation can be stored in the reaction products.

Properties that make solar radiation an advantageous source for producing chemicals and fuels are:

- radiation can be used at elevated temperatures within a clean, unpolluted atmosphere;
- radiation can provide high flux densities and high heating rates;
- radiation is a source of photons over a wide spectrum (300–2,100 nm) that can be used as a whole or in parts by spectral splitting or shifting (e.g. via gratings, filters, or frequency doublers, respectively).

To use the full potential of these solar attributes, it is important to examine alternative elementary methods for the collection and utilization of the concentrated solar radiation. The present status of research and development of converting solar energy into storable and transportable fuels and high value chemicals indicates that it is a new direction of solar application. Most of the reported activities are still in the laboratories, and engineering and fundamental research efforts are focused on several crucial points:

- coupling of the unsteady solar energy source with the usual need for steady-state chemical processing,
- development of solar adapted batch processes,
- development of receiver-reactors and window materials,
- investigation of photon-assisted thermochemical reactions using direct absorption techniques,
- investigation of thermal-assisted and photon-assisted electrochemistry.

---

[1] Manfred Fischer and Rainer Tamme, Deutsche Forschungsanstalt für Luft- und Raumfahrt (DLR), Pfaffenwaldring 38-40, D-7000 Stuttgart 80

Most of the investigations in solar chemistry were directed at using process heat, delivered by solar thermal conversion, for thermochemical reactions. This can be considered as the thermal path (sometimes also called the kT-path) of solar chemistry. Sects 9.2–9.6 deal with this straightforward approach, the *separated solution*, as well as with the more solar specific approach, the *intergrated solution*. Separated solution essentially means substitution of fossil energy by solar generated heat, adopting the operating conditions of the presently used process technology. Integrated solution means conducting the solar thermal conversion and the endothermic process within the same device, the receiver-reactor. The photochemical conversion (hν-path of solar chemistry) and its coupling with photothermal conversion leads to joint thermal-photochemical reactions realized by direct interaction of radiation with the chemically active absorbent agent. This approach can lead to unique solar processing and is presented in Sect. 9.7. The pure photochemical conversion is not treated in this chapter, since the research is still in the fundamental stages, and the presently available results cannot be directly extrapolated to large scale applications.

Section 9.8 deals with the electrolytic production of hydrogen with photovoltaic systems and with thermal solar power plants (eV-path of solar chemistry). Hydrogen is a clean fuel which is transportable and storable on a large scale in gaseous or cryogenic liquid form. It is technically capable of serving most, if not all, of the fuel functions served by today's natural gas, petroleum and coal products. Also in the future, fuels will remain the main source for the energy supply in industrialized and developing countries. In principle, hydrogen can be produced in unlimited amounts from the only needed material feedstock: water. The key resource needed for production is a suitable non-fossil primary energy source for splitting water into its elements: hydrogen and oxygen. Solar energy, including wind and hydro power converted into process heat and electricity are leading candidates for being inexhaustible clean primary energy sources for electrolytic hydrogen production.

9.2 Endothermal Chemical Processes Coupled with Solar Energy

Specific problems of using solar thermal energy for chemical processes are caused by the matching of the solar system and the needs of chemical engineering. The major constraint is the incompatibility of intermittently available process energy with the present requirement of definite process parameters for chemical processing. Possibilities of coupling solar thermal energy with chemical processes are demonstrated in Fig. 9.1. Three cases can be distinguished:

1. inclusion of thermal storage,
2. use of auxiliary heating,
3. direct coupling.

Requirements of steady state operating conditions, necessary for most of the established chemical processes, can be met by using thermal energy storage. For an optimized design of the chemical reactor and of the storage, continuous operation is feasible. Whether part of the solar energy is used to charge the storage and the rest to run the endothermic reaction, or solar energy is stored first and the chemical process is completely served by the thermal storage, depends on the choice of the receiver, the size of the storage system, and the solar multiple.

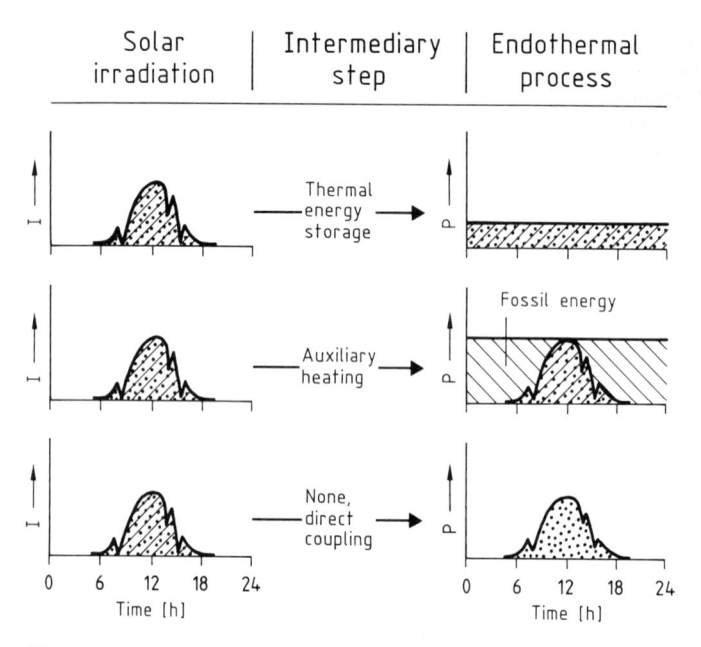

Fig. 9.1. Coupling of time dependent solar energy supply (I) with endothermal chemical processes (P).

The second case of Fig. 9.1 shows the use of auxiliary heating. The design of the reactor may be based on average or maximum irradiation. For continuous operation additional heating through a fossil fired burner is necessary. Solar/fossil hybrid configurations may not be attractive. The overall solar energy contribution to the process is small, a double layout is necessary, and the cycling of the backup system possibly causes adverse impacts on equipment lifetime. On the other hand, the solar/fossil hybrid operation represents a secure way of running endothermic chemical processes continuously with solar energy because an efficient high temperature storage technology is not yet available. The direct coupling of solar flux with chemical processes is solar unique. Gradients of incident solar flux may be partially smoothed by variation of mass flow. However, batch processing with non-constant process parameters is always the result. The advantages of direct coupling are:

- working at elevated temperatures and high energy densities as a result of radiant heating;
- utilization of strong temperature gradients for rapid heating;
- the possibility of combining thermal and photochemical effects.

Chemical reactors are classified in two ways, according to the type of operation and the design features [49]. The first classification comprises *batch processing* and the second *continuous processing*. A batch process operates in a cyclic discontinuous fashion; it includes charging the reactor with the reactants and removing the products. Batch processes usually employ standardized equipment which is more simply constructed and operated than most equipment used in continuous operation. Batch processing is preferable when:

- chemical reaction times are long;
- small quantities are produced;
- processing parameters must be varied with time;
- continuous transport of the reactants encounters difficulties.

The flexibility and cyclic operation of batch processing in which heat content, mass, temperature, reactants and product concentrations vary with time, are arguments for investigations of their suitability for operation with direct solar absorption technologies. New developments are necessary to improve batch processing and to reassess the role of batch versus continuous operation especially for solar applications.

9.3 Receiver-Reactors for Solar Chemical Applications

Direct coupling of solar energy with chemical processes implies that the chemical reactor must be integrated within the solar receiver. Such new reactor types are called receiver-reactors. For solar thermal applications several receiver-reactor types are under investigation. The following classification is useful [17]:

- tube receiver-reactors,
- indirect receiver-reactors, and
- direct absorption receiver-reactors.

Tube receiver-reactors contain reactor elements that are directly heated by solar energy within the receiver. Various geometries are possible. A spiral and an axial configuration are shown in Fig. 9.2. The spiral receiver-reactor has several circular elements connected together in parallel to form the walls of a cylindrical cavity receiver. A receiver-reactor of this type has been tested for methane/carbon dioxide reforming at the White Sands solar furnace, New Mexico, USA [42]. The axial receiver-reactor has elements that run parallel to the receiver axis. Figure 9.2(b) shows the reactor and heat exchanger elements consisting essentially of an internal tube arrangement with the headers at the cold end of the heat exchangers. Such an axial receiver-reactor[2] was tested at the Georgia Institute of Technology Advanced Component Test Facility (ACTF) in Atlanta/GA, USA, for thermal decomposition of sulfuric acid. The cavity was enlarged to avoid direct local illumination of the reactor tubes. Reflection and reradiation were used to smooth the local incident solar flux density [5].

Examples of *indirect receiver-reactors* are shown in Fig. 9.3. They all use an internal heat transfer fluid for the heat exchange between the absorption and reaction zones. Among many possibilities, developments have concentrated on molten salts and liquid metals. The heat pipe receiver-reactor is a bundle of individual heat pipes which interface with the reactors. The heat pipe approach makes it possible to decouple the heat transfer limitations of the reactor from the solar input. The result is a more compact receiver and a better heating of the reactor. However, this advantage is at the expense of a more complex and expensive receiver-reactor design [17].

The reflux receiver-reactor uses a boiling metal, such as sodium, as the heat transfer medium. The metal is contained in a vessel common to all the reactor elements. Temperature variations between the reactor tubes are eliminated by this arrangement. The basic concept is being investigated for application to dish/Stirling receivers [63] and liquid metal thermoelectric converters [41]. The molten material receiver-reactor uses an intermediate heat transfer medium such as molten salt or molten metal between the receiver cavity and the reactor elements. This buffers the reactor from short term fluctuations of the incident solar flux. It also has the advantage of incorporating thermal energy storage into the receiver-reactor. Heat

---

[2] built by General Atomics (GA) Technologies, San Diego/CA, USA

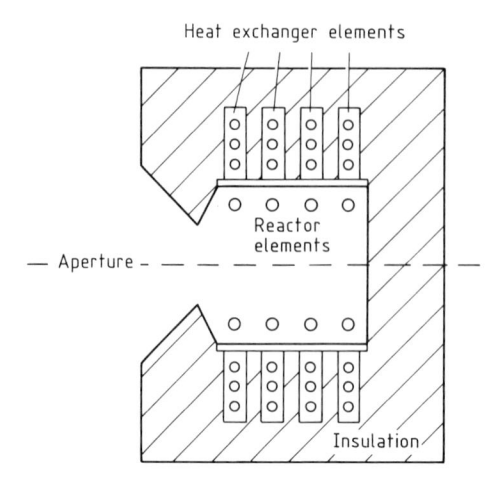

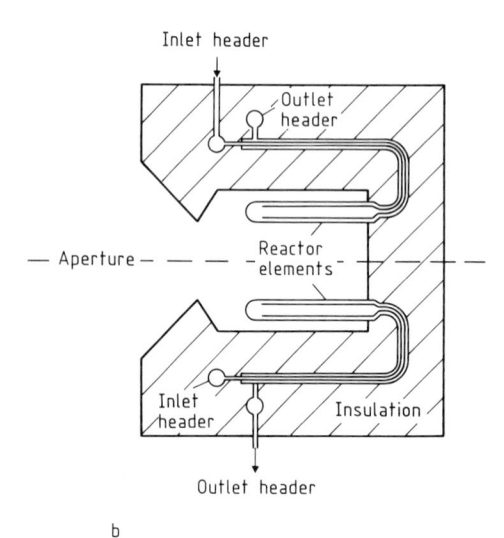

Fig. 9.2. Schematic diagram of tube receiver-reactors, type I: (a) spiral receiver-reactor; (b) axial receiver-reactor.

transfer through the molten material depends on thermal conductivity. Steam receivers based on this concept are used at Solarplant 1 [3], Warner Springs/CA, USA [14].

Direct absorption receiver-reactors (Fig. 9.4) absorb the solar radiation directly on the surface of the solid reactants. Absorption and chemical reaction are then no longer spatially separated. Direct absorption receiver-reactors in general require a transparent entrance window (aperture).

The matrix receiver-reactor features a solid matrix, e.g., a honeycomb or a mesh that serves as a catalyst support. Heat is delivered directly to the catalyst by radiation. Therefore, the chemical process is limited by the rates of diffusion or reaction rather than by heat transfer. Compact receiver-reactors are then possible. Using ceramic materials as catalyst

---

[3] built, owned and operated by La Jet Energy Co., Abilene/TX, USA

9.3 Receiver-Reactors for Solar Chemical Applications

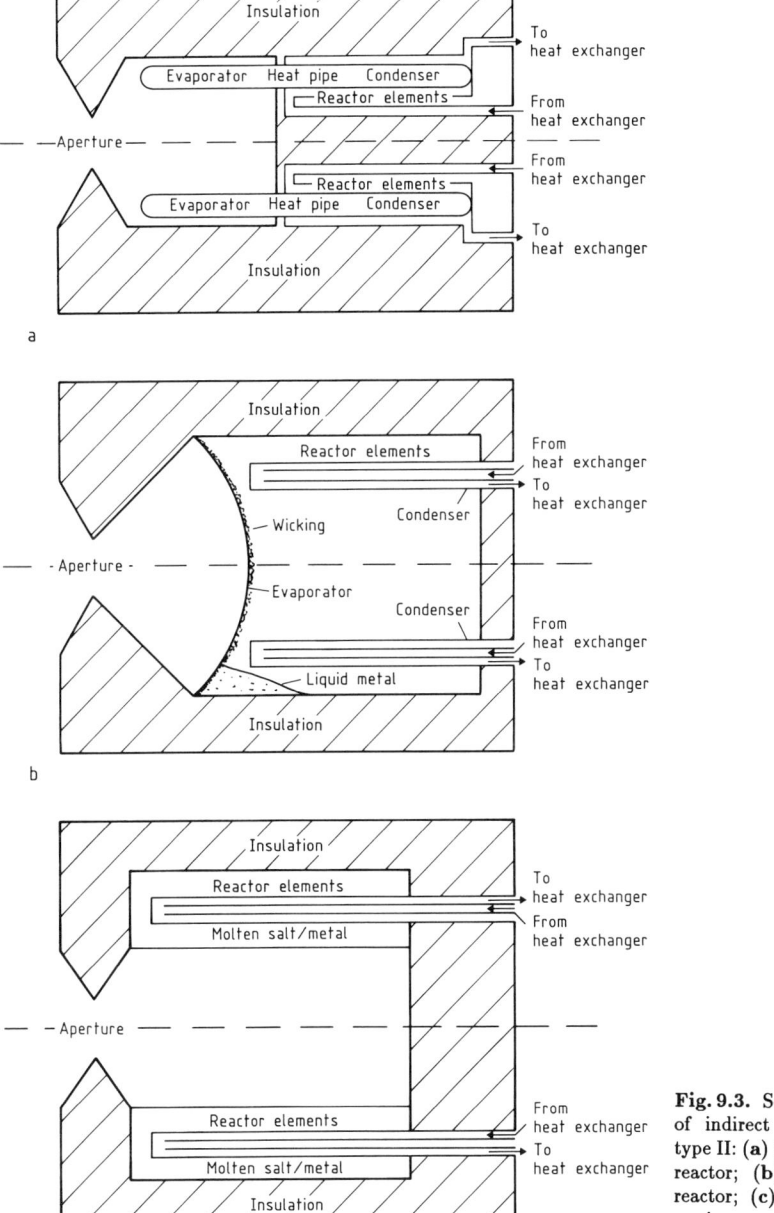

Fig. 9.3. Schematic diagram of indirect receiver-reactors, type II: (**a**) heat pipe receiver-reactor; (**b**) reflux receiver-reactor; (**c**) molten material receiver-reactor.

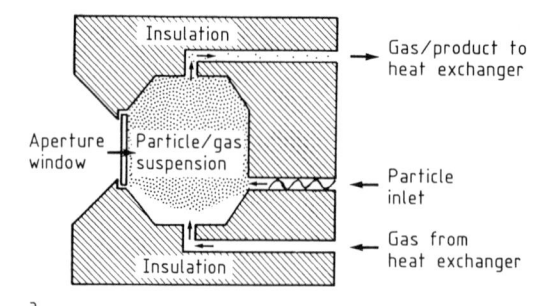

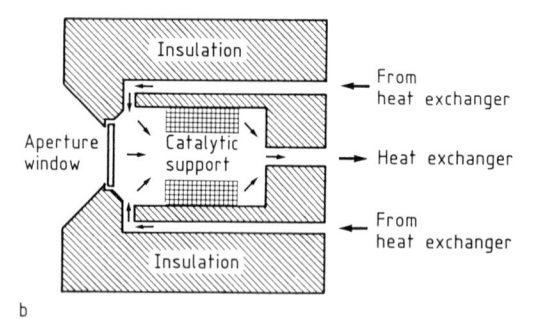

Fig. 9.4. Schematic diagram of direct absorption receiver-reactors, type III: (a) small-particle receiver-reactor; (b) matrix receiver-reactor.

support, operating temperatures may exceed 1,000°C. Sanders Associated designed a ceramic matrix receiver for dish applications [52]. Ceramic matrix receiver-reactors for low pressure thermal decomposition of water and hydrogen sulfide have been built and tested at the University of Minnesota, Minneapolis/MN, USA [18,37]. A ceramic mesh receiver-reactor is under development for solar enhanced methane/carbon dioxide reforming [10].

In the small particle receiver-reactor approach a suspension or a circulating fluidized bed of small solid particles is used. They can act as a heat absorbing agent for gas heating, as a catalyst, or as a feedstock for thermal decomposition processes or particle-gas reactions. For a catalytic reaction the small-particle receiver-reactor has all the advantages of the matrix receiver-reactor. By selecting an optimum combination of particle size and density (number of particles/unit volume), the designer can accommodate the three functions, absorption of solar radiation, heat transfer to the gas, and catalyst support to a higher degree than is possible with matrix support. A small-particle receiver that used soot particles for heating air was built by Lawrence Berkeley Laboratories, Berkeley/CA, USA, and successfully tested at the ACTF [34]. Decarbonation of calcium carbonate was studied in fluidized bed and rotary kiln receiver-reactors at the solar furnace, Odeillo Font Romeu, France [27]. Direct absorption receiver-reactors have the potential to be compact, relatively inexpensive, and are efficient high temperature thermochemical reactors. However, materials and design problems remain to be solved, in particular the manufacture of large transparent windows, stable at elevated temperatures and high pressures.

9.4 High Temperature Processes for Fuels and Chemicals Production

For conducting thermochemical reactions with solar energy, the traditionally used chemical reactors must be adapted to the specific properties of the solar energy heat source. An overview of important reactor types including examples of typical applications in the chemical industry is summarized in Tab. 9.1. The examples are given for the production of raw materials, inorganic and organic chemicals, storable and transportable fuels, and for several other non fuels or chemicals oriented processes.

Table 9.1. Chemical reactors for high temperature processes.

Reactor Type	Application	Examples
Blast furnace	Metallurgy; non-catalytic solid/gas reactions.	Iron melting; burning of limestone.
Converter	Metallurgy; non-catalytic reactions with solids.	Copper ore smelting; steel industry.
Multiple hearth roaster, suspension furnace	Non-catalytic solid/gas reactions; metallurgy.	Calcining processes; ore roasting; sulfur production.
Rotary kiln	Non-catalytic solid/gas reactions; solid-state reactions; metallurgy.	Cement production; calcining processes; barium sulfide production.
Fixed bed reactor	Catalytic solid/gas reactions.	Steam reforming of hydrocarbons; ethylene and styrene production.
Fluidized bed reactor	Non-catalytic and catalytic solid/gas reactions; physical processes.	Coal gasification; waste disposal; calcining processes; catalytic cracking.
Electrothermal furnace	Metallurgy; solid state reactions.	Production of calcium carbide; electrographite and silicon carbide.

For solar applications mainly fixed bed and fluidized bed reactors were proposed [3]. *Fixed bed reactors* play an important part in the chemical industry, especially in the field of heterogeneous catalytic processes. Most fixed bed reactors are of the tubular type with the catalyst in the tubes. When the catalyst is in pelleted or granular form, the catalyst particles are supported by a grid. In fixed beds, heat must flow through the packed catalyst to the heat transfer surface [38]. As examples for potential solar applications, various fixed bed catalytic processes are listed in Tab. 9.2.

Fluidized bed reactors are used for non-catalytic and catalytic solid/gas reactions and for several physical processes such as drying, heat treatment, adsorption/desorption and coating. The overview of today's relevant fluidized bed processes in Tab. 9.3 indicates the importance of this reactor type and presents examples for solar applications. Generally, fluidized bed reactors consist of a vertical cylindrical vessel containing fine solid particles that are either catalysts or reactants. The gas stream is introduced at the bottom of the reactor at such a rate that the solid particles become fluidized. The schematic of a small-particle receiver-reactor (Fig. 9.4(b)) is an example of a 'solarized' fluidized bed reactor.

A summary of important basic chemicals – inorganic and organic compounds – manufactured by high temperature endothermic reactions and generally suited for direct absorption

Table 9.2. Important fixed bed catalytic processes as examples for solar applications [29].

In the basic chemical and petrochemical industry	In the petroleum refining industry

• Steam reforming
• Hydrodealkylation
• Production of
  – Ethene
  – Butadiene
  – Styrene etc.
	• Catalytic reforming
	• Isomerization
	• Hydrosulfurization
	• Hydrocracking

Table 9.3. Fluidized bed processes with potential for solar applications.

Typical physical processes	Typical noncatalytic reactions	Typical catalytic reactions

• Drying of floatation ore and bulk articles
• Cooling of granulated fertilizer
• Heating of solids
• Defluorination of exhaust gas
	• Coal gasification
	• Pyrolysis of hydrocarbons
	• Reduction of iron ore and metal oxides
	• Waste combustion
	• Calcining of
	  – lime
	  – dolomit
	  – crude phosphate
	• Production of
	  – activated carbon
	  – metal fluorides
	  – phosphoryl chloride
	  – ammonium nitrate
	  – sodium nitrate
		• Cracking of hydrocarbons
		• Reforming of hydrocarbons
		• Dehydrogenation of isopropyl alcohol
		• Production of
		  – acetylene
		  – phthalic anhydride
		  – formaldehyde
		  – isoprene
		  – butadiene

techniques is given in Tab. 9.4. At present, only few of them have been investigated in a solar-driven reactor.

Upgrading of heavy crude oil, of tar sands, and oil shale has been proposed for a solar process heat application. Production of light hydrocarbons from heavy stocks requires upgrading by improving properties, especially the C/H ratio. This is equivalent to adding hydrogen or reducing the carbon content. To identify the potential of solar processing, it is important to remember that hydrocarbons, except methane, have a positive Gibbs free energy of formation (ΔG) at temperatures higher than 500 K so that thermodynamic equilibrium favors formation of carbon and hydrogen. Hydrogenation is exothermic, cracking is endothermic. The temperature dependence of reaction rates is such that the rate of cracking increases with temperature faster than the rate of hydrogenation [21,54]. Regarding these basic principles, thermal and catalytic upgrading processes are candidates for solar thermal applications. Commercially applied processes are visbreaking, delayed coking, fluidcoking and flexicoking. Visbreaking (viscosity breaking) is a mild thermal cracking (400–490°C, 1–5 MPa) with a low conversion rate. The main application of this widespread process is the cracking of heavy fuel oil. Coking processes are operated at higher temperatures (up to 600°C) and at moderate pressures of 0.2–0.5 MPa. Therefore, conversion rates are significantly higher.

Lurgi GmbH, Frankfurt, Germany, adapted its LR-process (Lurgi-Ruhrgas) to both conditions and developed a LR-coker for liquid heavy feeds, and a LR-carbonization unit for

9.4 High Temperature Processes for Fuels and Chemicals Production

Table 9.4. Endothermic chemicals selected for potential of solar energy input.

Base material	Thermal reaction	Main product	Processing
Alumina trihydroxide $Al(OH)_3$	Drying, dehydration, calcination	Anhydrous alumina Al_2O_3	Rotary kiln, Fluidized bed calciner; 1,000–1,300°C
Hydrous titanium dioxide $TiO(OH)_2$	Drying, dehydration, calcination	Titanium oxide TiO_2	Rotary kiln, fluidized bed calciner; 900–1,100°C
Sodium carbonate, Silica	Decarboxylation, calcination	Sodium metasilicate Na_2SiO_3	Rotary kiln, tunnel kiln; 850–1,050°C
		Sodium silicate $x\,Na_2O : y\,SiO_2$ $2 < x:y < 4$	Rotary kiln, regeneration furnace, open hearth furnace; 1,200–1,600°C
Limestone, dolomite $CaCO_3$, $Na_2Mg(CO_3)_2$	Decarboxylation	Calcium oxide CaO Dolomitic quicklime CaO/MgO	Rotary kiln, rotary hearth roaster, fluidized bed; 900–1,000°C
Crude barite, coal $BaSO_4$, C	Reduction	Barium sulfide BaS, feedstock for $BaCO_3$, $BaSO_4$	Rotary kiln; 1,000–1,200°C
Raw phosphate rock, Quartz, coal $Ca_3(PO_4)_2$, SiO_2, C	Reduction	Elemental phosphorus P, feedstock for phosphoric acid H_3PO_4	Electric furnace, nodulizing kiln, blast furnace; 1,300–1,450°C
Quartz, carbon SiO_2, C	Reduction, solid state reaction, carbothermal reduction	Silicon carbide SiC	Electric resistance furnace; 1,900–2,100°C
		Elemental silicon Si	Electric reduction furnace; 1,700–1,800°C
Alumina, carbon Al_2O_3, C	Carbothermal reduction	Elemental aluminium Al	Electric reduction furnace;[a] 2,000–2,300°C
Quicklime, carbon CaO, C	Carbothermal reduction	Calcium carbide CaC_2, feedstock for acetylene and fertilizer	Electric arc furnace; 2,000–2,500°C
Paraffins, naphthenes	Thermal cracking	Ethylene	Fixed bed reactor, fluidized bed reactor; 450–600°C
Ethanol (from biomass)	Catalytic dehydration	Ethylene	Fixed bed reactor; 300–350°C
Ethylbenzene	Thermal decomposition	Styrene	Flow reactor; 570–700°C
Ethylene dichloride	Catalytic cracking	Vinyl chloride	Fluidized bed reactor; 480–500°C
Alkyl fluoride Alkyl chloride	Pyrolysis	Dehalogenated Hydrocarbons	Fixed bed reactor, Fluidized bed reactor; 500–1,500°C

[a] Non-applied technology

tar sands and oil shale. The investigation of integrating solar energy in the LR-process for upgrading coke and tar sands has shown only a low potential [7]. A high potential was found for the gasification of lignite and biomass in a circulating fluid bed (CFB) gasification process [8]. A significant increase of the product yields was calculated for the advanced process that includes steam production and direct heating of the CFB (Fig. 9.5). A detailed evaluation of

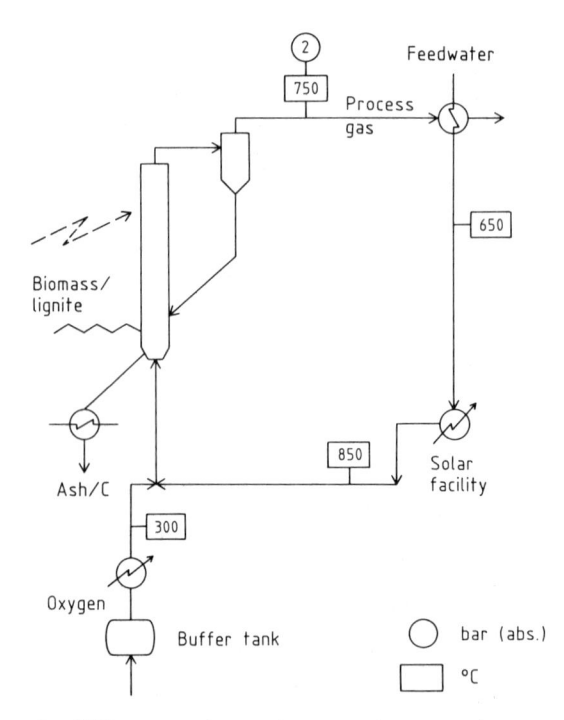

Fig. 9.5. Gasification of biomass and lignite in a solar-operated circulating fluidized bed (CFB) reactor [8].

the CFB-process for methanol production came to the conclusion that the specific methanol output – using lignite as feedstock – is increased by 16% for a CFB direct absorption process and by 57% for a proposed dual solar plant [8]. This configuration includes direct heating of the CFB and a separate solar plant for steam generation.

Thermochemical cycles in which the splitting of water for hydrogen production is performed in several consecutive reactions were investigated for nuclear process heat applications. While processing with nuclear heat is limited to reaction temperatures less than 900°C, solar application allows significantly higher operating temperatures. Therefore, the spectrum of the previously investigated cycles can be extended to include higher temperature reactions. For solar applications, the sulfuric acid-iodine process, developed by General Atomic, was proposed [6,30]. It is a multi step process, and the relevant high temperature reaction is the decomposition of sulfuric acid. Experimental investigations of this reaction were conducted in a tube receiver-reactor [5]. The flow sheet of a proposed 'solarized GA-process plant' is shown in Fig. 9.6 [39]. The solar part of the process, the sulfuric acid decomposition (section II), is operated discontinuously. The remaining processes, the formation of sulfuric acid (section I), the hydrogen iodide decomposition (section IV), and the sulfur trioxide formation (section V), are operated continuously.

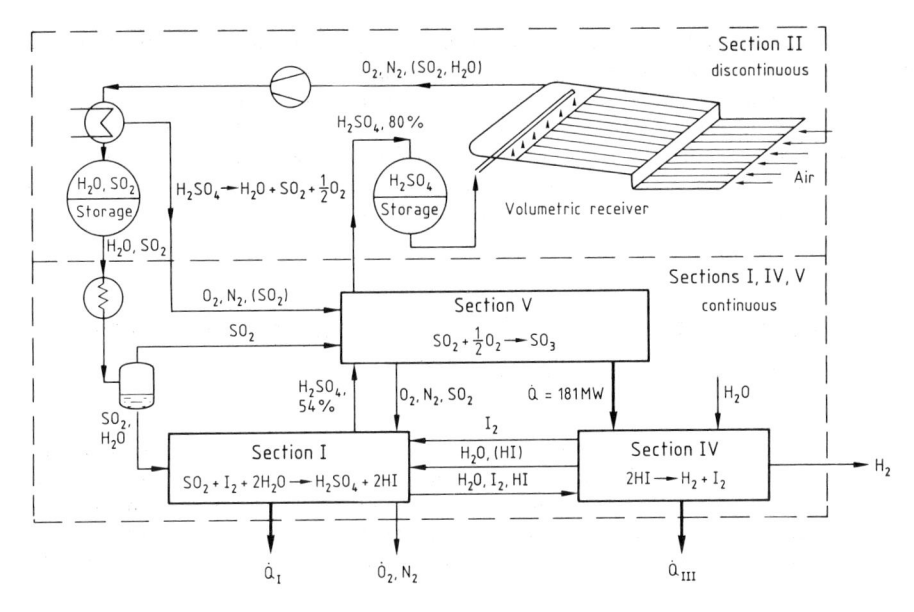

Fig. 9.6. Solarized sulfuric acid – iodine process plant [39].

9.5 Additional Chemical Processing Using Solar Energy

Several areas outside fuels and chemicals industries show promise for solar thermal applications:

- production of stone, clay and building materials;
- glass and ceramic manufacturing;
- minerals processing;
- cement production;
- metal fabricating;
- hazardous waste disposal.

Major products in the category 'stone, clay and glass' such as cement, lime, glass, mineral wool, brick and structural clay, require high temperature and direct heating. Therefore, they are potential candidates for direct absorption processing. For example, lime is produced by calcining of limestone or dolomite. For different raw materials, the decarboxylation reaction leads to different products, high-calcium quicklime CaO, or dolomitic quicklime CaO/MgO. Mineral processing and the manufacture of stone, clay and glass are accomplished in batch operations. Many of the processing temperatures required for the production of these materials are attainable with solar technology. Major problems include the transport of large volumes of solid materials to the solar heat source and the development of a solar technology which can heat the bulk solids in a reactor system located on ground level. Minerals, glass and ceramics are produced in furnaces that must operate continuously, even though materials are changed in batches. Failure of the refractory materials within the furnace occurs if the furnace is periodically heated and cooled. Consequently, high temperature thermal energy storage is needed to keep the furnace refractory at a constant temperature.

Hazardous waste disposal is another application where solar thermal energy can be employed in discontinuous operation. Many hazardous materials are pyrolyzed at elevated temperatures attainable by direct absorption techniques. Most of the waste materials are liquid and readily transportable to the location of the solar installation. This saves energy and, more importantly, leads to an improved waste disposal technology.

9.6 Steam/Carbondioxide Reforming of Methane – A Candidate Process

Synthesis gas is a gas mixture containing CO, H_2, and CO_2 with minor concentrations of other components such as methane. It can be used as the major basic feedstock in the production of chemicals and fuels. Regarding the subsequent synthesis of special chemical products, the value of the $H_2/(CO+CO_2)$ ratio is of importance. Natural gas, petroleum liquids, biomass and coal may all be completely reformed or partially oxidized to produce synthesis gas suitable for further processing.

Steam reforming of methane to produce synthesis gas for methanol production, hydrogen for ammonia synthesis or hydrogen for other processes is practiced worldwide commercially. Carbon dioxide and steam reforming of methane have been proposed for energy transport. Both reforming processes are catalytic endothermic reactions and can be described by several independent simultaneous reactions [60]:

(1a) $CH_4 + H_2O + 206$ kJ/mol $= 3H_2 + CO$
(1b) $CH_4 + CO_2 + 246$ kJ/mol $= 2CO + 2H_2$
(2) $CO + H_2O$ $= CO_2 + H_2 + 41$ kJ/mol
(3) $2CO + 172$ kJ/mol $= CO_2 + C$
(4) $CH_4 + 74$ kJ/mol $= 2H_2 + C$
(5) $C + H_2O + 131$ kJ/mol $= CO + H_2$

Reactions (1a) (steam reforming) and (1b) (CO_2 reforming) are the primary endothermic reactions of interest. The shift reaction (2) is a side reaction that runs either the exothermic path for the steam process (1a) or the endothermic path for the carbon dioxide process (1b). Therefore, the shift reaction reduces or increases the energy demand, necessary for the total reforming process. The carbon forming reactions (3) and (4) are to be avoided, to prevent catalyst deactivation and (fixed bed) reactor plugging. With catalysts used today (aluminium oxide supported nickel or rhodium catalysts) the reaction is fast enough that thermodynamic equilibrium can be obtained while suppressing carbon deposition. The influence of temperature and pressure variations on the product compositions can be deduced from equilibrium thermodynamics. Examples are given in Fig. 9.7 (for a steam-to-methane ratio of 2.5:1) and Fig. 9.8 (for a CO_2-to-methane ratio of 1.2:1) [59]. Generally, the methane conversion is more efficient by varying the process parameters towards high temperatures and low pressures.

For solar methane reforming several concepts have been investigated. Basically, they can be divided into a *separated system*, where the solar receiver is linked to the separate chemical reactor by using an intermediate gas loop. In the *integrated system*, the reforming reactor is placed directly within the solar receiver (receiver-reactor). Regarding the following different synthesis processes, the technically proven reaction conditions require continuous operation at 800–900°C and 1.5–4 MPa. For solar applications this can be adopted in a separated system by

9.6 Steam/Carbondioxide Reforming of Methane – A Candidate Process

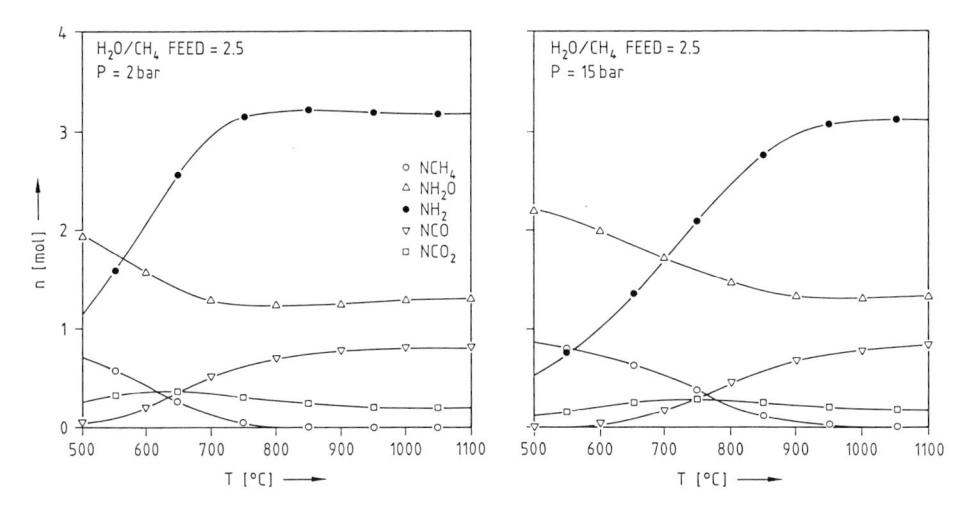

Fig. 9.7. Influence of temperature and pressure variation on steam reforming of methane (equilibrium assumed).

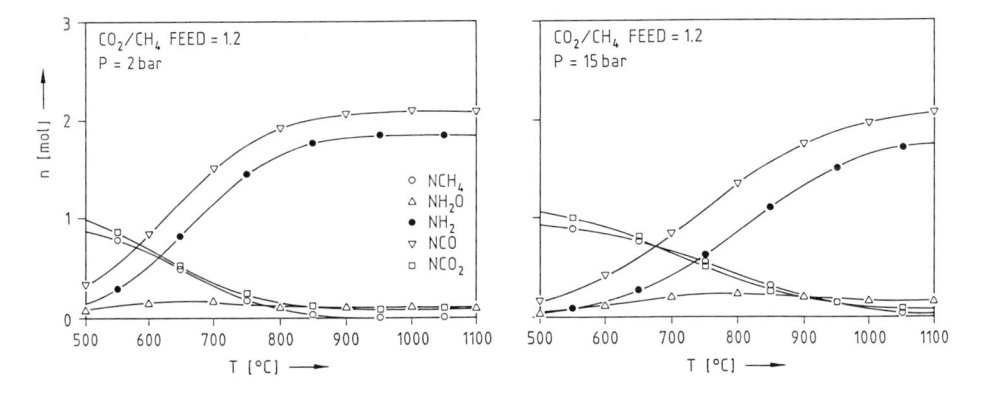

Fig. 9.8. Influence of temperature and pressure variation on CO_2-reforming of methane (equilibrium assumed).

using thermal storage or additional fossil firing. The possibility of steam reforming of methane in a solar heated tube reactor has been investigated for methanol, ammonia and oxoalcohol synthesis [4]. A detailed analysis of several plant configurations based on a separated system has shown that the highest potential case is a solar plant consisting essentially of a ceramic receiver operating at elevated temperatures in connection with a large thermal storage unit [45]. The schematic of a 40 MW_t solar steam reforming plant producing synthesis gas for methanol production is presented in Fig. 9.9. The influence of thermal storage on the potential of substituting fossil fuel through solar energy can be derived from Tabs. 9.5 and 9.6. With increasing solar multiple and storage capacity the solar plant shifts from a fuel saver operation to a monovalent system, where the total energy demand for the reforming process is supplied by solar energy (configuration S3, Tab. 9.6).

Carbon dioxide reforming of methane has been proposed mainly for energy transport. The feasibility of balancing the reaction system in a closed loop has been successfully demonstrated

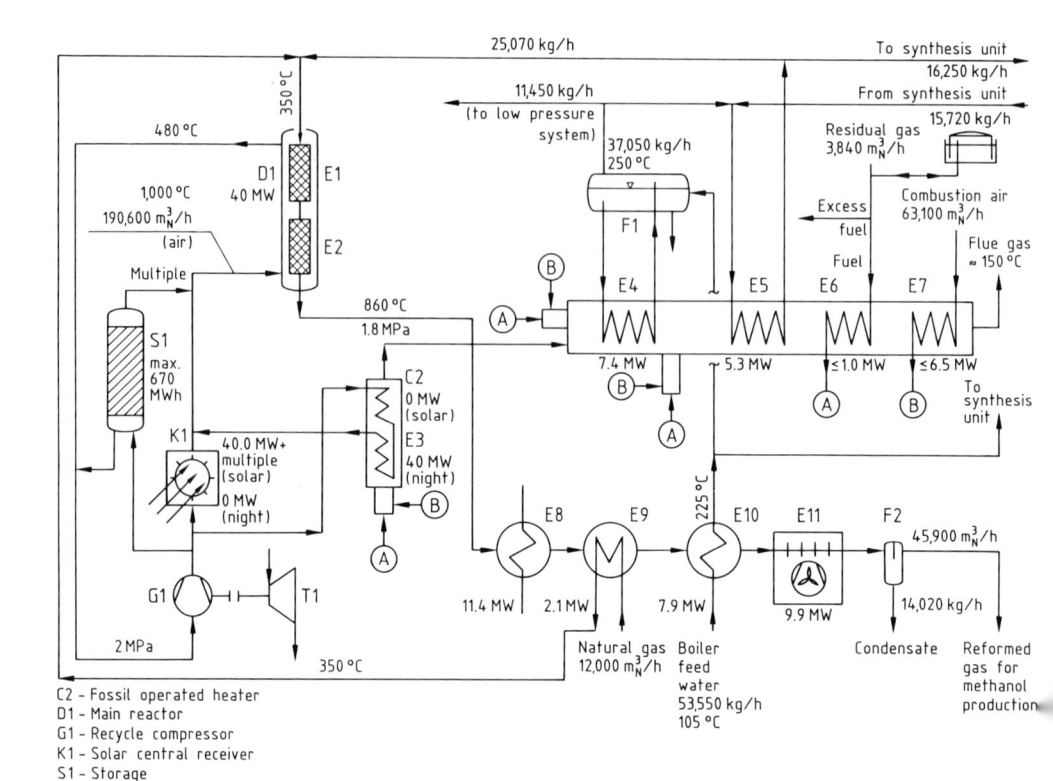

Fig. 9.9. Schematic of a 40 MW$_t$ solar steam reforming plant for methanol production [45].

C2 - Fossil operated heater
D1 - Main reactor
G1 - Recycle compressor
K1 - Solar central receiver
S1 - Storage

Table 9.5. Main plant data (40 MW steam reformer, 1,000°C, 2.0 MPa) [45].

Solar plant configuration	Receiver output (21.06, 12.00) MW$_t$	Solar multiple SM	Number of 115 MWh$_t$ storage modules	Annual solar energy GWh$_t$	Annual plant efficiency
S0, no storage	40.5	1	—	102.9	0.41
S1, with storage	61	1.5	1.5	150.6	0.43
S2, with storage	81	2.0	3	205.8	0.44
S3, with storage	2 × 65	2 × 1.6	2 × 3	2 × 169.6	0.46

[26]. Most activities are dealing with an integrated system and with investigating tube receiver-reactors [42,51] or direct absorption receiver reactors [10,40]. A schematic of a volumetric direct absorption receiver reactor is shown in Fig. 9.10. The foam absorber is coated with rhodium as the catalyst for the reforming reaction[4]. For large scale energy transport over long distances, carbon dioxide reforming has significant advantages over steam reforming [47,59]. Main reasons are the higher enthalpy of reaction for CO_2 reforming and the losses through

---

[4] This receiver reactor type is currently under test in a parabolic dish test facility as part of the joint DLR/SANDIA CAESAR-experiment within the IEA-SSPS project [10,50].

9.6 Steam/Carbondioxide Reforming of Methane – A Candidate Process

Table 9.6. Consumption data for methanol synthesis [45]. Feed: 12,000 m^3/h natural gas; main product: 13.75 t/h methanol; byproduct: 9,420 Nm^3/h hydrogen.

Plant configuration	Q_{solar}	Fuel gas for fossil firing, V_f, Nm^3/h			Fossil fuel substitution
		with irradiation	without		
	GWh/a	$V_{F,D}$	$V_{F,N}$	V_F	%
Fossil	–	3,578	3,578[a]	3,578	–
S0, no storage	103	0	3,273	2,312	29
S1, with storage	151	0	3,025	1,725	43
S2, with storage	206	0	2,532	1,045	59
S3, with storage	339	0	0	0	97

[a] Stored excess fuel included

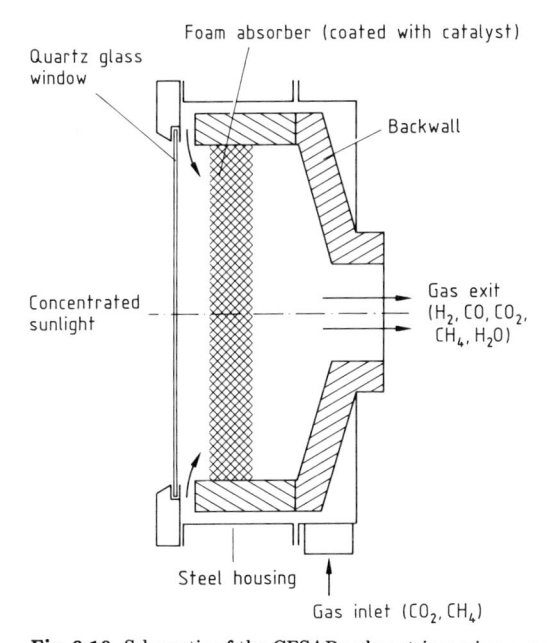

Fig. 9.10. Schematic of the CESAR volumetric receiver reactor [10].

steam generation which are, for the steam reforming system, about four times larger than for the CO_2 system.

9.7 High Temperature Processes by Direct Absorption of Solar Radiation

Concentrated sunlight is a resource with quite different characteristics than conventional energy sources, and offers new possibilities for fuels and chemicals production. Solar energy originates from the sun with a black body temperature of 5,800 K and arrives in the form of radiation energy. This high temperature means that most of the energy is carried by photons with significantly shorter wavelengths than those from conventional thermal energy sources. Existing thermal conversion techniques have been developed for less hot (and therefore longer wavelength) heat sources where photon energy is less important. Current solar thermal conversion has yet to exploit the short wavelenght nature of the radiation. The conventional methods of heat exchange using, for example pipes, are not necessarily the best of solar unique methods to convert sunlight to heat or other forms of energy.

Conventional heat transfer by absorption of radiation on a heat exchanger surface and subsequent transport through a wall into the heat transfer medium is limited by the stability of the structure materials. The highest temperatures are at the external absorber surface and a significant temperature drop and heat losses between the absorber and the thermal process result for technically relevant heat flux densities. Direct absorption can be defined as the process that takes place when the radiation absorbing material is either located within the working fluid or is directly acting as a component of a chemical reaction. Colored liquids, solid/gas suspensions and solid structures (mesh, foam, honeycomb) are examples of direct absorption systems [33]. Basically, they can act as:

- *inert material* heating the surrounding heat carrier gas without participation in any chemical process,
- a *catalyst* at the site of the catalytic or photocatalytic reaction and heat the surrounding gaseous components,
- *feedstock* for an endothermic chemical process; the entrainment gas acts only as a carrier of the solids and of the gaseous products, and as
- an *active component* of a solid-gas reaction, where the solid particles are subject to a chemical reaction with the entrainment gas [35].

When absorption occurs within a fluid by solid particles with small dimensions, there is only a small temperature difference between the absorber and the fluid due to the large surface area and the effectiveness of the direct contact heat transfer mechanism. An important consequence is that the highest temperatures occur within the fluid or on the surface of the chemically active particles and, because of this, higher temperatures of the respective applications are attainable than the given temperature limitation of the containment vessel. On the other hand, for the same output temperature as provided by conventional methods, there are lower temperature requirements for the receiver walls.

Calculations of gas and particle temperatures of single irradiated particles have shown that with concentration ratios achievable with a solar tower or a parabolic dish more than $1,000\,°C$ can be realized [24]. Figure 9.11 shows the equilibrium (stagnation) temperature of carbon particles with optically selective (radius $R_p < 0.1\,\mu m$) and non selective properties (radius $R_p > 1.0\,\mu m$). The solid curves indicate the case where the absorption area F_{abs} is equal to the cross section of the particles F_p, and the dotted curves represent the case that F_{abs} is equal to the total particle surface A_p caused by scattering effects. The possibility of realizing very high particle temperatures by direct absorption techniques can be derived from Fig. 9.12

9.7 High Temperature Processes by Direct Absorption of Solar Radiation

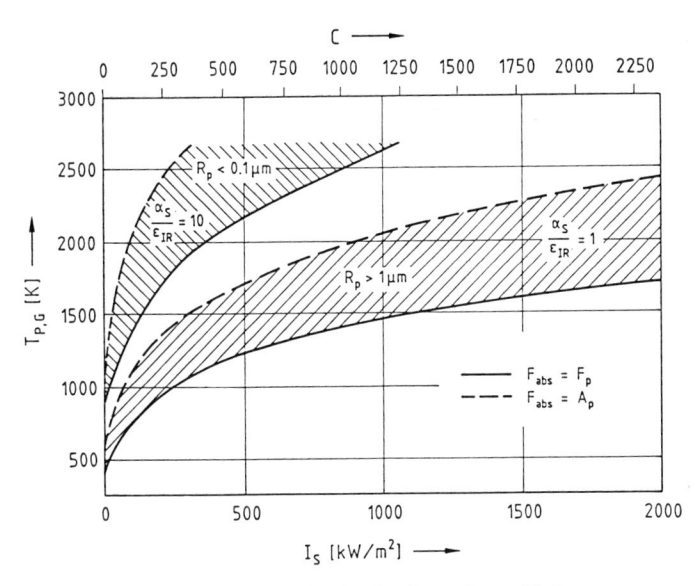

Fig. 9.11. Influence of the solar flux density E_s on the equilibrium temperature $T_{P,G}$ of carbon particles with selective ($R_P < 0.1\,\mu$m) and non-selective properties ($R_P > 1\,\mu$m) [24].

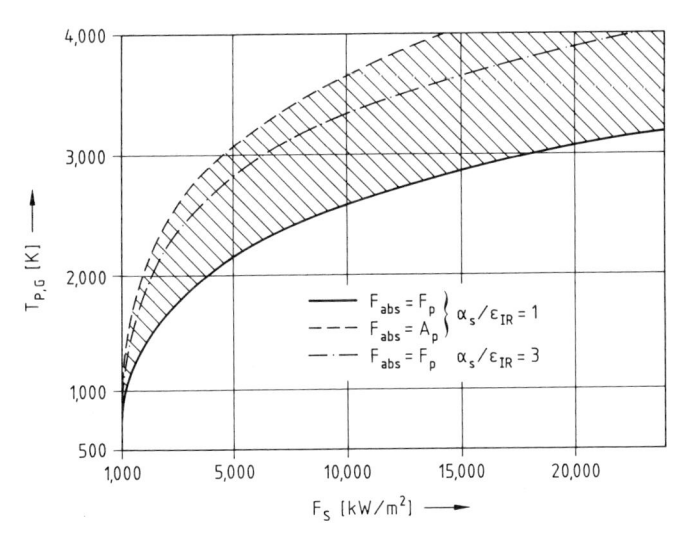

Fig. 9.12. Equilibrium temperatures of carbon particles $T_{P,G}$ for solar flux density E_s up to 20,000 kW/m² [24].

representing equilibrium temperatures of carbon particles calculated for concentration ratios up to 20,000 [23].

The second advantage of direct absorption systems occurs in applications, where it is desired to combine thermochemical and photochemical processes. The basic idea is that the photochemical reaction might be used as a *trigger* reaction to control the thermal reaction. The thermal energy can also make the photoreaction more efficient by reducing the photolytic energy necessary to complete the breaking of chemical bonds already highly excited by thermal energy. It is expected that the direct use of concentrated solar flux will enhance the conversion of basic chemicals and raw materials to high value products that are difficult or impossible to achieve by using fossil fuels. The enhancements are expected to be an improvement in product selectivity, or in catalyst activity, and a substantial increase of the reaction rates. Experimental research has been started to demonstrate such photo-enhancement effects and to determine the mechanisms and chemical kinetics of photo-assisted thermochemical reactions, concentrating on heterogeneous catalytic reactions dealing with methanation of CO [44], with decomposition of alcohols [61] and hydrocarbons [62], dissociation of SO_3 [36], and CO_2-reforming of methane [10,40,59].

Another important research effort is the investigation of the solar incineration of haloginated aromatic molecules. Specifically, these investigations try to separate photolytic effects from thermal effects with emphasis on determining the specific influence of high solar flux on solar incineration [31,32]. From the results, an assessment of the feasibility of using solar energy for hazardous waste detoxification is expected.

9.8 Electrolytic Production of Hydrogen with Photovoltaic and Solar Thermal Power Plants

The potential of hydrogen for the comprehensive utilization of solar energy is of particular importance. Hydrogen provides, among the new chemical fuels under consideration, an optimal mix of the following characteristics which are mandatory for a solar chemical energy carrier:

- efficient long-term storability and large-scale transportability of solar energy, i.e. of the most important conversion products of solar energy, thermal and electrical energy,
- practically unlimited feedstocks by closed cycles; in case of hydrogen only water,
- compatibility with existing energy supply structures and utilization techniques,
- minimal environmental impact,
- safe production, handling and use,
- potential to become economic.

Hydrogen is both a clean chemical fuel for direct use and a feedstock for synthesizing gaseous and liquid fuels.

To produce hydrogen from solar energy a number of processes are possible in principle, Fig. 9.13. The combination of various solar energy conversion technologies and electrolysis systems has attracted steadily increasing interest in recent years[5]. This section concentrates on the interconnected operation of solar power stations with water electrolysis plants. Figure 9.14 shows technically feasible and to some extent already proven conversion sequences.

9.8.1 Electrolytic Production of Hydrogen with Photovoltaic Systems

Both solar cells and water electrolysis represent well-known technologies of already proven reliability. Striking advantages of combining these technologies are (a) simple setup, (b) separation of electrical energy generation from the splitting of water, and (c) modular design of both PV generators and water electrolyzers, permitting variation of electrical power genera-

---
[5] e.g. [2,9,11,12,13,15,16,25,28,48,53,56]

9.8 Electrolytic Production of Hydrogen with Photovoltaic and Solar Thermal Power Plants

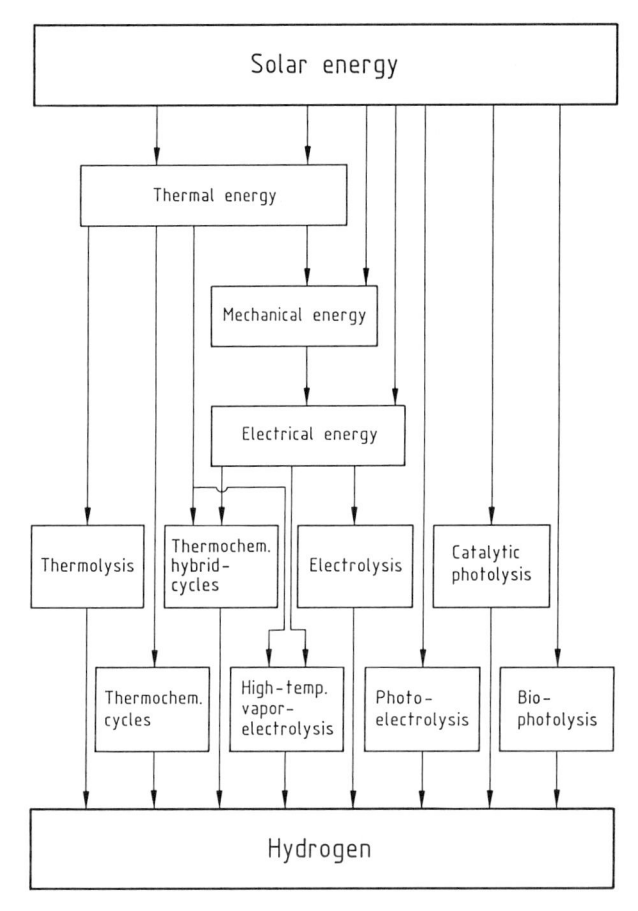

Fig. 9.13. Conversion steps and processes for the production of hydrogen from solar energy.

tion in a very broad range and adaption to the need of hydrogen production. Advantage (b) could be possibly decisive in comparison with single-stage photoelectrochemical and photocatalytic processes (at present still at the basic research stages). The following two aspects are, however, of key importance for the long-range perspective of water electrolysis systems:

- the cell voltage-current density characteristics of both components fit each other excellently, and
- both technologies still have considerable development potential.

Performance and dynamic characteristics of photovoltaic electrolysis systems. Photovoltaic electrolysis systems consist of PV arrays, electrolyzers, power conditioning units for matching the voltage-current characterstics of both, and H_2 storage facilities. The operating characteristic of PV arrays is an implicit current density-voltage relation which describes possible operating conditions of solar cells between open circuit voltage and short circuit current with insolation and cell temperature as parameters (Fig. 9.15). PV arrays should be operated at the maximum power point (MPP); the location of the MPP is strongly dependent on insolation and cell temperature, however (see Chaps. 2 and 8).

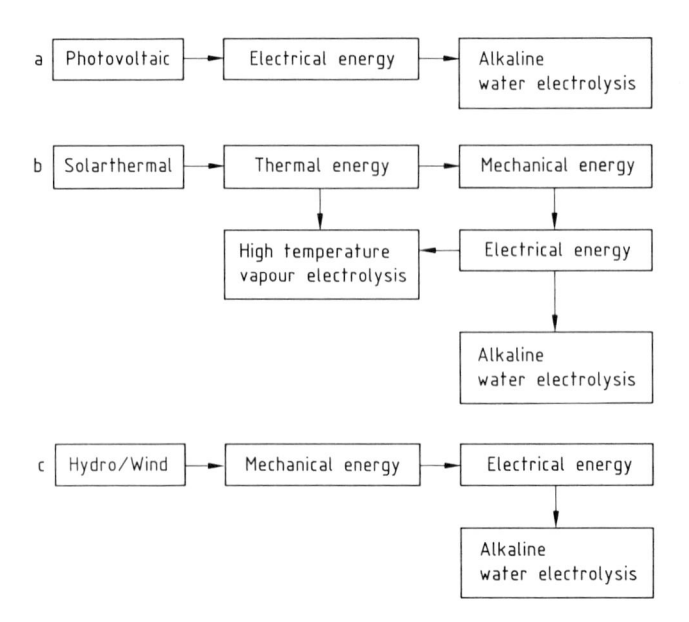

Fig. 9.14. Technically feasible conversion sequences for the production of hydrogen from solar energy. (a) PV-electrolysis combination, (b) solar thermal-electrolysis combination, (c) hydro/wind-electrolysis combination.

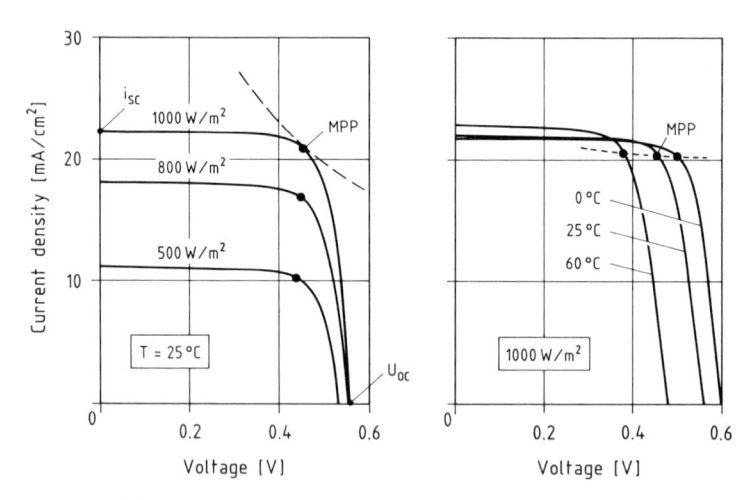

Fig. 9.15. Current density-voltage characteristics of a single solar cell with insolation and cell temperature as parameters (based on crystalline silicon).

Whereas conventional alkaline electrolysis units have been highly reliable in industry for decades, advanced electrolysis and high temperature vapor electrolysis with considerably improved efficiencies (Fig. 9.16) have only been tested on a pilot plant or laboratory scale. The higher the current density, the lower the capital cost; the lower the cell voltage, the lower the operating cost of electrolysis. The working conditions of PV arrays coupled with electrolysis are determined by the intersection of the current-voltage characteristics. Optimal operating

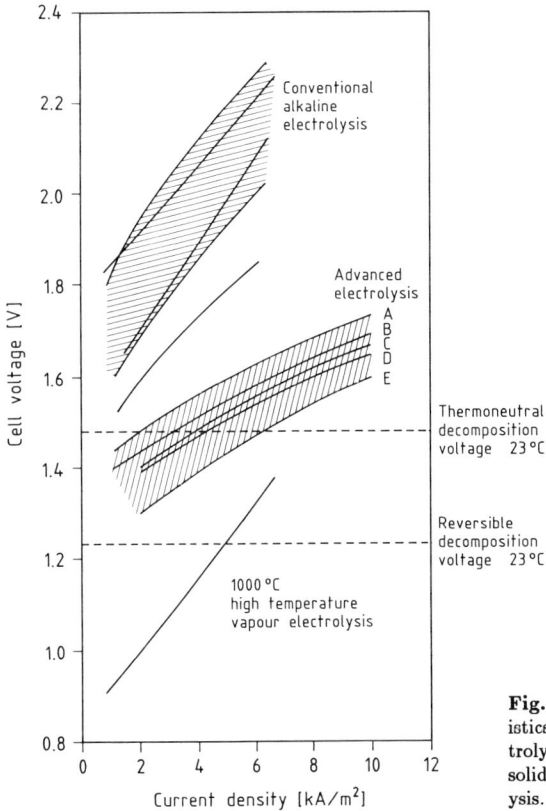

9.8 Electrolytic Production of Hydrogen with Photovoltaic and Solar Thermal Power Plants

Fig. 9.16. Cell-voltage/current-density characteristics of conventional and advanced water electrolyzers. A,B,D: advanced alkaline electrolysis; C: solid polymer electrolysis; E: molten salt electrolysis.

conditions are achieved when both characteristic curves intersect at the MPP of the PV array. There is, in principle, an intrinsic match between both characteristic curves due to the daily and seasonal change of the solar cell MPP as shown in Fig. 9.17. However, the characteristics of both PV array and electrolysis are not constant but vary with external parameters such as insolation, partial shading of large solar arrays and solar cell temperature, and system parameters such as electrolyte temperature, electrode polarization and degradation effects. The time constants for these dynamic effects range from seconds (insolation) to years (degradation effects).

The performance of PV electrolysis systems and components during stationary and dynamic operation modes, efficiency and performance of power conditioning, part-load and overload behaviour of electrolyzers as well as start-up and shut-down characteristics are important aspects of solar hydrogen production.

Figure 9.18 shows typical experimental results of an 10 kW$_e$ PV-electrolytic pilot plant. The system in this case is operated in a direct connection mode without power conditioning between solar generator and electrolyzer. The advanced alkaline electrolysis cell module consists of 25 single cells with an active area of 500 cm^2 in filter press type arrangement. The hydrogen production rate as given by the electrolytic current follows closely the global insolation even during radiation fluctuations caused by clouds in the afternoon hours. Immediately after start-up the electrolyzer voltage reaches production conditions, increases to a maximum value of 46 volts around 9 a.m., and then decreases steadily over the rest of the day.

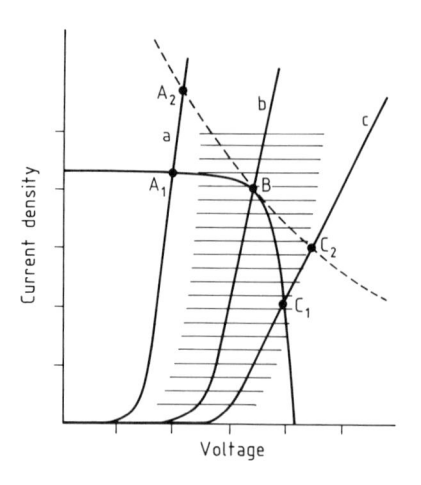

Fig. 9.17. Current-density/voltage-characteristics of solar cells and electrolyzers. Shaded region represents possible positions of point B, the maximum power point (MPP). Case b represents optimal operating conditions. Cases a and c show a mismatch in direct coupling (at points A_1 and C_1), illustrating the need for power conditioning, i.e. operation of the PV array in B and shifting of the electric power to points A_2 and C_2, respectively, along the constant power hyperbola depicted by the dotted line.

The electrolyte temperature rises from ambient in the early morning to 55°C around 3 p.m., and then decreases again. By changing the thermal conditions for all blocks and electrolyte circuits, an operation temperature of 95°C can be reached within 2 to 3 hours after start-up to improve electrolysis efficiency. It is evident from Fig. 9.18 that optimized PV-electrolytic systems are capable to continuously produce hydrogen over the entire solar day from sunrise to sunset. Start up and shut down require neither high levels of insolation nor time-consuming operational procedures.

Because of periodic intermittent operation, insolation and ambient temperature variation, actual solar cell temperature, electrolyte temperature, electrode polarization and degradation effects, a carefully designed coupling and control unit is needed to condition the energy flow from PV generator to the electrolyzer. Therefore, as in other power generation fields, automatic control strategies are the most viable technical means to continuously optimize plant operation. Such a power conditioning unit can in principle be realized either by a controllable switching interface (adaption of voltage and current levels by varying the serial and parallel interconnection of PV modules), or by full DC-DC conversion with maximum power tracking, or by partial DC-DC conversion with bypass control.

A laboratory test set-up consisting of a PV generator with a nominal rating of 100 W_e (AM 1, 1,000 Wm^{-2}, 25°C), a power conditioning unit, and an SPE (solid polymer electrolyte) electrolysis module has been used to investigate the dynamic behavior and the efficiency and performance of the power conditioning unit [56]. Figure 9.19 shows that power conditioning and the electrolyzer are able to follow rapid changes of insolation without significant delay. Electrolytic hydrogen production is proportional to the current of the electrolyzer. Continuous maximization of the electrolyzer current results, therefore, in a continuous optimization of system operation for maximum hydrogen production.

System simulation studies show that the mean efficiency of full or partial DC-DC conversion with MPP tracking must be well above 90% to realize essential advantages over direct connection with a controllable switching interface. The development potential involved is discussed in [43,55].

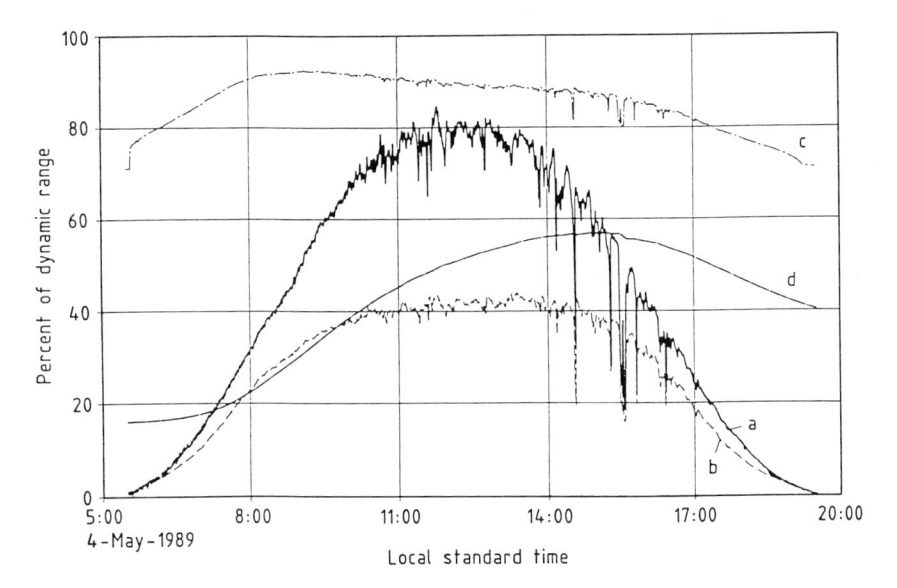

Fig. 9.18. Solar operation of a 10 kW$_e$ photovoltaic-electrolytic hydrogen production pilot installation. (a) global insolation $0\ldots1,250$ W/m^2 (at 35° tilt); (b) electrolyzer current; $0\ldots400$ A, (c) electrolyzer voltage; $0\ldots50$ V, (d) electrolyte temperature; $0\ldots100$°C.

Development potential to improve photovoltaic electrolysis systems. There is significant potential for further improvements of performance, efficiency, reliability and lifetime of components in PV electrolysis systems, which will lead to cost reduction and improved economy. In summary the following R&D tasks are of particular importance:

- advanced and new solar cell concepts including advanced thin film modules (Chap. 8);
- advanced alkaline and solid polymer electrolyte water electrolysis.

Concerning further improvements of water electrolyzer technology, development efforts concentrate on:

- zero-gap cell geometry;
- new diaphragm materials;
- new electrocatalysts;
- operation temperatures of 120–160°C.

Figure 9.20 shows the ohmic losses as well as the anodic and cathodic overvoltages of the alkaline water electrolysis. The most important objectives for technical improvements obviously are the reduction of overvoltages and of ohmic losses at high current densities. By coating the electrodes with electrocatalysts, the overvoltage losses of hydrogen and oxygen formation are reduced and a considerable improvement in water electrolysis is achieved. Numerous alloys and mixed oxides have already been investigated with respect to catalytic activity during electrochemical processes. However, in technical electrolyzers the application of these electrocatalysts has not yet been realized. New electrocatalysts for the cathode based on Raney-nickel (NiAl), with stabilizing additives, and for the anode based on complex mixed oxides with co-additives, promise further development possibilities.

Ongoing development efforts [57,58] will enhance the experience in operating such systems and will contribute towards reducing the cost of solar hydrogen production.

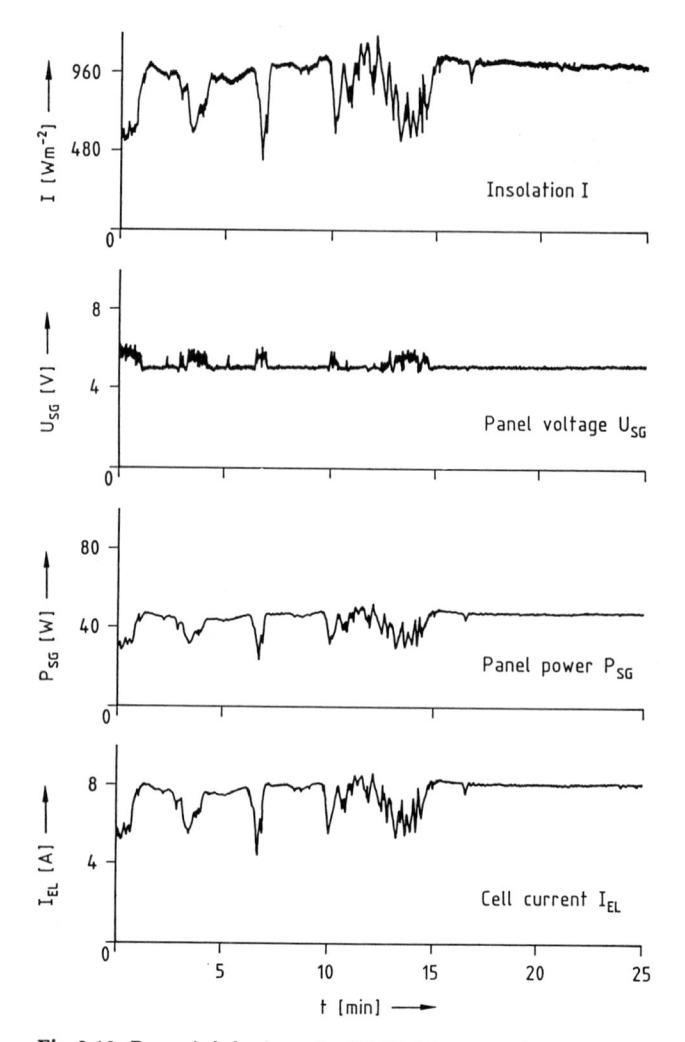

Fig. 9.19. Dynamic behaviour of a 100 W_e laboratory photovoltaic-electrolysis system during cloud occurrences [56].

The comparison of the different hydrogen production processes shows that hydrogen cost increases by about a factor of 4 for the processes starting from steam reforming of natural gas up to conventional water electrolysis, i.e. in the sequence of

- steam reforming of natural gas (cheapest H_2 process at present);
- gasification of lignite;
- partial oxidation of naphtha;
- gasification of heavy oil;
- gasification of hard coal;
- water electrolysis.

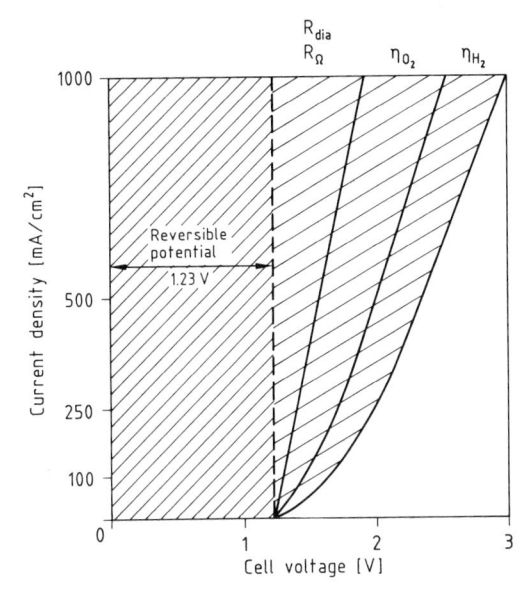

Fig. 9.20. Reversible potential, ohmic losses, and overvoltages of alkaline water electrolysis.

Only the availability of low-cost (non-fossil) electrical energy, considerable improvements in water electrolysis technology by reducing cell voltage and increasing current density, and/or the environmental necessity of reducing CO_2-discharge to the atmosphere, may change this sequence.

9.8.2 Electrolytic Production of Hydrogen with Thermal Solar Power Plants

Of the thermal solar power plant (SPP) technologies, the tower SPP promises high power (10 to some 100 MW_e) at low cost. Where only decentralized low power is required (20 to some 100 kW_e), the parabolic dish concept offers favorable possibilities. For electrolytic production of hydrogen, the well-proven alkaline electrolysis is available. In the future, advanced electrolysis systems are of interest for both thermal and photovoltaic SPPs. In the long term, high temperature vapor electrolysis offers specific opportunities for thermal SPPs because of the possibility of utilizing electrical energy in combination with high temperature process heat, resulting in high electrolysis efficiencies.

For the electrolytic production of hydrogen using thermal SPPs, both conventional alkaline electrolyzers and advanced concepts are of importance, in particular, the high-temperature vapor electrolysis [1,20].

Due to the thermodynamics of water splitting shown in Fig. 9.21, electrolysis of water vapor at high temperatures exhibits specific advantages [22]:

- total electrical energy requirement for water splitting is lower in the vapor phase than in the liquid phase, because the energy for vaporization can be produced thermally instead of electrically;
- minimum requirement for electrical energy ΔG, needed for electrolysis, decreases with increasing temperature. Thus the total efficiency η_{tot} can be substantially improved by providing a part of the splitting energy thermally instead of electrically (Fig. 9.23);
- improved reaction kinetics at elevated temperatures result in lower overvoltages.

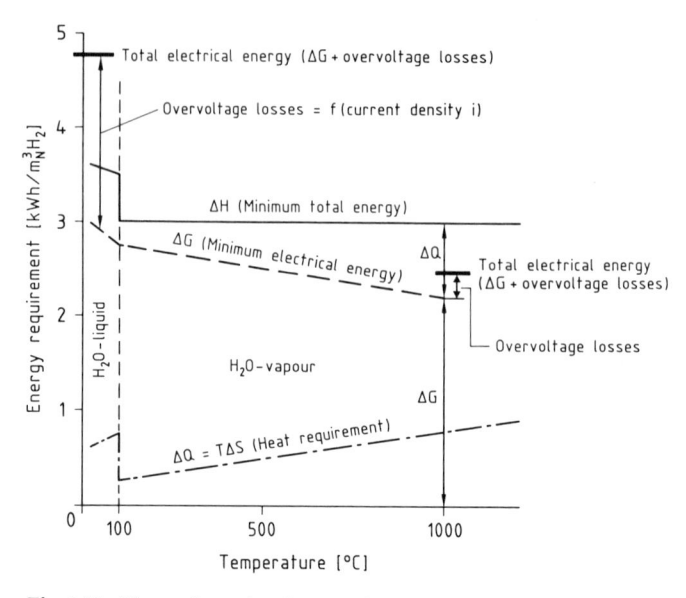

Fig. 9.21. Thermodynamics of water splitting as function of temperature.

Of outstanding importance for the coupling of high-temperature vapor electrolysis with thermal SPP are the possibilities for autothermal operation (i.e. the required high process temperature is produced electrically in the high temperature electrolysis itself) in combination with parabolic trough SPP, and for allothermal operation (i.e. the required high process temperature is produced thermally by the tower SPP) in combination with tower SPP, resulting in high total efficiencies for the conversion of solar radiation to hydrogen. Table 9.7 summarizes characteristic energy and efficiency data of electrolysis processes.

Table 9.7. Characteristic energy and efficiency data for hydrogen production, coupling thermal solar power plants with electrolysis processes [20]. Higher heating value of H_2 is 3.55 kWh/Nm³.

	High temperature vapor electrolysis		Conventional electrolysis
	autothermal	allothermal	
Electrical energy E_e, kWh/Nm³ H_2	3.2	2.6	4.6
Mid-temperature heat E_{t_m}, kWh/Nm³ H_2	0.6	0.6	–
High-temperature heat E_{t_h}, kWh/Nm³ H_2	–	0.5	–
Conversion factor $\left(\dfrac{\text{higher heating value } H_2}{\text{electrical energy } E_e}\right)$	1.11	1.37	0.77
Assumed power generation efficiency η_e – parabolic trough plant – tower plant	17%	25%	17% 25%
Total efficiency $\left(\dfrac{\text{higher heating value } H_2}{E_e/\eta_e+\sum E_{t_{m,h}}}\right)$	18%	31%	13% / 19%

9.8 Electrolytic Production of Hydrogen with Photovoltaic and Solar Thermal Power Plants

Autothermal high-temperature vapor electrolysis processes operate at a mean cell voltage of 1.32 V and can be combined with parabolic trough plants which deliver mid-temperature steam and electricity. The total efficiency for the conversion of solar energy to hydrogen could be raised further by allothermal high-temperature vapor electrolysis. In this case the cell voltage can be reduced below thermoneutral conditions. The high-temperature vapor electrolysis is of special interest for the combination with gas-cooled solar tower plants. Figure 9.22 shows the block diagram for such a plant which generates hydrogen via high-temperature vapor electrolysis together with electricity for the network [19,22]. Thermal energy available in the receiver of the gas-cooled thermal SPP is added directly to the high-temperature vapor electrolyzer as process heat; by bypassing the low efficiency in the conversion of thermal to electrical energy, a very high efficiency can be attained, see Fig. 9.23. The bulk of the receiver output is used for electricity production. However, high-temperature vapor electrolysis technology is still in the development stage. The process can be operated at lower (e.g. 3 bar) or higher (e.g. 25 bar) pressure levels. In this context it is important to avoid significant pressure differences between the inner and outer space of the ceramic electrolysis tubes.

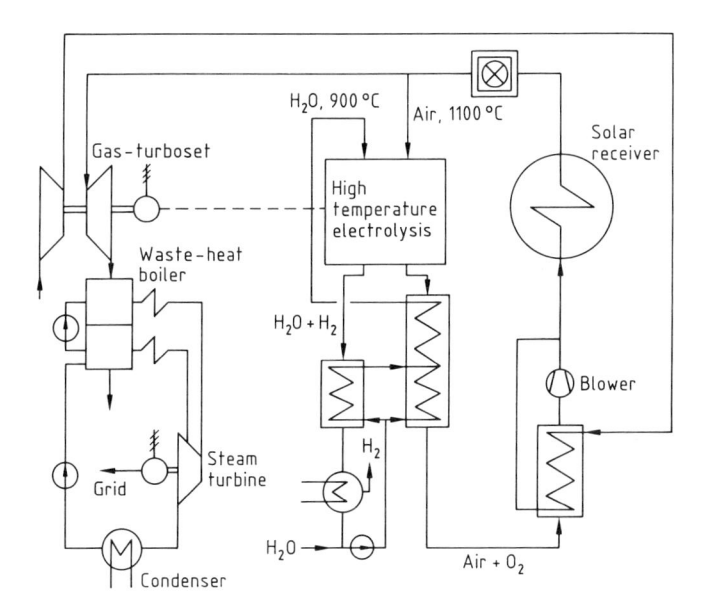

Fig. 9.22. Scheme of a gas-cooled solar tower plant with process heat utilization for high temperature vapor electrolysis [19,22].

In summary, by coupling thermal SPP with electrolysis plants, relatively high efficiencies for the production of hydrogen, the most important future secondary energy carrier, can be achieved. The high-temperature vapor electrolysis offers, in combination with thermal SPP, substantial further improvements in the conversion efficiency from solar energy to hydrogen (Tab. 9.7). It is of particular importance for such systems that high-temperature electrochemical processes can, in principle, be used in the reverse direction; for example, even during night electric power and high-temperature process heat can be produced with high system efficiencies, using hydrogen-fueled high-temperature molten carbonate fuel cells or solid oxide fuel cells.

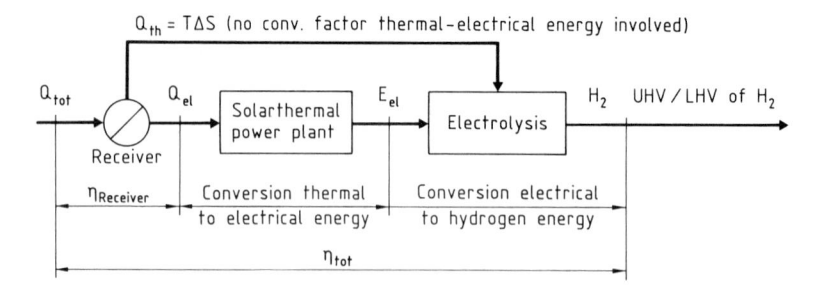

Fig. 9.23. Conversion factors and efficiency of hydrogen production in a solar thermal power plant coupled with high-temperature vapor electrolysis.

Bibliography

[1] *Gas-Cooled Solar Tower Power Plant GAST: Analysis of Its Potential* (in German). Study prepared for the BMFT, Bonn (D), BMFT, 1985
[2] *Hydrogen as Secondary Energy Carrier* (in German). Technical Report DLR-Mitteilungen 81-10, Köln (D), DLR, 1981
[3] *A Preliminary Assessment of the Potential for Integrating Solar Thermal Central Receiver Technology with Fuels and Chemical Processes.* Contractor Report SAND85-8183, Sandia National Laboratories, Albuquerque/NM, 1986
[4] Becker, M.; Harth, R.; Mueller, W. D.: Steam Reforming as a Key Process for Utilization and Transport of Solar Thermal Energy. In *Proc. 21st Intersoc. Energy Conv. Engg. Conf. (IECEC), San Diego/CA*, Washington/DC: American Chemical Society, 1986
[5] Besenbruch, G.: Thermochemical Water Splitting at GA Technologies. In *Proc. IEA-SSPS Experts Meeting on High Temperature Technology and Application, Atlanta/GA, IEA-SSPS TR 1/85*, pp. 407–449, Köln (D): DLR, 1985
[6] Besenbruch, G.; McGorkle, K. H.: *Thermochemical Water Splitting with Solar Thermal Energy.* Report GAA 16022, San Diego/CA, General Atomic, 1981
[7] Birke, G.; Reimert, R.: Integrating High-Temperature Solar Energy with Fuel Upgrading Processes. In *Proc. 3rd Int. Workshop on Solar Thermal Central Receiver Systems, Konstanz*, Becker, M. (Ed.), pp. 693–702, Berlin, Heidelberg, New York: Springer, 1986
[8] Birke, G.; Reimert, R.: Process Synthesis of a Gasification Process Modified for High Solar Energy Integration. In *Solar Thermal Energy Utilization*, Becker, M. (Ed.), pp. 547–620, Berlin, Heidelberg, New York: Springer, 1987
[9] Bockris, J. O.; Veziroglu, T. N.: A Solar-Hydrogen Economy for the USA. *Hydrogen Energy*, 7 (1982) 287–310
[10] Buck, R.: Volumetric Receivers: Potential and Problems. In *Proc. ISES Solar World Congress, Kobe 1989*, Oxford (UK): Pergamon Press, 1990
[11] Carpetis, C.: An Assessment of Electrolytic Hydrogen Production by Means of Photovoltaic Energy Conversion. *Hydrogen Energy*, (1984) 969–991
[12] Carpetis, C.: A Study of Water Electrolysis with Photovoltaic Solar Energy Conversion. *Hydrogen Energy*, 7 (1982) 287–310
[13] Carpetis, C.; Schnurnberger, W.; Seeger, W.; Steeb, H.: Electrolytic Hydrogen by Means of Photovoltaic Energy Conversion. In *Hydrogen Energy Progress IV, Proc. 4th WHEC, Pasadena/CA*, Veziroglu, T. N.; Van Vorst, W. B.; Kelley, J. H. (Ed.), pp. 1495–1512, Oxford (UK): Pergamon Press, 1982
[14] Carrol, D.: Solar Plant 1. *Sunworld*, 9 (1985) 10–11
[15] Costogne, N.; Yasni, R. K.: Performance Data of a Terrestrial Solar Photovoltaic Experiment. In *Proc. ISES Solar World Conference*, (Ed.), pp. 138–139, 1975
[16] Cox, K. E.: Hydrogen from Solar Energy via Water Electrolysis. In *Proc. 11th Intersoc. Energy Conv. Engg. Conf. (IECEC)*, pp. 926–932, 1976
[17] Diver, R. B.: Receiver/Reactor Concepts for Thermochemical Transport of Solar Energy. In *Proc. 21st Intersoc. Energy Conv. Engg. Conf. (IECEC), San Diego/CA*, Washington/DC: American Chemical Society, 1986

[18] Diver, R. B.; Pederson, S.; Kappauf, T.; Fletcher, E. A.: Hydrogen and Oxygen from Water – VI Quenching the Effluent from a Solar Furnace. *Energy*, 8 (1983)
[19] Doenitz, W.; Dietrich, G.; Erdle, E.; Streicher, R.: Electrochemical High Temperature Technology for Hydrogen Production or Direct Electricity Generation. *Hydrogen Energy*, 13 (1988) 283–287
[20] Doenitz, W.; Schmidberger, R.: Concepts and Design for Scaling-Up High Temperature Water Vapour Electrolysis. *Hydrogen Energy*, 7 (1982) 321–330
[21] Eickermann, R.: Thermal Cracking Processes (in German). *Chemie-Ing. Technik*, 55 (1983)
[22] Erdle, E.; Gross, J.; Meyringer, V.: Possibilities for Hydrogen Production by Combination of a Solar Thermal Central Receiver System and High-Temperature Electrolysis of Steam. In *Proc. 3rd Intl. Workshop on Solar Thermal Central Receiver Systems, Konstanz*, Becker, M. (Ed.), pp. 727–736, Berlin, Heidelberg, New York: Springer, 1986
[23] Erhardt, K.; Henne, R.; Köhne, R.; Tamme, R.: *Interaction of Highly Concentrated Solar Radiation with Chemical Compounds* (in German). DLR IB 441 484/84, Köln (D), DLR, 1984
[24] Erhardt, K.; Vix, U.: Direct Absorption of Concentrated Solar Radiation. In *Proc. 3rd Intl. Workshop on Solar Thermal Central Receiver Systems, Konstanz*, Becker, M. (Ed.), pp. 835–867, Berlin, Heidelberg, New York: Springer, 1986
[25] Esteve, D.; Ganibal, C.; Steinmetz, D.; Vialaron, A.: Performance of a Photovolatic Electrolysis System. In *Hydrogen Energy Progress III, Proc. 3rd WHEC, Tokyo/Japan*, Veziroglu, T. N.; Fueki, K.; Ohta, T. (Ed.), pp. 1593–1603, Oxford (UK): Pergamon Press, 1980
[26] Fish, J. D.; Hawn, D. C.: Closed Loop Thermochemical Energy Transport Based on CO_2 Reforming of Methane: Balancing the Reaction Systems. In *Proc. 21st Intersoc. Energy Conv. Engg. Conf. (IECEC), San Diego/CA*, Washington/DC: American Chemical Society, 1986
[27] Flamant, G.; Hernandez, D.; Bonet, C.: Experimental Aspects of the Thermodynamical Conversion of Solar Energy; Decarbonation of $CaCO_3$. *Solar Energy*, 24 (1980) 385–395
[28] Foster, R. W.; Tison, R. R.; Escher, W. J. D.; Hanson, J. A.: *Solar Hydrogen System Assessment*. Technical Report DOE/JPL-955492, US Department of Energy, 1980
[29] Froment, G.: Fixed Bed Catalytic Reactors, Technological and Fundamental Design Aspects. *Chemie-Ing. Technik*, 46 (1974) 374–380
[30] Funk, J. E.; Bowman, M. G.: Renewable Hydrogen Energy from Solar Thermal Central Receiver Systems. In *Proc. 3rd Int. Symp. Hydrogen from Renewable Energy, Hawaii*, University of Hawaii, 1986
[31] Graham, J. L.; Dellinger, B.: Solar Detoxification of Hazardous Organic Wastes. In *Solar Thermal Technology – Proc. 4th Intl. Symposium, Santa Fe/NM, 1988*, Gupta, B. P.; Traugott, W. H. (Ed.), p. 391, New York: Hemisphere Publ. Co., 1990
[32] Graham, J. L.; Dellinger, B.: Solar Thermal/Photolytic Destruction of Hazardous Organic Waste. *Energy*, 12 (1987)
[33] Hunt, A. J.: New Approaches to Receiver Design: Prospects and Technology of Using Particle Suspensions as Direct Thermal Absorbers. In *Proc. 3rd Intl. Workshop on Solar Thermal Central Receiver Systems, Konstanz*, Becker, M. (Ed.), pp. 835–842, Berlin, Heidelberg, New York: Springer, 1986
[34] Hunt, A. J.; Brown, C. T.: *Solar Testing of Small Particle Heat Exchange Receiver (SPHER)*. Technical Report LBL-15756, Lawrence Berkeley Laboratories, 1983
[35] Hunt, A. J.; et al.: *Solar Radiant Heating of Gas-Particle Mixtures*. Final Report FY 1985–86 LBL-22743, Lawrence Berkeley Laboratories, 1986
[36] Hunt, A. J.; Hodara, I.; Miller, F. J.; Noring, J. E.: Direct Absorption Receivers for Catalyzing Chemical Reactions. In *Solar Thermal Technology – Proc. 4th Intl. Symposium, Santa Fe/NM, 1988*, Gupta, B. P.; Traugott, W. H. (Ed.), p. 437, New York: Hemisphere Publ. Co., 1990
[37] Kappauf, T.; Murray, J. P.; Palumbo, R.; Diver, R. B.; Fletcher, E. A.: Hydrogen and Sulfur from Hydrogen Sulfide – IV Quenching the Effluent from a Solar Furnace. *Energy*, 10 (1985)
[38] Kirk-Othmer: *Encyclopedia of Chemical Technology*. Volume 12, New York: John Wiley & Sons, 3rd edition, 1982
[39] Knoche, K. F.: *Thermochemical Cycle Processes for Water Dissociation* (in German). Volume 729 of *VDI-Berichte*, Düsseldorf (D): VDI-Verlag, 1989
[40] Levy, M.; Levitan, R.; Rosin, H.; Adusei, G.; Rubin, R.: Storage and Transport of Solar Energy by Thermochemical Pipe. In *Solar Thermal Technology – Proc. 4th Intl. Symposium, Santa Fe/NM, 1988*, Gupta, B. P.; Traugott, W. H. (Ed.), p. 527, New York: Hemisphere Publ. Co., 1990
[41] Lukens, L. L.; Andraka, C. E.; Moreno, J. B.; Abbin, J. P.: Liquid Metal Thermoelectric Converter. In *Proc. 22nd Intersoc. Energy Conv. Engg. Conf. (IECEC)*, New York: American Institute of Aeronautics and Astronautics, 1987
[42] McCrary, J. H.; McCrary, G. E.; Chubb, T. A.; Nemecek, J. J.; Simmons, D. E.: *An Experimental Study of the CO_2-CH_4 Reforming-Methanization Cycle as a Mechanism for Converting and Transporting Solar Energy*. Volume 29, 1982

[43] Mehrmann, A.; Kleinkauf, W.; Pigorsch, W.; Steeb, H.: Dynamic of Small Photovoltaic Systems. In *Proc. 5th E.C. PV Solar Energy Conference, Athens 1983*, Palz, W.; Fittipaldi, F. (Ed.), p. 495, Dordrecht (NL): D. Riedl, 1984
[44] Mosfegh, A. Z.; Igantiev, A.: Photo-Enhancement of the Catalytic Methanation Reaction. *Energy*, 12 (1987)
[45] Mueller, W. D.; Fuhrmann, H.: Comparative Investigations and Ratings of Different Solar Systems Using Tubular Steam Reformers. In *Solar Thermal Energy Utiliztion*, Becker, M. (Ed.), Berlin, Heidelberg, New York: Springer, 1987
[46] Nix, G.; Sizmann, R.: High Temperature, High Flux Density Solar Chemistry. In *Solar Thermal Technology – Proc. 4th Intl. Symposium, Santa Fe/NM, 1988*, Gupta, B. P.; Traugott, W. H. (Ed.), p. 351, New York: Hemisphere Publ. Co., 1990
[47] Nix, R. G.; Bergeron, P. W.: Thermochemical Energy Transport for a Large Heat Utility. In *Proc. 21st Intersoc. Energy Conv. Engg. Conf. (IECEC), San Diego/CA*, Washington/DC: American Chemical Society, 1986
[48] Ohta, T.: *Solar Hydrogen Energy Systems*. Oxford (UK): Pergamon Press, 1979
[49] Perry, R. H.; Green, D.: *Chemical Engineers' Handbook*. New York: McGraw Hill, 6th edition, 1984
[50] Pritzkow, W.: The Volumetric Ceramic Receiver Potential of Ceramics for Solar Heat Exchangers. *Brit. Cer. Proc.*, 43 (1989)
[51] Rozenmann, T.: Energy Transport via a Direct Solar Reformer Reactor. In *Proc. 21st Intersoc. Energy Conv. Engg. Conf. (IECEC), San Diego/CA*, Washington/DC: American Chemical Society, 1986
[52] Sanders Associates, Inc.: *Parabolic Dish Module Experiment*. Final Test Report SAND85-7007, Sandia National Laboratories, Albuquerque/NM, 1985
[53] Sayigh, A. A. M.: The Use of Solar Energy – Photovoltaic in Hydrogen Production and Arid Zones like Saudi Arabia. In *Hydrogen Energy Progress III, Proc. 3rd WHEC, Tokyo/Japan*, Veziroglu, T. N.; Fueki, K.; Ohta, T. (Ed.), pp. 1431–1439, Oxford (UK): Pergamon Press, 1980
[54] Schuetze, B.; Hofmann, H.: How to Upgrade Heavy Feeds. *Hydrocarbon Processing*, 63 (1984)
[55] Steeb, H.; Kleinkauf, W.; Mehrmann, A.: Utilization of Solar Energy for Hydrogen Production. In *Proc. 4th Intl. Solar Forum, Berlin*, Auer, F. (Ed.), pp. 970–980, München (D): DGS-Sonnenenergieverlag, 1982
[56] Steeb, H.; Mehrmann, A.; Seeger, W.; Schnurnberger, W.: Solar Hydrogen Production: Photovoltaic System with Active Power Conditioning. In *Hydrogen Energy Progress V, Proc. 5th WHEC, Toronto/Canada*, Veziroglu, T. N.; Taylor, J. B. (Ed.), pp. 109–119, Oxford (UK): Pergamon Press, 1984
[57] Steeb, H.; Weiss, H. R.; Koshaim, B. H.: HYSOLAR, a Joint German Saudi Arabian Research, Development and Demonstration Program on Solar Hydrogen Production and Utilization. In *Hydrogen Energy Progress VI, Proc. 6th WHEC, Vienna/Austria*, Veziroglu, T. N.; Getoff, N.; Weinzierl, P. (Ed.), Oxford (UK): Pergamon Press, 1985
[58] Szyska, A.: *Realization of the Solar-Hydrogen Project at Neunburg vorm Wald (Germany)*. München (D): Solar-Wasserstoff Bayern GmbH
[59] Tamme, R.; Huder, K.: Production of Fuels and Chemicals by Solar Chemical Processing: Analysis of Methane Reforming Processes. In *Solar Thermal Technology – Proc. 4th Intl. Symposium, Santa Fe/NM, 1988*, Gupta, B. P.; Traugott, W. H. (Ed.), p. 425, New York: Hemisphere Publ. Co., 1990
[60] Ullmann: *Encyclopedia of Technical Chemistry* (in German). Volume 3 and 14, Verlag Chemie Weinheim, 4th edition, 1977
[61] Wentworth, W. E.; Batten, C. F.; Gong, W.: The Photo-Assisted Thermal Decomposition of Methanol and Isopropanol in a Fluidized Bed. *Energy*, 12 (1987)
[62] Wentworth, W. E.; Batten, C. F.; Hamada, M.: Photoassisted Hydrocarbon Reforming and Cracking Reactions. In *Solar Thermal Technology – Proc. 4th Intl. Symposium, Santa Fe/NM, 1988*, Gupta, B. P.; Traugott, W. H. (Ed.), p. 415, New York: Hemisphere Publ. Co., 1990
[63] Ziph, B.; Godett, T. M.; Diver, R. B.: Reflux Heat-Pipe Solar Receiver for a Stirling Dish-Electric System. In *Proc. 22nd Intersoc. Energy Conv. Engg. Conf. (IECEC)*, New York: American Institute of Aeronautics and Astronautics, 1987

10 Cost Analysis of Solar Power Plants

H. P. Hertlein, H. Klaiss and J. Nitsch [1]

The factors influencing the desirability of solar power plants (SPPs), and of SPP investment decisions, will be discussed in this chapter. The numerical details presented are based, as far as possible, on actual experience with SPPs but are also derived from study results whenever experimental system-level information has not yet become available. Some data may therefore become refined, even modified, in the future as a result of advances in and accumulating experience with SPP technologies. Value and benefit of SPPs are usually viewed differently by engineers, users, investors, or society at large; correspondingly, technical, economic, market, financial and environmental considerations may also lead to different conclusions.

10.1 SPP Technologies in Comparison

Minimizing the cost for installation, operation and maintenance, yet maximizing the annual energy output corresponding to local irradiation conditions, is the first-order engineering objective in any SPP design and technology development. On the other hand, the key parameters for large and long-range commercial investment decisions are the highest possible revenue and profit (in relation to investment risk) and the rate of return on investment. Governments set the regulatory framework for such SPP investment decisions, and in doing so, increasingly take environmental aspects on the national level into account.

Revenue potential and risk are different for the categories of thermal and photovoltaic SPPs for technical and maturity reasons. They depend on whether electricity, thermal energy or (chemical) products are the output objective. Hence, as a first step towards the discussion of SPP costs, some key technical and performance characteristics of the SPP technologies, as discussed in Chaps. 7 and 8, shall be summarized. In general terms, characteristics and features of the three major thermal and two photovoltaic SPP concepts are listed and topically compared in Tab. 10.1, assuming grid-connected electricity generation[2]; if thermal energy or solar fuels are the output objective, the list and the significance of differentiating characteristics would have to be reassessed.

---

[1] Hansmartin P. Hertlein, Deutsche Forschungsanstalt für Luft- und Raumfahrt (DLR), Linder Höhe, D-5000 Köln 90
Helmut Klaiss and Joachim Nitsch, Deutsche Forschungsanstalt für Luft- und Raumfahrt (DLR), Pfaffenwaldring 38-40, D-7000 Stuttgart 80

[2] The authors and editors incorporated all pertinent published data and records into this table, and their engineering judgment when recourse to study results had to be taken into account. Hence, any inaccuracies and bias in personal opinion and judgment cannot be excluded.

Table 10.1. Comparison of technical characteristics and features of thermal and photovoltaic solar power plants (SPPs) for grid-connected electricity generation.

Topic/Feature	Solar-thermal plants			Photovoltaic plants	
	Trough	Tower	Dish	Concentrating	Non-concentrating
Energy collection		concentration of irradiance by reflective surfaces		by Fresnel lenses	immediate absorption in solar cells
Radiation useable[a]		direct irradiation portion		> 40	total irradiation
Concentration	< 100	< 800	< 2,000		< 3
Collector movement	1-axis	is necessary for tracking the Sun in 2-axes	2-axes	2-axes	not required[b]
Radiation transformation into	thermal energy at temperatures			DC-electricity	
	> 350 C	≥ 500 C	< 800 C		
Transfer media	oil/steam	salt/metal/steam	gas	none	none
Energy conversion		thermodynamic/electromechanic conversion AC and/or heat (cogeneration)		photovoltaic conversion DC; AC via electronic inverter	
Energy output	2.0–4.5	3.0–6.0	> 2.8	8.0–14.0	3–8[c]
Land area required ha/MW$_e$	30–80	30–200 (dependent on SM)	6–60 kW$_e$		
Module size MW$_e$ (min/max)				1.0 kW$_e$–5 MW$_e$	0.1 kW$_e$–5MW$_e$ (modular)
Bulk storage	yes (by heat storage)	yes (from storage)	no	none	none[d]
Off-sunshine generation	yes (via fossil backup)	or via fossil backup)	yes (via fossil backup)		none[e]
Irradiation threshold kWh/m^2 d[f]	< 4.0	< 3.0	< 1.5		
Irradiance kW/m^2 threshold[g]	0.30	direct irradiation (SM = 1.0) 0.30	0.30	0.15	0.05
Net efficiency %/a[h]	10–15	direct irradiation (SM = 1.0) 12–15	15–25	10–15	total irradiation 9–12
Transient output response[i]	> 60min	> 60min	1–10min		instantaneous, 1-10 sec
Parasitics[j] %	8–10	10–12	12–17	low	negligible
Stand-by energy		necessary			negligible
Concentrator/collector soiling		is critical (needs regular cleaning)		0.2	2nd order influence
O&M (man-a/MW)		0.5–1.5			0.1
Availability (rep.)%/a	> 90	> 80	72	85	95–97
(expected) %/a	93	93	–	93–97	95–100[k]

10.1 SPP Technologies in Comparison

CF (reported) %	>5	11–24 (SM = 1)	8	24	26–35^l
(expected) %		40 (SM = 1.8)	25	–	>10
Lifetime					
(experienced) yrs	30	<5	<2	5	30
(expected) yrs	15	30	20	20	230
SPP's (installed)	8	7	15	7	220^n
(operating)^m		1	3	4	13
Cap.(installed) MW$_e$ ^{o}	289	21	7	1.2	12
(operating) MW$_e$	275	5	0.2	0.7	

SM = Solar multiple; CF = capacity factor; Total irradiation/irradiance $\hat{=}$ global irr. to oriented surface

^{a} Irradiation-concentrating plants require good irradiation, hence must be located at latitudes between $\pm 45°$.
^{b} Non-concentrating (flat-plate) photovoltaic modules are usually fix-mounted at latitude tilt; tilt may be adjusted seasonally, or modules may be mechanically moved in 1- or 2-axes sun-following mode for augmenting daily energy collection.
^{c} Land area requirements of non-concentrating photovoltaic plants depend on whether modules are mounted fixed/adjustable, or follow the sun in 1- or 2-axes.
^{d} Limited storage possible in electrochemical batteries, may be cost-effective under special circumstances.
^{e} Off-sunshine generation possible via backup Diesel-generator (up to multi-MW size) or via batteries (<1MW$_e$).
^{f} Minimum daily (direct or total) irradiation necessary before net electricity output can be generated after morning start-up (under SM = 1.0 condition); for SM = 1.2, the minimum irradiation for trough plants would be 2.5–3.5 kWh/m^{2}d, and 1–2 kWh/m^{2}d for tower plants.
^{g} Minimum irradiation needed to keep plant operational (under SM = 1.0 condition); for all concentrating systems, minimum irradiance level decreases with SM >1.0.
^{h} Calculated annual net electricity generation, based on 2,500 kWh/m^{2}a of (direct or global) radiation, 100% plant availability, and 1990 technology.
^{i} Valid for SM = 1.0; response capability to transient inputs decreases with plant size and energy inventory; increase in SM together with storage decouples output response from changes in irradiation input.
^{j} Calculated in % of annual gross energy yield, assuming 2,500 kWh/m^{2}a of (direct or total) radiation; values of the test and demonstration plants are higher.
^{k} Last figure refers to sun-following flat-module configurations.
^{l} Actual measured DC capacity taken as basis for reported CF of photovoltaic systems.
^{m} Number, 1988 status.
^{n} Of the about 220 operating grid-connected systems, only about 40 have a capacity >10 kW$_p$.
^{o} Only grid-connected plants; 1989 status.

For a correct interpretation of this list, some comments are necessary:

1. Numbers related to, or dependent on, solar radiation are expressed in terms of direct irradiance or irradiation for all concentrating systems, in terms of total (direct and diffuse) irradiance or irradiation in the case of non-concentrating SPPs.
2. All land- or area-specific numbers are a function of irradiation, conversion efficiency, and storage requirements; if nothing is stated to the contrary, a Barstow, southern California location and meteorology with 2,500 kWh/m^{2}a of irradiation annually, mature state-of-the-art technical system performance, and a solar multiple (SM) of 1.0 are assumed.
3. Capacity factors (CF) are stated in terms of nominal/nameplate alternating current (AC) output rating of thermal and photovoltaic SPPs and – for reported values – in terms of (measured) array direct current (DC) output rating for photovoltaic SPPs (determined at 20°C and 850 W/m^2 for concentrating systems, and 1,000 W/m^2 for non-concentrating systems).
4. All statements related to photovoltaic SPPs are based on crystalline solar cell technologies.

For all types of SPPs, annual net energy output is most significantly affected by annual values of (location-specific) irradiation, and by the actually achieved annual system output performance. Output performance is strongly influenced by net energy conversion efficiency and system operating availability during daylight hours. Plant operating mode, i.e. operation with or without storage and/or in conjunction with a backup fossil source, is another factor impacting annual net energy output. Correspondingly, some data in Tab. 10.1 show considerable spread. However, trough and tower thermal SPPs are expected to have SM > 1.0 in the future, with storage and/or a fossil backup source.

Presupposing that solar technology development advances steadily, that application and operation experience grows continually, and that an industrial volume manufacturing capability gradually emerges, SPP performance and capital investment requirements become a function of time. All data, trends or expectations about future SPP cost/performance quoted later are thus extrapolated from current state-of-the-art knowledge, based on such a scenario of steady development.

The situation (1989) of thermal and photovoltaic SPPs concerning market and development status can be summarized as follows:

- Parabolic trough SPPs, in fossil-hybrid configuration without thermal energy storage, achieved the breakthrough to commercialization in the sector of grid-connected bulk electricity generation in the service area of the Southern California Edison Co. (SCE); economy-of-scale considerations and increasing system-level operating experience favor system capacities > 30 MW$_e$; present annual manufacturing rate is about 80 MW$_e$/a.
- For economy-of-scale and performance reasons, tower SPPs tend towards large plant capacities, aiming for bulk electricity generation in grid-connection operation; need for further technology development and lack of adequate system-level operating experience are viewed as major barriers towards application in the (utility) market[3].
- A parabolic dish concentrator with Stirling converter demonstrated highest peak conversion efficiency of the thermal alternatives; but dish/Stirling technology has found little industrial support so far, mainly because of the high maintenance costs for Stirling engines.
- Small-scale PV systems are widespread and commercial for supplying electricity in stand-alone applications; large grid-connected systems, installed and currently operating, serve mainly demonstration purposes. Present annual solar cell production stands at about 45 MW$_p$/a (incl. 4–7 MW$_p$/a of amorphous Si-cells used in consumer products).

---

[3] It has been reported that 100 MW$_e$ are presently (end of 1989) under construction in the USSR.

10.1 SPP Technologies in Comparison

Major technical differences between the SPP alternatives are:

- In principle, all SPPs are suited for fossil-hybrid generation but in distinct modes and to a different degree; thermal SPPs can use a boiler and the (already existing) power conversion subsystem for this purpose; photovoltaic SPPs would need a separate Diesel-generator set used solely for backup.
- Only tower and parabolic trough SPPs are readily capable of intermediate thermal energy bulk storage and solar-derived cogeneration, thus offering the possibility to increase annual plant utilization and CF.
- Non-concentrating photovoltaic SPPs utilizing total irradiation are less sensitive to irradiance quality than concentrating thermal or photovoltaic SPPs, and hence are less sensitive with respect to site meteorology; no water resource on-site is required for photovoltaic and air-cooled thermal SPPs.
- For economic assessment studies, the yearly energy generated by a SPP can be approximately determined using energy input-output curves under conditions of good irradiation ($> 1,800$ kWh/m²a). The input-output relationships differ between the technology alternatives and are dependent on the operating mode (Figs. 10.1 and 10.2).
  - The amount of irradiation needed before achieving net output is lowest for photovoltaic SPPs, and becomes increasingly higher for dish/Stirling and tower/trough SPPs.
  - For tower/trough SPPs without intermediate energy storage, higher solar multiples (SM = 1.0–1.2) reduce the amount of (direct) irradiation needed before net output operating conditions are attained, but reduce the net output at high daily irradiation (Fig. 10.1); although some excess energy input on some high-irradiation days can thus not be utilized, annual energy generation may nonetheless be improved significantly (up to 20% with SM = 1.2).
  - Higher SMs in conjunction with intermediate thermal energy storage reduce daily irradiation input thresholds of thermal SPPs even further and annual energy yield improves (Fig. 10.1); obviously, fossil-hybrid operation renders net output generation increasingly independent of radiation input.

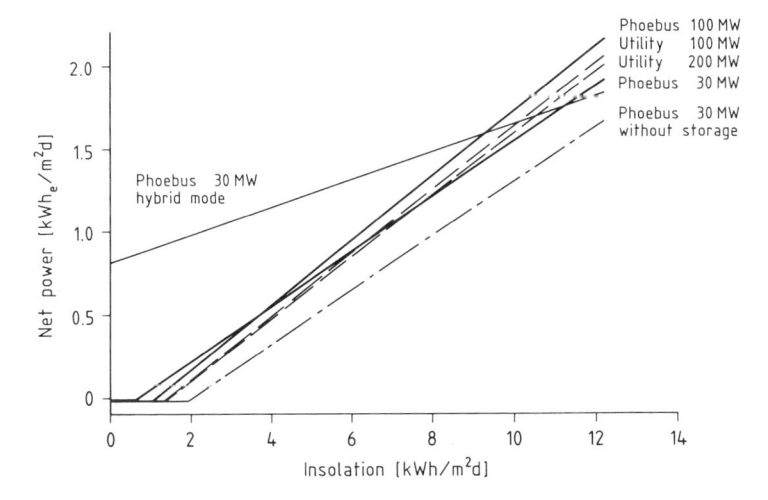

Fig. 10.1. Relationships of daily, collector-area-specific input of direct radiation to net electricity output of 30 and 100 MW$_e$ tower SPPs for varying solar multiples, storage capacity, and fossil hybrid operation mode (calculated using SOLERGY code and Barstow 1976 irradiation) [53].

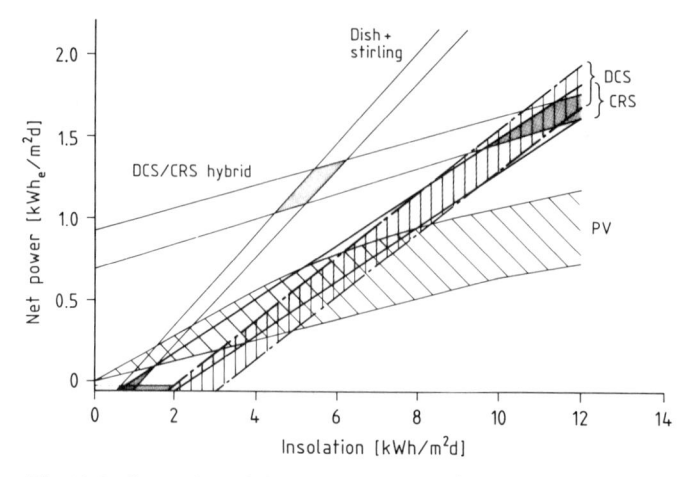

Fig. 10.2. Comparison of daily, collector-area-specific energy input-output characteristics of thermal and photovoltaic SPP alternatives (calculated using SOLERGY code and Barstow 1976 irradiation; grid-connected electricity generation; input in terms of global radiation for non-concentrating PV plants, of direct radiation for all other alternatives.) [53].

- At SM = 1.0 and without storage, dish/Stirling modules offer the lowest input energy threshold of all thermal SPP alternatives, while PV systems exhibit the lowest absolute threshold.

10.2 Investment, Operating and Maintenance Cost

If SPPs are to supply energy to a significant degree in the future, they must become economically competitive with established energy supply alternatives. Reduction of capital investment, operating and maintenance (O&M) costs – while simultaneously increasing energy output performance – are the key issues for rendering SPPs cost-competitive in relation to other power plant alternatives.

In the following, the present cost situation of thermal (tower; parabolic trough; dish/Stirling) and photovoltaic (flat-plate; concentrator) SPPs with ratings in the range of 0.02–200 MW_e (emphasis on ratings > 10 MW_e) will be discussed. Future costs for SPPs with mature technology are extrapolated on the basis of factors related to SPP market development.

It should be recognized, however, that firm data on cost and performance of SPPs are as yet few and heterogeneous. Actual data can be derived from only a limited number of experimental, demonstration and/or commercial SPP facilities either built or existing; some subsystem performance must still be inferred from conventional power plants utilizing similar technology. Considerable uncertainties exist as to the future technical, economic, financial and institutional developments. As availability, reliability, annual energy yield and efficiencies cannot be inferred from peak or short-term performance, these data must come from longer-term operating experience with new SPPs coming on line. Also, cost assumptions are frequently based on increasing production volume (learning curve effects); past predictions were often either too optimistic (e.g. PV) or too pessimistic (e.g. troughs).

For tower SPPs, only the 10 MW$_e$ Solar One facility in the U.S. provided firm cost data, but extensive cost studies have been performed [10,20,35]. For parabolic trough SPPs, cost figures are derived mainly from the commercial SEGS plants operating in the U.S. [44,52,74]. Dish/Stirling units provide the smallest data base for investment and O&M cost, as well as for availability, reliability and performance. For photovoltaic SPPs, cost data from operating systems as well as from cost assessment studies are available [13,73].

O&M costs, excluding any fossil fuel costs elements, are usually stated as a fraction of initial capital investment. In conventional power plants, O&M costs are of the order of 5–8%/a [4]. It is estimated that O&M costs for SPPs will be significantly lower with mature SPP technologies, but real numbers will depend on actual experience of repair and service requirements, and operating reliability.

For comparability, investment and O&M cost data are expressed in terms of plant electric output power, i.e. in $(US)/kW$_e$, and (levelized busbar) energy costs in terms of electricity generated, i.e. in $/kWh$_e$ [5]. Note that comparison solely on the basis of power-specific costs can be misleading for SPPs, especially when SPP alternatives with and without intermediate energy storage are compared. Storage usually requires SM $\gg$ 1.0 and a correspondingly higher capital investment, without output rating necessarily being altered. Also, cost figures depend to a considerable degree on the calculation method used, and are meaningful only within the boundary conditions specified or (tacitly) assumed (i.e. interest, depreciation rate, lifetime, manufacturing rate, learning curve influence over time, etc.). A difficulty in the evaluation of cost assessments is the difference in cost categories used. US cost studies usually differentiate between direct and indirect costs (i.e. costs for engineering, construction management, fees, financing, etc.). In European cost studies, indirect costs and contingencies frequently are factored into the direct costs[6].

10.2.1 Parabolic Trough Solar Power Plants

Investment costs. Real cost/price information is available from the commercial SEGS systems; their cost is about 100 Mio $ for a 30 MW$_e$ installation[7]. All operating SEGS plants are hybrid systems using natural gas as the fossil source. The most significant share of costs is contributed by the solar collector assemblies ($\approx$ 50%) and by the field installation ($\approx$ 25–30%) quite independent of solar collector type or configuration; the contribution from the power conversion unit (PCU), hybrid boiler and balance-of-plant components is comparatively minor ($\approx$ 25%).

The 1984 SEGS I *collector costs* including field piping and construction were 600 $/m^2 (using collector assemblies with 128 m^2 reflective surface); the SEGS III collector costs were 350 $/m^2 (for 235 m^2 reflecting surface), and reached 230–250 $/m^2 for the SEGS VII plant

---

[4] Of the total O&M cost, about 3.5–6% are caused by boiler operation/maintenance and by fuel transport, handling and treatment; the remainder of about 1.5–2.0% is attributable to power conversion system O&M requirements.

[5] All cost data are based on $(US) and exchange rates of 2 DM = 1 $(US) = 1 ECU (European Currency Unit), 1.2 DM = 1 sfr, 0.33 DM = 1 FF, 1.7 DM = 1,000 Lit, 1.1 DM = 100 Yen. If not stated otherwise, nominal interest rate is 8%, and plant lifetime is assumed to be 20 years. O&M costs are assumed to include all running costs (costs for operating personnel, service contracts, spares and equipment, consumables, rentals, etc.).

[6] If not stated otherwise, AFUDC (allowance for funds used during construction) are neglected in further discussions, although these project-specific expenditures may have considerable impact upon the real cost picture (in conventional power plants, AFUDC commonly amount to 10–15% of total investment costs).

[7] As the SEGS plants are marketed under commercial conditions, differentiation between prices offered and real cost for manufacture is problematic. Also, detailed cost break-downs are not divulged for reason of competition.

Table 10.2. Cost data of the SEGS parabolic trough SPPs (different price bases) [1,2,16,29,44,52,74,76].

	SEGS-Plants [a]					
	I	II	III	IV	V	VI
Status	operating 1984	operating 1985	operating 1986	operating 1987	operating 1988	operating 1988
Power MW_e	13.8	30	30	30	30	30
Annual output MWh/a	30,100	66,500	85,050	85,050	91,820	90,575
Size 10^3 m^2	82.9	165.4	204.0	204.0	233.0	188.0
Operation mode	hybr./stor.	hybrid	hybrid	hybrid	hybrid	hybrid
Total investment						
cost 10^6 $	62	96	101	104	122	116
$/kW$_e$	4,500[1)]	3,200	3,400	3,450	4,100	3,800
$/m^2	750	580	500	510	520	610
Collector						
type	LS-1/2	LS-1/2	LS-2	LS-2	LS-2	LS-2
cost incl. pipe $/m^2	600	n.a.	350	350	350	350
Fuel [d]						
consumption 10^3 m^3/a	4,829	9,452	10,131	9,622	9,764	8,150
costs 10^6 $	0.45	0.85	0.9	0.9	0.9	n.a.
Fossil share %	33	36	29	28	25	24

	SEGS-Plants [a]				Battelle	DOE
	VII	VIII	mid-term goal	long-term goal		
Status	operating 1988	operating 1989	project (1995)	study	study 1986	goal 1986
Power MW_e	30	80	100	100	100	100
Annual output MWh/a	94,410	253,000	324,000	383,000		
Size 10^3 m^2	183.0	464.0	475.0	855.0		
Operation mode	hybrid	hybrid	hybrid	storage [c]	storage [b]	storage
Total investment						
costs 10^6 $	117	233	180	289	820 (250)	700
$/kW$_e$	3,900	2,900	1,800	2,890	8200 (3500)	≈ 7000
$/m^2	630	500	380	338	n.a	n.a
Collector						
type	LS-2/3	LS-3	LS-4	LS-4	Acurex	
costs incl. pipe $/m^2	350/250	210	< 200	< 200	150	110
Fuel [d]						
consumption 10^3 m^3/a	8,150	23,659	30,400			
costs 10^6 $	n.a.	n.a.	n.a.	n.a.		
Fossil share %	25	25	25	0	0	0

[a] construction period 10–17 months; lifetime 30 years (SEGS I: 20 years)
[b] at CF = 0.52
[c] 6 h storage
[d] heating value 34.644 kJ/m^3

(with 546 m^2 reflecting surface). Costs of about 200 $/m^2 are envisaged for future plants. *Conversion subsystem costs* (including steam generator, (gas)heater/superheater and generator) amounted to 885 $/kW$_e$ for the SEGS I plant, and less than 600 $/kW$_e$ for the SEGS VII plant. For comparison, 630 $/m^2 (without auxiliary heater) are used in analytical studies.

Detailed *other cost* breakdown data are not available for the SEGS plants. In analytical studies [76] 30–48 $/m^2 are used for transportation costs, 7 $/kW$_t$ for storage (oil/rock con-

10.2 Investment, Operating and Maintenance Cost   375

figuration), and 20 \$/m² for balance-of-plant elements. The *total investment cost* and other cost data of the SEGS systems are summarized in Tab. 10.2. These figures show the considerable learning curve benefits which the LUZ Industries group has achieved, undercutting today already the DOE long-range goal for parabolic trough SPPs with storage. LUZ expects to achieve 2,500 \$/kW$_e$ total investment costs with 80 MW$_e$ systems (hybrid configuration, no storage), < 2,000 \$/KW$_e$ for plants of 100 W$_e$ rating, and ultimately about 1,800 \$/KW$_e$ for 100 MW$_e$ plants (contingent on new LS-4 collector assembly technology).

O&M cost. For the SEGS III plant, the O&M costs are calculated to be 3.2%/a of total investment in routine operation (equivalent to 16 \$/m²a of collector area), and are expected to be reduced to 1.2%/a (5 \$/m²a) for the 80 MW$_e$ and any future plants (these costs correlate with the 9 \$/m²a used in studies [76] and with the 6 \$/m²a US-DOE cost goal). Actual operating costs of the SEGS power plants are 1 Mio \$/a for SEGS III–VII (with a staff of 33) and 2.1 Mio \$/a for SEGS VIII (with staff of 53). However, the expenditure for natural gas used in hybrid/auxiliary operation mode must be added to total operating expenses.

10.2.2 Central Receiver (Tower) Solar Power Plants

Investment costs. The largest contributor to total tower SPP investment cost is the cost for the heliostat field (Fig. 10.3). At the beginning of tower SPP development ten years ago, the *heliostat field* cost share could have been 50% or more of total cost (for a plant of 20 MW$_e$ without storage). As a result of technological advances in the interim, this share would now be as low as 20–25%. Because economy-of-scale effects decrease the specific cost (\$/kW$_e$) of the power conversion subsystem, the cost share of the heliostat field and heliostat foundations increase with SPP size (assuming SM = 1.0). The inclusion of thermal storage necessitates a SM > 1.0, requiring more heliostats and leading to correspondingly higher heliostat field/foundation cost shares quite independent of plant rating. Hence, the percentage of heliostat subsystem costs is to be viewed in conjunction with plant rating and SM.

Figure 10.3 also shows that the cost of the receiver/tower subsystem is merely about 12% of total investment (25 \$/kWh$_t$ assumed), dependent on SPP rating and SM (at high SM values, the receiver/tower cost share is less sensitive to variations in SPP rating). A storage subsystems of 7 hrs capacity represents more than 10% of total investment, but depends strongly on capacity factor (CF) and storage medium.

Area-specific heliostat costs have been lowered substantially, as a historic review shows. The cost of the first 24 heliostats built for the Odeillo solar furnace was about 2,000 \$/m² (1987 value) in 1970, while the cost of heliostats for the Central Receiver Test Facility (CRTF), built in 1976, was about 850 \$/m² (Fig. 10.4). Today, heliostat costs for a 30 MW$_e$ plant are quoted in the range of 180–250 \$/m² (1987 value) [10]. All heliostats installed or currently quoted are of glass-metal construction with < 100 m² reflective area; prototype units up to 200 m² are being tested. Low-cost heliostat using high-reflectance stressed membranes are being investigated. Both glass-metal and stressed membrane heliostats hold promise of low specific costs [17,37,38]; long-term development goals aim at 40–60 \$/m² for stressed membrane heliostats, and at 60–80 \$/m² for glass-metal heliostats[8]. A break-down of the heliostat costs reveals the dominance of costs for foundation and, to a lower degree, of the

---

[8] Cost goals are contingent on the assumption of steady annual volume manufacture of about 0.5–1.0 Mio heliostats per year; however, so many heliostats represent, if deployed, about 10–20 GW$_e$ annual SPP capacity (assuming 150 m² heliostats, 15% annual plant efficiency and SM = 1.0). Smaller annual volumes would increase heliostat costs, but the requirement for steady volume manufacture remains essential.

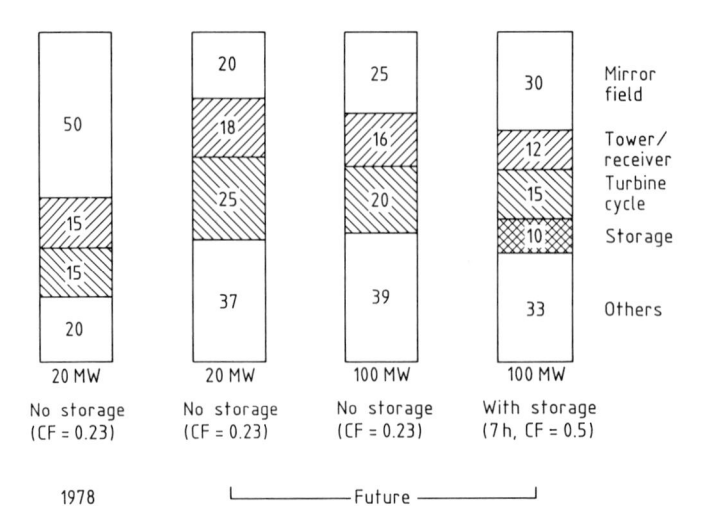

Fig. 10.3. Rough investment cost structure of tower SPPs as function of plant rating, capacity factor (CF) and start-up year [54]. The item 'Others' comprises all indirect costs such as construction management, equipment rental, spares, land and infrastructure, impact studies, contingencies, etc.

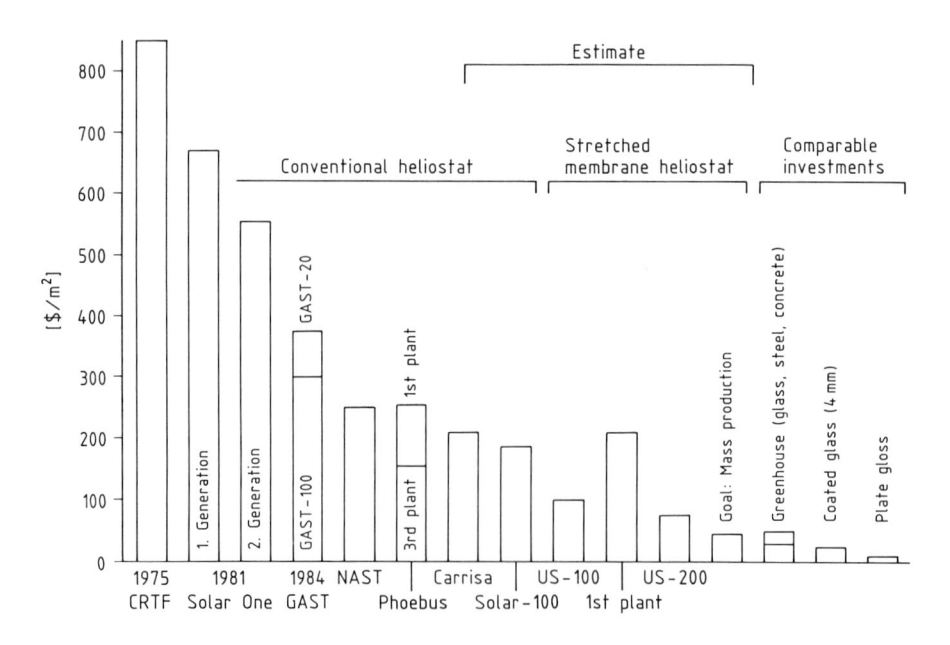

Fig. 10.4. Area-specific costs (realized, quoted or assumed) for glass-metal and stressed-membrane heliostats, and comparison with the cost for related technologies (1987 values, if not indicated otherwise) [10,17,20,30,35,40,54,59].

costs for the drive unit, mirror support and mirror protection. Actual cost shares depend on annual production volume. The share of foundation and site costs may reach 30% for a production volume of ≈ 1 Mio units/a.

Heat transfer medium, temperature and pressure, receiver size and configuration, are the key factors influencing *receiver/tower* costs (Fig. 10.5), the impact of volume manufacture is only minor. Receiver costs normalized in terms of thermal power (kW_t) vary widely for water/steam receivers but have a cost advantage over salt and sodium receivers at small plant power ratings. Because of physical volume and size, gas receivers appear to be restricted to plants of low power rating; multiple gas receivers are used in larger plants. Major cost differences between external and cavity receivers could not be identified.

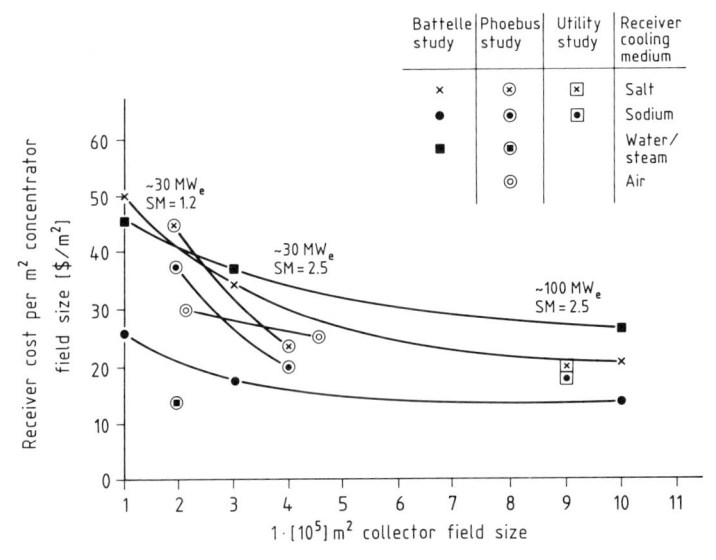

Fig. 10.5. Receiver costs in tower SPPs, normalized with respect to reflective heliostat field area (1987 values for Phoebus [10] and utility study [20]; 1984 values for Battelle [76]).

Actual tower costs depend on height, receiver bulk (which affects wind loads), receiver weight, and seismic conditions. Towers usually are costed on the basis of steel designs up to 120 m, and concrete construction for larger heights. In general, tower costs increase aproximately quadratically with height as does the size of the collector field as viewed by the supported receiver. Due to receiver bulk, towers with water/steam and air receivers cost more than those with salt or sodium receivers by a factor of 2 [38]. For example, typical *tower costs* are 5–10 Mio $ for a 200 m tower. Cost of the *storage subsystem* [43,45,58] depends mainly on storage medium and tank structure. Specific medium costs range widely from 0.05 $/kg or 1 $/kWh$_t$ (for reinforced concrete) to 5 $/kg or 80 $/kWh$_t$ (for silicone oil) [45,58]. Aggregate medium costs account for the largest share (40–60%) of total storage cost, approaching the high end with increasing volume (see Tab. 10.3).

A survey of past and projected *total investment* costs of tower SPPs is provided by studies which were carried out over the past 10 years. These studies include detailed cost and performance estimates and can thus provide insight into investment costs deemed achievable today, and those likely to be attainable in the future (see Tab. 10.4, 10.5). In all cases, total investment costs depend quite heavily on the assumed costs for the heliostat subsystem, on

Table 10.3. Estimates of total storage costs ($/kWh$_t$; 1986/87 values) as a function of storage capacity [10,20,76].

Technology	50 MWh$_t$ [a]	600 MWh$_t$ [b]	1,500 MWh$_t$	3,000 MWh$_t$
Sodium	180	70	19 [c]	18 [c]
Salt	90	35	16 [b]	12 [b]
Air	90	60	n.a.	n.a.

[a] [10]
[b] [20]
[c] [76]

Table 10.4. Investment cost (in Mio $) of U.S. tower SPPs and plant studies; US–100 and US–200 refer to the utility studies [20,35,64].

Project/study	Solar One	Carrisa	Solar 100	US–100	US–200	Phoebus 100 [b]
Price base year	1980	1984	1984	1987	1987	1987
Plant	1st	1st	1st	1st	n-th	
Power (MW$_e$)	10	30	100	100	200	100
Prime cooler	Steam	Sodium	Salt	Salt	Salt	Air
Solar multiple	1.1	1.26	2.42	1.8	1.8	≈ 2
Annual output GWh [a]	26	76	490	333	669	370
Site	4.7	7.5	–	4.3	7.0	–
Collector	49.2	37.6	163.6	92.2	142.5	92.2
Receiver/tower	22.6	10.9	42.9	33.2	50.7	31.3
Thermal transport	4.6	10.5	–	–	–	–
Storage	13.2	6.0	52.5	21.9	40.0	13.9
Steam generator	2.9	8.7	9.2	15.0	23.6	19.3
Turbine	10.1	16.8	14.2	–	–	–
EPGS/BOP	11.4	10.3	14.6	53.6	88.1	59.8
Miscellaneous	1.0	1.4	37.6	1.9	2.2	–
Indirects	21.5	28.2	59.2	49.7	53.1	48.7
Contingencies	–	9.5	54.5	–	–	–
Total $ Mio	141.2	147.4	448.3	271.8	407.2	265.2
$/KW$_e$	14,120	4,910	4,483	2,718	2,036	2,622
Heliostats $/m^2	685	210	187	100	75	100

[a] design value, Barstow weather condition (2,500 kWh(DNI)/m^{2}a)
[b] based on a US/European study

the solar multiple, and on plant size. Despite heliostat costs stipulated in these studies as low as 100 $/m^2, heliostat field costs remain dominant for total plant investment costs.

O&M costs. The studies on tower SPPs assume annual O&M costs in the range of 2–4% of total investment costs (Tab. 10.6). Nearly all O&M cost assumptions are based on a 3-shift, 7-days/week plant operation (including personnel expenses for heliostat washing). O&M costs are assumed not to increase linearly with plant size because the majority of O&M costs are attributed to control room personnel and to the maintenance of turbine and balance-of-plant subsystems, and because a high degree of plant automation is assumed.

10.2 Investment, Operating and Maintenance Cost

Table 10.5. Investment cost of the European tower SPP studies [10,12,19,27,54].

	GAST 20		Metaroz			Phoebus				
Price base	1984		1984	1987		1987		1987		1987
Plant	1st	5th	1st	1st	3rd	1st	3rd	1st	3rd	1st
Power MW$_e$	20	20	5	30	30	30	30	30	30	30
Prime cooler	gas	gas	gas	air		sodium		salt		steam
Solar multiple	1.0	1.0	1.0	1.2	2.5	1.2	2.5	1.2	2.5	1.2
Annual output GWh[a]	31.8	31.8	16.8	62	130	52	114	48	109	38
Collector	37.5	30.0	59.4	57.0	70.5	47.5	62.0	46.5	59.0	46.0
Receiver	26.5	17.5	2.2	6.5	11.0	7.0	7.5	8.0	9.0	2.5
Tower	17.0	15.0	–	3.5	4.5	1.5	3.0	1.5	3.0	1.5
Heat transfer	29.0	25.0	1.7	2.5	6.0	21.5	32.5	13.5	25.0	2.0
Steam generator	–[b]	–[b]	–[b]	7.0	6.5	4.5	4.0	7.0	6.5	–
Storage	–	–	3.3	5.5	40.0	8.5	40.0	4.0	20.0	–
Turbine	–[b]	–[b]	–[b]	12.0	11.0	12.5	11.5	13.0	12.0	13.5
Control system	–[c]	–[c]	–[c]	3.0	2.0	3.5	2.5	3.5	2.5	3.0
Electrical equip.	10.5	9.0	8.9	5.5	4.5	7.0	5.5	7.0	5.5	5.0
Buildings	–[d]	–[d]	–[d]	5.0	5.0	7.5	12.0	5.5	6.5	5.0
Contingencies	–[d]	–[d]	–[d]	10.0	16.0	12.5	18.0	10.5	15.5	8.0
Engineering	–[d]	–[d]	–[d]	12.5	12.0	16.0	14.0	15.0	13.0	11.0
Owners' cost	22.5	15.5	11.0	10.0	13.5	10.0	13.5	10.0	13.5	10.0
Total $ Mio	153	122	94.3	140	202	160	226	145	191	107
$/kW$_e$	7,650	6,100	18,860	4,666	6,750	5,330	7,533	4,833	6,366	3,580
Heliostats $/m^2	375	300	460	255	155	255	155	255	155	255

[a] design value, Barstow weather condition, 2,500 kWh(DNI)/m^{2}a; 1987 values
[b] included in heat transfer   [c] included in electrical equipment   [d] included in owners' cost

Table 10.6. Operation and maintenance costs for tower SPPs (different year price based on study assumptions, except for Solar One); personnel costs are based on a 3-shift 7-days/week operation and include the manpower cost for heliostat washing [10,20,35,64].

Operation and maintenance costs ($ Mio)			
GAST–20	2.7	Solar One (10 MW)	3.7
Proses–20	2.9	Carrisa (30 MW)	3.3
GAST–100	5.4	Solar–100	5.5
Metaroz (5 MW)	1.5	US–100	4.5
Phoebus (30 MW, SM = 1.2)	3.3	US–200	5.6
Phoebus (30 MW, SM = 2.5)	4.5		

O & M costs break-down ($ Mio)					
Solar One		US–100		Phoebus (SM = 1..2)	
Personnel[a]	2.56	Personnel[b]	2.95	Personnel[c]	1.6
Contract personnel	0.29	Contract personnel	0.82	–	–
Spares/material	0.47	Spares/material	0.47	Maintenance	1.7
Miscellaneous	0.05	Miscellaneous	0.27	–	–
Total	3.37	Total	4.52	Total	3.3

[a] 39 total personnel   [b] 46 total personnel   [c] 39 total personnel

10.2.3 Dish/Stirling Units

Investment cost. The collector represents the dominant cost element in dish/Stirling units, followed closely by the cost for balance-of-plant components [24]. The relative share of Stirling converter cost diminishes with size, but Stirling engines are not a mature technology and hence are costly components (no existing volume manufacturing capacity).

Dish collector. Costs depend primarily on design philosophy and weight. Early dishes for test purposes were designed on the basis of antenna structures, resulting in weights > 100 kg/m^2 and costs of about 1,000 \$/m^2. Today, costs of about 650 \$/m^2 (1987 value) are deemed achievable; weights as little as 30 kg/m^2 and costs of 120–150 \$/m^2 may be attainable using stretched-membrane technology in volume production (Fig. 10.6). Mirrors and mirror support structures contribute 50–60% to total collector cost, while the individual cost contribution for foundation, field installation, mechanical drives, wiring, etc., less than 10% each [66,70,76]

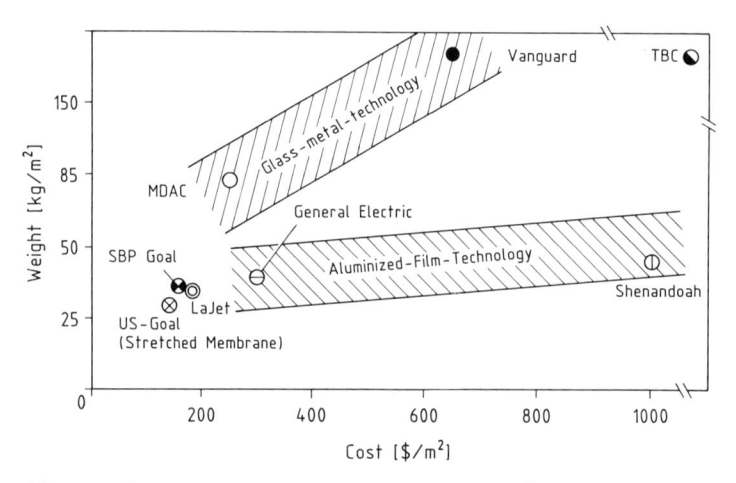

Fig. 10.6. Comparison of aperture-specific weights (kg/m^2) and costs (\$/m^2; 1987 value) of parabolic dish collectors technologies (TBC = Test Bed Concentrator; MDAC = McDonnel Douglas; SBP = Schlaich, Bergermann & Partner) [3,4,36,40,66,70,75].

Dish *receiver* costs depend strongly on assumed annual production rate since the subsystem receiver/Stirling engine still requires technological development and adaptation to solar (i.e. variable orientation) operating conditions. A Stirling receiver cost goal of 21 \$/kW$_t$ (or equivalently 21 \$/m^2 of aperture) has been stipulated, but near-term costs are still deemed closer to 70 \$/kW$_t$ [76]. Receiver costs for Brayton, Rankine and combined-cycle power conversion subsystems in the year 2000 are expected to range between 25-40 \$/m^2 (1987 value), depending on upper cycle temperature and technology (heat pipes at the high cost end) [36].

The receiver is commonly integral with the Stirling engine/generator power block. Approximately 30% of the power conversion subsystem cost can be attributed to the receiver (at least 25 \$/kW$_e$). With receivers, the cost of Stirling engines is currently > 20,000 \$/kW$_e$ (11 kW$_e$), and is estimated to be about 700-800 \$/kW$_e$ for 25 kW$_e$ (1987 value) at 1,000 units/a production rate, and about 160 \$/kW$_e$ at 10,000 units/a production rate (see Tab. 10.7) [40,75]. In this cost, generator costs of about 75 \$/kW$_e$ are already included. This figure compares well with US-DOE long-range goals (300 \$/kW$_e$) and with 450 \$/kW$_e$ long-range goals for a 11 kW$_e$ engine [66,70].

10.2 Investment, Operating and Maintenance Cost

Table 10.7. Breakdown of dish/Stirling investment costs [70,75,76].

	Vanguard 1 [a] 1984				SBP [f] 1987		DOE [h] 1986	Battelle [i] 1986
Units/Year	1	100	1,000	10,000	1 [c]	10 [d]	1,670 [e]	steady state prod.
Component								
Collector %	24	56	58	63	33	31	38	43	54
Conversion %	44	42	41	36	58	58	34	49	25
Others [b] %	32	2	1	1	9 [g]	11 [g]	28 [g]	8	21
Total Costs								
10^3 \$/unit	320	102	59	43	750	325	86	–	–
10^3 \$/kW$_e$	12.8	4.1	2.4	1.7	18.8	6.2	1.4	1.2	1.3

[a] 25 kW$_e$, diameter 11 m, $\eta = 26\%$
[b] Site equipment, insurance, transport etc.
[c] 40 kW$_e$, 1986, 17 m $\oslash$, $\eta = 14.2\%$
[d] 52.5 kW$_e$, 1990, 17 m $\oslash$, $\eta = 19.9\%$
[e] 60 kW$_e$, > 2000, 17 m $\oslash$, $\eta = 23.6\%$
[f] SBP = Schlaich, Bergermann & Partner
[g] includes transport from Germany to Saudi Arabia, transport insurance and buildings on site (assembly hall, workshop, education center, etc.)
[h] Goal, 100 MW$_e$, 395,000 m^2 collector field, $\eta = 27\%$
[i] 100 MW$_e$, 580,000 m^2 collector field, $\eta = 21\%$

Cost for Brayton turbines are expected to be similar to those of Stirling engines. Combined Brayton/Rankine converters and conversion cycles are technically complex and not mature in the < 1.0 MW$_e$ range, hence they are more expensive than any of the alternatives [40].

'Other' costs. Other costs depend strongly on aggregate plant capacity when several dish/converter units are combined. They range from 132 \$/m^2 of aperture area for a SPP of 0.5 MW$_e$ capacity, to a projection as low as 28 \$/m^2 for a 100 MW$_e$ plant. For dish/Stirling SPPs in the MW$_e$-range, a significant part of these costs is attributable to site and site preparation. Only few dish/Stirling *total investment* cost studies have been carried out. The results are listed in Tab. 10.7, contrasting the Vanguard 1 system at low and high production rate [75] against SBP costs and cost extrapolations [70], and against US/DOE goals [76] and other studies. It appears that about 2,000 \$/kW$_e$ is achievable at a 5,000 units/a manufacturing rate (equivalent to 125–300 MW$_e$/a of total SPP capacity), although (single prototype) costs today might still be as high as 15,000 \$/kW$_e$. Nonetheless, a discrepancy to US/DOE long-term cost goals and cost estimates (formulated on the assumption of constant volume manufacture and of a mature industry) for a 100 MW$_e$ dish/Stirling SPP is apparent (Fig. 10.7). All studies confirm the dominance of collector costs at high production rates, however, irrespective of absolute cost figures.

O&M costs. O&M cost of dish/Stirling systems is proportional to field size and unit production rate. 1.5%/a of collector and 'other' costs seems attainable. As the reliability of dish receivers and Stirling engines is yet unconfirmed, 20%/a of initial capital investment was assumed for O&M cost in studies [76]. SBP calculates with 3.5%/a (0.125 Mio \$ for a 525 kW$_e$ capacity SPP, and 7.4 Mio \$ for a 10 MW$_e$ capacity SPP; 3-shift operation). Taking all published cost data as a basis, O&M costs of about 50\$/kW$_e$ seem achievable. Nevertheless, O&M cost reduction must be the main objective for rendering dish/Stirling systems appropriate for rural areas.

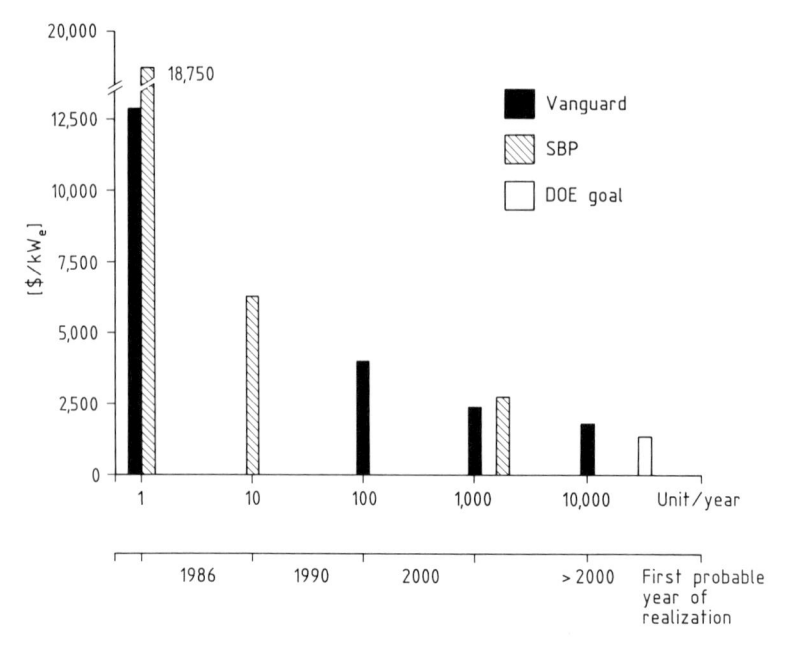

Fig. 10.7. Projected decrease of specific costs ($/kW$_e$; 1987 values) with annual manufacturing rate of dish/Stirling systems, and estimate of time by which such production volume might be attained [30,70,75,76].

10.2.4 Photovoltaic Solar Power Plants

Investment Cost. System investment is usually categorized in terms of module cost, balance-of-system (BOS) costs (subdivided into area- and power-related costs), and engineering cost. Area-related BOS costs include those for module support structures, foundations, electrical wiring, and for tracking/control elements in concentrating photovoltaic systems. Power-related BOS cost are for power conditioning (e.g. DC/AC inverters), metering and safety provisions [5]. In the past 10 years, *module cost* has decreased from 20 $/W$_p$ to less than 6 $/W$_p$ in commercial quotes for shipments > 1 MW$_p$ [Fig. 10.8] [9]. The manufacturer Siemens Solar assumes module costs of 3.27 $/W$_p$ by 1993, to be attained with a manufacturing capacity of 10 MW$_e$/a [13,61]. Costs as low as 2–2.5 $/W$_p$ are projected by the year 2000 for modules with crystalline-Si cells [5,7,46,55,60,63,65,68,73,77].

Long-term module costs as low as 0.5–1.0 $/W$_p$ are predicted for volume production and when novel technologies (e.g. thin-film cells) are employed [8] (Fig. 10.9). Long-term US/DOE goals are 0.3–0.5 $/W$_p$ (45–80 $/m^2) for flat-plate modules, and 0.25–0.4 $/W$_p$ (60–100 $/m^2) for concentrator modules [5,8]; mid-term goals for the mid-1990s are 0.9–2.4 $/W$_p$ (90–240 $/m^2) and 0.7–1.8 $/W$_p$ (110–275 $/m^2), respectively. A more detailed breakdown of module cost using crystalline-Si technology is provided in Fig. 10.10 [72,73]. The largest

---

[9] The low current cost quotes for photovoltaic modules are judged not to cover actual production costs; to some degree, these quotes anticipate future price levels which are expected to be attainable with single-source manufacturing capacities > 3 MW$_p$/a.

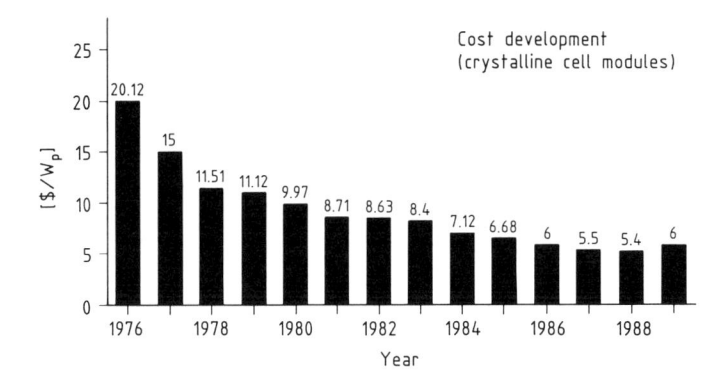

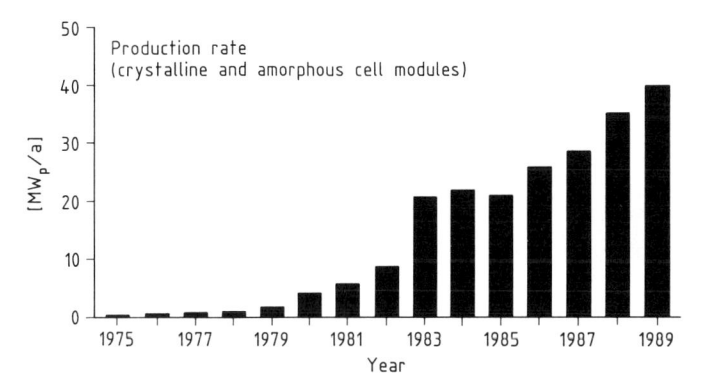

10.2 Investment, Operating and Maintenance Cost

Fig. 10.8. Development of photovoltaic module price and annual solar cell production (for crystalline and amorphous solar cells, at actual prices) [13,18,23,67,73].

relative and absolute contributors to total module cost[10] are wafer sawing/slicing during wafer production, cell manufacture during cell production, and cell interconnects/module assembly in the module production phase. Raw materials account for only about 25–30% of total cost.

With about 50–65% of total investment, module costs are dominant in the total cost for large photovoltaic SPPs. Near-term *balance-of-system* (BOS) cost are expected to be about 30% of net total cost; half of these are area-related and half power-related (see Tab. 10.8 and Fig. 10.11). Future BOS costs are estimated to range between 50 and 200 $/m² for area-related costs, and 0.15–1.3 $/$W_p$ for power-related costs (exact amount depending on flat-plate, tracking flat-plate, or 2-axes tracking concentrating system configuration). Effects of volume manufacture on BOS cost are expected to be less than for modules. But irrespective of actual figures, BOS costs for concentrating systems are judged to remain about twice as high as those for flat-plate systems. With decreasing module cost, the significance of the BOS cost fraction increases. On the other hand, the share of the area-related BOS costs decreases with increasing module efficiency.

---

[10] based on 23 Mio modules per annum (equivalent to 30 MW_p/a), 426 manyears/a for production, and ≈ 30,000 $/manyear.

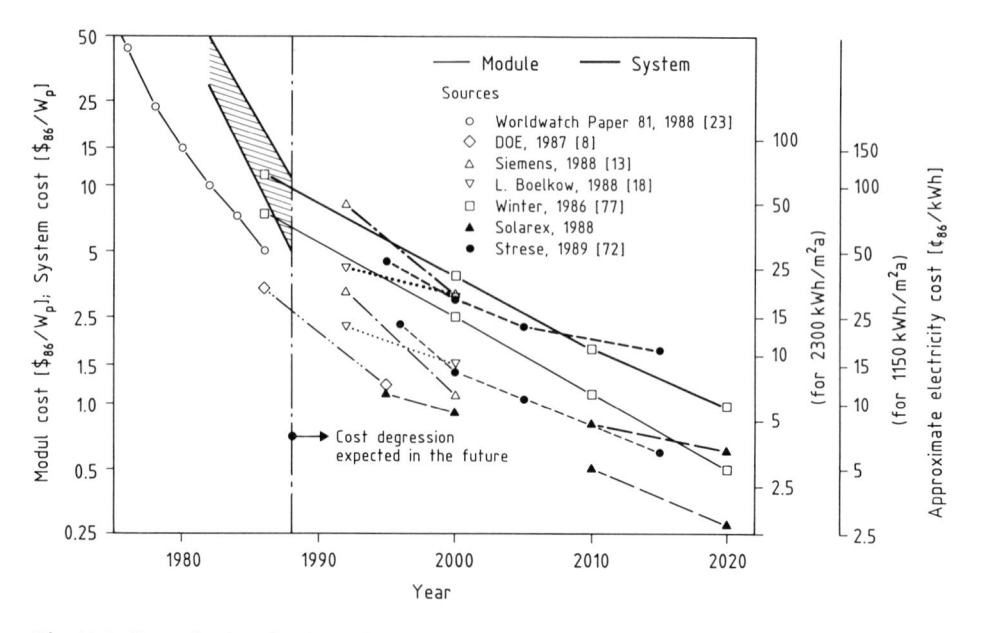

Fig. 10.9. Cost reductions for photovoltaic modules and systems (7% interest, 30 years depreciation, O&M costs 0.025–0.005 $/kWh$_e$) [8,13,18,23,73,77].

Total cost. Total cost estimates vary with assumptions made about production volume of modules, plant size, BOS cost estimates, system configuration, and the year the photovoltaic SPP takes up service (Tab. 10.8). While system costs in 1982 were still 30 $/W$_p$, present system costs are of the order of 10 $/W$_p$. In 1989 and 1990, installed system costs were reported of $\approx$ 9.50 $/W$_p$ for 192 kW$_p$ systems, and of 4.72 $/W$_p$ for a 419 kW$_p$ system (PVUSA project). It is estimated that 2.5–5 $/W$_p$ for system costs will be attained by the end of the century. Long-term (year 2000 and beyond) costs are predicted to become about 1 $/W$_p$ or less.

Recent cost analysis studies have been carried out mainly for photovoltaic SPPs employing flat-plate modules. The installed capacity is determined based on (manufacturer-supplied) DC power specifications. A correction factor of 1.33 for European conditions, and of 1.4 for sunbelt conditions, is introduced in the references to Tab. 10.8 to account for loss factors such as module mismatch, power conditioning and conversion, auxiliaries (i.e. for tracking), and cell degradation with temperature and time [73]. On this basis, photovoltaic SPP costs compare better with the AC-output-based specific investment costs of thermal SPPs.

O&M cost. Due to comparatively simple mechanical construction (no rotating machinery) and unattended operation, maintenance and operating requirements for fixed flat-plate photovoltaic SPPs are minimal; maintenance requirements increase with increasing complexity (i.e. mechanical tracking) and concentration (need for more frequent cleaning). Correspondingly, O&M cost of photovoltaic SPPs have been as low as 0.005 $/kWh$_e$ or 0.1%/a [69], but also as high as 1.5%/a (see also Chap. 8).

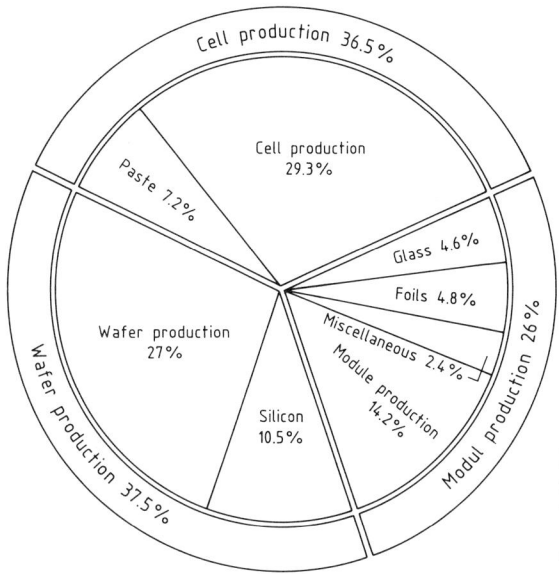

10.2 Investment, Operating and Maintenance Cost 385

Fig. 10.10. Structure of photovoltaic flat-plate module cost (based on a 1995 estimate of 2.3 $/W_p$ module cost (1988 value), under the condition of 30 MW$_p$/a single-source module manufacture) [72,73].

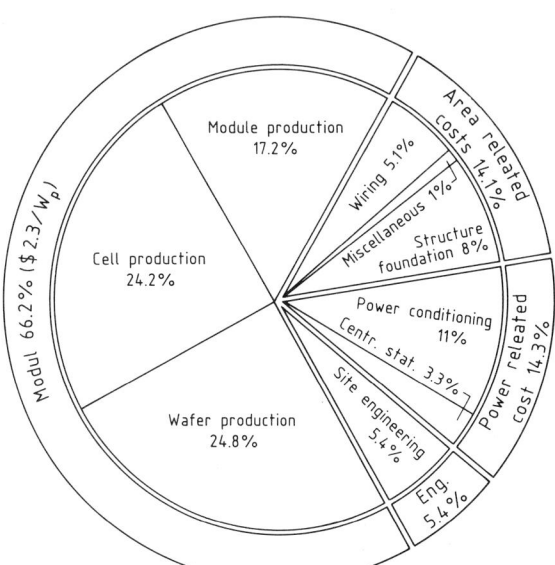

Fig. 10.11. Breakdown of photovoltaic SPP system costs (based on a 1995 estimate of 3.45 $/W_p$ total system cost (1988 value), under the condition of 30 MW$_p$/a single-source module manufacture) [72,73].

Table 10.8. Present and estimated future costs of photovoltaic SPPs [5,13,18,55,67,73,76].

		Average costs 1982 > 100 kW plants	Fraunhofer 1986	Siemens 1992 Crystalline flat plate system	Bölkow 1992 (pessim.)	1992 (optim.)
Net efficiency	%	5–8	10	12.5	12.5	12.5
Module costs	$/W$_p$	14.4	6.0	3.27	2.50	2.25
	$/m^2					
BOS costs	$/W$_p$	17.6	190	200	75	50
	$/m^2		1.3	0.65	0.75	0.35
Total costs	$/W$_p$	32	9.2	5.67	3.88	3.00
Correction [a]		12.8	3.6	2.02	1.28	1.0
Engineering	$/W$_p$	–[b]	0.8	0.50	0.52	0.17
Total costs	$/W$_{net}$	44.8	13.6	8.19	5.67	4.17

		> 2000 Amorphous, flat plate system	DOE long-term goal > 2000 Silicon, flat plate	> 2000 Concentrator, silicon	DLR 2020 Amorphous, flat plate
Net efficiency	%	15–20	25–30		
Module costs	$/W$_p$	1.50			0.5
	$/m^2		45–80	60–100	
BOS costs	$/W$_p$	50	50–100	125–200	80
	$/m^2	0.35	0.15	0.15	0.25
Total costs	$/W$_p$	2.22	≈ 0.8–1.4	≈ 0.9–1.5	1.4
Correction [a]		0.74	≈ 0.3–0.56	≈ 0.36–0.6	–[c]
Engineering	$/W$_p$	0.17	–[b]	–[b]	–[b]
Total costs	$/W$_{net}$	3.12	≈ 1.1–2.0	≈ 1.2–2.1	1.4

[a] correction factor = 1.4, to account for difference in ratio of module efficiency at STC (25°C) and annual system efficiency
[b] included in BOS costs
[c] included in lower efficiency

10.3 Power Plant Cost Analysis and Comparison

A correct and fair comparison between solar and conventional power plants is rendered difficult because of differing operating situations. Not only are type and character of input energy dissimilar, but also cost and availability of input energy differ with time and geographic region; in addition, technological maturity, operating constraints and financial conditions can significantly impact the result of economic analyses. The correct interpretation of economic results is further blurred by the fact that a conventional cost/benefit analysis does not reflect external costs (i.e. costs to society, such as CO_2/SO_2/nuclear follow-up costs). In the following, specific energy costs are calculated on the basis of life cycle costs (present value method) and are expressed as nominal levelized busbar energy costs (see also Chaps. 3 and 4) [22,34,49,73]. Because capital and all operating costs over the useful life of a facility are accounted for, the present value method is most commonly used by utilities in comparing costs of competing investments.

Often (country-specific) economic conditions/assumptions influence levelized energy costs more than technical performance parameters. Hence, the following simplified economic and technical boundary conditions are stipulated to apply [56,57]:

- taxes, inflation rates and end-of-life salvage values are zero;
- discount rate is 8%;
- plant operating lifetime is 20 years[11];
- annual fixed-charge rate is 10.2 %/a.

AFUDC are taken in consideration only if the construction period is longer than 6 months. As annual energy output of a SPP is a function of the amount, quality and distribution in time of local irradiation, specific energy costs are calculated using the excellent Southern California (Barstow) weather conditions (about 2,500 kWh/m²a of direct irradiation) as a basis, if not stated otherwise.

10.3.1 Conventional Power Plant Generating Costs

Total conventional plant investment costs differ over a wide range (see Tab. 10.9). The cost of a 1,200 MW$_e$ nuclear plant is about 1,700–3,000 \$/kW$_e$, and for a 400–600 MW$_e$ hard coal-fired plant about 900–1,100 \$/kW$_e$ (denitrogenization and desulfurization included). Plant investment varies in a wide range with plant rating. While in the past specific investment costs stayed fairly stable over time, excluding nuclear plants, fuel prices fluctuated widely. For example, in the decade from 1975–1985, fuel costs in Germany, and the contribution of fuel costs to specific energy costs, varied between the following extremes (in 1988 values, without taxes):

heavy oil	95–315\$/ton	or 0.022–0.075 \$/kWh$_e$
light oil	110–360\$/ton	or 0.027–0.090 \$/kWh$_e$
natural gas	0.5–2\$/kWh$_t$	or 0.017–0.070 \$/kWh$_e$
hard coal, imported	45–90\$/ton	or 0.015–0.030 \$/kWh$_e$
hard coal, domestic	69–130\$/ton	or 0.023–0.045 \$/kWh$_e$

The share of fuel cost in generating costs is quite different depending on fuel (about 70% for gas-fired plants, and 20% ≈ 0.012 \$/kWh$_e$ for nuclear plants). Levelized energy costs for fossil-fired power plants must therefore be quoted within a cost range [49].

The cost of electricity for coal and nuclear plants > 500 MW$_e$ varies with full-power operating hours and, in absolute terms, between countries (Fig. 10.12). With about 3,000 h/a of operation, generating costs of hard-coal-fired plants are higher than those of nuclear plants operating about 6,200-6,700 h/a both in Germany and France. The situation in the U.S. is the reverse, with nuclear generating costs significantly higher than those of hard coal-fired plants (excepting a few specific regions in the U.S.). Of all countries, France enjoys the lowest (nuclear) generating costs, as a result of high government subsidies.

Looking at the structure of generating costs (Fig. 10.13), the dominance of investment financing costs in solar and nuclear power plants – and of fuel costs in coal-fired (as well as in gas- or oil-fired) plants – emerges as the most differentiating element. Fuel costs for solar-only SPPs are of course zero, but do become a contributing factor in hybrid-operated SPPs. Presupposing an efficient, mature and reliable SPP system technology to be already available, generating costs of SPPs can be lowered further only by reducing the specific investment and

---

[11] A plant lifetime > 20 years, although customary for conventional power plants, is currently not verified for any SPP; demonstrated useful lifetimes range up to 6 years for parabolic trough SPPs, and up to 10 years for photovoltaic SPPs.

Table 10.9. Specific investment costs of conventional fossil- and nuclear-fueled power plants (different price bases) [9,15,49,77].

	Size	Costs ($/KW$_e$)
Hard coal power plants	400 MW	700–1,200
	600 MW	900–1,100
Nuclear power plants	1,300 MW	1,700–3,000
Oil and gas power plants	up to 1 MW	2,500–3,500
	1–4 MW	2,000–2,500
	4–10 MW	1,500–2,000
	10–30 MW	1,250–1,500
	30–100 MW	600–1,250
	100–1,000 MW	400–1,000
Gas turbine	5–30 MW	300–500
	30–80 MW	200–400
Diesel power plants	50–200 KW	1,000–1,500
	200–1,000 KW	1,000–1,200
	1–10 MW	800–1,100

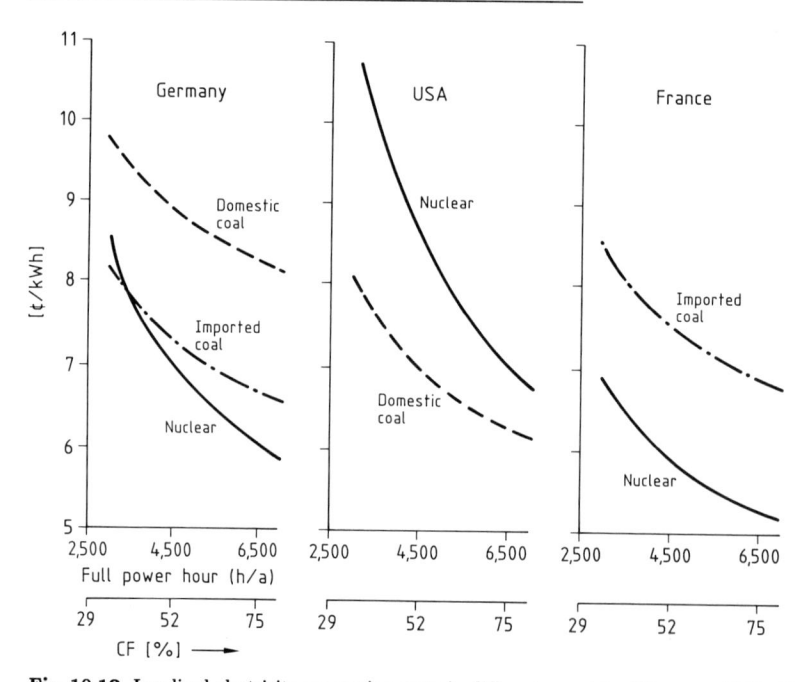

Fig. 10.12. Levelized electricity generating costs in different countries (Germany, USA, France) [71].

O&M costs, by increasing the SPP operational availability and useful lifetime, and by locating the SPP at sites with the best possible irradiation[12].

---

[12] As long as generating costs from fossil sources remain lower than those of SPPs in solar-only operation, SPP generating costs are reduced by the addition of a hybrid generating capability (capacity factor increase). In addition, hybrid capability enhances the assurance of generating capability, resulting in added capacity credit which may greatly increase the value of the SPP.

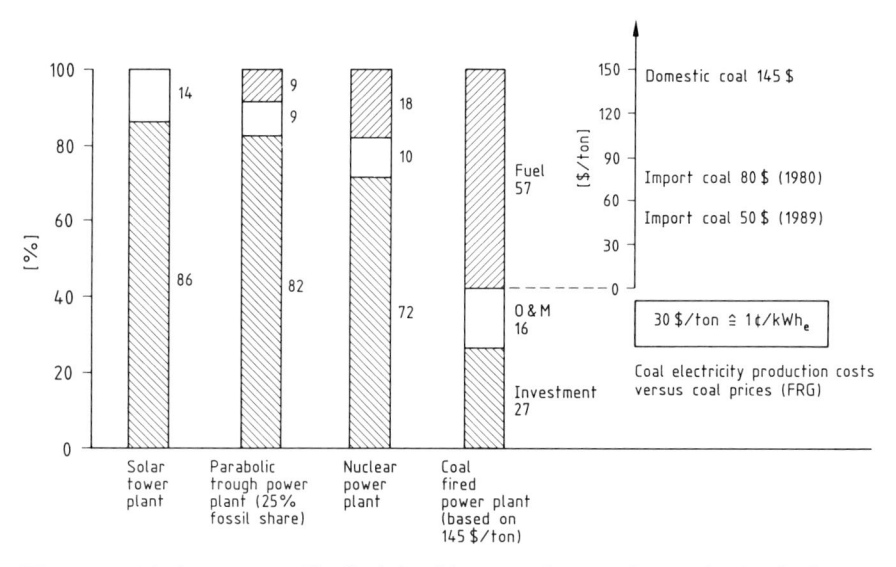

Fig. 10.13. Relative structure of levelized electricity generating cost of conventional and solar power plants (based on 1990 (1988) operation start-up for solar power (conventional) plants; the cost share of fuel in coal-fired plants is dependent on fuel price (30 $/ton of coal contributes about 0.01 $/kWh$_e$ to generating cost). A 25% contribution from fossil sources is assumed for parabolic trough SPPs [49,54].

10.3.2 Solar Power Plant Generating Costs

Generating costs[13] of *parabolic trough* SPPs are taken from the SEGS operating experience (Fig. 10.14). About 0.20 $/kWh$_e$ are quoted for the SEGS III facility in solar-only operating mode, and 0.17 $/kWh$_e$ in hybrid operating mode. Of these costs, 0.038 $/kWh$_e$ are attributable to O&M costs, and 0.011 $/kWh$_e$ to fuel costs. Generating costs < 0.10 $/kWh$_e$ are expected to be reached by 1995 after the last of the SEGS facilities has become operational in Southern California.

The electricity generating cost of *tower solar power plants* is calculated to be as low as 0.25 $/kWh$_e$ currently (1988 value) for a small first-generation plant of 30 MW$_e$ capacity with SM = 1.2, operating the equivalent of 2,066 h/a at nominal output capacity under an irradiation of 2,500 kWh(DNI)/m²a (cost would be ≈ 0.40 $/KW$_e$ at 1,900 KWh(DIN)/m²a and 1,800 h/a operation). With a 100 MW$_e$ tower SPPs operating under the same conditions, generating costs as low as 0.1 $/kW$_e$ are estimated for the future (Fig. 10.15). Increasing plant size to 200 MW$_e$ and capacity factor by adding storage, costs are extrapolated to fall as low as 0.06 $/kWh$_e$, assuming 3,600 h/a full-load operation.

It is predicted that electricity generating costs of *dish/Stirling* SPPs would be around 0.35 $/kWh$_e$ today (1990) at 525 kW$_e$ plant capacity [70]. Cost predictions under different conditions are illustrated in Fig. 10.16. Future generating costs as low as 0.10 $/kWh$_e$ are deemed achievable in the future with advanced technology, but cost elasticity with respect to plant size is low above 1 MW$_e$ aggregate plant capacity. For plant capacities < 5 MW$_e$, generating costs of dish/Stirling systems are lower than those of tower/trough SPPs.

---

[13] based on demand at temperatures < 500°C, which usually needs 10MW$_t$ or less; transport costs for high temperature heat are uncertain. As the use of solar radiation for the production of fuels, chemicals and solar hydrogen is still in the R&D phase, no cost estimates are attempted concerning fuels and chemicals.

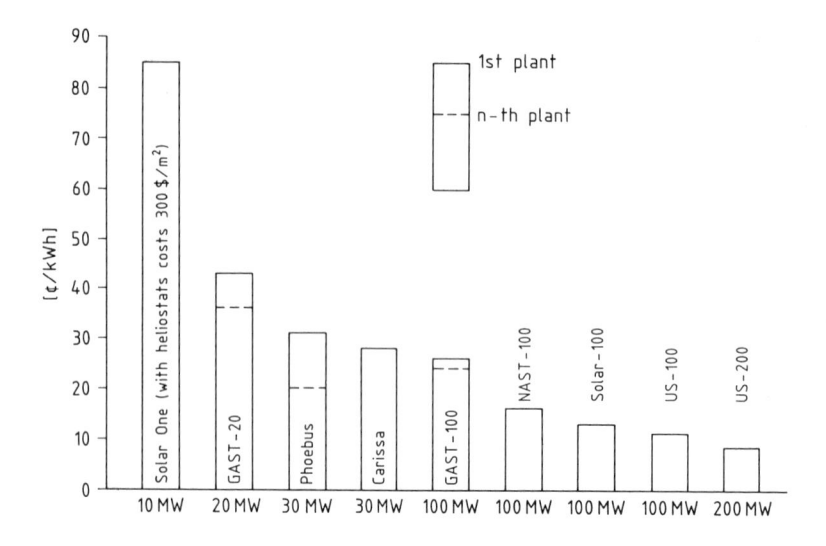

Fig. 10.14. Levelized electricity generating cost for the SEGS parabolic trough SPPs in hybrid operating mode (1987 values, 8% interest rate, lifetime 20 years).

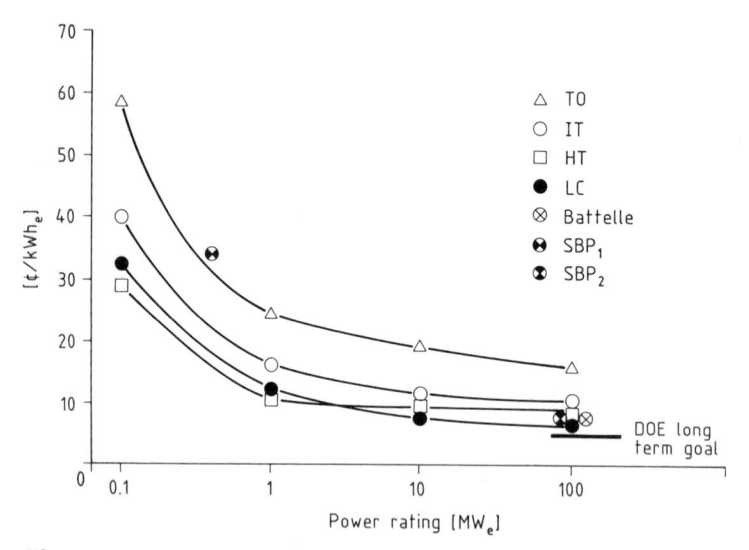

Fig. 10.15. Levelized electricity generating cost of tower SPPs (different price bases; costs calculated on the basis of 8% interest rate, lifetime of 20 years, buffer storage only, 2,500 kWh/m²a direct normal irradiation; Solar One costs calculated with a modified design and heliostat costs).

Photovoltaic SPPs are currently estimated to have generating costs of about 0.50–0.60 $/kWh$_e$ (1989 value) (Fig. 10.17). Assuming a 30 MW$_e$/a photovoltaic module production in a single manufacturing facility, generating costs as low as 0.20 $/kWh$_e$ are calculated under the condition of 2,400 h/a full-load operating hours. A goal of 0.06 $/kWh$_e$ has been stipulated by US-DOE, to be achieved by the year 2000.

10.3 Power Plant Cost Analysis and Comparison

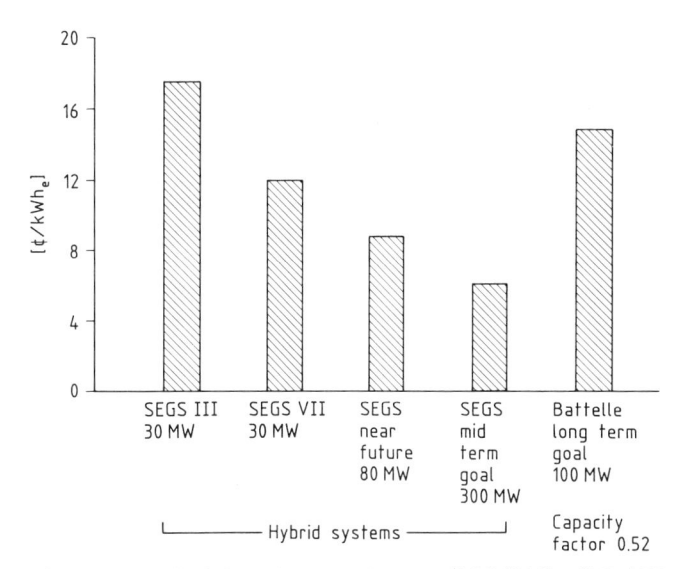

Fig. 10.16. Levelized electricity generating costs of dish/Stirling SPPs (1987 values; cost calculated on the basis of 8% interest rate, 20 years lifetime, 2,500 kWh/m^{2}a direct normal irradiation) (Remarks: TO: 1990, based on Vanguard Dish/USAB Stirling, 13% annual net energy yield; IT: 1995, improved Vanguard Dish/USAB Stirling technology, 19% annual net energy yield; HT: > year 2000 advanced technology, 26% annual net energy yield; LC: > year 2000 low-cost technology, 15% annual energy yield; Battelle: > year 2000 technology; SBP$_1$: 1990 technology, 525 kW$_e$ capacity; SBP$_2$: > year 2000 technology, 100 MW$_e$ capacity).

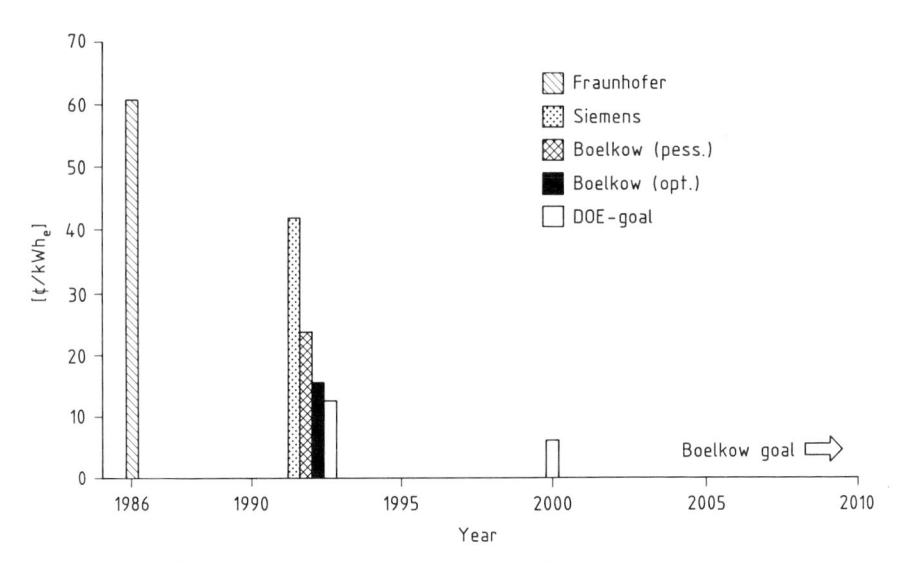

Fig. 10.17. Levelized electricity generating cost of photovoltaic SPPs (1986 values for US-DOE goal, 1988 values for other sources) [8,13,67,72].

10.3.3 Sensitivity Analysis of SPP Generating Costs

Sensitivity of SPP electricity generating costs to changes in the cost of contributing cost factors is illustrated in Fig. 10.18, using the tower SPP as an example. As expected, generating costs change nearly proportionally with discount rate and total investment cost. Changes in lifetime assumptions have a comparably large impact, particularly for short lifetimes. On the other hand, manpower costs as a fraction of total O&M costs are of smaller influence.

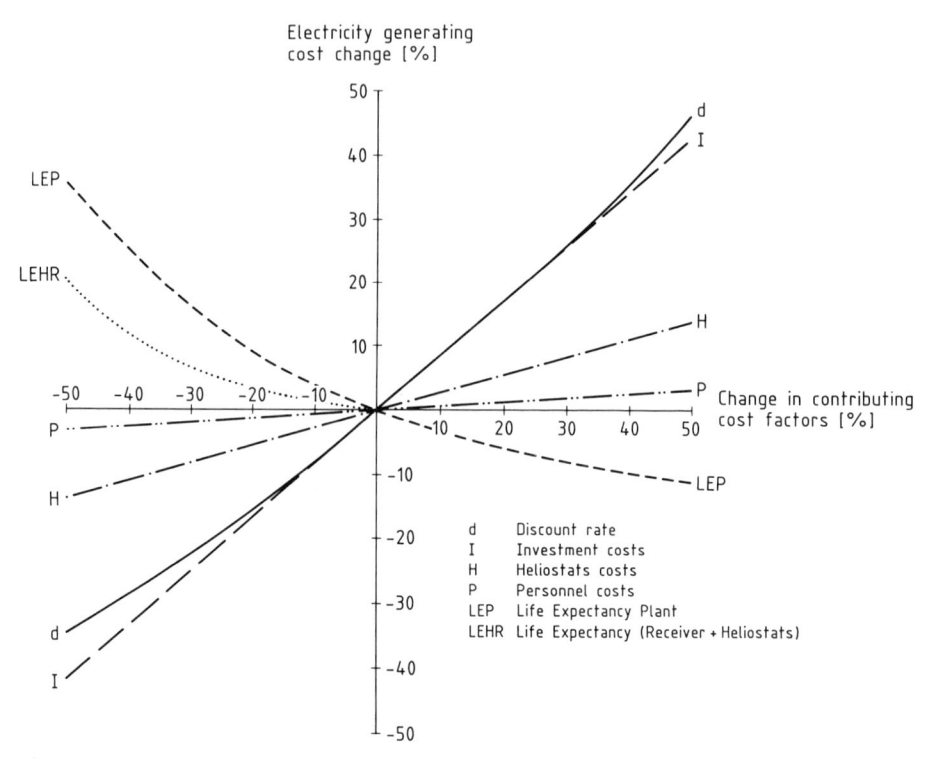

Fig. 10.18. Sensitivity of levelized electricity generating cost to changes in the cost of contributing elements typical for a SPP (based on a 100 MW$_e$ tower plant with storage and 3,815 full-power operating hours (CF = 46%)) [39].

Some other factors also influence SPP generating costs:

Operation start-up year. Assuming continuing advances in solar technologies with time (learning curve effects due to maturing of technologies, establishment of manufacturing capabilities, rising and stable production rates), extrapolations of plant performance data and economic results for the four major SPP technology lines, assumed to operate under identical conditions, are compiled in Tab. 10.10.

Up to 1987, data are taken from existing plants, state-of-the-art data are assumed for 1990 and beyond. Data for the year 2000 are calculated on the basis of an already established manufacturing capability for solar components. The year 2020 data express the performance deemed achievable when using today's advanced technologies and system concepts.

10.3 Power Plant Cost Analysis and Comparison

Table 10.10. Comparison of technical and economic data of the four major SPP alternatives, operating under identical conditions (1987 values; 2,500 kWh/m²a direct or global irradiation) [10,27,35].

		Tower				Trough			
		1985	1990	2000	2020	1985	1990	2000	2020
Performance									
Power	MW_{net}	10[d]	30[e]	100[f]	200[g]	30[h]	80[i]	300[j]	300
Net efficiency	% (solar)	8.1	11	15	15.1	–	–	–	–
	% (hybrid)	–	–	–	–	11.6	14.2	15.0	15.0
Reliability	%	92	93	95	95	95	95	95	98
Production[a]	GWh_e/a	13.6	62	333	669	85	250	900	1,080
Collector Area	10^3 m²	72	223	883	1,819	204	400	1,263	2,900
Operation mode		storage				hybrid		storage	
						gas	gas	gas	
Costs (1987)									
Investment[b]	10^6 \$	96.5	140	271.8	407.2	101	200	540	729
Specific costs[c]	\$/kW	9,650	4,670	2,718	2,036	3,370	2,500	1,800	2,450
	\$/m²	1,340	628	308	224	495	500	396	250
Collector only	\$/m²	300	255	100	75	175	125	115	95
							incl. pipes		
O&M	10^6 \$/a	3.35	3.3	4.5	5.6	3.2	3.6	6.5	8.7
	% of Inv.	3.5	2.2	1.7	1.4	3.2	1.8	1.2	1.2
Fuel	10^6 \$/a					0.9	2.0	7.5	

		Dish/Stirling				Photovoltaics[c]			
		1985	1990	2000	2020	1985	1990	2000	2020
Performance									
Power	MW_{net}	0.04	0.525	10	100	1[k]	10[l]	30[m]	100[n]
Net efficiency	% (solar)	13.5	19.5	21.0	23.4	8	12.5	13.5	14.0
Reliability	%	90	95	97	99	95	96	97	98
Production[a]	GWh_e/a	0.07	1.09	22	226	2.5	25.4	76	255
Collector Area	10^3 m²	0.227	2.27	42	379	19	116	320	1,020
Costs (1987)									
Investment[b]	10^6 \$	0.75	3.25	38	154	16.0	81.9	125	160
Specific costs[c]	\$/kW	18,750	6,190	3,800	1,540	16,000	8,190	4,170	1,600
	\$/m²	3,300	1,430	900	406	842	700	390	160
Collector only	\$/m²	2,200	660	420	320	9.0	3.27	2.25	0.5
O&M	10^6 \$/a	0.05	0.125	1.5	7.4	0.32	1.2	1.3	1.6
	%/of Inv.	6.7	3.8	4.0	4.8	2.0	1.5	1.0	1.0

[a] Barstow weather (about 2,500 kWhm² direct and global irradiation)
[b] Without AFUDC
[c] 1.4 W_p/1 $W_{AC,net}$ (Photovoltaics)
[d] Based on modified Solar One
[e] Based on modified Phoebus
[f] Based on modified Utility Study 100
[g] Based on modified Utility Study 200
[h] Based on modified SEGS III
[i] Based on modified SEGS VIII
[j] Based on future LUZ-projects
[k] Based on modified 1986 data (reduced efficiency)
[l] Based on modified Siemens data (1992)
[m] Based on modified optimistic Bölkow data (1992)
[n] Based on modified DLR goal (= 1\$/W module)
[o] No significant improvements up to 2000

Observations which may be drawn from this comparison are:

- current specific investment costs range from about 4,000 $/kW$_e$ for parabolic trough SPPs to about 10,000 $/kW$_e$ for photovolatic SPPs;
- specific investment costs will gradually be reduced in the future;
- most of the thermal SPP alternatives have not yet operated in optimal plant sizes;
- annual system energy efficiencies range between 8–23.4% and will improve with time;
- output generating performance is dependent on operating availability/reliability;
- O&M expenditures are lowest for photovoltaic SPPs (automatic operation), and highest for dish/Stirling SPPs (maintenance of Stirling engines);

Based on the performance data of Tab. 10.10, levelized electricity generating costs have been calculated (2,500 kWh/m²a irradiation); results are illustrated in Fig. 10.19. The figures give rise to the following observations:

- future levelized electricity generation costs become comparable for all SPP alternatives, converging around 0.1 $/kWh$_e$ with optimal system size but without storage, and under 2,500 kWh/m²a of irradiation;
- generation costs in the 0.05–0.10 $/kWh$_e$ are achievable only with storage, the lower value only with hybridization;

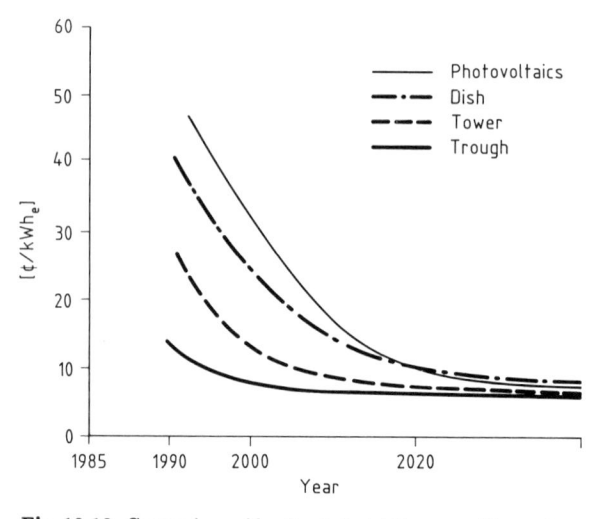

Fig. 10.19. Comparison of levelized electricity generating cost extrapolations for the four major SPP alternatives, operating under identical conditions (1987 values; 2,500 kWh/m²a direct or global irradiation).

Irradiation. Solar energy availability is crucial for the functional and economic performance of SPPs (see also Chap. 4). The data of Tab. 10.10 and Fig. 10.14 were calculated using Southern California (Barstow) irradiation data. However, few privileged regions of the world have such sunny weather, low cloud occurrence and clear air. Lower availability of direct radiation leads to lower annual energy output performance for concentrating SPPs (Tab. 10.11).

Non-concentrating photovoltaic SPPs using global radiation are equally dependent on radiation but less on the presence of direct radiation, i.e. on radiation quality. In addition, each type of SPP has a different input energy threshold for net output generation, and different stand-by parasitic energy needs (see Sect. 10.1).

10.3 Power Plant Cost Analysis and Comparison

Table 10.11. Variation of annual energy output performance of SPPs operating under Barstow, southern California irradiation conditions and Almería, southern Spain irradiation; results calculated, based on current (1990) technologies [31].

Barstow/Almería	Annual average efficiency [a] %			Energy output relationship Barstow/Almería [e]
	Barstow [b]	Almería [c]	Relationship [d] A/B	
Tower	11.0	8.8	0.80	1.70
Parabolic trough	9.0	7.4	0.82	1.65
Dish/stirling	19.5	17.0	0.87	1.56
Photovoltaics [f]	7.0	7.0	1.0	1.37

[a] Based on 1990 technology (same for both sites)
[b] Barstow weather 1985 (direct irradiation 2,585 KWh/m^{2}a)
[c] Almería weather 1983 (direct irradiation 1,896 KWh/m^{2}a)
[d] Relationship of Almería irradiation to Barstow irradiation
[e] Divisor for getting the energy output in Almería from Barstow data
[f] Global irradiation relationship Almería/Barstow assumed to be identical to direct irradiation relationship (0.73)

Local weather conditions, therefore, affect annual performance differently, depending on SPP type. Of all SPP types, large-capacity concentrating thermal SPPs are the most sensitive to irradiation and weather conditions, non-concentrating flat-plate photovoltaic SPPs the least sensitive.

For example, global and direct irradiation at Almería, southern Spain are each about 0.73 times that at Barstow. As a consequence, investment costs of a photovoltaic or a tower SPP would be about 1.37 times higher in Almería, provided the same plant output rating is to be maintained. Also taking into account that the presence of any input energy threshold results in overproportionally lower annual energy output performance at lower annual irradiation, investment costs for a tower SPP would be 1.25 times higher again in Almería than in Barstow. A non-concentrating photovoltaic SPP with low threshold, on the other hand, can still utilize the (higher) diffuse radiation portion at Almería. Hence, locating a SPP in Almería rather than in Barstow would lead to about 1.73 times higher investment costs for a tower SPP, but only to about 1.37 times higher investment costs for a photovoltaic SPP.

Storage and hybridization. Intermediate energy storage (in combination with SM > 1.0) and/or hybrid configuration permit SPP operation in cloudy/dark periods and extend the annual full-load operating hours. Output electricity generation can be better controlled and shifted to the on-peak load periods of a utility grid. Thus higher revenues, and often a higher capacity credit, can be gained. The economic optimum of storage capacity and SM depends on local climate and electricity rate structure. Assuming flat revenue for any electricity generated, minimum costs for a tower SPP can be expected with SM = 1.8 and a storage capacity equivalent to 4–5 h rated-power operation from storage (see Chap. 6). For the time being, it may be more cost-effective, however, to augment thermal SPP operation by small storage and a fossil burner. Critical points are the availability and cost of fossil fuel. As long as the cost of fossil-derived thermal energy is equal or less than the cost of solar-derived energy, hybrid operation is economically attractive.

Study results [54] corroborate that the addition of a fossil heater lowers the generating costs of a gas-cooled tower SPP over a wide range (within 1,800–6,000 h/a of full-load operation). On the other hand, studies based on conditions existing in the service area of the Arizona Public

Service Company in the U.S. contradict this finding [20]; in these circumstances, fossil-hybrid generation is less cost-effective than solar-only operation, due primarily to the relatively high local cost of natural gas (as compared to the low-cost coal and nuclear energy utilized in the utility grid).

Stand-alone photovoltaic systems often incorporate (electrochemical) battery storage for functional reasons. Such a configuration is not cost-effective for bulk energy storage in grid-connected configuration, either currently or in the forseeable future. This fact is the main disadvantage at the present time for bulk electricity generation by photovoltaic plants. Hydrogen, on the other hand, could eventually become the storage medium of choice for photovoltaic SPPs [77] (see also Chap. 9).

Plant size. The trend of the effect of power plant capacity on present (1988) and future (2020) levelized electricity generating costs is shown in Fig. 10.20 for conventional and solar power plants. These data are calculated, assuming mature solar technologies, reliable operation, and Southern California irradiation conditions. The following general observations can be made:

- electricity generating costs decrease with plant capacity (amount of decrease depending on plant type and regional conditions, and furthermore on the share of equal or similar modular components (i.e. PV module or heliostats) in the total plant investment);
- generating cost decrease is least for photovoltaic SPPs, and highest for thermal SPPs (highest economy-of-scale effects);
- at capacities < 0.01 MW_e, generating costs of photovoltaic SPPs are cost-competitive with those of conventional power plants even today;
- already in the medium term, generating costs of SPPs of all types show promise of becoming cost-competitive with those of conventional power plants (grid-connected operation mode).

The likelihood of the last observation is enhanced when considering that fossil fuel costs are expected to rise in the future, and that unforseen events are unaccounted which may affect the generating cost from, or the acceptability of, conventional supply sources (e.g. nuclear accidents, market or supply disruptions, environmental concerns). It must be kept in mind, however, that SPPs need long-term energy storage possibilities before they can functionally substitute conventional power plants entirely, even if they are already proving to be cost-effective in a fossil fuel saving operating mode.

10.3.4 Social Costs

For a true cost/benefit picture of energy supplies on the level of a national economy, it is necessary to look not only at internal but also external costs. The internal costs are those a producer must bear, or which are charged to the consumer. The external (or follow-up) costs are those which must be borne by the national economy (e.g. effects on employment rate or on national income) as well as the costs for avoiding, reducing or repairing environmental damages.

Only in certain cases is it possible to identify and quantify the external cost for electricity generation from conventional and/or renewable energy sources. Typical internalized costs are [50,79]:

- Desulfurization of a coal-fired plant   0.3–1.3 c/kWh
- Denitrogenization of a coal-fired plant   0.2–0.5 c/kWh
- Additional cost of dry vs. wet cooling   0.4–0.5 c/kWh

10.3 Power Plant Cost Analysis and Comparison

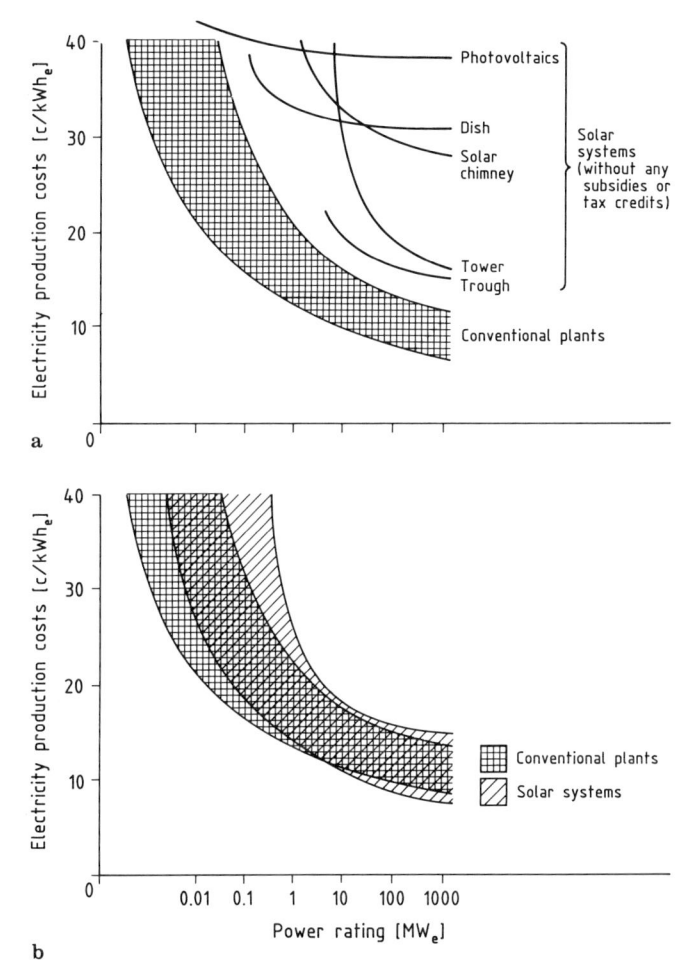

Fig. 10.20. Levelized electricity generating cost as a function of plant rating for conventional and solar power plants, (a) market and cost structure of electricity production costs versus installed plant sizes (1988), (b) market and cost structure of electricity production costs versus installed plant sizes (2020).

– Nuclear fuel reprocessing	0.5–5.0 c/kWh
– Dismantling of nuclear power stations	0.1–5.4 c/kWh

The cost of plant desulfurization or denitrogenization is fairly well known and can readily be internalized through regulation, forcing a higher investment. On the other hand, the assessment of cost for waste management and dismantling of nuclear power plants is subject to the influence of opinion or ideology, making internalization of these costs difficult. In other cases, either the cause/impact relationship is unknown, or effects such as environmental damage, health hazards, reduced quality of life, reduction of species of the flora and fauna, damage to forests, losses due to temperature changes, etc. have not been monetized. The result is that the principle cannot be enforced that originators of pollution shall defray the costs for damages.

Table 10.12 summarizes the results of an investigation which internalizes the external costs of electricity production in Germany [51]. The estimated cost ranges are based on $/kWh$_e$ of electricity produced in 1984 (price based on 1982). Without the inclusion of the macroeconomic effects of a modified gross value added, or of changes in saving and employment rates, a range of 0.02 to 0.045 $/kWh$_e$ for fossil fuel and 0.005 to 0.105 $/kWh$_e$ for nuclear electricity has been determined [51]. The weighting of these external costs relative to the share of electricity generated from fossil and nuclear energy results in average external costs of 0.025 to 0.065 $/kWh$_e$ for electricity produced in Germany in 1984. The cost of electricity production would be increased by more than 50% if the upper estimate of external costs is used as a basis. Other studies [25,26,42] show lower or higher values. In reality, most of the external impact when using fossil and nuclear sources is known only to a degree, or is not really quantifiable, for instance the effect of global warming associated with increased CO_2 levels.

External effects which cannot be monetized have to be internalized by government intervention in the market, since neglecting to internalize could lead to the misuse of funds and resources and serious losses in the achievable level of social welfare. Internalization leads to price escalation of conventional energy systems, increases the attractiveness of energy conservation, and increases competitivness of the renewable energies, and thus leads to higher end-use efficiency.

Table 10.12. Gross and average gross external costs of electricity generated from fossil and nuclear energy in Germany, excluding certain economic effects which are calculated as net effects for renewable energy sources (in cents/kWh$_e$; 1982 values; all figures are estimated minimal values) [51].

Gross external costs of electricity from	fossil fuels	nuclear energy
1. Environmental effects	0.57–3.05	0.6–6.0
2. Depletion surcharge (1985)	1.15	2.95–3.12
3. Goods and services publicly supplied	negligible	0.05
4. Monetary subsidies (incl. accelerated depreciation)	0.16	0.07
5. Public R&D transfers	negligible	1.17
Total	1.93–4.4	4.85–10.4

Average gross external costs of electricity due to	
1. fossil fuels (weighting factor 0.744)	1.43–3.28
2. nuclear energy (weighting factor 0.256)	1.24–2.66
Total	2.68–5.94

10.4 Market Considerations [14]

10.4.1 Introduction

For controlling global warming, the world conference 'The Changing Atmosphere: Implications for Global Security' closed in June 1988 in Toronto with the appeal to reduce worldwide CO_2 emissions by at least 20% by the year 2005. This goal can be attained quantitatively and in time only by an immediate stronger implementation of all possibilities for efficient energy use, and by the exploitation of available renewable energy sources.

Under such time constraints, only new technologies close to market introduction can be counted on to alleviate the situation, i.e. technologies which already have left the research stage and which have sufficiently demonstrated operating reliability and availability. Beginning market introduction is a presupposition for higher production rates and, therefore, cost reductions. Several times in the past, market introduction proved premature because of inadequate performance and cost-effectiveness of a new technology under real market dictates. The highest demand for new energy supplies exists today in the newly industrialized and the developing countries. Most of these countries are located within the sunbelt of the earth.

Investment in the energy supply infrastructure is a dominant economic factor for each country. Excepting countries with centrally planned economies (CPEs), these investments amount worldwide to about $320 \cdot 10^9$ \$/a today, which represents about 11% of the world's annual capital investments. Of this investment, about 60% stems from private sources, the remainder is defrayed by the public sector. The World Bank estimates the Third World demand for 80 GW per year for the next decades. Nearly 20 GW are needed in countries with an annual insolation of more than 2,000 kWh/m^{2}a of which solar plants could supply a large share.

Another important consideration for the assessment of an energy supply option is who owns and operates a power plant facility. SPPs of any size may be owned and operated in three different ways, i.e.

1. the SPP is privately owned and operated and the
   - electricity generated is self-consumed by the owner; a grid connection does not exist (utility-independent operation);
   - electricity generated is self-consumed; a grid connection exists and any excess demand is provided by the utility (fuel-saving mode of operation);
   - electricity generated is self-consumed; a grid connection exists, any excess demand is provided by the utility, but any surplus electricity generated is sold to the utility;
2. the SPP is owned and operated commercially, by utilities or by independent operators, and electricity generated is fed solely into the utility grid; in this case, SPPs add generating capacity in a mix with conventional generating plants;
3. the solar installation is owned and operated by utilities but is located on the premises of the utility customer; electricity generated is fed solely into the utility grid ('roof rent').

The amount of savings achieved from auto-generation and self-consumption is determined by the rate structure (for electric energy delivered and for generating capacity provided) applicable for the service area of the specific utility. The rate structure may differ between utilities, and, within one utility, for private households, commercial or industrial end-users. Concerning large SPPs, utility ownership and operation are the most relevant option for

---

[14] With contributions from Gerald W. Braun, Pacific Gas & Electric Co., San Ramon/CA.

utilities; consequently, information about SPPs is of interest for utility managers and engineers more than for any other group.

Assuming the traditional role of utilities in the energy economies of sunrich countries is maintained, this group of professionals will plan, specify, purchase, own and oversee operation of SPPs, and perhaps also design and manage their construction as well. Sometimes these functions are burdened by regulatory limitations or a strong influence of governments, predominantly in developing countries.

There are a few key considerations in the decision making process of a utility. First, there is the obligation to meet the customers' demand for power at every instant of every day of every year. Secondly, there is the imperative to do so at minimum costs, and thirdly to maximize value. Fourthly, there are no rewards for taking risks in either regard.

Necessarily then, utility plans are conservative, the planning methods thorough. The planning is concerned with costs, all the costs incurred over the lifetime of a power plant. The planning is concerned with value, the value of a SPP measured by the cost of fuels saved and of other types of capacity that are avoided by building it. And it is concerned with risks; risks that costs will be greater than estimated, risks that value will be compromised by failure to deliver energy in accordance with prediction, risks that the financing strategy goes wrong. Risks can never be reduced to zero. Risk is always a function of costs, value and financing strategy, and value is always interrelated with costs. Thus, the issue is optimization; optimization of the SPP based on independent variables like the solar resource, material and labor costs, etc.; optimization of the whole utility generating system to make best use of each power plant; achieving high reliability and stable, predictable costs through multiple power plants and resource diversity where possible.

10.4.2 Costs

Cost include initial costs and recurring costs (as for fuel and operations). Costs depend on plant size, site and concept. If the utility system is small, smaller plants have greater capacity value than larger plants. If a large plant is economic on a pure basis, its risk-adjusted cost may be significantly higher if there is no prior experience in the larger size range. Proper siting plays a critical role with all renewable options. The same amount of energy can, in principle, be captured at less cost where resource quality is high, but if the high quality site is remote from loads and infrastructure, higher construction, energy transmission and operating costs may be incurred. Site specific costs depend on choice of concept, i.e. the cost for flood protection, adequate seismic design, environmental compatibility, wind survival, etc..

If a series of plants is planned, economics achievable through learning, sequential construction, design standardization and bulk purchase of materials become important considerations. Another cost issue is associated with the incorporation of a storage or hybrid capability (for maximizing value). As discussed previously, these capabilities provide flexibility in relation to long-term changes in the utility system and its operating costs. It is important, therefore, to know fixed and variable costs for storage, and to carefully assess the long-term costs of fuel for hybrid operations.

10.4.3 Value

Aggregate customer demand varies with time of day, time of week and time of year, *when* solar power is delivered to the grid is comparable in importance to *how much*. The value of the energy delivered is thus time dependent, and this value is primarily determined by

10.4 Market Considerations

- the cost of the fuel that would be consumed to meet load if the SPP were not operating,
- the cost of additional power plants of other types that would have to be built to assure comparable levels of overall system reliability if the SPP did not exist.

For example, a naturally dispatching SPP would have a high value if

- it tended to be available during periods of maximum aggregate demand, thus reducing fuel consumption in the relatively inefficient power plants that are brought on line only during these peak periods,
- the utility needed to build power plants to meet increased demand, and the SPP can be expected to generate consistently during periods of maximum demand,
- in the future, the environmental threat from operating conventional plants of the utility system becomes unacceptably high.

Conversely as long as environmental considerations are only of minor public/political concern, a SPP would be much less valuable to a utility that needed no new generating capacity, had access to inexpensive fuel, and experienced peak demand periods at night. This means, a utility interest in power generation with SPPs will depend heavily on the extent of coincidence between SPP output and aggregate demand. This match depends strongly on local temporal solar resource characteristics and specific SPP design. The optimization of a SPP design for a maximum benefit-to-cost ratio is an interactive process involving long-term simulation of the operation of both SPP and the utility system as a whole. Models of the SPP and the utility system that permit this simulation are therefore of particular interest. Some other key variables that affect this optimization include the

- mix of other power generation resources in the utility system,
- integration of storage into SPPs,
- proportion of SPP capacity in the overall utility system,
- fuel costs for hybrid operation of SPPs and for operation of the other power plants.

Usually, all these factors are the basis for the formulation of power purchase agreements which contain the conditions for payments by utilities for electricity delivered to them. In these agreements, revenue payments for electricity delivered and for capacity provided by the SPP are distinguished. These payment factors differ between utilities and vary with season and the daily hours during which electricity is generated and delivered.

Electricity payments. The cost for generating electricity varies with demand over each day. For these reasons, utilities distinguish typically two or three periods in each day and associate different rates with each period: the on-peak period corresponding to the hours of each day when demand is greatest, the mid-peak period when demand is intermediate, and the off-peak period when demand is lowest. The tariff structure is further divided into Summer and Winter rate periods (Fig. 10.21).

The way the rates, i.e. the costs avoided by utilities by not generating but buying electricity (avoided costs), are determined differs widely between countries. In Germany, these costs are the average of production costs of the whole plant park of a utility; in the U.S., avoided costs are calculated in a complex formula stipulated by the Public Utility Regulatory Policies Act (PURPA) of 1978. This PURPA act establishes the right to produce power from renewable energy sources and from cogeneration facilities independent of utilities, and to sell power at avoided costs to utilities. Avoided cost is defined therein as 'the cost of energy to the utility which, but for the purchase, the utility would generate itself or purchase from another source' [14].

When new energy technologies are introduced into the market, data from expensive oil/gas plants have to be used to calculate the avoided costs. However, as more oil/gas is displaced,

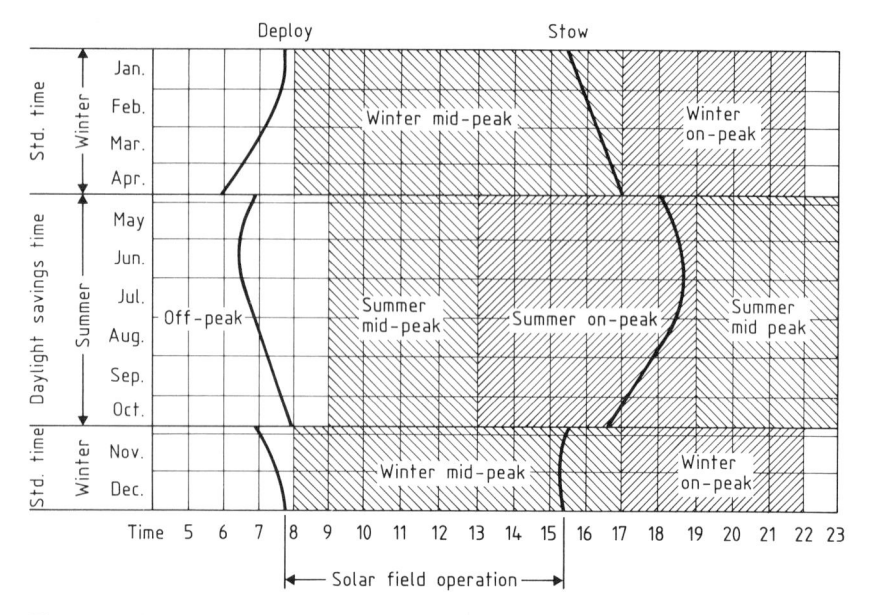

Fig. 10.21. Contract rate periods for the parabolic trough plants SEGS I-II in California [2]

the advantage of avoided-cost situations will become less significant and the allowable costs of renewable systems will drop [32], accompanied by decreasing tax credits, decreasing fuel costs and only slightly growing demand.

The difference in rates reflects also the different plant types and fuels used to generate electricity during each rate period. Commonly, baseload plants are run continuously day and night. The cost of electricity generated by peaking plants is often twice or more the cost of electricity generated by baseload plants, primarily because peaking plants characteristically operate with more expensive fuels and often with lower efficiency than baseload plants [21].

Capacity payments. Two reasons may account for interruptions in delivering electricity by utilities – forced outages and scheduled shutdowns of generating plants for service and maintenance. SPPs are burdened, in addition, by the unavoidable intermittance of solar radiation, necessitating a corresponding reserve generating capacity in the system (reserve margin). The capacity payment reflects the capital equipment cost that a utility may be able to avoid by being assured of power delivery by the independent producer, particularly during peak load periods. Capacity payments are therefore determined on the basis of availability or dependability of electrical power supply consistently during peak demand periods independent of season. If the supply of electricity were uncertain, the utility would have to provide reserve generation capacity within its own system, or would have to buy it from neighboring utilities. Since, apart from outage due to the weather, each SPP has additionally a specific outage probability, a cumulative availability curve for the whole utility system must usually be established for calculating the correct utility-specific capacity payment which can be offered in purchase contracts.

10.4.4 Financing

Financing of large power plants becomes increasingly difficult for utilities, especially in developing countries. Nuclear power plants with a capital requirement of more than $2\text{--}6\cdot10^9$ \$ each over a construction period of nearly ten years cannot be financed conventionally any more, i.e. by financing largely with equity and covering investment risks by assets [62]. Recently, nearly every new plant has been financed by a different method [33]. A typical capital requirement for a large tower SPP is currently $\approx$ 150 Mio\$ (30 MW$_e$). Total investment and the time needed for construction have decreased over the past years, but SPP lifetime, reliability and yearly production performance are not yet well established parameters. Hence, SPPs still represent considerable investment risks, even if revenues are calculated to exceed costs.

For any commercial SPP venture, a detailed feasibility study provides the technical, economic and commercial conditions under which the project can be carried out. The feasibility study defines and analyzes the critical elements such as plant size, production capability at a specific location, investment and production costs, and the revenues from sales, yielding a defined return on investment (ROI) related to one SPP configuration, or forming the basis for a trade-off between plant alternatives at the same site [6,11,28,41,47].

Some key items in the financing of a SPP are [35]:

- *Institutional, governmental, regulatory and utility incentives.* Incentives of some form are still necessary to accelerate the commercialization process of SPPs and to set the process for cost reduction into motion. Incentives may take the form of direct subsidies, favorable depreciation methods, federal and state tax benefits, acceptance of favorable rates for electricity by regulatory bodies, absorption of a portion of O&M costs by third parties, outright grants by governments or institutions, grace periods for capital repayment, loan guarantees, and demonstration programs undertaken by goverments.
- *Favorable long-term utility contracts.* It is difficult to predict the long-term avoided costs of a utility system. With uncertainty about future energy and capacity payments, long-term power purchase contracts become essential to enable third-party investors to quantify risk and the ROI position on a proposed project more accurately. Such power purchase agreements may, in addition, offer energy and capacity payments above the current avoided costs to provide revenues for adequate ROI. If government-backed, the value of purchase agreements to the investor would be even higher.
- *Financing strategy.* Non-conventional financing strategies include, for example, third party financing, corporate financing, multi-source financing or export financing. However, even simple financing strategies may pose, for various reasons, difficulties in many countries. Furthermore, the usually large number of participants in non-conventional financing schemes require more elaborate legal and administrative provisions which may render such financing strategies very expensive. For new first-of-its-kind technologies, the operational performance risk increases the financial risk. This often means that 'cash-flow-related lending' becomes difficult, and that the greatest portion of risk must be borne by the project proponents, i.e. the suppliers and constructors. Of course, for a given SPP technology the financing prospects improve with each additional plant in operation and the experience gained. An additional difficulty is that the financing strategies are usually not transferable from one country to another. The economic conditions between countries frequently vary, i.e. the taxes, tax structure, energy price levels, inflation rates, currency exchange rate and convertibility may differ.

Due to their more advanced stages of development and maturity, only parabolic trough SPPs and, with reservation, tower SPPs seem to be 'commercially' viable at present. Smaller

PVs and dish/Stirling SPPs may eventually also become commercially viable in selected circumstances.

The SEGS parabolic trough SPPs are, for example, third-party financed projects. The majority of funds is provided by institutional and corporate entities seeking investment opportunities, taking advantage of federal tax benefits and receiving the major portion of the cash flow derived from the operation of the plant. The second-largest investment segment are shares that are sold to private investors seeking primarily Californian tax benefits but receiving only minimal cash flow from the project [52]. To compensate for risk in cash flow in the early SEGS plants, a large investment portion was financed with equity. The interest rate for the borrower has been reduced continuously from one plant to the next because probability of performance improved as predicted, and confidence in the new technology increased. The LaJet Solarplant dish/farm project was financed by a limited partnership offer. Financing of the 30 MW_e Phoebus tower SPP project relies on an expected government subsidy amounting to nearly 50% of investment cost, while revenues from electricity generated are expected to finance most of the rest (project financing); a small remainder of investment is expected to be borne by the supplying industry.

10.4.5 Risk

An inherent risk for an investor is always associated with a larger investment and a long amortization planned. Long-term planning capital investment is customary for utilities because system expansion (expecially for generation) is extremely capital intensive, involving ROI periods measured in decades rather than years. A healthy and stable financial situation of a utility increases its credit worthiness and hence reduces its cost of capital, thus minimizing its cost of operation.

For new investments, the risk analysis is strongly correlated with the financing strategy. Assuming identical lifetime-based electricity generating costs, the return-of-investment period – and therefore the investment risk – for a fossil plant would be much lower than for SPPs because the time differences between investment and the rate revenues accumulate. A higher rate for the return on investment allows higher technical and/or performance risks to be acceptable which, contrary to need, could be higher for fossil power plants than for SPPs. Utilities deal with risks in two ways:

1. relying on proven techniques, and
2. quantifying uncertainties.

For example, a utility wishing to consider a SPP would first look to designs that had been successfully deployed and operated by other utilities. Likewise, an utility seriously contemplating SPPs would gather long-term irradiation data at likely sites in order to quantify the resource, reducing that risk to as low a value as possible. In the same vein, utilities will have considerably more interest in what costs and performance are guaranteed to be, rather than in what they are estimated to be. Without such guarantees, utilities will look to experience for validation of estimates. Actual initial costs and long-term performance of identical or highly comparable plants are of much more interest than actual costs and short-term performance of sub-scale prototypes.

Utilities gain much from each other's experience and tend to subject vendor claims to careful testing before assuming them to be valid. Their low tolerance for risk leads to a preference for incremental and only modest deviations from past practice. For example, limited scaling-up or technology changes in a key component would be acceptable, provided adequate testing had occurred. Introducing multiple new technologies along with major scale-ups would definitely not be acceptable.

Thus, experience, testing and vendor guarantees are the primary risk mitigation tools available to utilities. Since the dominant factor in SPP cost is the initial cost, and the dominant factor in value (typically) is energy production, utility interest will focus on experience, test results and guarantees, in this sequence.

10.4.6 Market Potential and Outlook

The most likely next stages in the evolution of the SPP technologies may be sketched in short terms as follows: The *tower SPP*, like all other concentrating thermal SPP technologies, requires an annual direct irradiation of $>1,800$ kWh/m^{2}a. A demonstration plant in the power range of 30–100 MW$_e$ would be needed to advance beyond the 10 MW$_e$ Solar One experiment and to demonstrate successful operation at near utility-scale with a high degree of reliability projected over a lifetime of 10 to 20 years. If, by this action, progress in the reduction of cost for components and, by automation, of O&M are achieved, and if plant parasitic energy needs and operating thresholds are lowered, commercialization of this plant type could start. The *parabolic trough* SPPs are mature and commercial. Further technology improvements are centered on direct evaporation and superheating of water in the receiver tubes (by enlarging the trough aperture and increasing CF and temperature $> 400°$C), as well as on inclusion of storage in the system to reduce the fossil fraction in generation (and the external costs).

In the sunbelt, the trough and tower SPPs will compete for the same market. The tower SPPs seem to be the choice for plant capacities $\geq$ 100 MW$_e$ and locations at considerable distance from the equator; the reverse holds true for trough SPPs. A utility system would have to have an installed capacity and peak load of more than 500 MW$_e$ in order to accommodate SPPs with a rating of $>$ 100 MW$_e$. With excellent direct irradiation conditions the energy cost for both systems may, for SPPs in the range of 50–100 MW$_e$, become as low as 0.1 \$/kWh$_e$; however, achieving such low costs depends strongly on plant rating and system operating mode, i.e. solar-only operation, or operation with storage and/or a hybrid source.

The marketing advantage of the *parabolic dish/Stirling* system is the small size of each modular unit. However, present investment costs are higher than for the other thermal alternatives, with considerable uncertainty about future investment and O&M costs, and about adequate reliability of the Stirling engine. The lowest-cost rating of SPPs consisting of a multitude of dish/Stirling units seems to lie between 500 kW and 10 MW, thus more readily being able to complement small existing energy supply systems in remote areas where, on the other hand, there is a probable shortage of skilled professionals. For plant units $>$ 10 MW$_e$, the economies in plant scale become poorer than for the other thermal alternatives due to the modular character of the technology.

Photovoltaic technology is close to maturity and its commercialization benefits from the application of PV modules in small power supplies. Nonetheless, the future cost situation for large photovoltaic SPPs appears uncertain in spite of steadily continuing basic R&D in cell development and advances in cell technology. The open question is whether these advances can be translated into dedicated cell and module volume production facilities for bringing module costs down. Stand-alone electricity supplies in remote or rural areas most probably are the earlier commercial market opportunities for photovoltaic SPPs of multi-kW size. As module prices fall, also plant sizes $>$ 1 MW$_e$ without storage capability may become commercially attractive. Irrespective of size, the advantage of photovoltaic SPPs persists, i.e. their ability to utilize also diffuse irradiation, and their low O&M needs and costs.

In a grid-connected utility environment, revenue considerations (in relation to cost and risks) will remain the dominant factor in thermal SPP investment decisions. If recent perfomance calculations are any guide, the revenue position improves for thermal SPPs with SM > 1.0 and with storage, i.e. higher revenues can be achieved through higher capacity payments and by shifting generation to peak demand/high tariff periods while specific generating costs remain equal or are even lower. As long as fossil energies are low-priced, additional revenues from capacity credits may be gained by incorporating fossil input sources in hybrid SPP configurations, requiring only $\approx 4\%$ additional investment in thermal SPPs.

Current world primary energy needs are about 12 TWa/a. Today, more than two-thirds of total primary energy is used by the industrialized countries. Of this total, renewable energies account for only a small fraction (about 7% hydropower and 10% non-commercial energy, most of it biomass used as fire wood). The fraction attributable to solar is currently negligible, except for certain regions such as California. Future global energy demand will be determined by world population growth, the rise in the standard of living, and the primary energy saving efforts. Different projections assume an energy demand between 11 and 28 TWa/a by 2020.

An important consequence of these scenarios is the structural change between the developing and industrialized countries. Many developing countries are located in the sunbelt, and it can be expected that solar energy will play a growing role in their energy supply future. The developing countries primarily need reliable SPP technologies, and the greater part of SPP investments would have to be supplied by the country itself.

In the industrialized countries, the current energy market is characterized by strong competition, especially due to a predicted rather stable energy demand in the near future. Therefore, a 'newcomer' must be better in every respect than, or at least equal to, the established energy supply competitors (fossil, nuclear). Indirect solar energy sources alone (wind, biomass, hydropower, geothermal, etc.) do not have the potential to provide the total global primary energy demand. Therefore, the direct utilization of solar energy is inescapable if the future energy supply is to be shifted towards renewable energy sources. The amount of solar energy available to the continental land areas is about 25,000 TWa/a; hence, merely about 0.4% of the Earth's surface would theoretically suffice to supply the whole world with primary energy. The market for large SPPs is commonly seen in grid-connected operation augmenting a centralized electricity supply structure. For political and infrastructural reasons, a decentralized electricity market would seem preferable but would be of less significance in relation to global primary energy needs. In developing countries, most power plants are fossil-fired, and one driving motive for the development of SPPs is to replace the fossil plants in these countries. If CO_2 emissions have to be restricted in the long term, SPPs have to substitute fossil power plants in other countries as well.

The SPP market for the four technologies discussed in this book is growing. Long-term energy production costs are favourable for all technologies. Whether SPPs will become commercial in a specific country depends not only on further technical advances and entrepreneutral vigor but even more on forward-looking national energy policies in countries which properly take account of the total costs of conventional energy supplies, and consequently support the increased exploitation of renewable and ecologically benign energy sources in concert with a responsible and efficient end-use of energy.

Bibliography

[1] *Under the Sun, the Newsletter of LUZ International Ltd.*, 4 (1988)
[2] Flachglas Solartechnik, private communication, Köln (D), 1987/88/89
[3] SPS-Stirling Power Systems Corporation, private communication, 1987
[4] Presentation Material, Ann Arbor/MI, Stirling Power Systems Corp. (SPS), 1987/88
[5] *Annual Progress Report: Photovoltaics, FY 1986*. DOE/CH/10093-H3, 1987
[6] *Contents, Set-up and Data Requirements for the Various Investment Studies for Industrial Projects*. Frankfurt a. M. (D), Kreditanstalt für Wiederaufbau (KfW), 1975
[7] *Design of a PV Central Power Station*. Technical Report SAND82-7149, Denver/CO, Martin Marietta Corp., 1984
[8] Five Year Research Plan, 1987–1991: USA's Energy Opportunity. National Photovoltaics Program, 1987. DOE/CH/10093-7
[9] *Generating Costs of Large Nuclear and Coal Power Plant Units Starting Operation in 1990* (in German). Technical Report, Frankfurt a.M. (D), Vereinigung Deutscher Elektrizitätswerke (VDEW), 1987
[10] *International 30 MW_e Solar Tower Plant, Feasibility Study – Phase I: Presentation of Results*. Technical Report, Madrid, Miner/DLR, 1987
[11] *Manual for the Preparation of Industrial Feasibility Studies*. UNIDO/United Nations, New York, 1978
[12] *METAROZ: Study about the Possibilities of a Solar-Thermal Power Plant in the Metaroz Valley* (in German). Technical Report, SOTEL Consortium, Baden (CH), 1984
[13] *Photovoltaics – A Comment on the Costs of Crystalline PV-Modules in Large-Volume Production, and of Electricity Generated with PV Power Plants* (in German). Technical Report, Siemens AG, 1988
[14] Public Law 95-617, Public Utility Regulatory Policies Act of 1978 (PURPA), Sect. 210
[15] *Relative Cost of Electricity Production*. Technical Report P 300-86-006, California Energy Commission, 1987
[16] SEGS-1, Solar Electric Generating System–1. Fact Sheet by The Electric Energy Information Center, Daggett/CA, 1986
[17] Solar Thermal Technology Annual Report, Fiscal Year 1986 (prepared for U.S. DOE). Sandia National Laboratories, Albuquerque/NM and Livermore/CA; Solar Energy Research Institute, Golden/CO, 1987
[18] *Summary Report on Solar Hydrogen Energy Economy*. Ad-hoc-Committee appointed by the BMFT, Bonn (D), 1988
[19] *Technology Program GAST: Gas-cooled Solar Tower Power Station* (in German). Technical Report, Bergisch-Gladbach (D), Interatom GmbH, 1988
[20] *Utility Solar Central Receiver Study*. Midterm Review April 1987; Final Review, August 1987; Summary Report (Draft), 1987
[21] The Value of Reliability. *EPRI Journal*, (1985)
[22] *World Energy Conference 1983: Energy 2000–2020 World Prospects and Regional Stresses*, London: J.-R. Frisch, Graham & Trotmann, 1983
[23] World Watch Paper 81. World Watch Institute, Washington/DC, 1988
[24] Alvis, R. L.: *Some Solar Dish / Heat Engine Design Consideration*. Technical Report SAND84-1698, Sandia National Laboratories, Albuquerque/NM, 1984
[25] Awad, A. H.; Veziroglu, T. N.: Environmental Damage due to Fossil Fuel Use. *Int. J. Hydrogen Energy*, 15 (1990)
[26] Awad, A. H.; Veziroglu, T. N.: Hydrogen versus Synthetic Fossil Fuels. *Int. J. Hydrogen Energy*, 9 (1984) 355–366
[27] Becker, M.; Dunker, H.; Sharan, H.: *Technology Program "Gas-Cooled Solar Tower Power Station (GAST)", Analysis of its Potential* (in German). T 86-087, Bonn (D), BMFT, 1986
[28] Binz, R.: New Instruments for Export Financing (in German). *Sulzer News Bulletin*, (1986)
[29] Braun, G. W.; Doyle, J. F.; Iannucci, J. J.: Solar Power Receives New Impetus in USA. *Solar Energy*, (1986) 63–69
[30] Butler, B. L.; Beninga, K.; Royval, P.: Stressed Membrane Research – SAIC. In *Proc. 8th ASME Solar Energy Div. Conference, Anaheim/CA*, Ferber, R. R. (Ed.), American Society of Mechanical Engineers, New York: 1986
[31] Chiang, C. J.: *SUNBURN: A Computer Code for Evaluating the Economic Viability of Hybrid Solar Central Receiver Electric Power Plants*. Technical Report SAND86-2165, Sandia National Laboratories, Albuquerque/NM, 1987
[32] DeMeo, E. A.; Taylor, R. W.: Solar Photovoltaic Power Systems: An Electric Utility R&D Perspective. *Science*, 224 (1984)

[33] Diehl, R.: Financing of Nuclear Projects (in German). *Atomwirtschaft*, (1983) 631–636
[34] Doane, J. W.; O'Toole, R. P.; Chamberlain, R. G.; Bos, P. B.; Maycock, P. D.: *The Cost of Energy from Utility-Owned Solar Electric Systems*. JPL 5040-29, Pasadena/CA, Jet Propulsion Laboratory, 1976
[35] Doyle, J. F.; Box, P. B.; Weingart, J. M.: *Solar Thermal Central Receiver Integrated Commercialization Analysis*. Technical Report SAND86-8176, Sandia National Laboratories, Livermore/CA, 1986
[36] Droher, J. J.: *Performance of the Vanguard Solar Dish-Stirling Engine Module*. Technical Report EPRI AP-4608, Electric Power Research Institute, Palo Alto/CA, 1986
[37] Epstein, M.: Problems of Mass Production. In *Solar Thermal Central Receiver Systems*, Becker, M. (Ed.), Berlin, Heidelberg, New York: Springer, 1986
[38] Falcone, P. K.: *A Handbook for Solar Central Receiver Design*. Technical Report SAND86-8009, Sandia National Laboratories, Livermore/CA, 1986
[39] Faninger, G.; Bucher, W.; Geyer, M.; Klaiss, H.; Thornton, J.: *A Model for the Economic Assessment of Solar Power Plants*. DOE Report /DR/00789-T109, Stuttgart (D), 1984
[40] Fortgang, H. R.; Mayers, H. F. Technical Report JPL Publication 80-42, 1980
[41] Francois, P.: Marketing Strategies and Financing of Solarthermal Power Plants in the USA. In *Technical and Economic Aspects of Solar Energy* (in German), Aringhoff, R. (Ed.), Regensburg (D): Transfer Verlag, 1987
[42] Friedrich, F. R.; Kallenbach, U.; Thoene, E.; Voss, A.: *External Costs of Electricity Generation* (in German). Frankfurt (D): VWEW Verlag, 1989
[43] Geyer, M.: *High-Temperature Storage Technology* (in German). Berlin, Heidelberg, New York: Springer, 1987
[44] Geyer, M.; Klaiss, H.: 194 MW of Solar Electricity with Trough Collectors (in German). *Brennstoff-Wärme-Kraft*, (1989) 288–295
[45] Geyer, M.; Tamme, R.; Klaiss, H.: High Temperature Thermal Storage in Solar Plants. In *Proc. 2nd IEA-SSPS Task IV Status Meeting, Denver/CO*, 1987
[46] Goldemberg, J.; Johansson, T. B.; Reddy, A. K. N.; Williams, R. H.: An End-Use Oriented Global Energy Strategy. *Annual Rev. Energy*, 10 (1985) 613–688
[47] Grosse, P. B.: Project Financing – Application Possibilities for the Financing of Investments in the Energy Sector (in German). In *Technical Investments in the Energy Sector – Economics and Financing* (in German), Düsseldorf (D): VDI-Verlag, 1988
[48] Hansen, J.: *TDSA Project Joint Test and Operation (August 1986–1989 / 1986–December 1989)*. Technical Report, Köln (D), DLR, 1990
[49] Hansen, U.: Comparison of Fossil, Nuclear and Solar Generating Costs (in German). In *Technical and Economic Aspects of Solar Energy* (in German), Aringhoff, R. (Ed.), Regensburg (D): Transfer Verlag, 1987
[50] Hansen, U.; Mussenbrock, K.; Schoen, R.: Reduction of Toxic Emissions from Power Plants – Cost Modelling (in German). In *Documentation Flue Gas Scrubbing* (in German), Düsseldorf (D): VDI-Verlag, 1985
[51] Hohmeyer, O.: *Social Costs of Energy Consumption*. Berlin, Heidelberg, New York: Springer, 1988
[52] Kearney, D.; Jaffe, D.: Bright Future for Californian Solar Plants. *Modern Power Systems*, (1988)
[53] Kiera, M.; Meinecke, W.; Wehowsky, P.; (M. Geyer); (J. Kern); (H. Klaiss): *Comparison Study of Solar Tower and Farm Power Plants* (in German). Bergisch-Gladbach (D), Interatom GmbH, 1990
[54] Klaiss, H.; Geyer, M.: Economic Comparison of Solar Power Electricity Generating Systems. In *GAST – The Gas-Cooled Solar Tower Technology Program*, Becker, M.; Boehmer, M. (Ed.), Berlin, Heidelberg, New York: Springer, 1989
[55] Klaiss, H.; Nitsch, J.: *Renewable Sources of Energy for Baden-Württemberg: The Import of Solar-Generated Energy Carriers* (in German). Technical Report, Stuttgart (D), DLR, 1987
[56] Klaiss, H.; Nitsch, J.; Geyer, M.: Economic Analysis of Large Solar Tower Power Plants (in German). *Spektrum der Wissenschaft*, (1988)
[57] Klaiss, H.; Nitsch, J.; Geyer, M.: Solar Tower Power Plants – An Option for the Export Market (in German). *Zeitschrift für Energiewirtschaft*, 1 (1986)
[58] Kolb, G.; et al.: Evaluation of the Molten Salt Thermal Storage System of CRTF. In *Proc. 2nd IEA-SSPS Task IV Status Meeting, Denver/CO*, 1987
[59] Krawiec, F.; Thornton, J.; Edesess, M.: *An Investigation of Learning and Experience Curves*. Technical Report SERI/TR-353-459, Solar Energy Research Institute, Golden/CO, 1980
[60] Luque, A.: Photovoltaics in 1986 – Routes to Low Cost. In *Proc. 7th E.C. PV Solar Energy Conference, Sevilla*, Goetzberger, A.; Palz, W.; Willeke, G. (Ed.), pp. 9–18, D. Reidel, 1987
[61] Magid, L. M.: Prospects for Large Power Plants. In *Proc. 5th E.C. PV Solar Energy Conference, Athens 1983*, Palz, W.; Fittipaldi, F. (Ed.), pp. 437–444, Dordrecht (NL): D. Riedl, 1984

[62] Mueller, W. D.: Financing – a New Bottleneck (in German). *Atomwirtschaft*, (1976) 397
[63] Neal, R. W.; DeDuck, P. F.; Marshall, R. N.: *Assessment of Distributed Photovoltaic Electric Power Systems*. Technical Report EPRI AP-2687, Electric Power Research Institute, Palo Alto/CA, 1982
[64] Norris, H. F.: *Total Capital Cost Data Base – 10 MW_e Solar Thermal Central Receiver Pilot Plant*. Technical Report SAND86-8002, Sandia National Laboratories, Livermore/CA, 1986
[65] Palz, W.: Photovoltaic Power Generation – a 1987 Review. *Solar Energy*, 5 (1988)
[66] Pauschinger, T.; Schiel, W.; Klaiss, H.: *Economic Assessment of Solar Dish-Stirling Facilities* (in German). IB 444 007/88, Stuttgart (D), DLR, 1988
[67] Raeuber, A.; Holland, M.; Holder, B.: Photovoltaic Generation of Energy (in German). In *Energiegutachten Baden-Württemberg*, Freiburg (D): FhG-Institut für Solare Energiesysteme, 1987
[68] Risser, V. V.; Fuentes, M. K.: Linear Regression Analysis of Flat-Plate Photovoltaic Systems Performance Data. In *Proc. 5th E.C. PV Solar Energy Conference, Athens 1983*, Palz, W.; Fittipaldi, F. (Ed.), p. 623, Dordrecht (NL): D. Reidel, 1984
[69] Schaefer, J.: Photovoltaic Operating Experience. *EPRI Journal*, (March 1988) 40
[70] SBP, private communication and information material
[71] Schmitt, D.: Electricity Generating Costs – Coal and Nuclear as Primary Energy Carriers in Comparison (in German). *Handelsblatt*, 95 (1987) 23–24
[72] Strese, D.: The Ludwig Bölkow Study: Solar Electricity Becomes Cost-Effective (in German). *Bild der Wissenschaft*, (1988) 49–56
[73] Strese, D.; Schindler, J.: *Cost Degression of Photovoltaics, Phase 1: Manufacture of Multi-Crystalline Solar Cells and Their Application in the Utility Sector* (in German). München (D): Ludwig-Bölkow-Systemtechnik GmbH, 1989
[74] Verhey, J. F.: Luz Experiences in Project Development: A Minefield of Opportunity. Solar Thermal Technology – 4th Intl. Symposium, Santa Fe/NM, oral presentation, 1988
[75] Washom, B. J.: *Vanguard I Solar Parabolic Dish-Stirling Engine Module*. Technical Report DOE/AL-16333-2, Advanco Corp., 1984
[76] William, T. A.; Dirks, J. A.; Brown, D. R.; Drost, M. K.; Antoniac, Z. A.; Ross, B. A.: *Characterization of Solar Thermal Concepts for Electricity Generation*. Technical Report PNL-6128, Richland/WA, Battelle Pacific Northwest Laboratories, 1987
[77] Winter, C.-J.; Nitsch, J. (Ed.): *Hydrogen as an Energy Carrier – Technologies, Systems, Economy*. Berlin, Heidelberg, New York: Springer, 1988
[78] Winter, C.-J.; Nitsch, J.; Klaiss, H.: Solar Energy – Its Contribution to the Future Energy Supply of the Federal Republic of Germany (in German). *Brennstoff-Wärme-Kraft*, 35 (1983) 243–254
[79] Winter, C.-J.; Nitsch, J.; Klaiss, H.; Voigt, C.: Solar Energy Utilization – A Technical, Economical and Political Task for an Industrialized Country in Middle Europe (in German). In *Proc. 5th Intl. Solarforum, Berlin*, Auer, F.; Lanz, T. (Ed.), pp. 3–25, München (D): DGS Sonnenenergie-Verlags GmbH, 1984

Appendix A: Glossary of Terms

A.1 Solar Resource Terminology

Albedo. The ratio of radiation diffusely reflected by a surface, to the radiation falling on it.

Angle of incidence. The angle between the central ray of direct irradiance incident upon a surface and the normal to the surface at the point of incidence.

Atmospheric attenuation. The attenuation of flux density of solar radiation by absorption and scattering (i.e. by extinction) as a result of atmospheric conditions.

Black body. A term denoting an ideal body which would absorb all (reflect none) of the radiation falling upon it.

Cloud cover. The fraction of the sky hemisphere which is covered with clouds, usually estimated by a trained observer in eigths or tenths of sky covered.

Declination angle. Angle between the celestial equator along the great circle passing through the Sun and the celestial north pole. By convention, the solar declination angle is positive when the Earth-Sun vector points northward relative to the equatorial plane. For the northern hemisphere, the declination angle varies from -23.44 degrees on December 21 (winter solstice) to $+23.44$ degrees on June 22 (summer solstice).

Elevation angle. Angle subtended by the line of sight to the center of the Sun and the horizon. The sum of elevation angle and zenith angle is 90 degrees.

Insolation. The solar radiation available terrestrially. The maximum flux density at sea level and clear sky is about 1000 W/m^2.

Insolation, direct. Solar radiation incident within the solid angle subtended by the solar disk. It should be distinguished from diffuse or multidirectional irradiation. Clouds, fog, haze, smoke, dust, and molecular scattering and absorption attenuate the direct insolation.

Insolation, direct normal (DNI). The direct beam radiation or insolation on a surface perpendicular to the Sun's rays.

Intensity (radiant intensity). The radiant flux leaving a source per unit solid angle (Units: Wsr^{-1}).

Irradiance (flux density). The radiant flux incident upon unit area (Units: W/m^2).

Irradiance, diffuse. The irradiance upon a receiving surface by solar radiation which has become scattered during its passage through the sky, is called diffuse irradiance or sky irradiance. It is the downward scattered and reflected solar radiation coming from the entire hemisphere excluding the part of the hemisphere covered by the Sun's disc.

Irradiance, direct (beam irradiance). The radiant flux density from the Sun's disc and taken from a small circumsolar region of the sky within a subtended angle of 5 degrees.

Irradiance, global (global solar flux density). The direct plus diffuse solar irradiance upon a surface of horizontal orientation.

Irradiance, total. The global solar irradiance plus irradiance of terrestrial and atmospheric origin falling upon a surface of horizontal orientation (in case of meteorological measurements), or falling upon a surface (in photovoltaic engineering terms).

Irradiation (radiant exposure). The radiant energy incident per unit area. The product of irradiance (flux density) and its duration (Units: J/m^2).

Irradiation, diffuse. The diffuse solar energy (diffuse irradiance) incident upon a specified surface and integrated over a specified period of time.

Irradiation, direct (beam irradiation). The direct (beam) irradiance indident upon a specified surface and integrated over a specified period of time (e.g. daily, monthly).

Irradiation, global. The global irradiance incident upon a specified surface and integrated over a specified time period. Monthly Mean Daily Global Radiation refers to the daily mean global irradiation on a horizontal surface exposed to a hemispherical sky, averaged over a specified month.

Irradiation, total. The total irradiance, integrated over a specified time period, and incident upon a surface of horizontal orientation (for meteorological measurements), or upon a surface (in photovoltaic engineering terms, called also plane-of-array (POA) irradition).

Pyranometer. An instrument for the measurement of solar global irradiation received from the entire hemisphere.

Pyrheliometer. An instrument for measuring the amount of direct (beam) irradiation at normal incidence, usually with a subtended cone angle of $\approx 5°$.

Pyrradiometer. An instrument for the measurement of both solar and terrestrial (total) irradiation received from the entire hemisphere.

Radiance. The radiant flux leaving or arriving at a surface in a given direction per unit solid angle and per unit of surface area projected orthogonal to that direction. (Units: $Wm^{-2}sr^{-1}$)

Radiant energy. Electromagnetic energy emmanated by a radiation source (Units: J).

Radiant flux density. Radiant flux crossing a unit area element (Units: W/m^2).

Radiant flux (radiant power). Radiant energy passing through an area in unit time.

Radiation. The emission and propagation of electromagnetic energy through space or material.

Radiation, circumsolar. Solar radiation scattered by the atmosphere into an angular annulus around the Sun. It produces the solar aureole, whose angular extent is directly related to the atmospheric turbidity, increasing with turbidity. Usually the angular width of the Sun with its aureole is taken to be 5 degrees.

Radiation, direct solar (beam). The solar energy incident within the solid angle subtended by the solar disk on a surface.

Radiation, extraterrestrial solar. Solar radiation received outside of the Earth's atmosphere. This solar radiation would reach the ground in the absence of the Earth's atmosphere.

Solar constant. The extraterrestrial insolation incident upon a surface normal to the Sun. The value accepted in 1980 by the IEA is 1,367 W/m^2.

Solar energy. The electromagnetic energy emitted from the Sun, normally perceived as sunshine, but over a wider spectrum than is seen by the human eye or transmitted by the atmosphere.

Solar time. The time related to the position of the Sun. Solar noon occurs when the Sun reaches its zenith.

Sunshine, duration of. The time period (hours or minutes) during which direct sunshine is available. Measurements using this imprecise criterion have been taken for over 140 years in Europe, and serve both as a general measure of regional cloudiness and as a rough measure of global irradiance. Models have been developed that relate sunshine duration reasonably well with global irradiance.

Sunshine hours. (see Sunshine, duration of)

Zenith angle (zenith distance). The angle subtended by the zenith and the line of sight to the Sun. The sum of elevation angle and zenith angle is 90 degrees.

A.2 Solar Thermal Terminology

Absorber. The blackened surface which absorbs solar radiation and produces heat.

Absorptance. The ratio of the radiant flux absorbed in a material to the total incident radiant flux.

Absorptive coating. A coating which improves the absorptance of a material to incident radiation.

Absorptivity. The ratio of radiant energy absorbed by a surface, to that incident upon the surface.

Availability, operating. The percentage of time the unit was available for service, whether operated or not. It is equal to available hours divided by the total hours in the period under consideration.

Base load plant. A power plant in operation on an almost continuous basis; a plant with a capacity factor greater than 0.6.

Blocking. The interception of part of the reflected sunlight from one heliostat by the backside of another heliostat.

Buffer storage. The use of short term energy storage (typically less than one-half hour of storage) for smoothing transients associated with fluctuations of the energy source.

Capacity. The amount of net energy that can be delivered from a fully charged storage system and be used as a source of energy to generate electricity (Units: J or MWh_t).

Capacity factor. Energy produced over specified time period(s), e.g. weekly, monthly, during peak periods or annually, divided by the product of nameplate power rating times the aggregate hours of the time periods(s) chosen.

Cavity receiver. A solar radiation receiver in the form of a cavity. The radiation enters through one or more openings (apertures) and becomes absorbed on internal heat exchanger surfaces.

Central receiver system. A solar power plant system in which the available solar radiation is concentrated by an array of heliostats onto a tower-mounted receiver.

Cogeneration. The production of electricity or mechanical energy, or both, in conjunction with the production and use of process heat.

Collector efficiency. The ratio of the energy collection rate of a solar collector to the radiant flux intercepted by it.

Concentration ratio. The ratio of radiant flux density output to radiant flux density input.

Concentrator. A device that increases the flux density.

Convection. Heat transfer resulting from forced or natural fluid motion.

Design point. Radiation available for driving a process on a particular hour and day of the year. The nominal output (rating) of a system or component is specified for this design point.

Distributed collector system (DCS). A solar thermal system comprised of arrays of line- or point-focus concentrating collectors. Thermal energy is collected via heat transfer media and is transported to central points for processing.

Dual-axes tracking. A system capable of rotating independently about two not parallel axes.

Efficiency, annual average. The useful annual energy of a system, divided by the annual insolation available for use by the solar system.

Efficiency, collector. The ratio of the energy collection rate of a solar collector to the solar radiant flux intercepted by the collector.

Efficiency, heliostat field. The ratio of the solar flux directed to the receiver cavity aperture or onto an external receiver area, to the product of the solar irradiance and total heliostat field reflective area.

Efficiency, receiver. The ratio of the heat absorbed by the receiver working fluid and delivered to the base of the tower, to the solar radiant flux delivered to the receiver under reference conditions.

Emissivity. The ratio of radiant energy emitted by a surface, to that emitted by a black body of the same temperature and area.

Farm system. (see Distributed Collector System)

Flux (radiant power). The time rate of flow of radiant energy.

Flux density. The radiant flux crossing a unit of area.

Fresnel lens. Usually a focusing lens or mirror; the focusing is achieved by a series of concentric or parallel zones of engraved or molded structures on a flat surface of glass or plastic.

A.2 Solar Thermal Terminology

Heat transfer fluid. A fluid used to absorb heat in one region, e.g. at a receiver, and to transport and deliver it to a different region, e.g. to thermal storage.

Heliostat. A device consisting of an assembly of mirrors, support structure, drive mechanism, and mounting foundation, which is continously moved so that the Sun's rays are kept reflected in a fixed direction.

Heliostat field efficiency. The ratio of the solar radiant flux into the receiver cavity aperture or onto an external receiver under specified reference conditions, to the product of incident solar flux and total heliostat field reflective area.

Hours of storage. The number of hours a plant can produce power at a stated output level, normally at full-rated system load, when operating exclusively from an initially fully charged storage unit.

Hybrid system. In general, any energy system which operates on two or more energy input sources, or which provides more than one form of energy output. In particular, an energy conversion system that can be operated from solar energy or fossil fuel either interchangeably or simultaneously, or that combines different technologies for the utilization of solar and wind energies.

Intercept factor. The fraction of radiant flux incident on the receiver which reaches the absorber.

Line-focus collector. A solar collector that concentrates and absorbs solar radiation along a strip of focus.

Nameplate rating. The continuous operation of a power plant under specified conditions as designated by the manufacturer.

Optical efficiency. For a thermal solar power plant, the product of the four parameters reflectance of the heliostats, transmittance, intercept factor, and absorptance of the receiver. For a collector, the fraction of incident solar radiation absorbed.

Parasitic power. Power required to operate an energy conversion system (e.g., to operate pumps motors, computers, lighting, air conditioning).

Peak load. The maximum load during a given time interval.

Peaking plant. A power plant operated predominantly to cover peak demand periods; generally plants with capacity factor less than 0.18, but available on demand on short notice.

Point-focus collector. A solar collector that concentrates and absorbs solar radiation at a disk of focus.

Pointing error. The standard deviation (RMS), usually expressed in milliradians, of the difference between the desired aimpoint direction and the beam centroid direction.

Process heat. Heat used in agricultural, chemical, or industrial operations, excluding space heating.

Receiver. A radiation absorbing system that accepts radiation and delivers heat to a heat transfer fluid.

Receiver efficiency. The ratio of the thermal output delivered by the receiver heat transfer fluid, to the incident solar radiant flux under reference conditions.

Reflectance. The ratio of radiant flux reflected from, to incident on, a surface. In a focusing system, this is the specular reflectance of the mirrors.

Retrofit (repowering). The redesign and equipment of an existing fossil-fueled power plants with solar energy collection systems in order to replace a portion or all of the fossil fuel normally used.

Single-axis tracking. A system tracking the Sun's position by rotating about one axis only, e.g. a polar axis, a north-south or an east-west axis.

Solar furnace. A solar radiation concentrating device used to obtain absorber temperatures over 2,760°C.

Solar multiple. Defined at the system design point as the ratio of the input absorbed at the input end, to the fraction of input required to deliver rated output at the output end. Specifically for thermal solar systems, the ratio of thermal input absorbed by the heat transfer fluid of the receiver, to the fraction of input required to operate the turbine generator at rated net electrical output (with only solar radiation as input).

Spillage (radiation). The fraction of concentrated solar radiation which misses the absorber of the receiver.

Stand-alone. Any power system that operates on a local energy source only, with no off-site backup power system.

Stow. A position, or act of reaching a position, of storage for heliostats or other movable collectors.

Thermocline. The thermocline is the zone or layer in a thermal storage volume in which the vertical temperature profile changes rapidly.

Thermocline storage. The storage of thermal energy with hot and cold media contained in the same vessel (tank), employing the mechanical stability of lower density hot fluid atop the higher density cooler fluid.

Working fluid. A fluid which can be heated, cooled, pressurized, and expanded to do work, e.g., drive a turbine in a power cycle.

A.3 Photovoltaic Terminology

Absorption edge. Rapid rise of optical absorption as the wavelength of incident radiation is reduced.

Air mass. (AM) The path length of direct solar beam through the Earth's atmosphere, expressed as a multiple of the path length to a point at sea level with the Sun directly overhead.

Amorphous silicon. A type of solid silicon with its atoms not regularly arranged in a crystal lattice; usually only available as thin films.

Anti-reflection coating. The employment of interference in thin film(s) of dielectric material at a surface to reduce its reflection of radiation.

Array. A mechanically integrated assembly of modules or panels together with support structures (but without foundation), together with tracking, thermal control and other components as required, which forms a DC power producing unit.

Band gap. Range of forbidden electron energies between two adjacent bands of allowed energy states; in particular, energy gap between valence and conduction bands in a semiconductor.

Blocking diode. A diode connected in series with solar cell(s), module(s) or panel(s) to prevent reverse current in such solar cell(s), module(s) or panel(s).

Conduction band. Lowest electron energy band in a solid not completely filled with electrons at absolute zero, representing energy states in which electrones are not bound to atoms.

Conversion efficiency. The ratio of maximum electrical power output to the product of photovoltaic device area and incident irradiance measured under defined test conditions and expressed as a percentage.

Current voltage characteristic. The output current of a photovoltaic device as a function of output voltage (I = f(V)) at a particular temperature and irradiance.

Czochralski process. Method of growing single crystals by vertical pulling of a seed crystal from a melt of silicon.

Fill factor (FF). The ratio of maximum power to the product of open-circuit voltage and short-circuit current.

Float zone process. Method of growing a single crystal by creating a small molten zone at the seed crystal end of a polycrystalline feed rod, and moving this zone slowly up the rod.

Gettering. Method of producing and keeping high vacuum in closed containers such as a vacuum tube. Alternatively, method of fixing and neutralizing adverse impurities by chemical reactions, e.g. in semiconductors.

Inverter. A electromechanic or electronic device which changes direct current (DC) into an alternating current (AC).

Maximum power point. The point in the I-V-diagram of a solar cell or PV module where, for a given irradiance and cell temperature, the product of current I and voltage V is a maximum.

Module. The smallest complete assembly of interconnected solar cells protected against the environment.

Nominal operating cell temperature (NOCT). The equilibrium mean solar cell temperature within a module in a standard reference environment (800 W/m^{-2} irradiance, 20°C ambient temperature, 1 m/s wind speed), electrically open-circuited and open-rack mounted at normal incidence to the Sun at solar noon.

Open circuit voltage (V_{OC}). The voltage across an unloaded (open) photovoltaic device at a particular temperature and irradiance.

Panel. A group of modules fastened together, preassembled and wired, designed to serve as an installable unit in an array and/or subarray.

Peak watt. The power supplied by a solar generator when exposed, under natural incidence, to a maximum of solar irradiance under conditions of air mass 1.5 and 25°C cell temperature, or, in a test environment, to standard test conditions (see below).

Photovoltaic cell. (see Solar Cell).

Photovoltaic effect. Direct conversion of radiant energy into electrical energy.

Photovoltaic efficiency (or conversion efficiency). The ratio of maximum electrical power to the product of PV device area and irradiance, expressed as a percentage.

Short circuit current. The (output) current of a photovoltaic device in short-circuited condition at a particular temperature and irradiance.

Solar cell. The basic photovoltaic device which generates electricity when exposed to solar radiation.

Spectral response. The dependence of output (of electrical power, thermal power, chemical flux densities) on frequency (or wavelength) of the incident radiation.

Standard operating conditions (SOC). Irradiance of 1000 Wm^{-2} with reference solar spectral irradiance distribution ($\approx$ AM 1.5) and an ambient temperature of 20°C.

Standard test conditions (STC). Irradiance of 1000 Wm^{-2} with reference solar spectral irradiance distribution ($\approx$ AM 1.5) and a cell temperature of 25°C.

Tandem arrangement. A stack arrangement of two or more simultaneously irradiated solar cells differing in bandgap.

Valence band. Band of highest electron energies, completely filled with electrons in a semiconductor or insulator at absolute zero.

A.4 Financial Terminology

Avoided costs. The incremental costs that result (a) from investing in additional generating capacity in a utility system, or (b) from spending input fuels and energy to generate electricity, both of which a utility can avoid by buying electricity from a third party.

Busbar energy cost. The cost of producing electricity, including plant capital and operating and maintenance expenses. Does not include cost of transmission or distribution.

Capacity credit. The amount of generating capacity displaced by a solar power plant, expressed in MW_e or as a fraction of the nominal solar plant output. Determined by individual utilities.

Capacity payment. The portion of payment granted to a third-party producer by utilities which is based on costs avoided for installing additional generating capacity in a utility system.

Cost/performance ratio. A measure used in evaluating system design alternatives wherein both cost and system performance are taken into account.

Cost/value ratio. A measure used in evaluating the cost of a system over its lifetime compared with the value of its product (e.g. energy output).

Direct costs. The portion of investment costs attributable to deliverable hardware and equipment.

Discount rate. The annual rate used in present worth analyses that takes into account inflation and the potential earning power of money while moving the present worth forward or backward to a single point in time for comparison of value.

Energy payment. The portion of payment granted by utilities to a third-party producer which is based on costs avoided in a utility system for not expending input fuels/energies.

Fixed charge rate. The amount of revenue per dollar of capital expense that must be collected annually to pay for the fixed charges associated with the plant ownership, e.g., return on equity, interest payment on debt, depreciation, income taxes, property taxes, insurance, repayment of initial investments, etc. It may also include operations and maintenance expenses expressed as a fraction of the capital cost.

Indirect costs. No uniform definition exists. Indirect costs may include, for instance, the costs for engineering, construction management, spares, fuel inventories, fees, financing, AFUDCs (allowance for funds used during construction), and contingencies for various reasons.

Internal rate of return (IRR); return on investment (ROI). The true rate at which an investment is repaid by proceeds from a project. It is the discount rate at which the incremental cash inflows expected from a project (after taxes, but before allowance for depreciation) have a present value which is equal to the discounted present value of all incremental cash outflows required to implement the project, i.e. the discounted net present value is zero.

Levelized busbar energy cost (or: levelized energy cost). The constant annual revenue per unit of energy required over the lifetime of a plant to compensate for its fixed and variable costs, interest costs and shareholder return.

Levelized fixed charge rate. The fixed charge rate that produces a constant level of payments over the life of a plant whose present worth is the same as the present worth of the actual cash flow.

Net present value. Present value of revenues minus present value of expenditures/costs. This difference must be ≥ 0 if an investment is to be profitable.

Payback period (or: payback time). Time period after which the capital expended for investments has flown back and is recouped by the financing entity (i.e. the utility, the investor, etc.).

Present value. For a given interest rate, the discounted value of past expenditures/costs and/or of future revenues (over lifetime) relative to a specific point in time (in energy projects, usually the year of facility start-up).

Specific costs. Costs related to a physical unit (power, energy, area, etc.).

Standard offers. Different options for long-term energy and capacity credit rates contained in power purchase agreements between utilities and third-party electricity suppliers.

Total costs. The sum of direct and indirect costs.

Appendix B: Abbreviations and Acronyms

B.1 Radiation, Solar

AM	air mass
C	concentration ratio
CFLOS	cloud free line of sight
DNI	direct normal insolation
EOT	equation of time
IR	infrared
LDS	limb darkened Sun
LST	local standard time
TE	transverse electric
TM	transverse magnetic
UTC	universal time coordiate
UV	ultraviolet
VIS	visibility
VR	visual range

B.2 Thermal

BOP	balance-of-plant
CFB	circulating fluidized bed
CFR	conical flux density redirector
CPC	compound parabolic collector
CRS	central receiver system
DCS	distributed collector system
HTF	heat transfer fluid
ORC	organic Rankine cycle
OTEC	ocean thermal energy conversion
PCM	phase-change material
PCU	power conversion unit
PET	polyethylene terephtalate
PMMA	polymethyl metacrylate
SBC	shading, blocking and cosine (losses)
SM	solar multiple
SPP	solar power plant
SS	second-surface
SR	slant range
TBC	test bed concentrator
TC	terminal concentrator

B.3 Photovoltaic

AC	alternating current
BOS	balance-of-system
BSF	back surface field
BSR	back surface reflector
CVD	chemical vapor deposition
CZ	Czochralski
DC	direct current
EG-Si	electronic-grade silicon
EOL	end of life
FF	fill factor
FZ	float zoning
HVDC	high voltage direct current
IBC	interdigitated back contact
IR	infrared
MIS	metal-insolator-semiconductor
mono-Si	monocrystalline silicon
MPP	maximum power point
NOCT	nominal operating cell temperature
OC	open circuit
PC	point-contact
PESC	passivated emitter solar cell
PF	power factor
poly-Si	polycrystalline silicon
PV	photovoltaic
RFI	radio frequency interference
rms, r.m.s.	root-mean-square
SC	short circuit
SoG-Si	solar-grade silicon
SOC	standard operating conditions
STC	standard test conditions
UV	ultraviolet

B.4 Cost/Economic

AFUDC	allowance for funds used during construction
CAP	capita, inhabitants
CF	capacity factor
GNP	gross national product
IRR	internal rate of return
LEC	levelized energy cost
LLEC	least levelized energy cost
NPV	net present value
O&M	operations and maintenance
PEC	primary energy consumption
PVM	present value multiplier
ROI	return on investment

B.5 Acronyms

ACTF	Advanced Component Test Facility
ANU	Australian National University
ASI	Arco Solar Inc.
BMFT	Bundesministerium für Forschung und Technologie
CEC	Commission of the European Communities
CESA	Central Solar de Almeria
CNRS	Centre National des Recherches Scientifiques
CRTF	Central Receiver Test Facility
DLR	Deutsche Forschungsanstalt für Luft- und Raumfahrt
DOE	Department of Energy
DRTF	Distributed Receiver Test Facility
GAST	Gas-Cooled Solar Tower
GIRT	Georgia Institute of Research and Technology
IEA	International Energy Agency
IER	Instituto de Energias Renovables
LaJet	LaJet Energy Corporation
LBL	Lawrence Berkely Laboratories
MAN	Maschinenwerke Augsburg-Nürnberg
MBB	Messerschmitt-Bölkow-Blohm
MDAC	McDonell Douglas
MOB	Mobil Solar Corp.
MRI	Midwest Research Institute
MSSTF	Medium-Scale Solar Test Facility
NEDO	New Energy and Industrial Technology Development Organization
OPEC	Organization of Petroleum Exporting Countries
PG&E	Pacific Gas and Electric Co.
PKI	Power Kinetics Inc.
PURPA	Public Utility Regulatory Policies Act
SBP	Schlaich, Bergermann und Partner
SCE	Southern California Edison Co.
SEGS	Solar Energy Generating System
SERI	Solar Energy Research Institute
SKI	Solar Kinetics Inc.
SMUD	Sacramento Municipal Utility District
SNL	Sandia National Laboratories
SNLA	Sandia National Laboratories, Albuquerque
SNLL	Sandia National Laboratories, Livermore
SOL	Solarex Corp.
SSI	Siemens Solar Inc.
SSPS	Small Solar Power System
STEP	Solar Total Energy Project
STIP	Solar Thermal Irrigation Project
TDSA	Testing of Dishes in Saudi Arabia
UNSW	University of New South Wales
USAB	United Stirling AB
WMO	World Metorological Organization
ZSW	Zentrum für Solarenergie- und Wasserstoff-Forschung

Subject Index

aberration 99
  chromatic 85, 118
absorptance 18
absorption 27
  coefficient 286, 287
absorptivity 168, 169
actual solar time 23
aerosol 27, 31
air mass 21
albedo
  atmospheric 31
  ground 31
  surface 31, 34, 288
albedometer 78
aureole 32, 88, 89
availability 157, 320

backup 199, 371, 396
Beer's law 286
biomass 1, 2
Boltzmann's constant 17, 18
Brewster's angle 86

capacity 13
  base load 154
  factor 13, 154, 199, 201, 249, 319, 370
  installed 11
  payments 402
capital, investment of 4, 6, 9, 157
Carnot 54
  efficiency 140, 176
  relation 36
Cartesian frame 22
Cassegranian 96, 97
Central Limit Theorem 92, 95
characteristics, input-output 151, 248
chemicals 343
circulator 70
circumsolar 33, 88
  ratio 89
cirrus clouds 27–29, 31
cloudfree lines of sight (CFLOS) 29
clouds 287
coal 1
cogeneration 12
collector 135, 220
  cleaning 117, 220, 225, 246
  equation 44, 48

  field 136, 216
  fluorescence 67
  heliostat 95
  line-focussing 11, 95, 215
  mirrors 223
  parabolic dish 110, 229, 234
  parabolic trough 192, 215
  point-focussing 11, 95, 136, 229
  subsystem 154, 223
  vacuum tube 136
concentration 11, 41, 136, 285
  factor 87, 176, 254
  optics 11, 84
  ratio 43
  ratio, flux density 96
  ratio, geometric 96, 106, 115
concentrator 41, 135
  compound elliptical 101
  compound parabolic 99
  dispersive 307
  facet 116
  flow-line 102, 103
  fluorescent 307
  Fresnel 305
  ideal 98
  line-focussing 43
  parabolic 104, 250, 307
  parabolic dish 101
  parabolic trough 104
  point-focussing 43
  terminal 98, 99, 108
  trumpet 102, 103
  V-trough 306
conical flux density redirector (CFR) 103
convection 45
conversion 3
  chemical 336
  DC-AC 142, 299, 311
  DC-DC 297, 358
  electrochemical 354
  of solar energy 136
  of solar radiation 5
  photochemical 336, 337, 353
  photothermal 336, 337, 347, 352
  photovoltaic 354
  radiation-to-thermal 164
  thermodynamic 139
converter 3, 35

Subject Index

cost 400
  -to-benefit ratio 157
  assumption 12
  avoided 401
  balance-of-system 383
  collector 123, 373, 380
  conversion subsystem 374
  direct 373
  energy unit 158
  external 9, 386, 396
  financing 403
  generating 14, 387, 389, 392, 396, 401
  heliostat 125, 375
  indirect 373
  investment 13, 122, 157, 367, 372, 373, 375, 380, 382, 387, 394
  levelized energy 122, 157, 390, 394
  maintenance 157
  media 203
  module 382
  of support structures 301
  optimization 125
  receiver 375, 380
  social 9, 12, 396
  storage 203
  storage subsystem 375, 377
  total 308, 375, 378, 381, 384
  tower 377
  tracking 123
Cowper 211
current, short circuit 61, 284, 296
cycle 219
  Brayton 139, 142, 190, 254
  fatigue 165, 178
  organic Rankine 136, 139, 190
  Rankine 139, 142, 254
  Stirling 139, 190, 254
  thermochemical 346
Czochralski technique 289

declination 22
design
  parameters 158
  point 154
  process 158
diagram
  block 142
  loss-tree 147
  Sankey 147
  system 144
dilution factor 38, 43
diode
  blocking 296
  bypass 296
  equation 59
  shunt 296
direct coupling 338
dispersion 85
distribution

annual load 202
bivariate Gaussian 90
brightness 90
  function 90
  Gaussian 91, 178
  limb darkened 88, 91
  moments of 90
  of power levels 202
  uniform 88
doping 55
downhill reaction 70

efficiency 148
  24-hours 150
  annual 150, 201
  annual average 13, 319
  array 314
  conversion 150
  cycle 140, 200, 228, 249
  energy 257
  gross 150
  heliostat 120
  net power 150
  operating hours 150
  part-load 142, 201
  power plant 142
  solar cell 285
  subsystem 149
  system 164
  thermal 46
electrochemistry 75
electrolysis 337, 354, 355
  advanced alkaline 357, 359
  alkaline 355
  electrocatalytic 359
  high temperature vapor 361, 363
  solid polymer electrolyte 359
elevation angle 20, 22
emittance 18
energy
  carrier 70
  demand 1
  depletable 135
  end-use 134
  flux density 73
  gain factor 5
  gap 54, 284, 286
  input 199
  needs 406
  nuclear 1, 2, 8
  output 313, 387
  pay-back time 5
  rational use of 3
  renewable 4, 135
  resource 1, 4, 406
  self-consumption 399
  solar 4
  supply 399
  supply heptagon 3

value 400
environment 9
  greenhouse effect 399
  penalties 5
  pollutants 2
  pollution 1, 4, 397
equation of time EOT) 23
error
  beam 95, 121
  function 92
  mirror surface 93, 94
  slope 93, 121
  tracking 95, 171
Etendue 20, 42, 88, 102
exergy 35, 40, 64, 74

f-factor 38
field-receiver ratio 155
fill factor 63, 106, 285
fin factor 45
financing 403
float-zoning technique 289
flux 18, 20, 176, 194
  density 5, 20, 95, 112, 127
  density distribution 96, 120, 179
  peak density 96
  redirector, conical 103
focal length 87
focus 87, 113
Fresnel, lenses 85, 87, 117
Fresnel, reflector 115
Fresnel, relations 85

gas, natural 1
gasification 345
geographic latitude 22
geothermal resource 1
Grasshoff number 172

heat
  dissipation 147
  distribution of 165, 178
  loss coefficient 45
  transfer fluid (HTF) 136, 164, 204, 232, 237, 339
  transfer media 179
  transfer subsystem 144
heliostat 115, 124, 237, 243
  array 115, 122
  field 119, 126, 244
  parameters 120
  spacing 119
Hermite polynomials 91
horary circle 22
hour angle 22
hydrogen 3, 12, 337, 354, 361, 396
hydropower 3

inner photoeffect 54

ionization 54
irradiance 19, 20, 286, 313
  beam 29
  diffuse 29
  direct 29, 370
  direct normal 155
  extraterrestrial 88
  global 45
  spectral 287
  terrestrial 88
  total 45, 370
irradiation 3, 394
  characteristics 151
  diffuse 11, 13
  direct 3, 11, 12
  direct normal 151, 153
  global 13, 155, 394
  quality 156
I-V-characteristic 284

Kirchhoff's law 18

Lambert's law 19
land area requirement 5, 130, 135, 310
least energy cost (LEC) 122, 157
limb darkening 19
losses 86, 167
  atmospheric 122, 147
  conductive 147, 171
  convective 46, 124, 147, 171, 172
  conversion 147, 313
  electrical 311
  ohmic 147
  optical 147
  radiation 171
  radiative 124
  receiver 164, 226
  reflection 168
  resistive 313
  specularity 92
  spillage 90, 124, 147, 164
  thermal 164, 169
  tracking 313
local standard time (LST) 23

market
  competitive 9
  consumer-electronics 288
  electricity 406
  potential 405
material degradation rate 164
maximum power point 62, 284
  tracking 298
mean solar time 23
measurement
  non-intrusive 192
  of radiation 76, 193
  optical 193
  techniques 192

Subject Index

media
  chemical storage 204
  phase change 204
meridian 22
Mie scattering 27, 86
mirror, rotated 113
module 135

Nusselt number 172

oil
  mineral 1
  silicone 204, 206
  synthetic 204, 206, 216
  thermal 206, 216
optimum temperature 176, 177

parasitics 147, 156, 201, 228
performance
  conversion 317
  electrolysis system 357
  input-output 225
  long-term 261
  modelling 128
  output 154, 370
  simulation 177, 263
  solar power plant 128, 225, 246, 259, 318
period
  mid-peak 401
  off-peak 401
  on-peak 228, 401
photochemical
  potential 64
  processes 70, 71
photochemistry 11, 12, 14, 336
  exergy yield 74
  maximum yield 74
photodiode 57
photoeffect 54, 55
photoionization 54
photosphere 18
photovolatic modules 293
photovoltaic, array 284
photovoltaic, conversion 55, 138, 142, 283
photovoltaic, module 283
photovoltaic plant 138, 310, 370, 382, 389, 405
  concentrating 289, 304
  grid-connected 309
  non-concentrating 283
  rating 314
  simulation 324
  tracking 303, 304
Planck's constant 18
pollutor-pays-principle 397
potential, chemical 73, 74
power
  conversion unit 142, 220
  rated output 199
  virtual rated 202

power plant
  capacity 396
  coal 387
  conventional 134, 387
  grid-connected 283, 370, 399
  monitoring 312
  nuclear 387
  tidal 1
Prandtl number 172
process
  catalytic 343, 348
  chemical 337
  direct absorption 352
  heat 14, 44, 46, 49, 337
  high temperature 343
  methane reforming 348, 349
  photo-initiated 14
  photocatalytic 14, 355
  upgrading 344
pyranometer 76, 78
pyrgeometer 77
pyrheliometer 76, 77
pyrradiometer 77, 79

radiance 20
  temperature 37, 49, 64
radiant
  excitance 17, 20
  intensity 20
  power 20
radiation 3
  black body 18, 88
  diffuse 32
  diffuse solar 76
  direct 163
  direct solar 76
  global 45
  global solar 76
  Lambertian 88, 99
  quality 35, 36
  total 45, 76
radiosity 17, 20
Rayleigh scattering 27, 31, 86
reaction
  endothermal 337
  photo-assisted 354
reactor 190
  axial 339
  batch processing 338
  cavity receiver 190
  chemical 11
  continuous processing 338
  direct absorption receiver 340, 342
  fixed bed 343
  fluidized bed 343
  indirect receiver 339
  integrated 348
  liquid metal 339
  matrix receiver 342

molten salt 339
    receiver 336, 337, 350
    separated 348
    spiral 339
    tube receiver 339
receiver 135, 163
    air 181
    cavity 164, 165, 170
    central 11, 119, 136
    cylindrical 106
    direct absorption 165, 189, 192
    distributed 11
    external 164, 165, 190
    fixed 114
    flat 106, 109
    fluid film 192
    gas-cooled 124
    heliostats 11
    operating temperature 163, 164
    parabolic dish 190
    parabolic trough 164, 184
    sodium 180, 210, 243
    spherical 106, 110
    tube 164, 179, 184, 191
    volumetric 165, 170, 185, 191, 192
    water/steam 180
reflectance 86
    hemispherical 92
    specular 92
reflection 85
    coefficient 86, 286
reflectivity 168
refraction 85, 117
refractive index 84
Reynolds number 172
rim angle 101, 103, 106
risk 367, 400, 404

selective absorption/transmission 48, 53
semiconductor 54
    direct 286
    indirect 286
silicon
    amorphous 288
    electronic grade 289
    solar grade 290
Snell's law 85, 98, 115
solar
    chemistry 70, 71
    constant 18, 19
    cyclic energy systems 6
    disc 88
    energy 135
    energy utilization 3, 215
    fuel 11, 70, 336
    multiple 154, 200, 201, 249, 371, 395
    zenith 287
solar cell 56
    amorphous 288
    amorphous thin film 289, 291
    characteristic 284, 296
    compound semiconductor 289
    crystalline silicon 288, 289
    efficiency 63
    encapsulation 294
    equation 58, 60
    ideal 57
    interconnection 293, 295, 311
    monocrystalline 285, 287, 288
    multicrystalline 288
    multijunction 287
    multiple-stack tandem 70
    point contact 290
    polycristalline thin film 292
    sandwich stack 67
    tandem 67, 69, 287, 289, 292
    temperature 285, 294
solar energy, density 5
solar power plant 134, 367
    characteristic 371
    chimney 138
    concepts 2, 12, 205
    dish/Stirling 139, 190, 250, 254, 257, 370, 380, 405
    farm 136, 190, 215
    hybrid 136, 143, 219, 222, 231, 371, 395
    individual dish 250
    parabolic trough 370, 373, 389, 405
    photovoltaic 7
    thermal 12
    tower 136, 237, 370, 375, 405
spectral
    distribution 33
    matching 67
    radiance 19
Staebler-Wronski effect 291
stagnation temperature 47
standard deviation 90, 94
standard spectrum 38
storage 13, 136, 199, 220, 337
    218
    bulk 371
    bulk electricity 13
    bulk energy 144
    capacity 156, 201, 203, 395
    dual medium 205, 206, 237, 243
    electrochemical 199, 299
    hour 199
    loss factor 200
    single medium 204
    subsystem 144, 208
    thermocline 204
storage media 204
    ceramic 212
    Cowper-type 212
    latent heat 204
    molten salt 207, 242
    oil 204, 206, 207

Subject Index

rock/sand   207
sensible heat   203
sodium   204, 210
Sun
  degraded   90, 95, 121
  photometry   78
  radiosity   18
  radius   18
  temperature   18, 176
sunrise   24, 199
sunset   24, 200
sunshape   89, 90, 91, 121
sunshine
  recorder   77, 80
  hours   9
supporting structures   293, 300, 301
surface
  azimuth   26
  tilt   26
system
  central receiver   207, 237
  design optimization   123
  line-focussing   114
  non-imaging   43
  optimization   125, 126
  point-focussing   114
  scaling   119

sizing relations   129
tracking   88

test sites   262
thermal inventory   156
thermochemistry   75, 177
tilt angle   32
time
  local solar   23
  local standard   23
tracking facet   114, 115
transmission coefficient   27, 86
transmittance   86, 293

utilization factor   199

value   400
VIS   28
visibility   28
voltage, open circuit   61, 284, 296

Wien's law   48, 73
wind-heater   201

zenith
  angle   21
  distance   21

C.-J. Winter, J. Nitsch (Eds.)

Hydrogen as an Energy Carrier

Technologies, Systems, Economy

1988. XII, 377, pp. 188 figs. Hardcover DM 158,- ISBN 3-540-18896-7

Contents: Significance and Use of Hydrogen: Energy Supply Structures and the Importance of Gaseous Energy Carriers. Technologies for the Energetic Use of Hydrogen. Hydrogen as Raw Material. Safety Aspects of Hydrogen Energy. – Production of Hydrogen from Nonfossil Primary Energy: Photovoltaic Electricity Generation. Thermomechanical Electricity Generation. Water Splitting Methods. Selected Hydrogen Production Systems. Storage, Transport and Distribution of Hydrogen. – Design of a Future Hydrogen Energy Economy: Potential and Chances of Hydrogen. Hydrogen in a Future Energy Economy. Concepts for the Introduction of Nonfossil Hydrogen. Energy-economic Conditions and the Cooperation with Hydrogen Producing Countries. - Index.

The book deals with the possiblities of an energetic utilization of hydrogen. This energy carrier can be produced from the unlimited energy sources solar energy, wind energy and hydropower, and from nuclear energy. It is also in a position to one day supplement or supersede the fossil energy carriers oil, coal and gas.

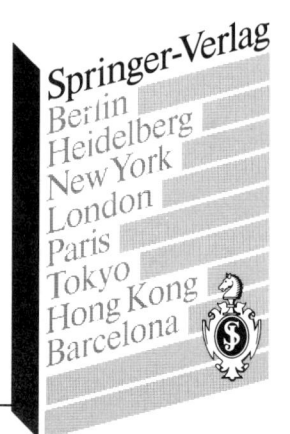

Springer-Verlag
Berlin
Heidelberg
New York
London
Paris
Tokyo
Hong Kong
Barcelona